Biosimulation in Drug Development

Edited by
Martin Bertau, Erik Mosekilde,
and Hans V. Westerhoff

Related Titles

U. Marx, V. Sandig (Eds.)

Drug Testing in vitro

Breakthroughs and Trends in Cell Culture Technology

2007
ISBN 978-3-527-31488-1

M. Reddy, R. S. Yang, M. E. Andersen, H. J. Clewell III

Physiologically Based Pharmacokinetic Modeling

Science and Applications

2005
ISBN 978-0-471-47814-0

E. Klipp, R. Herwig, A. Kowald, C. Wierling, H. Lehrach

Systems Biology in Practice

Concepts, Implementation and Application

2005
ISBN 978-3-527-31078-4

T. I. Oprea (Ed.)

Chemoinformatics in Drug Discovery

2005
ISBN 978-3-527-30753-1

P. M. Carroll, K. Fitzgerald (Eds.)

Model Organisms in Drug Discovery

2003
ISBN 978-0-470-84893-7

A. Avdeef

Absorption and Drug Development

Solubility, Permeability, and Charge State

2003
ISBN 978-0-471-42365-2

H. van de Waterbeemd, H. Lennernäs, P. Artursson (Eds.)

Drug Bioavailability

Estimation of Solubility, Permeability, Absorption and Bioavailability

2003
ISBN 978-3-527-30438-7

M. Lutz, T. Kenakin

Quantitative Molecular Pharmacology and Informatics in Drug Discovery

1999
ISBN 978-0-471-98861-8

Biosimulation in Drug Development

Edited by
Martin Bertau, Erik Mosekilde,
and Hans V. Westerhoff

WILEY-VCH Verlag GmbH & Co. KGaA

The Editors

Prof. Dr. Martin Bertau
Freiberg University of Mining and
Technology
Institute of Technical Chemistry
Leipziger Strasse 29
09599 Freiberg
Germany

Prof. Dr. Erik Mosekilde
Technical University of Denmark
Department of Physics
Systems Biology Group
Fysikvej 309
2800 Kgs. Lyngby
Denmark

Prof. Dr. Hans V. Westerhoff
Vrije Universiteit
Faculty of Earth and Life Sciences
Molecular Cell Physiology
De Boelelaan 1087
1081 HV Amsterdam
The Netherlands

Library of Congress Card No.:
applied for

British Library Cataloguing-in-Publication Data
A catalogue record for this book is available
from the British Library.

**Bibliographic information published by
the Deutsche Nationalbibliothek**
Die Deutsche Nationalbibliothek lists this
publication in the Deutsche
Nationalbibliografie; detailed bibliographic
data are available in the Internet at
<http://dnb.d-nb.de>.

Typesetting   VTEX, Vilnius, Lithuania
Printing   betz-druck GmbH, Darmstadt
Binding   Litges & Dopf GmbH, Heppenheim
Cover Design   WMX-Design, Bruno Winkler,
Heidelberg

Printed in the Federal Republic of Germany
Printed on acid-free paper

ISBN: 978-3-527-31699-1

Contents

Preface

Drug development is far from a straightforward endeavor. It starts with the identification of a pharmaceutically promising substance, the potential of which is further
investigated in near-exhausting physiological analyses (Chapter 1). If it is found
effective, this does not necessarily mean that we know all the molecular consequences when administered. Each patient is an individual, with unique features,
which may mean patient treatment at the level of the individual; so-called personal
medicine. This illustrates the complexity of drug development and discovery.

Does it get any simpler when we consider the physiological and molecular events
upon drug ingestion? A perorally administered drug first of all enters the gastro-
intestinal tract where it encounters the biotransformatory activity of the intestinal
microflora. Our knowledge of which compounds finally enter the blood stream is
fragmentary at best, not to mention the interactions of the drug with the different
organs and the resulting organ–organ interplays (Chapter 3). And this is not the
only question arising. How do drugs interact with human physiology? Whether we
can simulate drug effects then becomes very important (Chapters 6 and 8). Further, how do the organs interact with each other upon the ingestion of a drug?
Can we achieve a whole-body simulation? Or virtual populations? Can we, and to
what extent, treat individuality and gender in drug administration? What advantages lie behind technologies that can simulate drug action and effect (Chapters 11
and 14)? If we can do all this, how can we use these simulations to our advantage in
drug discovery and development? To what are undesired side-effects attributable?
What do we know of drug–drug interactions, drug–drug metabolite interactions,
and their outcomes on organs or on the interplay between different organs? Can
we, by applying novel methodologies, shorten package inserts, even make them
superfluous?

In other words, drug development is increasingly characterized by the requirement to understand highly complex biological processes and to exploit the rapidly
growing amount of biological information. The methods that are currently applied
in the development of new medicines require new and effective means to evaluate,
integrate, and accumulate this biological knowledge. Essential improvements will
result from the use of computational models that can provide a dynamic and more
quantitative description of the relevant biological, pathological, and pharmacoki-

Biosimulation in Drug Development. Edited by Martin Bertau, Erik Mosekilde, and Hans V. Westerhoff
Copyright © 2008 WILEY-VCH Verlag GmbH & Co. KGaA, Weinheim
ISBN: 978-3-527-31699-1

netic processes. These ambitious goals will in fact be achieved by the integration of a new methodology, biosimulation, into the drug development process.

Why biosimulation? The high development costs and long lead times of new drugs are associated with numerous clinical trials that a drug must undergo to document its function and show that adverse side-effects do not occur. The use of professional simulation models can simplify the drug development process enormously by exploiting the information available in each individual trial much more effectively. Besides, this approach is used in many industries where computer simulation allows new concepts and designs to be tested and optimized long before the first example of the product is manufactured. Consequently the simulation approach is strongly recommended by the United States Food and Drug Administration, both because of its potential for reducing the development costs, and because of the associated reduction in the use of laboratory animals and test persons. Through the establishment of "virtual populations", as information is gradually accumulated, variations in the efficacy and possible side-effects of a new drug can be predicted on a gender- and age-specific basis and/or for patients with specific gene modifications (Chapter 16).

The establishment of a more quantitative description of biomedical systems as the foundation for a disease-driven drug development process requires an unusually broad range of insights and skills in biological as well as in technical and mathematical realms. This approach is highly interdisciplinary and commands outstanding expertise in bioinformatics, biochemistry, cellular biology, electrophysiology, intercellular communication, physiology, endocrinology, neurology, nephrology, pharmacology, pharmacokinetics, systems biology, complex systems theory, and *in silico* modeling and simulation techniques (Chapters 2 and 15). It also requires close collaboration between experimental and theoretical partners. Modeling is the translation of (often) experimentally derived scientific hypotheses into a formal mathematical framework. In order to improve, the models must continuously be confronted with new experimental data; and precisely this process of generating new hypotheses and formulating critical experiments represents the most effective way of expanding our knowledge about drug function.

The use of biosimulation and mathematical models in drug development is based on the circumstance that, contrary to the conventional assumption of homeostasis, many biological systems have unstable equilibrium points and operate in a pulsatile or oscillatory mode. This is the case, for instance, for the release of pituitary hormones that typically occurs at more or less regular intervals of two hours, or for the optimal point-of-time of drug administration (Chapters 10 and 12). Or, in several cases it has been reported that the cellular response to an oscillatory signal is stronger than the response to a constant signal of the same average magnitude, suggesting that the oscillatory dynamics play a role in the regulatory effect of the hormone. Rhythmic and pulsatile signals are also encountered in intracellular processes as well as in communication between cells. Many nerve cells are excitable and respond in an unusual fashion to small external stimuli. Other cells display complicated patterns of spikes and bursts (Chapters 7 and 13).

Biosimulation can also contribute to the development of a methodology for the prediction of drug-likelihood. Neural networks can be used to prescreen drug candidates and to predict absorption rates, binding affinities, metabolic rate constants, etc., from knowledge about previously examined compounds. This allows the subsequent biochemical screening to be performed on a reduced set of candidates that have a significantly increased likelihood of possessing the desired functionality. In this context reports from the European Federation for Pharmaceutical Sciences (EUFEPS) and from other European organizations have repeatedly stressed the need for a targeted effort to speed up the development of new and safe drugs. In fact, the introduction of new technologies such as high-throughput screening (HTS), computational chemistry, and combinatorial and automated chemistry has made research and discovery in the pharmaceutical industry significantly more effective. These technologies allow the screening of compound libraries for potential drugs at rates that are hundreds of times faster than even the best skilled chemist can do. Thus, a major improvement in drug development will require a more knowledge-oriented process that builds on a detailed and quantitative understanding of the biological and pathological processes associated with the functioning of the drug.

Specifically the biosimulation approach to drug discovery provides efficient means to: (1) define and control the conditions under which experiments and clinical tests are performed, (2) extract the information available in the individual trial and validate it in terms of current knowledge in cell biology, medicine, etc., (3) accumulate information from trial to trial and redesign the trial procedure to become an adaptive process where information acquired in one trial is immediately used to improve the process, (4) extrapolate results obtained from experiments on cell cultures and from animal experiments to applications for human patients, (5) predict the variation of drug efficacy and the occurrence of side-effects, taking account of genetic modifications, and gender, age, or weight characteristics, and (6) predict the likelihood that a particular chemical compound will function as a drug on the basis of knowledge about related compounds.

How can we implement biosimulation within the drug development process? For these purposes, let us first take a look at what information can be taken from simulation models. They describe the temporal variation of a system in terms of the processes and interactions that are presumed to be at work. If the information obtained from these models is viewed in connection with the development of a drug, then the model will combine a pharmacokinetic description of the absorption, distribution, metabolism, and excretion of the drug with a detailed representation of the mechanisms responsible for its function and for the development of side-effects and possible synergetic interactions with other drugs (Chapters 3 and 4). To the extent that they are important, this description will include a representation of the drug's interaction with cellular receptors of the intracellular reaction cascades, and of the drug's effects on the intercellular communication or cellular energy metabolism (Chapters 8, 9, 11 and 14). It may also be important to examine interactions with specific organs (heart, liver, kidney, etc.) as well as with hormonal and immune regulations (Chapters 6, 9 and 12). These combined efforts will finally result in a simulation model that so to speak translates our knowledge about the biological system into mathematical equations. In the initial stage of the drug development process, one can use the simulation model to test any hypothesis one might have regarding the function of the drug vis-à-vis the established biological understanding. With approaches from bioinformation, one can change the product in order to optimize its function, and even before the first molecule is produced one can estimate the likelihood that a given agent will function as a drug (Chapter 17).

During the subsequent trial phase, the simulation model can be used as a vehicle to define an effective test protocol. The model can be used to check that the information obtained from the tests is consistent. Provided that it represents a proper representation of the underlying pharmacokinetic and biological mechanisms, the model can be used to predict the metabolism of a drug compound or its effect outside its normal physiological regime and under conditions not previously experienced (Chapter 2). To the extent that these predictions are substantiated, our understanding of the drug's action is gradually extended. If the predictions are false we must examine both the hypotheses and the experimental procedures. The advantage is here that any discrepancy between hypotheses and experimental results appear right away and in a clear manner.

At the present very few models exist that can describe a disease longitudinally, i.e. from its very start to the final cure. The advantage of simulation models is that they can represent available information about the relevant biological and pathological processes on any time scale. Hence, it is possible to describe how a disease develops over years with respect to the performed treatment (the so-called disease life-cycle). This is of particular interest in connection with modern life-style diseases where active collaboration and compliance by the patient plays an important role. By means of a simulation model health care providers can examine the long-term consequences of possible adjustments in the treatment, and from simplified versions of the models the patients may learn to appreciate the significance of following a prescribed treatment and/or keeping a specific diet. Insights gained from such models can also be used to tailor health care policies, public campaigns, or educational programs. Examples could include the models developed some 15 years ago to design policies against the spread of AIDS or the treatment of diabetes (Chapter 6).

The *3R Declaration of Bologna on Reduction, Refinement and Replacement Alternatives and Laboratory Animal Procedures* adopted by the Third World Congress on Alternatives and Animal Use in the Life Sciences (31 August 1999) calls for a reduced use of laboratory animals in cosmetic, chemical, and medical research through the use of alternative methods and through the use of methods that can produce the same information with fewer animal experiments. The United States Food and Drug Administration strongly advocates the use of computer simulation models, and except for the replacement of animal experiments by human experiments, it is unlikely that any other technique than biosimulation can accomplish the goals of the *3R Declaration*. Application of computer simulations in the drug development process can reduce the use of laboratory animals significantly through a more rational exploitation of the information acquired in each test and through a better planning of the experiments. As information is accumulated using *in silico* models, the use of laboratory animals can gradually be replaced by computer models. However, establishing the information needed in the biological simulation models cannot go without a minimum number of independent animal experiments. But their number will be significantly smaller than the number of animal experiments required by standard test procedures. In this way the biosimulation can contribute to the fulfillment of the goals of the *3R Declaration*. Precisely the same arguments for the use of a more rational test procedure also apply to experiments on volunteers. Biosimulation may therefore contribute towards the establishment of a *3R Declaration* pertaining to volunteer experiments.

All these issues are addressed by activities which are currently being conducted in part by the BioSim Network (www.biosim-network.net) which is a *Network of Excellence* under the *Life Sciences, Genomics and Biotechnology for Health Thematic Priority Area* of the Sixth European Framework Programme. The Network was initiated on 1 December 2005 through a grant from the European Commission. The immediate goal of the network is to develop a modeling and simulation approach that can render the drug development process more rational to the benefit of the patient. In a broader perspective the network aims at contributing in an essential manner to the development of an integrated and quantitative understanding of biological processes in accordance with the 3R principles.

BioSim includes a total of 40 partners, out of which 26 are academic research groups from various universities throughout Europe and nine are industrial partners. The network also includes the regulatory agencies from Denmark, Holland, Sweden, and Spain, since the introduction of simulation models in the drug development process will require significant changes in the way the regulatory authorities evaluate new drug candidates, and the regulatory authorities need to establish their own expertise in the use of simulation models. Industrial partners are involved in a variety of different activities from the development of new simulation software to the test of drug candidates. Half of the academic groups are primarily experimentally oriented, and the other half cover many forms of modeling expertise. The network also involves hospital departments that perform experimental treatments of depression, various forms of tremor, and cancer. In this way a platform for fruitful interdisciplinary collaboration has been established.

In this way biosimulation aims at developing insights, concepts, and illustrative examples that can support drug development by an approach in which mathematical modeling and simulation are used as essential tools. The goal is to introduce this methodology as a natural tool in the initial selection of new drug candidates, in the planning of *in vitro* and *in vivo* trials, and in the interpretation of the obtained results. Correctly applied, this approach should allow pointless lines of research to be exposed – and ineffective or toxic drug candidates to be abandoned – at an early stage in the process.

With this collection of 17 contributions this book reflects the broad variety of methods, technologies, and fields of applications that biosimulation adopts within the drug development process. What is further illustrated by these chapters is the high complexity of biological systems. This book also shows that enormous progress has already been achieved now in successfully converting the language of biology, biochemistry, and physiology into terms of mathematics, i.e. the translation of *in vivo* into *in silico*. Yet, even with this research already begun, it has become clear that everything that can be summed up today as a quantum leap in the mathematical treatment of biological systems can be only the starting point for much more complex research which cannot be realized except by more intensive interdisciplinary approaches than ever to understand how living systems work. With respect to drug development this book demonstrates that both much has been done and much remains to be done to improve drug development processes. All the approaches and strategies depicted here point clearly in one direction: toward the benefit of the patient.

It is, above all, especially this feature that characterizes the envisaged impact of biosimulation on drug development. The diverse research activities in neighboring scientific areas have too long been conducted in parallel. In addition each discipline has accumulated a wealth of knowledge, the combination of which with others is much more a must than an interesting option for patients' benefit. This was also the impetus for us to combine the biosimulation activities of different international research groups into one book. As will be obvious to the reader, biosimulation is a fascinating multi-faceted research field with a direct link to application. For this reason it is not only academic groups speaking in this book, but also industry whose job it is to translate scientific results in terms of commercially available drugs.

The editors of this book would like to thank all contributors for the professional manner in which they have demonstrated the usefulness of biological modeling and simulation in drug development. It is their commitment and their investment of effort and time we are indebted to, without which this book never would have been accomplished. Also we gratefully acknowledge the integrating activities of the European Commission who realized the potential of biosimulation in drug development with regard to the benefit of the patient and who invested additional effort in providing the research framework BioSim under the roof of which international activities in the field of biosimulation in drug development are being combined. Special thanks also go to Dr. Waltraud Wüst and Dr. Frank Weinreich from the publishing house Wiley-VCH who from the very beginning greatly supported our activities in preparing this book and who always had an open mind for our ideas and special wishes along the publishing process. Last but not least we would like to thank the assistance of Dr. Frank Bruggeman and Dr. Olga Sosnovtseva.

It is our concern to spread our enthusiasm for the fascinating and facet-rich field of biosimulation to a broad scientific community and share it with them. If this has been accomplished, then the goal of this book will have achieved its purpose.

Freiberg, Lyngby, Amsterdam
September 2007

Martin Bertau
Erik Mosekilde
Hans Westerhoff

List of Contributors

Atilla Altinok
Université Libre de Bruxelles
Unité de Chronobiologie Théorique
Faculté des Sciences
Campus Plaine, C.P. 231
1050 Brussels
Belgium

Marcus Belke
Philipps University of Marburg
Institute of Physiology
Deutschhausstrasse 2
35037 Marburg
Germany

Martin Bertau
Freiberg University of Mining and
Technology
Institute of Technical Chemistry
Leipziger Strasse 29
09599 Freiberg
Germany

Marival Bermejo
Universidad de Valencia
Facultad de Farmacia
Vicente Andrés Estellés s/n
46100 Burjassot (Valencia)
Spain

Anne Beuter
Université Bordeaux 2
Institut de Cognitique
146, rue Léo Saignat
33076 Bordeaux Cedex
France

Hans A. Braun
Philipps University of Marburg
Institute of Physiology
Deutschhausstrasse 2
35037 Marburg
Germany

Frank J. Bruggeman
Centrum voor Wiskunde en
Informatica (CWI)
Multiscale Modelling and Nonlinear
Dynamics
Kruislaan 413
1098 SJ Amsterdam
The Netherlands
and
Vrije Universiteit
Faculty of Earth and Life Sciences
Molecular Cell Physiology
De Boelelaan 1085
1081 HV Amsterdam
The Netherlands

Biosimulation in Drug Development. Edited by Martin Bertau, Erik Mosekilde, and Hans V. Westerhoff
Copyright © 2008 WILEY-VCH Verlag GmbH & Co. KGaA, Weinheim
ISBN: 978-3-527-31699-1

Lutz Brusch
Dresden University of Technology
Centre of Information Services and
High Performance Computing
01062 Dresden
Germany

Vicente Casabó
Universidad de Valencia
Facultad de Farmacia
Vicente Andrés Estellés s/n
46100 Burjassot (Valencia)
Spain

Gunnar Cedersund
University of Linköping
Department of Cell Biology
58185 Linköping
Sweden

Morten Colding-Jørgensen
Novo Nordisk A/S
Development Projects Management
Novo Allé
2880 Bagsværd
Denmark

Eugene Cox
Johnson & Johnson Pharmaceutical
Advanced PK/PD Modeling &
Simulation
Clinical Pharmacology
2340 Beerse
Belgium

Filip De Ridder
Johnson & Johnson Pharmaceutical
Research & Development
Advanced Modeling & Simulation
Biometrics
2340 Beerse
Belgium

Gemma L. Dickinson
University of Sheffield
School of Pharmacy and
Pharmaceutical Sciences
3.119 Stopford Building
Oxford Road
Manchester M13 9PT
United Kingdom

Christine Erikstrup Hallgreen
Novo Nordisk A/S
Development Projects Management
Novo Allé
2880 Bagsværd
Denmark

Juris Galvanovskis
University of Oxford
OCDEM, Department of Physiology
Churchill Hospital, Old Road
Oxford OX3 7LJ
United Kingdom

Albert Goldbeter
Université Libre de Bruxelles
Unité de Chronobiologie Théorique
Faculté des Sciences
Campus Plaine, C.P. 231
1050 Brussels
Belgium

Isabel González-Álvarez
Universidad de Valencia
Facultad de Farmacia
Vicente Andrés Estellés s/n
46100 Burjassot (Valencia)
Spain

Hanne Gürtler
Consultant to Novo Nordisk A/S
Living United Consult
Kongevejen 2
3450 Allerød
Denmark

René Normann Hansen
Novo Nordisk A/S
Development Projects Management
Novo Allé
2880 Bagsværd
Denmark

Hanna M. Härdin
Vrije Universiteit
Faculty of Earth and Life Sciences
Molecular Cell Physiology
De Boelelaan 1085
1081 HV Amsterdam
The Netherlands
and
Centrum voor Wiskunde en
Informatica (CWI)
Scientific Computing and Control
Theory
P.O. Box 94079
1090 GB Amsterdam
The Netherlands

Ulrich Hemmeter
University of Marburg
Department of Psychiatry and
Psychotherapy
Rudolf-Bultmannstrasse 43
35033 Marburg
Germany

Niels-Henrik Holstein-Rathlou
Department of Medical Physiology
Panum Institute
University of Denmark
2200 Copenhagen N
Denmark

Martin T. Huber
University of Marburg
Department of Psychiatry and
Psychotherapy
Rudolf-Bultmannstrasse 43
35033 Marburg
Germany

Mats Jirstrand
Fraunhofer-Chalmers Research Centre
for Industrial Mathematics
Chalmers Teknikpark
41288 Gothenburg
Sweden

Hui Kimko
Johnson & Johnson Pharmaceutical
Research & Development
Advanced PK/PD Modeling &
Simulation
Clinical Pharmacology
Raritan, NJ 08869
USA

Carsten Knudsen
The Technical University of Denmark
Department of Physics
Systems Biology Group
Fysikvej 309
2800 Kongens Lyngby
Denmark

Thomas Vagn Korsgaard
Novo Nordisk A/S
Development Projects Management
Novo Allé
2880 Bagsværd
Denmark

Ursula Kummer
Department of Modelling of
Biological Processes
Institute for Zoology/BIOQUANT
INF 267
69120 Heidelberg
Germany

Jürgen Kurths
University of Potsdam
Institute of Physics
Am Neuen Palais 10
14469 Potsdam
Germany

Jakob Lund Laugesen
The Technical University of Denmark
Department of Physics
Systems Biology Group
Fysikvej 309
2800 Kgs. Lyngby
Denmark

Thorsten Lehr
Boehringer Ingelheim Pharma
GmbH & Co. KG
Department of Drug Metabolism
and Pharmacokinetics
Birkendorfer Strasse 65
88397 Biberach an der Riss
Germany

Francis Lévi
INSERM U776
Université Paris Sud XI
Rythmes Biologiques et Cancers
Hôpital Paul Brousse
94800 Villejuif
France

Fabio Luciani
University of New South Wales
School of Biotechnology and
Biomolecular Sciences
2026 Sydney
Australia

Virginia Merino
Universidad de Valencia
Facultad de Farmacia
Vicente Andrés Estellés s/n
46100 Burjassot (Valencia)
Spain

Julien Modolo
Université Bordeaux 2
Institut de Cognitique
146, rue Léo Saignat
33076 Bordeaux Cedex
France

Surya Mohanty
Johnson & Johnson Pharmaceutical
Research & Development
Advanced Modeling & Simulation
Biometrics
1125 Trenton Harbourton Rd
Titusville, NJ 08560-1504
USA

Erik Mosekilde
The Technical University of Denmark
Department of Physics
Systems Biology Group
Fysikvej 309
2800 Kgs. Lyngby
Denmark

Harald Murck
Novartis Pharmaceuticals Corporation
US Clinical Development and
Medical Affairs
One Health Plaza, Building 701, 642B
East Hanover, NJ 07936-1080
USA

Denis Noble
Department of Physiology
Anatomy and Genetics
Parks Road
Oxford, OX1 3PT
United Kingdom

Ferenc Orosz
Hungarian Academy of Sciences
Institute of Enzymology
Biological Research Center
P.O. Box 7
1518 Budapest
Hungary

Judit Ovádi
Hungarian Academy of Sciences
Institute of Enzymology
Biological Research Center
P.O. Box 7
1518 Budapest
Hungary

Juan José Perez-Ruixo
Johnson & Johnson Pharmaceutical
Research & Development
Advanced Modeling & Simulation
Clinical Pharmacology
2340 Beerse
Belgium

Svetlana Postnova
Philipps University of Marburg
Institute of Physiology
Deutschhausstrasse 2
35037 Marburg
Germany

Patrik Rorsman
University of Oxford
OCDEM, Department of Physiology
Churchill Hospital, Old Road
Oxford OX3 7LJ
United Kingdom

Amin Rostami-Hodjegan
University of Sheffield
Academic Unit of Clinical
Pharmacology
M Floor Medicine and Pharmacology
Royal Hallamshire Hospital
Sheffield S10 2JF
United Kingdom

Mahesh Samtani
Johnson & Johnson Pharmaceutical
Research & Development
Advanced PK/PD Modeling &
Simulation
Clinical Pharmacology
Raritan, NJ 08869
USA

Hans Günter Schäfer
Boehringer Ingelheim Pharma
GmbH & Co. KG
Department of Drug Metabolism
and Pharmacokinetics
Birkendorfer Strasse 65
88397 Biberach an der Riss
Germany

Horst Schneider
Philipps University of Marburg
Institute of Physiology
Deutschhausstrasse 2
35037 Marburg
Germany

Bo Söderberg
Complex Systems Division
Department of Theoretical Physics
Lund University
22834 Lund
Sweden

Olga V. Sosnovtseva
Systems Biology Group
Department of Physics
The Technical University of Denmark
Fysikvej 309
2800 Lyngby
Denmark

Alexander Staab
Boehringer Ingelheim Pharma
GmbH & Co. KG
Department of Drug Metabolism
and Pharmacokinetics
Birkendorfer Strasse 65
88397 Biberach an der Riss
Germany

Peter Strålfors
University of Linköping
Department of Cell Biology
58185 Linköping
Sweden

Jan H. van Schuppen
Centrum voor Wiskunde en
Informatica (CWI)
Scientific Computing and Control
Theory
P.O. Box 94079
1090 GB Amsterdam
The Netherlands

An Vermeulen
Johnson & Johnson Pharmaceutical
Research & Development
Advanced PK/PD Modeling &
Simulation
Clinical Pharmacology
2340 Beerse
Belgium

Karlheinz Voigt
Philipps University of Marburg
Institute of Physiology
Deutschhausstrasse 2
35037 Marburg
Germany

Hans V. Westerhoff
Vrije Universiteit
Faculty of Earth and Life Sciences
Molecular Cell Physiology
De Boelelaan 1085
1081 HV Amsterdam
The Netherlands

and
Manchester Interdisciplinary Biocentre
(MIB)
131 Proncess Street
Manchester M1 7ND
United Kingdom

Bastian Wollweber
Philipps University of Marburg
Institute of Physiology
Deutschhausstrasse 2
35037 Marburg
Germany

Alexey Zaikin
University of Potsdam
Institute of Physics
Am Neuen Palais 10
14469 Potsdam
Germany

Part I
Introduction

1
Simulation in Clinical Drug Development

Juan Jose Perez-Ruixo, Filip De Ridder, Hui Kimko, Mahesh Samtani,
Eugene Cox, Surya Mohanty, and An Vermeulen

Abstract

Modeling techniques can be efficiently used to integrate and quantify the information collected throughout the drug development process, and to put in perspective other relevant external information with certain underlying assumptions. Stochastic models provide a way of integrating the available information, quantifying the relationship between an outcome of interest and underlying explanatory variables, and accounting for the inherent between subjects variability. The proper management of the knowledge generated with the modeling techniques implies the exploration of different scenarios of interest via simulation techniques, which can be utilized as a guiding tool at every decision making stage of a drug development. This process, traditionally termed as a modeling and simulation exercise, is illustrated in four different examples, which discuss the diverse situations at different stages of drug development. Each of these examples are real situations with key decisions made based on the knowledge obtained from these models. The best service that modeling and simulation can provide to the drug development engine is to elicit discussion on the unverifiable assumptions at each stage and guide in decision making with calculated risk. This approach will facilitate the development of better dosing regimens for new medicines, which ultimately will help to maximize the benefit of the drug therapy and improve the quality of life of the patients.

> *"The greatest thing in this world is not so much where we are, but what*
> *direction we are moving"*   Oliver Wendell Holmes (1809–1894)

1.1
Introduction

Over the past decade, the pharmaceutical industry as a whole has suffered from a scarcity of successful new drug molecules that have made it to the market. Despite the major advances in basic science and its clinical application, the success rate of

Biosimulation in Drug Development. Edited by Martin Bertau, Erik Mosekilde, and Hans V. Westerhoff
Copyright © 2008 WILEY-VCH Verlag GmbH & Co. KGaA, Weinheim
ISBN: 978-3-527-31699-1

drug development is decreasing over the last decades. The current medical product development path is becoming increasingly challenging, inefficient, and costly [1, 2]. A new compound entering Phase I development today has an 8% chance of achieving approval of a new drug application and, on average, the process will take 10–12 years and cost more than US$ $0.8–1.7 \times 10^{12}$ [3]. Several initiatives have been taken by pharmaceutical industry, drug regulators and the academic community in an attempt to change and improve the paradigm of drug development. The *Critical Path Initiative* taken by the FDA [1] and the *Innovative Medicine Initiative* of the European Union [2] highlight that the substantial evolution in basic science has not led to a similar innovation and improvement in the drug development process. If the costs and hurdles of medical product development continue to increase, innovation probably would stagnate or decline. The *Critical Path Initiative* identifies those phases of drug development where there is room for improvement by embracing new scientific and technological tools and by implementing model-based drug development. The concept of model-based drug development consists of building pharmaco-statistical models of drug efficacy and safety using all available pertinent data, and offers an important approach to improve drug development knowledge management and the decision-making process [3]. Modeling and simulation techniques applied to drug development were also recognized by the pharmaceutical industry as one of the emerging technologies that contribute to improve the business model for developing new therapeutically and commercially successful drugs, which ultimately will benefit the patients [4–6].

Over recent years, the field of modeling and simulation has clearly moved from using empirical functions to describe and summarize the data to the utilization of mechanism-based pharmacokinetic and pharmacodynamic (PK/PD, [7]) models. Mechanism-based modeling provides better quantification and prediction of drug disposition and dynamics as it reflects the essential underlying principles of pharmacology, physiology, and pathology. Models should be built to quantitatively characterize the time course of drug effects ('quantitative pharmacology') and generate hypotheses that are the objectives of subsequent studies. Mathematical and statistical expressions for the time course of the disease and the drug effects should be developed, linking measures of drug exposure (dose, plasma concentration, biophase concentration) to therapeutic or side-effects, and including relevant patient covariates [8]. At the start of drug development, models mostly are largely mechanistic in nature, and they tend to become more empirical as the drug moves through the different development phases. At the same time, the focus changes from understanding how the drug works to helping design studies that provide the highest probability of showing efficacy while at the same time optimizing safety. Modeling and simulation should thus be at the center of drug development and support the end-to-end process. Ultimately, the goal of drug development consists of bringing compounds to the market that show some benefits over existing therapy and have their place in the competitive landscape. Realizing that goal should be the overall objective of any modeling and simulation exercise.

Clinical trial simulation or computer-assisted trial design or biosimulation is not a revolutionary technique. Computer simulation has been used in the automotive and aerospace industry for more than 20 years, improving the safety and durability of new products at a reduced development cost. This concept of simulating clinical trials, taking careful note of the intricacies of the trials at hand is a very powerful tool. Implementation of clinical trial simulation requires the implementation of various models developed in a systematic way, mimicking the real clinical trial that is yet to be executed (see Section 1.2). This exercise increases the efficiency of drug development, i.e. minimizing cost and time, while at the same time maximizing the information contained in the data generated from the trial [8, 9]. It allows the possibility of being explicit about the assumptions and testing them out, generating a hypothesis and assessing its importance, and exploring numerous what-if scenarios, which would otherwise not be feasible in the real world. But most importantly, modeling and simulation techniques provide the scientific basis to select the optimal dosing regimen that will most benefit the patients, in terms of efficacy, safety, and quality of life.

In the implementation stage, it is crucial that the aim of the simulation exercise as well as all the assumptions behind the models be well defined. A challenge is to find what aspects are important in the model and which approximations can be safely made so that the complexity of the model will basically be determined by the availability of relevant data and the final goal of the modeling exercise. In order to allow for generalization, the model should not be more complex than needed or, in other words, as simple as possible, but not simpler. Clinical trial simulations can be used to quantitatively and objectively evaluate the chance of success of a plethora of trials (or drug development strategies) with different assumptions and thus optimize the decision-making process. It is key to note that variability in outcome and uncertainty associated with previous information must be incorporated appropriately in these simulations. Key design features such as doses, dosing schedules, inclusion criteria, study duration, number and timing of measurements, cross-over versus parallel treatments, superiority versus non-inferiority design can be explored and the sensitivity of the study outcome towards these variables could then easily be quantified. In essence, this technique formalizes the integration of subjective elements invariably used in drug development decisions, trial design used, and other information, in a quantitative manner and tests the sensitivity of the assumptions.

The integration of information via the utilization of modeling techniques is an efficient mean to quantify information, which not only aids in designing a clinical trial, but also provides the necessary information to supplement a knowledge-based decision-making process for drug development strategies. In this way, a quantitative risk–benefit assessment can be obtained by determining the probability of success of a new clinical trial, given a certain experimental design. Integration of clinical trial simulation with decision theory in a systematic way can provide an excellent platform for decision-making at any stage of drug development and could serve as an invaluable tool for any compound development strategy [10].

All of the above should ultimately result in applying the 'learning and confirming' paradigm as laid out by Lewis Sheiner as the basis of drug development, where in the early phases of drug development (Phases I and IIa) the focus lies on learning how the drug works, and in the later phases (Phases IIb and III/IV) on confirming that certain dosing regimen works in a specific patient population [11]. In this way, the drug development process is split in two major learn–confirm cycles. In the first, one learns in normal subjects what dose is tolerated (Phase I) and confirms that this dose has promising pharmacological activity in a selected group of patients (Phase IIa). In the second cycle, the goal of the learning step (Phase IIb) is to understand how to use the drug in a representative patient population to come to an acceptable risk/benefit ratio. The drug efficacy and safety are subsequently confirmed in a large and representative patient population (Phase III).

Modeling the pharmacokinetics of a drug using population approaches has become a commonplace, and population simulation is often used to predict the expected exposure of multiple doses using single-dose administration data, of different doses and dosing regimens, and how the pharmacokinetics of a drug will behave in patients with certain characteristics such as renal or liver impairment, among others [12, 13]. In many dossiers submitted to regulatory authorities, population pharmacokinetic analyses are an integral part of the submission, both in the United States and in Europe, and the results of covariate analysis are often used to optimize the drug label [14]. Population pharmacokinetic and pharmacodynamic analyses have seen a real evolution over the past few years, with many more mechanistic models being developed, specifically in the field of oncology, diabetes, hematology, and internal medicine, among others. The bulk of these models are developed using clinical data, though some models try to combine both preclinical (*in vitro*) and clinical data. However, still too little of that work is part of submission dossiers or provided to regulatory authorities in support of the decision-making processes about Phase II/III clinical study design, though that is only a matter of time [15].

The main objective of this chapter is not to provide an exhaustive review of the publications related to simulation in drug developments as the "*Simulation for Designing Clinical Trials*" was the topic of a book recently edited by Hui C. Kimko and Stephen B. Duffull [16], and the excellent articles about "*Model-Based Drug Development*" recently published by Lalonde et al. and Grasela et al. in "*Clinical Pharmacology and Therapeutics*" [17, 18]. Therefore, the reader is referred to the literature for further detailed discussion on technical and philosophical aspects of the clinical trial simulation. After a brief description of the different types of models used for simulations in clinical drug development, this chapter focuses on real-life examples where modeling and simulation techniques have been efficiently used in different stages of drug development. The examples presented encompass the full course of clinical development from Phase I to Phase III. They include the application of modeling and simulation techniques: to integrate the information obtained from preclinical stages of drug development and to predict the outcome of Phase I studies of erythropoietin receptor agonists; to select an optimal dosing regimen of an antimicrobial entering Phase II, on the basis of the information obtained from Phase I studies conducted in healthy subjects and *in vitro* information about the pathogen sensitivity to the drug; to design a Phase II dose-finding study to get support for the initiation (or not) of the Phase III program; and to predict the outcome of a Phase III study in light of the information gathered during the Phase II program.

1.2
Models for Simulations

Realistic predictions of study results based on simulations can be made only with realistic simulation models. Three types of models are necessary to mimic real study observations: system (drug–disease) models, covariate distribution models, and study execution models. Often, these models can be developed from previous data sets or obtained from literature on compounds with similar indications or mechanisms of action. To closely mimic the case of intended studies for which simulations are performed, the values of the model parameters (both structural and statistical elements) and the design used in the simulation of a proposed trial may be different from those that were originally derived from an analysis of previous data or other literature. Therefore, before using models, their appropriateness as simulation tools must be evaluated to ensure that they capture observed data reasonably well [19–21]. However, in some circumstances, it is not feasible to develop simulation models from prior data or by extrapolation from similar drugs. In these circumstances, what-if scenarios or sensitivity analyses can be performed to evaluate the impact of the model uncertainty and the study design on the trial outcome [22, 23].

System models, also known as input/output models, include pharmacokinetic/pharmacodynamic (PK/PD) models and disease progression models. The time course of drug concentration and its relationship with the response can be described using mathematical equations to illustrate what the body does to the drug (PK) and what the drug does to the body (PD). PK/PD modeling is widely used in drug development and there is extensive literature about this topic [8]. However, the importance of a disease progress model was recognized only recently [24–26]. Mathematical expressions for disease progression models may be derived from the data on placebo treatments included in clinical trials and in some (rare) circumstances databases may be available that provide data on the natural course of disease progression. System models are usually built using nonlinear mixed effects modeling, resulting in a fixed effects structural model that represents a theoretical typical value of the population and random effects models that describe between-subjects and within-subject variability, as well as the residual unexplained variability. In addition, the precision of the model parameter estimates must be obtained in order to account for statistical model uncertainty when simulations are performed [27].

Some factors or covariates may cause deviations from the population typical value generated from system models so that each individual patient may have different PK/PD/disease progression profiles. The relevant covariate effects on drug/disease model parameters are identified in the model development process. Clinical trial simulations should make use of input/output models incorporating covariate effects. The virtual patient population generated for the simulation should therefore display physiologically reasonable covariate distributions, which should be representative of real subjects participating in clinical trials. The distributions of these covariates need to be characterized in order to reflect the heterogeneity in the patient population. Some covariates may be correlated (e.g. age, weight) and the correlation between covariates should be considered as a joint distribution in the simulation in order to avoid generating unrealistic representations of virtual patients. The joint covariate distribution model may include demographic (e.g. body weight, age, sex), physiologic (e.g. renal function, hepatic function) and pathologic (e.g. baseline disease severity, minimum inhibitory concentration of microbes) aspects that are both representative of patients that will be enrolled in the trial as well as pertinent to the simulation model [28–30]. The particular interest on this regard is the new discipline, pharmionics, which is concerned with quantitative assessment of what patients do with prescription drugs [31].

Finally, it is inevitable that patients and investigators may deviate from the protocol during execution of a real clinical trial (e.g. by not taking a study drug, dropping out of the study, or not sampling according to a protocol). Such uncontrollable deviations yield different observations from a perfectly performed study according to a study protocol. Therefore, study execution models need to be introduced into the virtual study to let it mimic reality as closely as possible. Typically, study execution model(s) include patient-specific effects such as dropout and adherence of the patient to the dosing regimen and investigator effects such as measurement errors. The dosing algorithm may be straightforward as in a parallel fixed-dose study, but with a dose titration study more detailed dosing algorithms are needed to adjust doses according to simulated observations during the virtual study [32–34].

1.3
Simulations in Clinical Drug Development: Practical Examples

1.3.1
Predicting the Outcome of Phase I Studies of Erythropoietin Receptor Agonists

In this case study a simulation strategy, based on a mechanistic PK/PD model, was developed to predict the outcome of the first time in man (FTIM) and proof of concept (POC) study of a new erythropoietin receptor agonist (ERA). A description of the erythropoiesis model, along with the procedures to scale the pharmacokinetics and pharmacodynamics based on preclinical *in vivo* and *in vitro* information is presented. The Phase I study design is described and finally the model-based predictions are shown and discussed.

A mechanism-based PK/PD model for rHu-EPO was used to capture the physiological knowledge of the biological system. An open, two-compartment disposition model with parallel linear and nonlinear clearance, and endogenous EPO at baseline, was used to describe recombinant human erythropoietin (rHu-EPO) disposition after intravenous administration [35]. The pharmacodynamic effect of rHu-

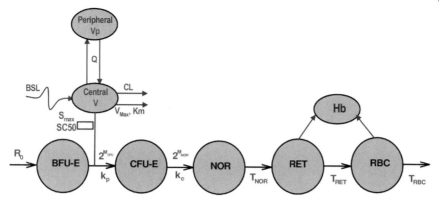

Fig. 1.1 Pharmacokinetic and pharmacodynamic model for rHu-EPO in healthy subjects.

EPO on hemoglobin was characterized using a previously developed model, based on the precursor-dependent indirect response model and cell lifespan concepts [36]. Figure 1.1 displays the scheme of the PK/PD model. Briefly, it is assumed that the progenitor cells BFU-E are generated at a constant rate of R_0 and the differentiation of BFU-E cells into CFU-E is controlled by processes with a first order rate of k_p. This maturation process is stimulated by rHu-EPO serum concentration according to a sigmoid function, characterized by the maximum stimulation of BFU-E to differentiate to CFU-E (S_{max}), and rHu-EPO serum concentration eliciting 50% of S_{max} (SC_{50}). The factor $2^{M_{CFU}}$ reflects the fact that one BFU-E cell gives rise to $2^{M_{CFU}}$ CFU-E cells. The CFU-E cells proliferate on average M_{NOR} times and transformed to normoblasts (NOR) according to the first-order rate constant k_c. It is further assumed that normoblasts are transformed to reticulocytes (RET) after their maturation time T_{NOR}. The rate of NOR elimination to the RET pool becomes the NOR production rate delayed by the time T_{NOR}. An analogous mechanism controls the transformation of RET to mature RBC (RBC) after their lifespan T_{RET} expires. The RBC production equals the RET transition rate and RBC are removed from the circulation due to senescence after their lifespan T_{RBC}. Hemoglobin concentration in blood is proportional to the absolute amount of RET and RBC, and the proportionality constant is the mean corpuscular hemoglobin.

A population-based interspecies allometric scaling was used to predict the expected pharmacokinetic profile in normal healthy subjects accordingly to the methodology described by Jolling et al. [37]. Data collected in rat, rabbit, dog, and monkey were used to extrapolate to humans the model parameters of an open, two-compartment disposition model with linear elimination using body weight and maximum lifetime potential as a scaling factor. With this approach the coefficients and exponents of the standard allometric scaling equations were estimated directly from the raw plasma concentration versus time data, which avoids the inherent bias arising from the classic allometric scaling procedure that does not consider within-species variability by using the estimation of mean pharmacokinetic parameters. Nevertheless, the standard interspecies allometric scaling has been successfully used recently to extrapolate the pharmacokinetics of rHu-EPO alpha and beta, and the parameters of the erythropoietic system [38]. Based on *in vitro* UT-7 cells proliferation assays [39], the same intrinsic activity and a 45-fold lower potency of the new ERA relative to rHu-EPO was assumed. This ratio was used to scale the potency of the new ERA in humans given the rHu-EPO potency in healthy subjects obtained previously [36]. In addition, a competitive agonist model was implemented via Gaddum's equation to quantify the pharmacodynamic interaction between the new ERA and endogenous EPO.

The Phase I study was designed as a randomized, double-blind, placebo-controlled, single ascending dose study to evaluate the pharmacokinetics and pharmacodynamics of the new ERA. Cohorts of nine subjects with baseline hemoglobin levels less than 14.5 g/dl were randomized to receive ERA treatment ($N=6$) or matched placebo ($N=3$) at each dose level. The various dose levels considered in this study were $1\times$, $3\times$, $10\times$, $30\times$, $100\times$, and $300\times$. The objective of the study was to identify the pharmacological effective dose (PED), defined as the dose level where four or more treated subjects achieved more than 1 g/dl increase from baseline in hemoglobin within 28 days.

Simulations were performed to evaluate the probability of identifying the PED with the current study design. One hundred cohorts of six subjects receiving ERA treatment were simulated per dose level according to the Phase I study design and the mechanistic PK/PD model described above. Body weight and endogenous EPO at baseline were obtained by resampling from a population of healthy subjects that were included in rHu-EPO Phase I studies [35]. Figure 1.2 displays the observed and the model-based predicted time course of the change in hemoglobin from baseline as a function of dose. From the simulated data, the number of subjects per cohort that achieved >1 g/dl increase in hemoglobin within 28 days was calculated and the probability of achieving the PED was derived (Fig. 1.3). Notably, the observed PED (vertical bold line) falls in the predicted dose range that suggests a 95% probability of achieving the PED.

This example shows that: (1) the mechanistic PK/PD model developed based on literature data of rHu-EPO and preclinical information of a new ERA is suitable to provide a better quantification and prediction of the drug disposition and the time course of hemoglobin in adult healthy subjects, and (2) this model can be used to optimize the design of the Phase I studies of new ERAs, with respect to key design features (number of dose levels, selection of dose levels, number of subjects per dose level, PK/PD sampling times). In this way, a quantitative risk–benefit assessment can be obtained by determining the probability of success of a Phase I study with new ERA, conditional on a certain experimental design.

1.3.2
Simulations for Antimicrobial Dose Selection

In this case study a simulation strategy was developed to select the dosing regimen of an antibiotic. A description of the critical factors for the interaction between drug, pathogen and host is presented, along with the procedures to integrate all

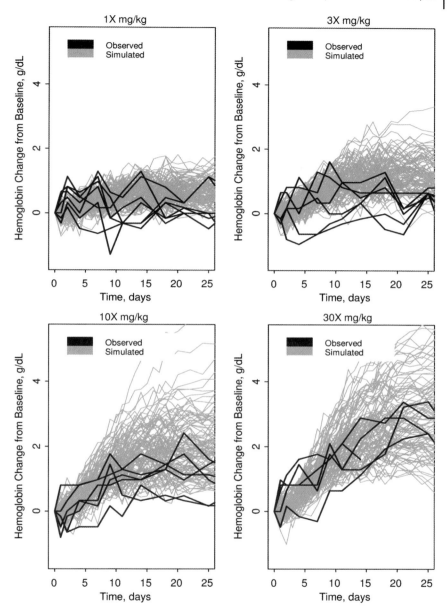

Fig. 1.2 Observations and model-based predictions for hemoglobin as a function of dose.

this information using modeling techniques. Finally, the probability of target attainment of the new anti-microbial agent against a specific pathogen of interest is
presented as a function of dose and the dose selection process is discussed.

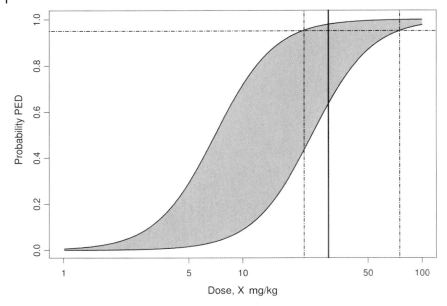

Fig. 1.3 Probability of achieving the pharmacological effective dose as a function of dose. The shaded area represents the uncertainty in the probability of achieving the pharmacological effective dose. Vertical dotted lines represent the dose range where more than 95% of the replicate cohorts have four or more subjects with hemoglobin increases of more than 1 g/dl within 4 weeks. The vertical bold line represents the observed pharmacological effective dose.

Application of biosimulations that take into account variability in pathogen susceptibility and human pharmacokinetics for bridging preclinical/early development and clinical dose selection is relatively new. The utility of modeling and simulations techniques for dose optimization of antimicrobials was first illustrated in 1998 to the FDA Anti-Infective Drug Products Advisory Committee for the antibiotic evernimicin [39]. Modeling and simulation techniques allow the integration of prior information regarding the pharmacokinetics of the drug in the target population and the differences in pathogen susceptibility to forecast the likelihood of success of a chosen dosing regimen in clinical trials.

The use of modeling and simulation techniques in the early drug development stages for antimicrobials has shown that a successful dose selection may be driven by four critical factors that describe the interaction between drug, pathogen and host [40]. These factors include: (1) the PK/PD target, (2) the distribution of pathogen susceptibility to the drug, (3) clinical pharmacokinetics (including interindividual variability), and (4) the drug's protein binding characteristics. The PK/PD target serves as a predictor of drug efficacy and is a drug exposure metric normalized to the pathogenicity of the virulent organism. The degree of pathogen virulence is obtained from *in vitro* experiments that measure the minimum inhibitory concentration (MIC) required to suppress bacterial growth. Drug exposure is represented by metrics such as area under the curve (AUC), peak concentration (C_{max}) or the percent of a dosing interval during which concentrations stay above the MIC (%T>MIC).

It has been shown that PK/PD targets are drug class-specific and fixed for particular drugs of interest. Despite the large number of antimicrobial agents, there is an increasing consensus that PK/PD targets for anti-infectives are similar across species [40]. This may be a realistic assumption because the PK/PD target reflects the drug's mechanism of action that is responsible for the *in vivo* interaction between the drug and the pathogen, and should therefore be independent of the host species. During early clinical development the choice of the PK/PD target and its magnitude are obtained from dose-fractionation experiments in the murine model of infection or from *in vitro* hollow fiber systems mimicking the *in vivo* dynamics [40]. Pathogens reside in the interstitial space between cells and the fraction of drug that is accessible to this effect site is the free concentration in plasma [41]. The PK/PD target obtained from preclinical experiments is therefore corrected for differences in protein binding between humans and the animal/*in vitro* setup. It is then recognized that in the clinical situation there exists between subject variability in human pharmacokinetics and there is a range of MIC values for pathogen susceptibility to the drug. Pathogen susceptibility and pharmacokinetic variability are brought together in a simulation model and each factor is described by a distribution of values.

Pharmacokinetic data reveals that depending on the variability in drug absorption and disposition, the future patient population will have a certain distribution for drug exposure metrics (AUC, C_{max}, %T>MIC; Fig. 1.4a). It should be realized that, even though the PK/PD target is fixed, the target exposure to be achieved at each MIC changes. As an illustration, if the AUC/MIC target is 100 h, then with each doubling of MIC the target AUC needed for successful treatment also has to double so that the target stays fixed. The concept, commonly referred to as dual individualization, implies that as the pathogens become more virulent greater drug exposure is needed to suppress their growth. The fraction of the exposure metric distribution attaining the target therefore changes with MIC and is usually reported as percent fractional target attainment versus MIC (Fig. 1.4b).

Similar to the variability in the human population, the pathogen of interest also displays variability in its susceptibility to the drug. The probability of being inhibited at a certain MIC is therefore obtained from a large collection of pathogen strains (usually several 100s to 1000s of strains) and the results are displayed as the percent fraction of strains inhibited at specific MICs (Fig. 1.4a). The probability of attaining the PK/PD target for the virtual population of patients infected by the specific pathogen is then obtained by multiplying the fractional target attainment at each MIC by the proportion of strains with that MIC. The product of fractions at each MIC is added and the overall probability of achieving the PK/PD target is derived. This computation explicitly answers the question about the probability of target attainment at specific candidate doses of the new anti-microbial agent against a specific pathogen of interest.

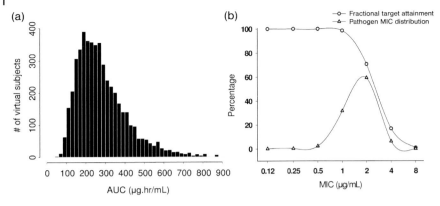

Fig. 1.4 (a) AUC distribution for a hypothetical drug for 5000 virtual subjects at a daily dose of 1 g. (b). Fractional target attainment against the intended pathogen and the distribution of pathogen virulence for the hypothetical drug.

When evaluating these results, one needs to be cognizant of the limitation that the number of subjects available for pharmacokinetic model development is usually small and the underlying population pharmacokinetic model is often derived from healthy subjects. PK/PD target attainment therefore commonly reflects what one would expect for a patient population with drug disposition similar to that of healthy subjects. Thus, Phase II/III studies are needed to understand how target attainment predicts microbiological efficacy and for validating the PK/PD target that is most appropriate for predicting clinical success [42]. Despite these limitations, information derived from these analyses is useful as an objective criterion for go/no go decisions regarding transition of a new anti-microbial agent into full development. In addition, modeling and simulation techniques are useful for comparing target attainment for a new compound versus comparators since the relevant information for marketed products is commonly available in the public domain. In summary, a method has been delineated for using microbiological, preclinical animal data, and PK information from early Phase I studies for the selection of doses with high probability of success in Phase II/III studies.

1.3.3
Optimizing the Design of Phase II Dose Finding Studies

In general, the objective of Phase II dose finding studies is to learn about and identify the dosing regimen that is expected to have the highest probability of achieving the appropriate attributes to meet the target drug profile. This target drug profile is typically defined in terms of obtaining sufficient/maximal efficacy and acceptable/minimal risk of adverse events and/or safety issues. These target dosing regimens may or may not be studied in the Phase II trials. After a successful Phase II program, the drug development program usually migrates to a Phase III program where a number of pivotal trials are conducted to confirm the expected target product profile of this target dosing regimen.

Typically, Phase IIb dose finding studies recruit about 50 to 200 subjects in a relevant population, which are randomized to one of the three to six treatment arms. The studies are nearly always placebo-controlled and sometimes include an active comparator arm. Notably, Phase IIb studies enable insight into other variables than drug exposure/dose that may explain the magnitude of the treatment response. For example, as illustrated in the next section, the severity of disease may be reflected in the pretreatment (baseline) value of the efficacy endpoints used in the evaluations of the Phase IIb study outcome. Also a relationship may be identified with the magnitude of drug response or other explanatory variables of treatment response that are related to either patient population (demographic variables, prior and concomitant drug treatment) or the drug treatment (such as different drug formulations). As such the term "dose finding" does not fully cover the potential and the objective of the Phase IIb study. Rather, the Phase IIb program is the key learning Phase in the clinical drug development program, prior to embarking on an elaborate, expensive and often lengthy Phase III development program and subsequent preparation of regulatory filing. The learning potential of these trials is therefore crucial to its success.

Recently, it has been suggested that Phase IIb clinical trials might help in supporting a single Phase III confirmatory trial. The Food and Drug Administration Modernization Act of 1997 established that the determination of substantial evidence of effectiveness as required for approval of a new drug needs to be based on data from one adequate and well controlled investigation and confirmatory evidence. Given adequate empiric confirmation of clinical benefit with one well designed Phase III clinical study and a known mechanism of pharmacologic action for the new candidate, an alternative route to demonstrate causal confirmation is establishing that the drug elicits the expected pharmacologic action. Confirmatory evidence is evidence that convincingly establishes pharmacologic causation of benefit, which could not necessarily be evidence from a replicate confirmatory trial, but could be additional evidence from one or more independent dose–response Phase IIb studies. Dose–response constitutes the strongest possible positive evidence of a pharmacologic mechanism of action and the Phase IIb trials that follow the temporal progress of biomarkers of drug effect can supply both empiric and causal confirmation of drug effectiveness. Although positive evidence of a dose–response relation in a single well designed and successful randomized-control trial does eliminate confounding, it does not eliminate transience or interaction, both of which depend on pharmacologic action plus some other factor. Independent evidence is needed to eliminate those possibilities and a single Phase III offering empiric confirmation of both effectiveness and pharmacologic mechanism of action, coupled with independent evidence of pharmacologic action from other studies provides persuasive support for the conclusion that drug effectiveness is independent of time and place and should be sufficient to provide substantial evidence of effectiveness [43].

The success of the trial should always be defined as the ability to learn and understand the quantitative relationship between treatment (dose) and nontreatment-related explanatory variables and the magnitude of drug response. This should result in the identification of the dosing regimen that is expected to result in the target drug response. Therefore the design of the trial(s) should be such that the likelihood of achieving these objectives is maximized. It is important to distinguish the expected success of the Phase II trial from the expected clinical success of the drug. In the context of the Phase II studies, the expected clinical success can be defined as the likelihood that a trial design can correctly identify the dose that causes the response of interest [44, 45]. If the drug is truly not expected to meet the required target product profile, that should be known after completion of the Phase II program (minimizing false-positive trial outcomes). In that situation, an optimized learning trial may allow to understand why the drug did not meet the target performance objective. As such, it may provide directions for improvement of clinical response, such as improving the formulations or the dosing regimen, or by focusing on a patient population that is more or less severely diseased. Conversely, if the drug is truly (intrinsically) successful, the trial needs to identify this drug's intrinsic potential (minimizing false-negative trial outcomes). Finally, the trial should enable the identification of the optimal treatment in the right patient population with a sufficient level of: (a) precision, and (b) absence of bias. This minimizes the risk of identifying a sub-optimal treatment in a random and a structural way, respectively. These metrics of trial success and how they can be utilized to optimize Phase IIb trial designs have already been discussed in the literature [44, 45] and are further explored in the illustration below.

Quantitative pharmacometric models provide powerful tools to optimize the design of Phase II trials in a constructive, objective, and transparent way. These models describe the relationship between the magnitude of expected drug/treatment response in a patient population based on the available preclinical and clinical data of the compound in development. For instance, initial signals of clinical efficacy may be obtained from a prior POC study. In addition, relevant information may be utilized with respect to expected placebo and drug responses in the patient population. The compound may be n-th in class and information on similar compounds in the patient population may be available either within a company or in the public. With these models, virtual trials are simulated and the outcome evaluated on the basis of the performance metrics discussed above. Trial parameters such as size, number of treatment groups, patient allocation, dose strengths, trial duration, patient inclusion criteria, etc., may now be changed in order to evaluate the impact of these parameters on the trial performance. It is important to realize that, in general, at this stage of drug development, the amount and specificity of prior information used to construct the models is fairly limited. Therefore the expected pharmacometric relationships come with a high degree of uncertainty and a substantial number of assumptions. Modeling and simulation techniques force to be explicit about the assumptions and allow to address the possible impact of these uncertainties and assumptions on the expected trial outcome in a systematic and comprehensive way.

Table 1.1 Expected trial performance (%) for various trial scenarios on the basis of precision and the probability of false- positive and false-negative trial outcomes.

Scenario	Dose levels (mg)	N (per dose level)	Precision[a]	False-positives[b]	False-negatives[c]
1 Reference	0, 50, 200, 400, 800	40	85	8	3
2 Skip 50 mg arm	0, 200, 400, 800	50	80	8	6
3 Reduce trial size by 50%	0, 50, 200, 400, 800	20	60	15	13
4 Skip 800 mg arm	0, 50, 200, 400	50	70	8	20

a) $0.5 < D_{20\%,est}/D_{20\%,true} < 2.0$
b) $P(D_{20\%,est} < 800 \mid D_{20\%,true} > 800)$
c) $P(D_{20\%,est} > 800 \mid D_{20\%,true} < 800)$

Drug X is in development for indication Y with an hypothetical arbitrary clinical response (ACR) used as primary endpoint. The target drug profile for X is to achieve at least a 20% reduction in ACR versus placebo. The maximum tolerated dose for Drug X, as obtained from Phase I tolerability studies, is 800 mg/day. A dose finding trial needs to be designed that maximizes the ability to identify the minimal dose that is expected to result in this target response ($D_{20\%}$). On the basis of available preclinical data, public domain data (placebo response), and preliminary clinical data from the POC study, a model is developed that relates expected ACR to various relevant explanatory variables, such as dose, time, baseline, etc. For a particular trial design (see Table 1.1), the expected ACR is simulated on the basis of the model parameters. The simulated data are analyzed and the estimates of the model parameters are obtained. Both the "true" $D_{20\%}$ ($D_{20\%,true}$) and the estimated $D_{20\%}$ ($D_{20\%,est}$) are calculated from the true model parameters (used for the simulation) and the estimated model parameters (obtained from the analysis of the simulated data), respectively, and compared with each other (Fig. 1.5). The closer the $D_{20\%,est}$ is to the $D_{20\%,true}$, the more successful is the trial. This exercise is replicated, for instance 1000 times, in order to account for uncertainty in model parameter, to evaluate the expected trial performance.

Figure 1.6 illustrates how the distribution of the distance of $D_{20\%,est}$ versus $D_{20\%,true}$ is used to evaluate the performance of the trial under a particular design. One possible simulated trial outcome is that the $D_{20\%,true}$ is less than 800 mg/day, but the trial identifies the $D_{20\%,est}$ to be higher than 800 mg (dark grey upper left rectangle). Given the unacceptable tolerability profile at doses higher than 800 mg/day, it would be decided to stop the program (false-negative outcome). Another trial outcome may be that the $D_{20\%,true}$ is higher than 800 mg/day, but the trial identifies the $D_{20\%,est}$ to be lower than 800 mg (dark grey lower right rectangle). The decision now would be to move forward into an elaborate Phase III program with a potentially non-efficacious compound (false-positive outcome). Clearly, one would like to minimize the probability of both false-positive and false-negative trial outcomes, although in certain instances, there may be incentives for false-positive

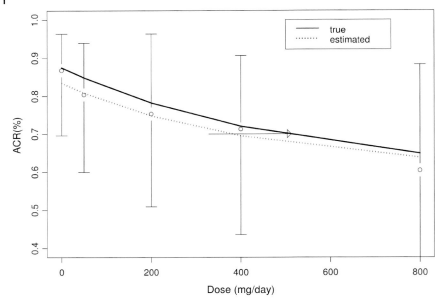

Fig. 1.5 Illustration of the simulation and analysis of a virtual trial outcome. The solid line represent the "true" dose–response relationship based on a sampled set of parameters from the joint posterior distribution of the model parameters. The circles represent the simulated drug effects in the patients included in the trial on the basis of the "true" model parameters and the errors bars represent the 95% confidence interval of the treatment effects in the various dose groups. The dashed line represents the estimated dose–response relationship on the basis of the model parameters as obtained form the analysis of the simulated trial data. The arrow indicates the deviation of the estimated $D_{20\%}$ from the true $D_{20\%}$.

outcomes (for instance a potential blockbuster compounds). After minimizing the probability of false-positive/negative outcomes, what remains is the objective to identify the target dose with an acceptable level of precision and accuracy (absence of bias). This minimizes the risk of identifying a sub-optimal dose (either too high or too low). Precision here is defined as the variance of the distribution of the ratio of $D_{20\%,est}$ over $D_{20\%,true}$. The level of precision that would be considered acceptable may be different from one program to the other, but it is important to note here that these definitions may easily be adapted. For illustration purposes, the precision being the ratio of $D_{20\%,est}$ over $D_{20\%,true}$ has been identified to be between 0.5 and 2.0. Finally, the trial parameters (such as overall trial size, number of treatment groups, patient allocation, dose) can be alternated in order to identify the design with the optimal expected performance according to the above-defined definitions of trial success (see Table 1.1).

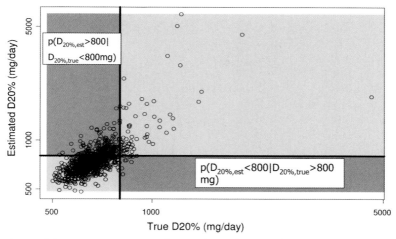

Fig. 1.6 Schematic representation of expected trial performance. For each simulated trial replicate (circles) the true $D_{20\%}$ derived from the model parameters used to simulate the replicate trial is compared with the estimated $D_{20\%}$ as obtained from the model parameters coming from the analysis of the simulated trial data. The dark grey upper left and lower right rectangles represent the probability of a false-negative and false-positive trial outcome, respectively. The light grey rectangles represent the probability of a true positive or negative trial outcome, regardless of the precision and bias of the estimated $D_{20\%}$ value.

1.3.4
Predicting the Outcome of Phase III Trials Using Phase II Data

Entry into Phase III is a very important step in clinical drug development, as it requires a large investment of both time and resources. Unfortunately, Phase III failure rates have been unacceptably high over the past decade. It is estimated that, out of 300 Phase III trials of major pharmaceutical companies, about 50% have failed. It is now generally recognized that modeling and simulation provide one of the new tools that can improve the efficiency of the drug development process by facilitating the quantitative decision-making process [46]. This section discusses how modeling and simulation may contribute to the success in Phase III, illustrated with a case study.

Obviously success in Phase III requires an (intrinsically) successful drug. Apart from commercial criteria, an (intrinsically) successful drug is one which yields medically meaningful effects in the target patient population when administered at a safe dose level. Whether a drug candidate is suitable for entry into Phase III needs to be assessed at the end of Phase II. This requires a thorough integration and interpretation of all available knowledge about the drug. Most of this information has been gathered in the preceding phases of drug development. Typically, this knowledge base consists of a multitude of data sources. Besides Phase II patient data, a lot of other data regarding safety and efficacy may be of interest, such as preclinical data, healthy volunteer data, or data from biomarkers. In addition, knowledge about the target patient population and the disease are also important to fully understand what will happen in Phase III. This information can come from previous clinical trials with the drug candidate or (unsuccessful) predecessors or public domain competitor data. Traditionally, this large, heterogeneous collection of data is interpreted in a global qualitative way, which is not always straightforward. Key features of this information are its variability (at various levels, e.g. among patients in a trial, between trials, etc.) and its uncertainty. Quantitative models can help to integrate data from a variety of sources in a rational way, and specifically include stochastic components to address variability and uncertainty. Of course, building useful models at this stage highly depends on the information that is available. All modeling and simulation activities in the earlier phases of drug development, which are described in the previous sections of this chapter, are therefore crucial to guarantee that optimal data are collected and that the right knowledge is generated for later use.

In order to decide to enter Phase III or not, all available knowledge has to be used to predict the clinical outcome (efficacy and safety) in Phase III. Phase III clinical trials differ in several aspects from Phase II trials, which makes this step not always easy. These aspects are basically: (1) the patient population, (2) the dose range studied, and (3) the primary endpoint.

The target population may differ from the Phase II population in baseline characteristics such as disease status and severity, age, and medical history. Phase IIb studies provide the setting to explore the role of these factors on the clinical outcome and account for the significant effects when predicting the outcome of Phase III. Although typically fewer doses are studied in Phase III than in Phase IIb, the choice of these doses is crucial in designing the Phase III trial and very much depends on the availability of an appropriate Phase IIb study design, study conduct, analysis, and interpretation. In addition, changes in formulation or dosing regimen might be considered for a variety of reasons. Again, quantitative models can be used to predict Phase III outcomes under these circumstances; always quantifying both the expected effects as well as the uncertainty. In many cases, the definition of the primary endpoint to measure clinical efficacy is different in Phase III trials. Frequently, the Phase III endpoint is a simple summary measure, such as the percentage of patients that respond to treatment, which is derived from more elaborate measurements taken during the trial. The most extreme situation is where the Phase II endpoint is a surrogate marker. In any case, a model that relates the different endpoints can be helpful here.

When the decision is made to move into Phase III, adequate clinical trials need to be designed and executed. The models as used before, supplemented with specific components for trial execution, can provide a better understanding of the performance of different trial designs. This is the realm of what is usually labeled as clinical trial simulation (CTS). The performance of a Phase III trial is usually measured by its statistical power. Given the trial design and assuming that the drug, as used in the trial, has a certain effect in the patient population studied, what is the probability of showing a statistically significant benefit over a control group? In principle, the statistical power of a trial depends on three quantities: (1) the magnitude of the postulated treatment benefit, (2) the expected variability of the outcome among patients in the population, and (3) the number of patients. Of these three factors, only the last one is under quantitative control when the trial is designed. The first two quantities are not precisely known and depend on a variety of other factors, including trial design features such as doses, in-/exclusion criteria, dropouts, trial duration, etc.

De Ridder [47] described the application of modeling and simulation techniques in the transition from Phase II to Phase III for a drug for symptom relief in a chronic condition. A dose–response model for the clinical endpoint, in this case the proportion of patients achieving a predefined level of symptom relief, was developed using data of two different Phase II dose finding studies in patients. The model was based on the longitudinal profiles of changes in symptom scores, but all predictions were transformed to percent responders, which was the planned Phase III clinical endpoint. The two Phase II studies differed, among other things, in the range of baseline disease severity of patients that were included. As a consequence, a profound difference in the dose–response of the percent responders between studies was observed (Fig. 1.7a). In the study with less severe patients, the placebo response was higher and the additional effect of treatment limited. This

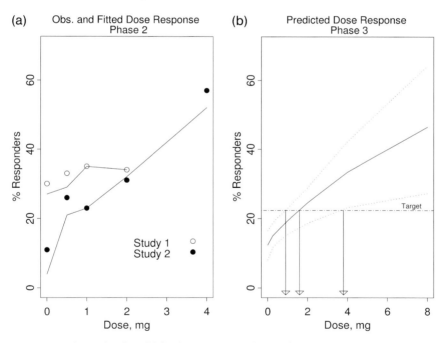

Fig. 1.7 (a) Observed and model fitted percent responders in Phase II trials. (b) Predicted dose–response of response (% responders) in a Phase III population based on a population simulation of the Phase II model. The dotted lines are the 5th and 95th percentiles of the prediction distribution and represent uncertainty in the model parameters.

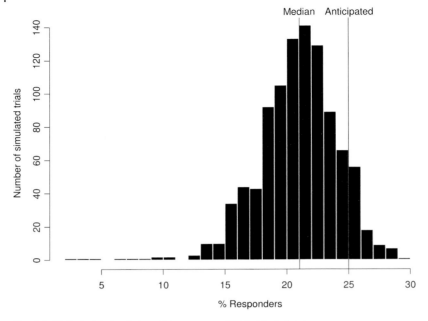

Fig. 1.8 Clinical trial simulation of the response (%) in blinded data after enrolment of 50% of the patients.

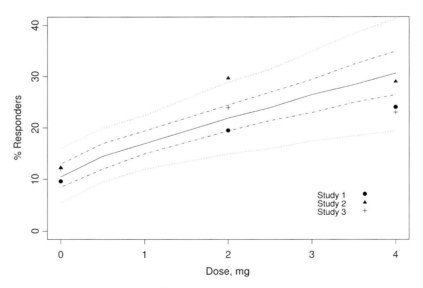

Fig. 1.9 Clinical trial simulation of the dose–response of the response (% responders) in Phase III trials compared with the actual observed outcome in the three trials. Solid line: median. Dotted lines: 5th, 25th, 75th, and 95th percentiles. Each arm of a study enrolled about 200 patients.

result, however, could be explained by the difference in baseline severity between the studies. A dose–response model including a component describing the relationship between baseline severity and drug effect adequately described the data of both studies. The model also allowed predicting the dose–response in the Phase III population, with restricted inclusion criteria (Fig. 1.7b). From this predicted dose–response, the target dose which would yield a clinically relevant placebo-subtracted response of 10%, along with its uncertainty interval, was estimated, supporting the inclusion of 2 mg and 4 mg in the Phase III trials. In addition, the model was also used to assess the robustness of the ongoing Phase III trials with respect to uncertainty of the true dose–response, patient variability in baseline severity, and drug–response.

Once the Phase III clinical trials were started, a blinded data survey was done when 50% of the patients were enrolled. This is not an uncommon practice. Blinded data can be used to assess the appropriateness of assumptions that were made when designing the trials, such as the overall response or variability, among others. However, to judge the observed data it is necessary to have an idea of what reasonably can be expected. Due to the multitude of factors involved, clinical trial simulations can help here. The observed blinded response rate was lower than the antipicitated value, but was well within what could be expected from the simulation results (Fig. 1.8). Finally, when the true outcome of the Phase III trial was compared with the predictions from the clinical trial simulations, an overall good agreement was observed (Fig. 1.9).

1.4
Conclusions

Information collected throughout the process of drug development can be systematically integrated and quantified using modeling techniques, which also allow the incorporation of other relevant external information with certain underlying assumptions. Models provide a way of describing and quantifying the relationship between an outcome of interest and underlying explanatory variables so that, at every decision-making stage of drug development, the models incorporating available information can be utilized as a guiding tool. Due to the inherent variability in the subjects, these models have a stochastic element. The clinical outcome of an individual or a group of individuals is often the interest at a decision-making juncture. The extreme cases in the realm of assumptions in any model can provide us with the boundary of the outcome space. It is crucial to understand the realm of possible outcomes and the distribution thereof before embarking on a decision. Simulations, based on these stochastic models can easily provide the distribution of outcome space. These distributions, with appropriate decision-making criteria can provide reasonable confidence for each subsequent step. This process, traditionally termed as a modeling and simulation exercise is well illustrated in four different examples in this chapter. These examples discuss the diverse situations and differing stages of drug development. Each of these examples is a real situation with key decisions made based on the stress test of these models. The best service that modeling and simulation can provide to the drug development engine is to elicit discussion on the unverifiable assumptions at each stage and guide in decision making with calculated risk. This approach facilitates the development of better dosing regimens for new medicines, which ultimately helps to maximize the benefit of the drug therapy and improve the quality of life of the patients.

Acknowledgments

This chapter is in honor to Vladimir Piotrovsky, who devoted a big part of his career to the application and implementation of modeling and simulation tools in pharmaceutical drug development at Johnson & Johnson Pharmaceutical Research & Development and beyond. We all miss his guidance and support.

References

1 U.S. Food and Drug Administration: *Innovation/stagnation. Challenge and opportunity on the critical path to new medical products. View from the U.S. Food and Drug Administration (FDA)*. U.S. Department of Health and Human Services, Food and Drug Administration, Washington, D.C., **2004**.

2 EFPIA: *Innovative medicine for Europe*. EFPIA, Brussels, **2004**.

3 Tufts Center for the Study of Drug Development: *Backgrounder: how new drugs move through the development and approval process*. Tufts Center, Boston, 2001; and Gilbert, J., P. Henske, and A. Singh: Rebuilding big pharma's business model. *InVivo* (The Business & Medicine Report), *Windhover Information*, 21, No. 10, **2003**.

4 Zhang, L., Sinha, V., Forgue, S. T., Callies, S., Ni, L., Peck, R., Allerheiligen, S. R.: Model-based drug development: the road to quantitative pharmacology. *J Pharmacokinet Pharmacodyn* **2006**, 33:369–393.

5 Miller, R., Ewy, W., Corrigan, B. W., Ouellet, D., Hermann, D., Kowalski, K. G., Lockwood, P., Koup J. R, Donevan, S., El-Kattan, A., Li, C. S., Werth, J. L., Feltner, D. E., Lalonde, R. L.: How modeling and simulation have enhanced decision making in new drug development. *J Pharmacokinet Pharmacodyn* **2005**, 32:185–197.

6 Gieschke, R., Steimer, J. L.: Pharmacometrics: modelling and simulation tools to improve decision making in clinical drug development. *Eur J Drug Metab Pharmacokinet* **2000**, 25:49–58.

7 Danhof, M., de Jongh, J., De Lange, E. C., Della Pasqua, O., Ploeger, B. A., Voskuyl, R. A.: Mechanism-based pharmacokinetic-pharmacodynamic modeling: biophase distribution, receptor theory, and dynamical systems analysis. *Annu Rev Pharmacol Toxicol* **2007**, 47:357–400.

8 Sheiner, L. B., Steimer, J. L.: Pharmacokinetic/pharmacodynamic modeling in drug development. *Annu Rev Pharmacol Toxicol* **2000**, 40:67–95.

9 Lockwood, P., Ewy, W., Hermann, D., Holford, N.: Application of clinical trial simulation to compare proof-of-concept study designs for drugs with a slow onset of effect; an example in Alzheimer's disease. *Pharm Res* **2006**, 23:2050–2059.

10 Holford, N. H. G., Kimko, H. C., Monteleone, J. P. R., Peck, CC.: Simulation of clinical trials. *Annu Rev Pharmacol Toxicol* **2000**, 40:209–234.

11 Sheiner, L. B.: Learning versus confirming in clinical drug development. *Clin Pharmacol Ther* **1997**, 61:275–291.

12 Piotrovsky, V.: Pharmacokinetic and pharmacodynamic variability: estimation and appraisal of its impact on dose optimization with an example of gender differences. In: Krishna, R., ed. *Dose optimization in drug development*. (Drugs and the

pharmaceutical sciences, volume 161) Marcel Dekker, New York, **2006**.

13 Piotrovsky, V., Van Peer, A., Van Osselaer, N., Armstrong, M., Aerssens, J.: Galantamine population pharmacokinetics in patients with Alzheimer's disease: modeling and simulations. *J Clin Pharmacol* **2003**, 43:514–523.

14 Sun, H., Fadiran, E. O., Jones, C. D., Lesko, L., Huang, S. M., Higgins, K., Hu, C., Machado, S., Maldonado, S., Williams, R., Hossain, M., Ette, E. I.: Population pharmacokinetics. A regulatory perspective. *Clin Pharmacokinet* **1999**, 37:41–58.

15 Gobburu, J. V., Marroum, P. J.: Utilisation of pharmacokinetic-pharmacodynamic modelling and simulation in regulatory decision-making. *Clin Pharmacokinet* **2001**, 40:883–892.

16 Kimko, H. C., Duffull, S. B.: *Simulation for designing clinical trials. A pharmacokinetic–pharmacodynamic modeling perspective.* (Drugs and the pharmaceutical sciences, volume 127) Marcel Dekker, New York, **2003**.

17 Lalonde, R. L., Kowalski, K. G., Hutmacher, M. M., Ewy, W., Nichols, D. J., Milligan, P. A., Corrigan, B. W., Lockwood, P. A., Marshall, S. A., Benincosa, L. J., Tensfeldt, T. G., Parivar, K., Amantea, M., Glue, P., Koide, H., Miller, R.: Model-based Drug Development. *Clin Pharmacol Ther* **2007**, 82:21–32.

18 Grasela, T. H., Dement, C. W., Kolterman, O. G., Fineman, M. S., Grasela, D. M., Honig, P., Antal, E. J., Bjornsson, T. D., Loh, E.: Pharmacometrics and the transition to model-based development. *Clin Pharmacol Ther* **2007**, 82:137–142.

19 Yano, Y., Beal, S. L., Sheiner, L. B.: Evaluating pharmacokinetic/ pharmacodynamic models using the posterior predictive check. *J Pharmacokinet Pharmacodyn* **2001**, 28:171–192.

20 Williams, P. J., Ette, E. I.: Determination of model appropriateness. In: Kimko, H. C., Duffull, S. B., eds. *Simulation for designing clinical trials. A pharmacokinetic–pharmacodynamic modeling perspective.* (Drugs and the pharmaceutical sciences, volume 127) Marcel Dekker, New York, **2003**.

21 Mentre, F., Escolano, S.: Prediction discrepancies for the evaluation of nonlinear

mixed-effects models. *J Pharmacokinet Pharmacodyn* **2006**, 33:345–367.

22 Nestorov, I. A.: Sensitivity analysis of pharmacokinetic and pharmacodynamic models in clinical trial simulation and design. In: Kimko, H. C., Duffull, S. B., eds. *Simulation for designing clinical trials. A pharmacokinetic–pharmacodynamic modeling perspective.* (Drugs and the pharmaceutical sciences, volume 127) Marcel Dekker, New York, **2003**.

23 Kraiczi, H., Frisen, M.: Effect of uncertainty about population parameters on pharmacodynamics-based prediction of clinical trial power. *Contemp Clin Trials* **2005**, 26:118–130.

24 Chan, P. L. S., Holford NHG: Drug treatment effects on disease progression. *Annu Rev Pharmacol Toxicol* **2001**, 41:625–659.

25 Piotrovsky, V.: Drug efficacy analysis as an exercise in dynamic (indirect-response) population PK-PD modelling. *Population Approach Group in Europe (PAGE) Meeting*, Paris, **2002**. Available at: http://www.page-meeting.org/default.asp?abstract=305.

26 Post, T. M., Freijer, J. I., DeJongh, J., Danhof, M.: Disease system analysis: basic disease progression models in degenerative disease. *Pharm Res* **2005**, 22:1038–1049.

27 Holford, N. H. G.: Input–output models. In: Kimko, H. C., Duffull, S. B., eds. *Simulation for designing clinical trials. A pharmacokinetic–pharmacodynamic modeling perspective.* (Drugs and the pharmaceutical sciences, volume 127) Marcel Dekker, New York, **2003**.

28 Mould, D. R.: Defining covariate distribution models for clinical trial simulation. In: Kimko, H. C., Duffull, S. B., eds. *Simulation for designing clinical trials. A pharmacokinetic–pharmacodynamic modeling perspective.* (Drugs and the pharmaceutical sciences, volume 127) Marcel Dekker, New York, **2003**.

29 Tannenbaum, S. J., Holford, N. H., Lee, H., Peck, C. C., Mould, D. R.: Simulation of correlated continuous and categorical variables using a single multivariate distribution. *J Pharmacokinet Pharmacodyn* **2006** [Epub ahead of print].

30 Tannenbaum, S. J., Holford, N. H., Lee, H., Peck, C. C., Mould, D. R.: Simulation of correlated continuous and categorical variables using a single multivariate distribution. *J Pharmacokinet Pharmacodyn* **2006**, 33:773–794.

31 Urquhart, J.: Pharmionics: research on what patients do with prescription drugs. *Pharmacoepidemiol Drug Saf* **2004**, 13:587–590.

32 Kastrissios, H., Girard, P.: Protocol deviations and execution models. In: Kimko, H. C., Duffull, S. B., eds. *Simulation for designing clinical trials. A pharmacokinetic–pharmacodynamic modeling perspective.* (Drugs and the pharmaceutical sciences, volume 127) Marcel Dekker, New York, **2003**.

33 Kenna, L. A., Sheiner, L. B.: Estimating treatment effect in the presence of noncompliance measured with error: precision and robustness of data analysis methods. *Statist Med* **2004**, 23:3561–3580.

34 Girard, P.: Clinical trial simulation: a tool for understanding study failures and preventing them. *Basic Clin Pharmacol Toxicol* **2005**, 96:228–234.

35 Olsson Gislekog, P., Jacqmin, P., Perez-Ruixo, J. J.: Population pharmacokinetics meta-analysis of recombinant human erythropoietin in healthy subjects. *Clin Pharmacokin* **2006** (in press).

36 Krzyzanski, W., Jusko, W. J., Wacholtz, M. C., Minton, N., Cheung, W. K.: Pharmacokinetic and pharmacodynamic modeling of recombinant human erythropoietin after multiple subcutaneous doses in healthy subjects. *Eur J Pharm Sci* **2005**, 26:295–306.

37 Jolling, K., Perez-Ruixo, J. J., Hemeryck, A., Vermeulen, A., Greway, T.: Mixed-effects modelling of the interspecies pharmacokinetic scaling of pegylated human erythropoietin. *Eur J Pharm Sci* **2005**, 24:465–475.

38 Woo, S., Jusko, W. J.: Interspecies comparison of pharmacokinetics and pharmacodynamics of recombinant human erythropoietin (rH-EPO). *Drug Metabolism and Disposition* **2007** (in press).

39 Erickson-Miller, C. L., Pelus, L. M., Lord, K. A.: Signaling induced by erythropoietin and stem cell factor in UT-7/Epo cells: transient versus sustained proliferation. *Stem Cells* **2000**, 18:366–373.

40 Ambrose, P. G. Monte Carlo simulation in the evaluation of susceptibility breakpoints: predicting the future: insights from the society of infectious diseases pharmacists. *Pharmacotherapy* **2006**, 26:129–134.

41 Drusano, G. L., Preston, S. L., Hardalo, C., Hare, R., Banfield, C., Andes, D., Vesga, O., Craig, W. A.: Use of preclinical data for selection of a phase II/III dose for evernimicin and identification of a preclinical MIC breakpoint. *Antimicrob Agents Chemother* **2001**, 45:13–22.

42 Craig, W. A.: Pharmacokinetic/pharmacodynamic parameters: rationale for antibacterial dosing of mice and men. *Clin Infect Dis* **1998**, 26:1–10.

43 Preston, S. L., Drusano, G. L., Berman, A. L., Fowler, C. L., Chow, A. T., Dornseif, B., Reichl, V., Natarajan, J., Corrado, M:. Pharmacodynamics of levofloxacin: a new paradigm for early clinical trials. *JAMA* **1998**, 279:125–129.

44 Peck, C. C., Rubin, D. B., Sheiner, L. B.: Hypothesis: a single clinical trial plus causal evidence of effectiveness is sufficient for drug approval. *Clin Pharmacol Ther* **2003**, 73:481–490.

45 Lockwood, P. A., Cook, J. A., Ewy, W. E., Mandema, J. W.: The use of clinical trial simulation to support dose selection: application to development of a new treatment for chronic neuropathic pain. *Pharm Res* **2003**, 20:1752–1759.

46 Mandema, J. W., Hermann, D., Wang, W., Sheiner, T., Milad, M., Bakker-Arkema, R., Hartman, D.: Model-based development of gemcabene, a new lipid-altering agent. *AAPS J* **2005**, 7:E513–E522.

47 Burman, C. F., Hamren, B., Olsson, P.: Modelling and simulation to improve decision-making in clinical development. *Pharmaceut Statist* **2005**, 4:47–58.

48 De Ridder, F.: Predicting the outcome of phase III trials using phase II data: a case study of clinical trial simulation in late stage drug development. *Basic Clin Pharmacol Toxicol* **2005**, 96:235–241.

2
Modeling of Complex Biomedical Systems

E. Mosekilde, C. Knudsen, and J. L. Laugesen

Abstract

The purpose of this chapter is to provide a first introduction to some of the main concepts and ideas of mechanism-based modeling of biomedical systems. We present three examples of the use of modeling and simulation techniques to establish a quantitative description of different pharmacological and physiological phenomena related to insulin absorption from the skin and to the regulation of pancreatic insulin secretion. Our first example describes some of the considerations underlying the formulation and test of a model of ultradian oscillations of insulin secretions in man. The second example describes the formulation of a model that can explain an observed initial slow absorption of soluble insulin following a subcutaneous bolus injection. Finally the last example analyses a simplified model of a bursting pancreatic beta-cell. The idea is here to underline the importance of nonlinear dynamic phenomena in the regulation of intracellular processes as well as of the cell-to-cell communication and the processes that control the secretion of insulin. The ultradian oscillations observed at the systemic level may similarly be involved in the regulation of the pancreatic activity.

> *"The success we have had in the medical treatment of many diseases by far outstrips our understanding of the underlying biological processes."*
> D. Marsh, School of Medicine, Brown University

2.1
Introduction

Opportunities in the pharmaceutical industry are fantastic, and so are the challenges: There are large and rapidly growing needs for improved medical treatments of cancer, diabetes, depression, and a variety of other chronic progressive disorders and lifestyle diseases. We need drugs to fight infections caused by HIV, SARS, and avian flue, and to treat the many deadly diseases that plague the developing countries. New types of antibiotics must be developed to overcome the spreading

Biosimulation in Drug Development. Edited by Martin Bertau, Erik Mosekilde, and Hans V. Westerhoff
Copyright © 2008 WILEY-VCH Verlag GmbH & Co. KGaA, Weinheim
ISBN: 978-3-527-31699-1

of resistant bacteria. "Personalized medicine" is required to take hand of the large inter-patient variability in drug action and "biological drugs" to improve specificity.

Yet, as explained in the preceding chapter [1], the immediate prospects for the pharmaceutical industry are fairly bleak: While the expenditures for research and development continue to rise almost exponentially, the outcome in terms of new successful drugs has shown a clear tendency to decline, and today the development costs for a single drug can easily run into the one billion Euro bracket. Moreover, two out of three drugs fail in the last phase of the development process, at a stage where enormous investments in time and money have already been made.

The large development costs are associated with the significant number of tests that a drug candidate must undergo in order to prove both its efficacy and its lack of adverse side-effects. A major concern in the pharmaceutical industry is, therefore, to develop methods that can be used to screen out poorly functioning or unacceptable drug candidates as early as possible in the development process. Molecular dynamics [2] and various forms of bioinformatics [3] can be used to design and select compounds with specific molecular structures, physicochemical properties (solubility, diffusion constants, etc.), affinities to various cellular receptors, or metabolic pathways in the organism.

Development of biomarkers [4] represents another approach in the search for indicators and useful predictive methods that can provide early information about the efficacy and toxicity of a drug candidate. The rise in blood glucose concentration after a meal may be considered as a "biomarker" for diabetes, and the ability of a particular insulin variant to quickly regulate the glucose level could be used as a biomarker for its efficacy. It is also possible, for instance, that one can test the extent to which different food components can prevent specific forms of cancer within a few weeks by using urine samples to determine their ability to repair DNA damage [5]. One of the challenges in this area is to find a good biomarker that can distinguish, for instance, between the relief of noncausal symptoms and the treatment of the actual disease. An alleged biomarker may also turn out to be associated with the immediate response to the drug and fail to account for adaptation and other more long-term processes. It is obvious that a precise understanding of the biological processes underlying the considered disease will be important to build confidence in a particular biomarker. At the same time efforts are being made to use bioinformatics and molecular approaches in the identification of new biomarkers [6].

Drug development will continue to require an extensive test program in order to precisely delineate dose ranges and patient groups and to ensure that unwanted side-effects are minimized. The purpose of the present book is to argue for the use of modeling and simulation techniques as a third component in the establishment of the predictive methods that are necessary for a rational drug development process. Over the past two or three decades, the application of computer simulations made it possible for industries in most of the classic engineering fields to test new concepts and designs long before the first prototype of a product was manufactured. The result of this approach has been significant improvements in product quality with parallel reductions in the development costs.

The continuous confrontation of model predictions with experimental results and *vice versa* is also a hallmark of most areas of natural science. Advanced mathematical concepts and methods have traditionally been considered less useful in the life sciences, partly, of course, because living systems are so much more complicated than most of the systems that we have been concerned with in physics and chemistry. However, this state of affairs is coming to an end. Today *Systems Biology* is considered as one of the most exciting areas of science, and many professional biological and physiological journals publish detailed model studies, side by side with purely empirical studies.

As described in several chapters of the present book [1, 7], the application of pharmacokinetic (PK) and pharmacodynamic (PD) methods is widely accepted in the pharmaceutical industry. A PK model typically predicts the availability of a drug in the blood and interstitial spaces at different times after the drug has been administered. The model is used to determine characteristic parameters of the absorption, distribution, metabolism, and excretion processes from experimentally observed time courses, or the model follows the rates of formation and removal of various metabolites. PD models describe the effects of the drug (and its metabolites) as a function of time, again based on statistical fits to experimental results.

The PK/PD approach is rapidly developing in the direction of so-called physiologically based PK/PD models [7, 8] where the availability and effects of a drug are calculated in different organs in dependence, for instance, of the blood perfusion, the binding affinities, and the metabolic rates of the various tissues. At the same time, as the models increase in size, some of the parameters are most appropriately fixed through independent experiments. In this way, the physiologically based PK/PD models start to bridge the gap between classic PK/PD and mechanism-based modeling.

The generic model structure may be pictured as a set of coupled differential equations:

$$\frac{d\mathbf{x}}{dt} = \mathbf{F}(\mathbf{x}, \mathbf{p}, t) \tag{1}$$

representing the material conservation equations for the considered chemical species and specified compartments. Here, $\mathbf{F}$ describes the structure of the system, i.e. the interactions among the various variables in terms of biological regulatory mechanisms, physical or chemical laws, etc. The vector $\mathbf{x}$ represents the state variables, $\mathbf{p}$ the parameters, and t time. One can assume that the structure $\mathbf{F}$ of the system is known. Using appropriate statistical methods, one can then determine the values of the various parameters by fitting the model predictions to observed time courses.

Alternatively, one can consider determination of the structure as the objective of the analysis. In this case, the parameters must be known from independent measurements, and one can then use the response of the system to different external challenges to determine the most appropriate structure. We refer to this approach as mechanism-based or, in contrast to the empirical and data-driven methods, as hypothesis-driven.

In the mechanism-based approach, a model is validated by its ability to reproduce observed temporal behaviors, i.e. wave forms, phase relationships, parameter dependences, and stability properties under many different conditions. For oscillatory phenomena, prediction of amplitudes and frequencies (using independently determined parameters) play a significant role. Further validation of the model is based on its ability to predict the outcome of new experiments, performed under conditions not previously examined. Among the advantages of this approach are that the model can be gradually expanded without changing already consolidated parts and that the model, in principle at least, allows translation by replacement of, e.g. parameters from animal studies by parameters relevant to man.

In the present chapter we try to illustrate different aspects of mechanism-based modeling through three simple examples related to the absorption, production, and control of insulin. Section 2.2 describes some of the considerations underlying the establishment and test of a model of ultradian oscillations of insulin secretion in man [9, 10]. In light of today's understanding, the model is clearly incomplete. However, this is not our concern. Rather, the emphasis of the discussion is on the many different types of experiment required to discriminate among different hypotheses concerning the biological processes involved.

The second example (Section 2.3) describes the formulation of a model that explains an observed initial slow absorption of soluble insulin [11]. The model resembles to some extent a typical PK model, except that instead of determining all parameters simultaneously, the parameters are deduced one by one from different features of the absorption curve. More importantly, however, when the model was first formulated, it served to examine different hypotheses about the processes that took place in the tissue. The model also involves simulation of the diffusion of insulin in subcutis. Finally, it is worth mentioning that, in contrast to typical PK models, our model allows slower parts of the absorption process to play out before the more rapid parts.

Finally, Section 2.4 analyses a simplified model of a bursting pancreatic β-cell [12]. The purpose of this section is to underline the importance of complex non-linear dynamic phenomena in biomedical systems. Living systems operate under far-from-equilibrium conditions. This implies that, contrary to the conventional assumption of homeostasis, many regulatory mechanisms are actually unstable and produce self-sustained oscillatory dynamics. The electrophysiological processes of the pancreatic β-cell display (at least) two interacting oscillatory processes: A fast process associated with the K^+ dynamics and a much slower process associated with the Ca^{2+} dynamics. Together these two processes can explain the characteristic bursting dynamics in the membrane potential.

The stability properties of a time-continuous system depend on the eigenvalues of its equilibrium point. As long as the real parts of all the eigenvalues are negative, the equilibrium point is stable. The transition in which the equilibrium point, under variation of a parameter, loses its stability is known as a bifurcation [13, 14]. From a mathematical point of view, this can happen in two different ways: A single real eigenvalue can pass from the negative real halfplane into the positive real halfplane (a saddle-node bifurcation), or a pair of complex conjugated eigenvalues can cross the imaginary axis (a Hopf bifurcation). Before the Hopf bifurcation, trajectories approach the equilibrium point from all different initial conditions in a spiraling manner. After the bifurcation, trajectories spiral away from the equilibrium point, and formation of a stable oscillatory dynamics of finite amplitude then depends on the existence of nonlinear limitations and damping mechanisms.

With the development of *Nonlinear Dynamics* and *Complex Systems Theory* over the past two to three decades, we have come to understand many aspects of the complicated phenomena that can take place when a system is forced to operate deeper in the unstable regime, or when many oscillatory processes interact. It was demonstrated, for instance, that there are three main routes to deterministic chaos (through an infinite cascade of period-doubling bifurcations, through different types of intermittency, or through torus destabilization), and each of these routes display universal, quantitative phenomena [13, 14]. As we try to illustrate in Section 2.4, *Nonlinear Dynamics* provided us with many of the concepts and tools we need to understand and describe the complicated phenomena we observe in cellular electrophysiological dynamics. A more detailed presentation of the role of nonlinear dynamic phenomena in physiological control systems is given in connection with our discussion of kidney pressure and flow regulation in a later chapter of this book [15].

2.2
Pulsatile Secretion of Insulin

Investigations performed during recent decades revealed a great variety of rhythms of significance for the regulation and function of normal physiological systems. Hormonal secretion provides several illustrative examples with the release of luteinizing hormone [16], growth hormone [17], and insulin [9, 18] displaying pronounced ultradian oscillations with periods of 2–3 h. The release of insulin also displays a faster rhythm, with a period of 9–15 min [19]. These rapid oscillations are likely to be of significance for the metabolic control processes in the liver, and there are several other examples to show that rhythmic administration of a hormone can be more effective than constant administration at the same average rate. It is also known that disruption of certain biological rhythms can lead to a state of disease [20] while, in other cases, synchronization of ultradian release processes can cause abnormal biological conditions such as, for instance, hot flashes [21].

Many hormonal secretion processes also exhibit strong circadian components. This is true, for instance, for cortisol, antidiuretic hormone, and growth hormone. The secretion of growth hormone is markedly increased during the early periods of sleep, and the secretion of antidiuretic hormone also reflects the sleep–wake cycle. The mechanisms underlying these oscillations can often be traced back to cyclical variations in the activity of the central nervous system. At the same time, the circadian rhythm modulates the above-mentioned ultradian oscillations.

Besides in the characteristic secretion patterns, the circadian rhythm is also revealed in several other parameters. Under daily-life conditions, a reduced food intake causes the blood concentrations of both glucose and insulin to decrease during the night. In certain cases, this may activate the hepatic glucose release. In experiments with constant glucose infusion, however, one finds that the glucose concentration is particularly high during night time [9], indicating that the cellular response to insulin is lower. This is accompanied by a particularly strong ultradian pulse of insulin secretion when the person wakes up. In many cases, the combined effects of the various rhythms produce a strikingly complex hormonal release pattern, and the question may arise whether this pattern can be given a causal explanation [22].

From a physical point of view, the rhythmic phenomena are related to the fact that biological systems are maintained under far-from-equilibrium conditions through a continuous dissipation of energy [23]. However, non-equilibrium conditions can also give rise to more complicated behaviors. Chaotic dynamics, for instance, can arise either as a regular rhythmic process is destabilized and develops through a cascade of period-doubling bifurcations [24], by torus destruction in connection with the interaction of two or more rhythms, or via different types of intermittency [25].

Non-equilibrium systems may thus display modes of behavior that are completely different from the simple types of dynamics we know for linear systems. Moreover, while in a linear system the effect of a drug will always be the same, the precise timing of the administration may be significant for a rhythmic (or chaotic) system. During certain phases of the oscillation, even a major dose may have little effect while a smaller dose may be highly effective (or even toxic) during other phases. Attempts to take advantage of this phenomenon are made, for instance, in experiments with chronotherapy of cancer and other diseases [26, 27]. In the chronotherapy of cancer one makes use of the fact that the circadian rhythm interacts with the cell cycle so that the rate of cell division displays a 24-h component. By virtue of the difference between normal and cancerous cells it is possible to adjust the timing of the daily drug administration such that the toxic effects on the cancer cells are optimized relative to the effects on the normal cells.

Synchronization is a universal phenomenon in nonlinear dynamic systems by which two (or more) oscillators tend to adjust their motions relative to one another so as to attain a state where they are completely entrained or, alternatively, a state in which there is a rational relation between their periods [28, 29]. The synchronization of our circadian rhythm to the local time and the afore-mentioned role of hormonal entrainment in the generation of hot flashes are typical examples. It is also well known that the beating of the heart and the respiration may synchronize, typically with a 1:3 or 1:4 relation between the two periods. By forcing an oscillating system with an external periodic signal and detecting the various synchronization regions, one can obtain significant information about the nonlinearities at play in the system. An example is the investigation by Sturis et al. [10] of how the endogenous insulin secretion synchronizes with an externally forced periodic glucose infusion.

As illustrated in Section 2.4, the significance of nonlinear dynamic phenomena seems to be even more pronounced at the cellular level. The insulin-producing

β-cells in the pancreas, for instance, are known to show variations in their hormonal release that are directly correlated with complicated patterns of bursts and spikes in their membrane potentials [30]. The release of insulin in response to a sudden increase of the blood glucose concentration also displays a bi-phasic behavior with a fast initial release followed by a slower and more continuous response [31]. This phenomenon is believed to be associated with the existence of two (or more) pools of insulin-carrying vesicles in the individual cells. Vesicles in a pool close to the cell membrane are prepared for immediate release, while a pool father away from the membrane provides a more continuous supply.

Communication between the cells takes place via a variety of different mechanisms, including the short-range diffusive exchange of ions and small molecules through gap junction [32] and the response of the individual cell to variations in the intercellular Ca^{2+} concentration produced by the bursting activity of neighboring cells [33]. Hence, one can observe synchronization of the bursting activity between adjacent cells [34] as well as waves of cytoplasmic calcium propagating across islets of Langerhans [35]. The above-mentioned rapid oscillations in the secretion of insulin may be associated with a modulation of the bursting activity of the individual cells that arises from such intercellular interactions.

Synchronization of cellular activity is known from many other types of tissue, and transitions between different types of synchronization and between smaller and larger clusters of synchronized units may represent an important component in the overall regulation of a physiological system. In their inactive state, smooth muscle cells, for instance, are found to exhibit incoherent waves of cytoplasmic Ca^{2+} produced by an instability associated with so-called calcium-induced calcium release [36]. When activated, however, the cells synchronize in a regular oscillatory pattern [37]. In contrast, several cases are known where a similar transition is related to the development of a state of disease. It has long been recognized, for instance, that the onset of epileptic seizures is associated with a synchronization of the firing activity for a group of cells in the brain [38].

Let us try to illustrate the mechanism-based modeling approach through the process of formulating a model of the ultradian oscillations in human insulin secretion [9]. A better understanding of the role and underlying mechanisms of these oscillations is clearly of interest in the design of an optimal treatment of diabetes.

With a sufficient temporal resolution, the ultradian oscillations in human insulin secretion can be detected as a "ringing" in the response to a meal [18]. However, as illustrated in Fig. 2.1, the oscillations persist during constant enteral nutrition or constant glucose infusion, and they are found to be accompanied by significant oscillations in the blood glucose concentration. For experiments conducted at an average glucose concentration of 100–120 mg/dl, the amplitude of the ultradian oscillations observed in healthy young volunteers is of the order of 10–20 mg/dl. The oscillations tend to grow in amplitude and become more regular as the rate of glucose infusion is increased.

We conclude that the ultradian insulin oscillations cannot be related to an intermittent supply of glucose. Neither do the oscillations appear to be generated through interaction with counter-regulatory hormones, since analysis of simulta-

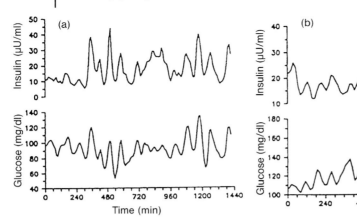

Fig. 2.1 Examples of ultradian oscillations in human insulin secretion and blood glucose concentration: (a) during continuous enteral nutrition and (b) during constant glucose infusion. Closer inspection shows that the glucose oscillations lead the insulin oscillations by a few minutes. Redrawn from [39, 40].

neous variations in glucagon and cortisol concentrations fails to show correlations with the insulin oscillations [10]. Moreover, experimental results indicate that the ultradian insulin oscillations persist following a pancreas transplant, indicating that the oscillations are independent of central neurogenic signals [41].

At this stage of the process we have already made use of a variety of independent experimental results in order to narrow down the possible mechanisms. Two fundamentally different types of explanation still remain: (1) the ultradian oscillations could reflect the activity of a pancreatic pacemaker, or (2) the oscillations could arise from an instability in the insulin–glucose feedback regulation. Experimentally, one can distinguish between these two possibilities by putting a clamp to the glucose concentration. If the insulin oscillations were caused by an intrapancreatic pacemaker, one would expect them to persist in spite of a fixed glucose concentration. However, if the oscillations were related to an instability in the glucose–insulin feedback regulation they should stop when the glucose concentration is clamped. Experiments by Sturis et al. [9] suggest the second possibility to be correct and this, therefore, forms the basis of our hypothesis.

Before we can start to develop a model we also have to decide how to interpret the behavior observed in Fig. 2.1. The variations in insulin and glucose concentrations could be generated by a damped oscillatory system that was continuously excited by external perturbations (e.g. through interaction with the pulsatile release of other hormones). However, the variations could also represent a disturbed self-sustained oscillation, or they could be an example of deterministic chaos. Here, it is important to realize that, with a sampling period of 10 min over the considered periods of 20–24 h, the number of data points are insufficient for any statistical analysis to distinguish between the possible modes. We need to make a choice and, in the present case, our choice is to consider the insulin–glucose regulation to operate close to a Hopf bifurcation where the transition from damped oscillatory dynamics to self-sustained oscillations takes place.

The insulin–glucose regulation represents a negative feedback where a rising glucose concentration by stimulating the secretion of insulin leads to an increase in the cellular glucose uptake and, hence, reduces the plasma glucose concentration. This type of control is generally stable. Moreover, the finite lifetime of insulin in the organism and the dependence of cellular glucose utilization on the plasma glucose concentration represent effective damping mechanisms. For such a system to become unstable and generate an oscillatory dynamics, there must be delays in the feedback regulation. One such delay could be associated with the finite equilibration rate for the insulin concentration between the plasma and interstitial compartments. However, with reasonable values of the diffusion capacity across the capillary wall, this delay does not suffice to generate self-sustained oscillations. Other possible delays could be associated with the response of muscle and adipose tissue cells to insulin or that of pancreatic β-cells to glucose. Such delays are characteristic of patients with type II diabetes, and the oscillatory dynamics could, perhaps, be considered a preclinical indicator (a biomarker) of diabetes. However, oscillations are also observed in normal young men and we, therefore, tend to rule out the significance of these types of delay in the present context.

In the version of the model to be considered here we assume the existence of a delay of 20 min in the response of the hepatic glucose release to changing insulin concentrations. The release of glucose from the liver is primarily controlled by glucagon. However, the release is suppressed by the presence of insulin, and this suppression involves a number of intermediate processes that could produce a delay. With this extension, the model can generate self-sustained oscillations, and our formulation of an initial dynamic hypothesis is completed. Note, however, that by neglecting glucagon as a regulatory factor, application of the model is restricted to the intermediate blood glucose concentrations at which the experiments with ultradian oscillations in insulin secretion were performed. To reduce the significance of internal insulin secretion, glucose clamp experiments to test the effect of new insulin analogues are often performed at fasting glucose concentrations. Here, additional or different regulatory factors may come into play.

As previously noted, the idea is that all parameters and relationships must be determined from independent experiments, i.e. experiments designed explicitly to provide each specific parameter or relationship. Alternatively, the parameters may be taken from the literature, or they may be estimated on the basis some semiquantitative argumentation. As discussed in Section 2.1, the advantage of this approach is that is allows us to test the *structure* of the model. Over the years, as the model is expanded and subjected to more and more tests, the procedure also allows us to accumulate and solidify a significant quantitative knowledge about the functioning of the system. In the present case, the parameters include the degradation of insulin in the plasma and intercellular compartments, the diffusion capacity for insulin between these compartments, and the delay in hepatic glucose release. The nonlinear functions include the relations between insulin secretion and blood glu-

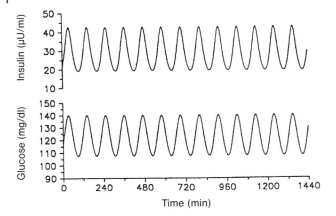

Fig. 2.2 Simulation of a mechanism-based model of ultradian insulin–glucose oscillations. Using independently determined parameters and nonlinear relations, the model displays self-sustained oscillations of the correct period with proper amplitudes and phase relationships. The model also responds correctly to a meal as well as to changes in the rate of glucose infusion.

cose concentration and those between cellular glucose uptake and (intercellular) insulin concentration.

With such independently determined parameters, the model must now be able to reproduce the experimentally observed characteristics of ultradian oscillations in human insulin secretion, e.g. the observed period, the oscillation amplitudes and phase relations, and the "ringing" of the insulin secretion in response to a meal. If the model can only produce the desired behavior after minor adjustments of the parameters, such adjustments may be performed. In most cases, however, the model either produces temporal variations in qualitative and semi-quantitative agreement with the empirical results or it does not produce anything that resembles them at all. In the latter case we conclude that the model structure is incorrect, and that a new hypothesis must be formulated.

Figure 2.2 presents the results obtained with our mechanism-based model of the ultradian insulin–glucose oscillations [9]. Although clearly only a preliminary model of the phenomenon, the applied model passes all of the above tests. The model produces self-sustained oscillations of the correct period and proper amplitudes, and the model also responds correctly both to a meal and to changes in the rate of glucose infusion. The next step is to use the model to predict the outcome of experiments that have not previously been performed. To the extent that the model is successful in such predictions, the hypothesis underlying the model structure gains additional support.

A first experiment to test the model predictions could be to replace the constant glucose infusion rate GIR_0 by a harmonically oscillating infusion rate [10]:

$$GIR(t) = GIR_0(1 + A \cdot \cos(2\pi t/T)) \tag{2}$$

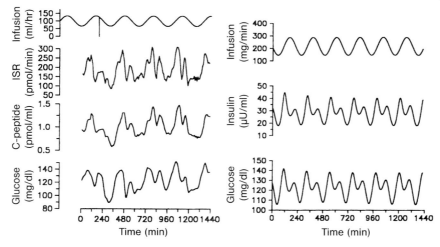

Fig. 2.3 Comparison of experimental (left) and simulation results (right) for the case of 2:1 synchronization of the hormonal control system to the external forcing signal. Reproduced from [10]. Note the similarity between the two set of curves.

Here, A represents the amplitude and T the period of the external forcing. In accordance with the results of nonlinear dynamic analysis [28, 29, 42], one expects the ultradian oscillations in insulin secretion to entrain with the external forcing signal such that, for a given amplitude of the forcing signal, there exists a finite interval of forcing periods where 1:1 synchronization occurs. Due to the presence of nonlinear interactions, the hormonal regulation adjusts its rhythm so as to produce precisely one pulse of insulin secretion per period of the forcing signal. This is clearly in contrast to the behavior of linear systems where, by virtue of the principle of superposition, two oscillatory processes can exist independently of one another.

The interval for the forcing period T in which 1:1 synchronization can be observed generally increases with the forcing amplitude, and the rate of this increase provides information about the nonlinearities in the system. Moreover, synchronization can also occur at other ratios of the two periods. Hence, for a given amplitude A, there is an interval of T in which 2:1 synchronization can be observed, i.e. two pulses of insulin secretion in each oscillation period of the infused glucose. Figure 2.3 provides an example of this type of behavior. Here, the left panel shows experimentally obtained results, and the right panel displays the corresponding simulation results. We immediately acknowledge the similarity between the two sets of curves. 2:1 synchronization can be considered to arise through phase-locking between the rate of insulin secretion and the second harmonic of the externally forced plasma glucose variation.

As a final comment to this section, let us note that experiments have also been performed to examine the possible physiological function of the oscillatory insulin secretion. An obvious question of interest was whether pulsatile release of insulin is more effective in eliciting glucose uptake by the muscle and adipose tissue cells than a constant release at the same average rate. In the experiments, the internal secretion of insulin (and glucagon) was suppressed by administering somatostatin to healthy young volunteers. Glucose was infused at a constant rate, and the observed variations in blood glucose concentrations over a 24-h period were compared between the cases of oscillatory and constant insulin infusion. A subsequent modeling study [43] showed that the expected higher efficiency in the case of oscillatory insulin infusion is difficult to explain in terms of an interaction of the insulin concentration oscillations with the regeneration dynamics of the cellular insulin receptors. The model shows that the amplitude of the insulin oscillations in the intercellular space are too small to give a noticeable effect.

2.3
Subcutaneous Absorption of Insulin

The previous section presented a simplified (and perhaps somewhat idealized) picture of the mechanism-based modeling approach. To provide a more concrete example let us consider the problem of modeling the absorption kinetics of subcutaneously injected soluble insulin [11].

Figure 2.4 shows a set of absorption curves obtained for seven type I diabetic patients treated at the Steno Memorial Hospital. Each patient received a bolus injection of 7 IU (40 IU/ml) radiolabeled soluble insulin (Velosulin, Nordisk Gentofte) in the thigh, and the fraction of insulin remaining in the subcutaneous depot was measured over an 8-h interval by means of a scintillation counter.

The absorption curves in Fig. 2.4 show a certain interpatient variability, with one patient displaying a particularly slow absorption process and another patient demonstrating a relatively fast absorption. However, inspection of the figure also reveals a consistent reduction of the absorption rate during the initial phases of the process, and only after 3–4 h does this rate reach its full magnitude. Absorption curves of the type shown in Fig. 2.4 do not provide information that can be used to explain this phenomenon. Fortunately, a single study exists [44] in which the experimental conditions were varied enough to give us the necessary insight.

The experimental procedure used to obtain the absorption curves in Fig. 2.5 is similar to that described in connection with Fig. 2.4, only the experiments have now been conducted over a range of different insulin concentrations and different injection volumes. Several of the experiments fall outside the range of pharmacological interest. Nevertheless, they can contribute significantly to our understanding of the processes involved in subcutaneous insulin absorption. From the absorption curves in the top row (insulin concentrations of 40 IU/ml) we notice that the delayed initial absorption phase exists for relatively high injection volumes only. For an injected volume of 0.01 ml, the effect is not clearly visible. From the absorption curves displayed in the left column (injection volumes of 1 ml), we conclude that the reduced absorption rate occurs for relatively high insulin concentrations. For concentrations of 4 IU/ml or less, the effect is not visible. Finally, the absorption

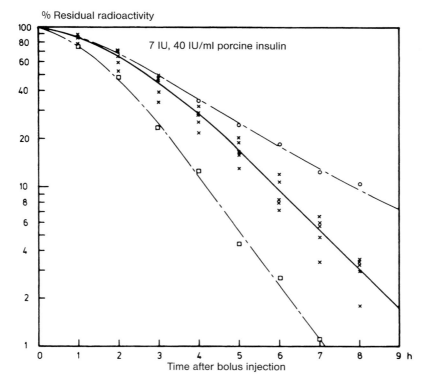

Fig. 2.4 Absorption curves for seven type I diabetic patients following a bolus injection of approx. 0.2 ml of 40 IU/ml soluble insulin. Note the initial reduction of the absorption rate. (Reproduced from [11]).

curves for low insulin concentrations and low injection volumes show a characteristic tail phenomenon with a reduced absorption rate in the last part of the process.

To explain the complex of phenomena observed in Fig. 2.5, we postulate that insulin is present in the subcutaneous depot in three different forms: free insulin on dimeric form, free insulin on hexameric form, and insulin bound to various substances and surfaces in the tissue. Moreover, we assume that only dimeric insulin can diffuse across the capillary wall. In qualitative terms, our explanation is then:

1. At high insulin concentrations most of the insulin is free, and the chemical balance between the two-multimeric forms is shifted towards hexameric insulin. Since hexameric insulin is assumed to not be absorbed, the fractional absorption rate is small. However, as the insulin concentration in the depot gradually falls, the chemical balance is shifted towards the dimeric form, and the absorption rate increases.

2. If the injected volume is large, diffusion of insulin in the tissue has little significance for the size of the depot and, hence, for the concentration of insulin. However, at the smallest injection volumes, the insulin molecules can rapidly

spread over areas comparable to the injected depot, and the associated reduction of the insulin concentration shifts the balance towards dimeric insulin and speeds up the absorption process.

3. At low concentrations a notable fraction of insulin may be bound in the tissue. The absorption rate will then be limited to the rate at which insulin is released from the bound state.

   The above description provides a coherent explanation to the overall picture. There are clearly details in Fig. 2.5 that are not accounted for. In panel F, for instance, only one of the curves displays the tail phenomenon. In panel H, none of the absorption curves show a tail, even though one would expect such tails on basis of the systematic aspects of the figure. At the end of this section, we make a few comments on the role of such deviations in connection with mechanism-based modeling.

To quantify our hypothesis and determine the parameters we note that:

- The absorption rate $B \approx 1.2 \times 10^{-2}$/min observed during the final phases for high concentrations and high injection volumes is interpreted as the absorption rate for dimeric insulin. This is approximately the same rate that we observe at lower concentrations and injections volumes before the tail develops.

- The reduced absorption rate observed in the initial phases at high insulin concentrations is the dimeric absorption rate multiplied by the fraction of insulin on dimeric form. This allows us to determine the equilibrium constant $Q \approx 0.06$ $(\text{ml/IU})^2$ for the mass balance between dimeric and hexameric insulin in the tissue.

- The slope of the tail observed at low insulin concentrations determines the lifetime $\tau \approx 800$ min of insulin in the bound state, and the intersection of the (backwards extrapolated) tail and the vertical axis determines the binding capacity $C \approx 0.08$ IU/ml in the tissue.

The set of (partial) differential equations that represent the above hypotheses is:

$$\frac{\partial C_H}{\partial t} = P(QC_D^3 - C_H) + D_H \nabla^2 C_H \tag{3}$$

$$\frac{\partial C_D}{\partial t} = -P(QC_D^3 - C_H) - BC_D - SC_D(C - C_B) + \frac{C_B}{\tau} + D_D \nabla^2 C_D \tag{4}$$

$$\frac{\partial C_B}{\partial t} = SC_D(C - C_B) - \frac{C_B}{\tau} \tag{5}$$

Here, C_H, C_D, and C_B denote the local concentrations of hexameric, dimeric, and bound insulin, respectively, and C is the binding capacity for insulin in the tissue. Only dimeric and hexameric insulin is supposed to diffuse in the tissue. The corresponding diffusion terms are $D_D \nabla^2 C_D$ and $D_H \nabla^2 C_H$, with D_D and D_H being the two diffusion constants and $\nabla^2 = \frac{\partial^2}{\partial x^2} + \frac{\partial^2}{\partial y^2} + \frac{\partial^2}{\partial z^2}$ the Laplacian operator. For the

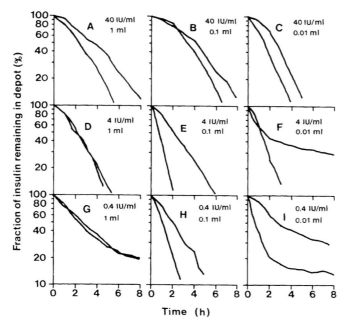

Fig. 2.5 Absorption curves for soluble insulin injected into the thigh
in combinations of three different insulin concentrations (40 IU/ml,
4 IU/ml, 0.4 IU/ml) and three different volumes (1 ml, 0.1 ml,
0.01 ml). The delayed initial absorption phase displays both a volume
and a concentration dependence.

diffusion constants we use $D_D = D_H = D = 1.6 \times 10^{-4}$ cm^2/min corresponding to
the diffusion constant for insulin in water at 37 °C. The two remaining parameters
P and S control the transition rates between the various forms of insulin. We have
assumed that these rates are high compared to the absorption rate, so that the depot
always maintains a state of quasi-equilibrium between the three forms of insulin.
This implies that we do not need to know P and S, as long as they are sufficiently
large. It may be of interest to note that similar conclusions hold for many other
biomedical systems: Although physiologically meaningful, a significant number of
parameters may not be of importance for the observed quantitative phenomena.

In the region of pharmacological interest, both the diffusive spread of the subcutaneous depot and the binding of insulin in the tissue can be neglected. This
simplifies the model considerably, and the resulting set of two coupled first-order
ordinary differential equations can easily be solved numerically. This allows us to
reproduce the absorption kinetics observed in Fig. 2.4. It also allows us to calculate
the fraction of insulin on dimeric form as a function of time during the absorption
process, as well as other variables not directly amenable to experimental investigation. It is important to emphasize, however, that information about the behavior of
the absorption process outside the domain of physiological relevance provides the
key to our hypotheses about the underlying biochemical processes.

Let us finally note that even though we have determined most of the parameters from the observed absorption curves, the dynamic hypotheses have allowed us to identify each parameter with a separate feature of these curves. By choosing $D_D = D_H = D$ as the diffusion constant for insulin in water we have also been able to make an independent test of that part of the hypotheses that concerns the diffusive spread of the depot.

Figure 2.6 displays a set of calculated curves for the residual insulin in the depot for different insulin concentrations and injection volumes. Inspection of the figure shows that both the experimentally observed volume and concentration effects are reproduced.

The full model can reproduce all the systematic features of the absorption curves in Fig. 2.5 [11]. In the present form, the model does not account for the interpatient variability observed in Fig. 2.4 or for the nonsystematic phenomena discussed in connection with Fig. 2.5. Ideally, in our mechanism-based modeling approach, any form of unsystematic variation in the experimental results should be given a separate explanation, i.e. specific studies should be undertaken to explain the observed variability in absorption rates and binding capacities. Statistical outliers should be considered as potential sources of new information.

To complete this discussion, let us finally note that our model (or the underlying hypotheses) allows us to predict the absorption curves for insulin concentrations outside the range of our experimental curves (Fig. 2.5). For a bolus injection of 0.1 ml of 100 IU/ml soluble insulin, for instance, we expect the initial fraction of hexameric insulin to be even higher than for 40 IU/ml. Hence, the fractional absorption rate is very low at the beginning, and it takes an even longer time before the insulin concentration reduces to such a level that dimeric insulin dominates,

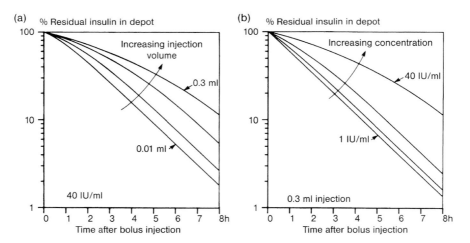

Fig. 2.6 Simulated absorption curves for different injection volumes (a) and concentrations (b) of the bolus injection. Note that the reduced initial absorption rate is observed only for large injection volumes and concentrations.

and the absorption rate attains its full value. To realize a faster initial absorption rate, one must develop insulin variants that have a lower tendency for polymerization. This has also been done, partly with the help of molecular dynamics and bioinformatics.

2.4
Bursting Pancreatic β-Cells

The purpose of the previous section was to illustrate how mechanism-based modeling can be applied to a relatively simple problem of drug absorption. We used a similar approach to examine the enhancement and inhibition of HIV infection of monocytes by antibodies against HIV [45], a problem of direct relevance to the development of an HIV vaccine. We also tried to establish the criteria for a successful treatment of HIV through gene modification of T helper cells and/or hematopoietic progenitor cells [46]. In collaboration with Li. Mosekilde and J.S. Thomsen [47], we considered the potential of various treatments of osteoporosis by integrating information about the influence of different drugs on the remodeling parameters associated with renewal of bone tissue during menopause.

However, as discussed in the introduction, living systems can exhibit much more complicated temporal behaviors. The purpose of the present section is to illustrate how different forms of complex dynamics can arise in the electrophysiological activity of pancreatic β-cells, and to show how these phenomena can be accounted for in a simple nonlinear dynamic analysis.

Living cells operate far from thermodynamic equilibrium by virtue of the continuous activity of ion pumps, and the enormous differences between the intra- and extracellular ion concentrations they maintain. Hence, many biological cells display an excitable electric activity, or the membrane potential exhibits complicated patterns of slow and fast oscillations associated with variations in the ionic currents across the membrane. This activity plays an essential role for the function of the cell as well as for its communication with neighboring cells. It is well known, for instance, that pancreatic β-cells under normal circumstances display a bursting behavior with alternations between an active (spiking) state and a silent state [48, 49]. It is also established [50, 51] that the secretion of insulin depends on the fraction of time that the cells spend in the active state, and that this fraction increases with the concentration of glucose in the extracellular environment. The bursting dynamics controls the influx of Ca^{2+} ions into the cell, and calcium is considered an important trigger for the release of insulin. In this way, the bursting dynamics organizes the response of the β-cells to varying glucose concentrations. At glucose concentrations below 5 mM, the cells do not burst at all. For high glucose concentrations (>22 mM), however, the cells spike continuously, and the secretion of insulin saturates [52].

A number of experimental studies have shown that neighboring β-cells in an islet of Langerhans tend to synchronize their membrane activity [53], and that cytoplasmic Ca^{2+} oscillations can propagate across clusters of β-cells in the presence of glucose [35]. The precise mechanism underlying this interaction is not fully known. It is generally considered, however, that the exchange of ions via low impedance gap junctions between the cells plays a significant role [54]. Such synchronization phenomena are important. Not only do they influence the activity of the individual cell, but they also affect the overall insulin secretion. Actually, it appears that isolated β-cells in general do not burst but show disorganized spiking behavior as a result of the random opening and closing of potassium channels [55, 56]. A single β-cell may have in the order of a few hundred such channels. However, with typical opening probabilities as low as 5–10%, only a few tens open during a particular spike. Organized bursting behavior arises for clusters of 20 or more closely coupled cells that share a sufficiently large number of ion channels and a sufficient membrane capacity for stochastic effects to be less important.

Models of pancreatic β-cells are usually based on the standard Hodgkin–Huxley formalism with elaborations to account, for instance, for the intracellular storage of calcium, for different aspects of the glucose metabolism, or for the influence of ATP and various hormonal regulators. Over the years, several models of this type have been proposed with varying degrees of detail [57]. At the minimum, a three-dimensional model with two fast variables and one slow variable is required to generate a realistic bursting behavior. The slow dynamics are often considered to be associated with changes in the intracellular Ca^{2+} concentration. The fast variables are usually the membrane potential V and the opening probability n of the potassium channels.

Although the simple models have been around for quite some time, their bifurcation structure is so complicated as to not yet be understood in full. Conventional analyses [53, 58] are usually based on a separation of the dynamics into a fast and a slow subsystem, whereafter the slow variable is treated as a (bifurcation) parameter for the fast dynamics. How compelling such an analysis may appear, particularly when one considers the large ratio of the involved time constants, the analysis fails to account for the more interesting dynamics of the models. Simulations with typical β-cell models display chaotic dynamics and period-doubling bifurcations for biologically interesting parameter values [59] and, according to the theory of nonlinear dynamic systems, such phenomena cannot occur in a two-dimensional, time-continuous system such as the fast subsystem.

Wang [60] proposed a combination of two different mechanisms to explain the emergence of chaotic bursting. First, the continuous spiking state undergoes a period-doubling cascade to a state of chaotic firing, and this state is destabilized in a boundary crisis. Bursting then arises through a homoclinic connection that serves as a reinjection mechanism for the chaotic saddle created in the boundary crisis. Belykh et al. [61] presented a qualitative analysis of a generic model structure that can reproduce the bursting and spiking dynamics of pancreatic β-cells. They considered four main scenarios for the onset of bursting, emphasizing that each of these scenarios involves the formation of a homoclinic orbit that travels along the route of the bursting oscillations and, hence, cannot be explained in terms of bifurcations in the fast subsystem. Lading et al. [62] studied chaotic synchronization (and the related phenomena of riddled basins of attraction, attractor bubbling, and on–off intermittency) for a model of two coupled, identical β-cells, and Yanchuk et al. [63] investigated the effects of a small parameter mismatch between the coupled chaotic oscillators.

Most recently, Belykh et al. [64] performed a detailed qualitative analysis of a prototype of a bursting biological cell model. This analysis showed that bursting cells represent perhaps the first example of a realistic system that satisfies the conditions for structurally stable chaotic dynamics. Bertram et al. [65] developed a model of pancreatic β-cells which includes equations for the key mitochondrial variables. They used the model to explain experimental observations of the opposite effects of raising cytosolic calcium at low and high glucose concentrations, and to predict the effect of a mutation in the mitochondrial enzyme nicotinamide nucleotide transhydrogenase. Finally, Pedersen and Sørensen [66] studied the effect of noise on the cellular bursting period.

The purpose of the present section is to give a somewhat simpler account of the bifurcation structure of the individual β-cell. Our analysis reveals the existence of a squid-formed area of chaotic dynamics in parameter plane, with period-doubling cascades along one side of the arms and saddle-node bifurcations along the other. The transition from this structure to the so-called period-adding structure involves a subcritical period-doubling bifurcation and the emergence of type III intermittency. The period-adding transition itself is found to be nonsmooth and to consist of a saddle-node bifurcation in which stable (*n*+1) spike behavior is born, overlapping slightly with a subcritical period-doubling bifurcation in which stable *n* spike behavior ceases to exist [62].

Bursting behavior similar to the dynamics that we have described for pancreatic β-cells is known to occur in a variety of other cell types as well. Pant and Kim [67], for instance, developed a mathematical model to account for experimentally observed burst patterns in pacemaker neurons, and Morris and Lecar [68] modeled the complex firing patterns in barnacle giant muscle fibers. Braun et al. [69] investigated bursting patterns in discharging cold fibers of the cat, and Braun et al. [70] studied the effect of noise on signal transduction in shark sensory cells. Although the biophysical mechanism underlying the bursting behavior may vary significantly from cell type to cell type, we expect many of the basic bifurcation phenomena to remain the same.

The starting point for our analysis of bursting phenomena in pancreatic β-cells is the following model suggested by Sherman et al. [53]:

$$\tau \frac{\mathrm{d}V}{\mathrm{d}t} = -I_{Ca}(V) - I_K(V, n) - g_S S(V - V_K) \tag{6}$$

with:

$$\tau \frac{\mathrm{d}n}{\mathrm{d}t} = \sigma[n_\infty(V) - n] \tag{7}$$

$$\tau_S \frac{\mathrm{d}S}{\mathrm{d}t} = S_\infty(V) - S \tag{8}$$

$$I_{Ca}(V) = g_{Ca}m_\infty(V)(V - V_{Ca}) \tag{9}$$

$$I_K(V, n) = g_K n(V - V_K) \tag{10}$$

$$w_\infty(V) = \left[1 + \exp\left\{\frac{V_w - V}{\theta_w}\right\}\right]^{-1} \tag{11}$$

Here, $w = m$, n, and S. V represents the membrane potential, n is the opening probability of the potassium channels, and S accounts for the presence of a slow dynamics in the system. I_{Ca} and I_K are the calcium and potassium currents, $g_{Ca} = 3.6$ and $g_K = 10.0$ are the associated conductances, and $V_{Ca} = 25$ mV and $V_K = -75$ mV are the respective Nernst (or reversal) potentials. The ratio τ/τ_S defines the relation between the fast (V and n) and the slow (S) time scales. The time constant for the membrane potential is determined by the capacitance and typical conductance of the cell membrane. With $\tau = 0.02$ s and $\tau_S = 35$ s, the ratio $k_S = \tau/\tau_S$ is quite small, and the cell model is numerically stiff. The calcium current I_{Ca} is assumed to adjust immediately to variations in V. For fixed values of the membrane potential, the gating variables n and S relax exponentially towards the voltage-dependent steady-state values n_∞ (V) and S_∞ (V). Together with the ratio k_S of the fast to the slow time constant, V_S is used as the main bifurcation parameter. This parameter determines the membrane potential at which the steady-state value for the gating variable S attains one-half of its maximum value. The other parameters are assumed to take the following values: $g_S = 4.0$, $V_m = -20$ mV, $V_n = -16$ mV, $\theta_m = 12$ mV, $\theta_n = 5.6$ mV, $\theta_S = 10$ mV, and $\sigma = 0.85$. These values are all adjusted to fit experimentally observed relationships. In accordance with the formulation used by Sherman et al. [53], there is no capacitance in Eq. (6), and all the conductances are dimensionless. To eliminate any dependence on the cell size, all conductances are scaled with the typical conductance. Hence, we may consider the model to represent a cluster of closely coupled β-cells that share the combined capacity and conductance of the entire membrane area.

Figure 2.7a shows an example of the temporal variations of the variables V and S as obtained by simulating the cell model under conditions where it exhibits a characteristic bursting dynamics. Here, $k_S = 0.32 \times 10^{-3}$ and $V_S = -38.34$ mV. We notice the fast spikes in the membrane potential are associated with the extremely rapid opening and closing of some of the potassium channels. The opening probability n changes from nearly nothing to about 10% at the peak of each spike, and the membrane potential rises from -56 mV to -22 mV in a fraction of a second. We also notice how the slow variable increases during the bursting phase to reach a value around 553, whereafter the cell switches into the silent phase, and S gradually relaxes back. If the slow variable is considered to represent the intracellular Ca^{2+} concentration, this concentration is seen to increase step by step during the spikes until it reaches some threshold value, and the bursting phase terminates. Hereafter, S decreases as Ca^{2+} is continuously pumped out of the cell.

Let us start our bifurcation analysis with a few comments concerning the equilibrium points of the β-cell model. The zero points of the vector field Eqs. (6) to (11) are given by:

$$g_{Ca}m_\infty(V)(V - V_{Ca}) + g_K n(V - V_K) + g_S S(V - V_K) = 0 \qquad (12)$$

$$n = n_\infty(V) = \left[1 + \exp\left(\frac{V_n - V}{\theta_n}\right)\right]^{-1} \qquad (13)$$

$$S = S_\infty(V) = \left[1 + \exp\left(\frac{V_S - V}{\theta_S}\right)\right]^{-1} \qquad (14)$$

so that the equilibrium values of n and S are uniquely determined by V. Substituting Eqs. (13) and (14) into Eq. (12), the equation for the equilibrium potential becomes:

$$f(V) \equiv g_{Ca}m_\infty(V)(V - V_{Ca}) + g_K n_\infty(V)(V - V_K)$$

$$+ g_S S_\infty(V)(V - V_K) = 0 \qquad (15)$$

with $m_\infty(V)$ as given by Eq. (11). Assuming $V_{Ca} > V_K$ and considering the conductances g_{Ca}, g_K, and g_S to be positive by definition, we observe that any equilibrium point of the β-cell model must have a membrane potential in the interval $V_K < V < V_{Ca}$. Moreover, there must be at least one such point.

In order to examine the stability of the equilibrium points it is customary to separate the three-dimensional system Eqs. (6) to (11) into a fast subsystem involving V and n and a slow subsystem consisting of S. The z-shaped curve in Fig. 2.7b shows the equilibrium curve for the fast subsystem, i.e. the value of the membrane potential in the equilibrium points ($dV/dt = 0$, $dn/dt = 0$) as a function of the slow variable S, which is now to be treated as a parameter. In accordance with common practice, those parts of the curve in which the equilibrium point is stable are drawn with full lines, and parts with unstable equilibrium points are drawn as dashed curves. Starting from the top left end of the curve, the equilibrium point is a stable focus. The two eigenvalues of the fast subsystem in the equilibrium point are complex conjugated and have negative real parts, and trajectories approach the point from all sides in a spiraling manner.

With increasing values of S, as we pass the point marked by the black square, the fast subsystem undergoes a Hopf bifurcation. The complex conjugated eigenvalues cross the imaginary axis and attain positive real parts, and the stable focus is transformed into an unstable focus surrounded by a limit cycle. The stationary state, which the system approaches as initial transients die out, is now a self-sustained oscillation. This state represents the spiking behavior.

As S continues to increase we reach a point marked by an open circle. Here, the equilibrium point undergoes a saddle-node bifurcation. Somewhere before this bifurcation, the unstable focus point has turned into an unstable node with two positive real eigenvalues. In the saddle-node bifurcation, one of these eigenvalues

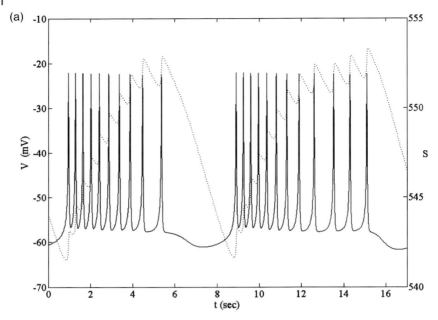

Fig. 2.7 (a) Temporal variation of the membrane potential *V* and the intracellular calcium concentration *S* in the considered simple model of a bursting pancreatic cell. (b) Bifurcation diagram for the fast subsystem: the black square denotes a Hopf bifurcation, the open circles are saddle-node bifurcations, and the filled circle represents a global bifurcation. (c) Trajectory plotted on top of the bifurcation diagram. The null-cline for the slow subsystem is shown dashed.

crosses back into the negative half-plane, and the unstable node is transformed into a saddle point with one positive and one negative eigenvalue.

We can now follow the saddle point backwards in the diagram (i.e. for decreasing values of *S*) until the dotted branch reaches another point marked with an open circle. Here, the second real eigenvalue turns negative, and the saddle point is transformed into a stable node with two attracting directions. This branch of stable node points continues to exist for increasing values of *S* in the lower right corner of the diagram.

The dynamics of the β-cell model now depends on the point of intersection between the equilibrium curve for the fast subsystem [the solution to Eqs. (12) and (13)] and the so-called null-cline for the slow subsystem [the solution to Eq. (12)]. If this point falls on one of the fully drawn branches of the equilibrium curve, the equilibrium for the three-dimensional model is stable, and the model produces neither bursting nor spiking dynamics. If, as sketched in Fig. 2.7c, the two curves intersect in a point of unstable behavior for the fast subsystem, the equilibrium point for the full model is also unstable. The null-cline in Fig. 2.7c is drawn as a dashed curve. Below the null-cline for the slow subsystem, $dS/dt < 0$, and the slow

(b)

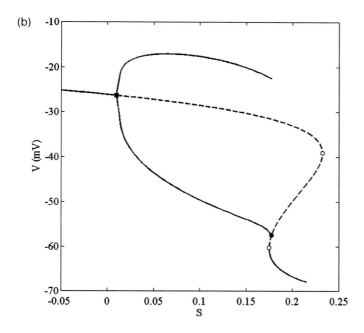

(c)

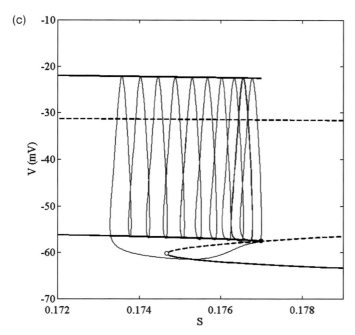

Fig. 2.7 (continued)

dynamics carries the system to the left. Above the null-cline,$dS/dt > 0$, and the point of operation for the fast subsystem gradually moves to the right.

To explain how the bursting dynamics arises let us start in the lower right corner of Fig. 2.7c. Here, the fast subsystem is stable while S gradually decreases. This corresponds to the conditions during the silent phase of the bursting period. When the system reaches the saddle-node bifurcation, the stable node solution ceases to exist, and the fast subsystem jumps to the alternative stable solution, namely the limit cycle oscillation around the upper branch of the equilibrium curve. Here, $dS/dt > 0$, and the system moves slowly to the right as S increases. This corresponds to the conditions during the bursting phase.

The bursting dynamics ends in a different type of process, referred to as a global (or homoclinic) bifurcation. In the interval of coexisting stable solutions, the stable manifold of (or the inset to) the saddle point defines the boundary of the basins of attraction for the stable node and limit cycle solutions. (The basin of attraction for a stable solution represents the set of initial conditions from which trajectories asymptotically approach the solution. The stable manifold to the saddle point is the set of points from which the trajectories go to the saddle point). When the limit cycle for increasing values of S hits its basin of attraction, it ceases to exist, and

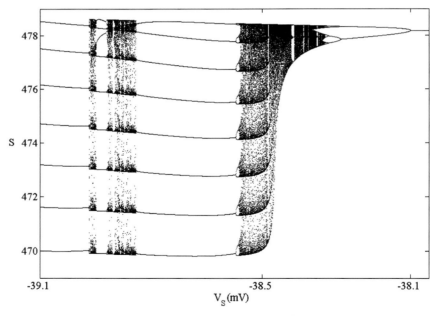

Fig. 2.8 One-dimensional bifurcation diagram for $k_S = 0.37 \times 10^{-3}$. The β-cell model displays chaotic dynamics in the transition intervals between periodic spiking and bursting and between the main states of periodic bursting. A careful description of the bifurcation diagram involves a variety of different transitions, including Hopf and saddle-node bifurcations, period-doubling bifurcations, transitions to inter-mittency, and homoclinic bifurcations.

the system again jumps back to the stable node solution. While the local Hopf and saddle-node bifurcations are associated with eigenvalues that cross the imaginary axis, the global bifurcation involves a collision of two structures in phase space. It is interesting to note that the characteristic slowing down of the spiking dynamics as the system approaches the end of the bursting phase observed in Fig. 2.7a is also found experimentally, lending credibility to the postulated mechanisms.

Figure 2.8 shows a one-dimensional bifurcation diagram for the cell model with V_S as the control parameter. Here, $k_S = 0.37 \times 10^{-3}$. The figure resembles figures that one can find in early papers by Chay [57]. The diagram was constructed from a set of Poincaré sections: For each value of the bifurcation parameter we intersected the trajectory in phase space with a plane at $n = 0.04$ and recorded the corresponding values of S. With the chosen section, all spikes performed by the model are detected. For $V_S > -38.1$ mV, the model exhibits continuous periodic spiking. As V_S is reduced, the spiking state undergoes a period-doubling cascade to chaos with characteristic periodic windows. Each window is terminated by a saddle-node bifurcation to the right and by a period-doubling cascade to the left. Around $V_S = -38.45$ mV, a dramatic change in the size of the chaotic attractor takes place. This marks the transition to bursting dynamics through the formation of a homoclinic connection in three-dimensional phase space [61].

Below $V_S \approx -38.5$ mV the bifurcation scenario is reversed, and for $V_S \approx -38.57$ mV a backwards period-doubling cascade leads the system into a state of periodic bursting with five spikes per burst. The interval of periodic bursting ends near $V_S = -38.84$ mV in a saddle-node bifurcation leading to chaos in the form of type-I intermittency [24]. With further reduction of V_S, the chaotic dynamics develops via a new reverse period-doubling cascade into periodic bursting with seven spikes per burst. It is clear from this description that chaotic dynamics tends to arise in the transitions between continuous spiking and bursting, and between the different bursting states.

To establish a more complete picture of the bifurcation structure, we applied a large number of such one-dimensional scans to identify the main periodic solutions (up to period 10) and to locate and classify the associated bifurcations. The results of this investigation are displayed in the two-dimensional bifurcation diagram shown in Fig. 2.9. To the left in this figure, we observe the Hopf bifurcation curve delineating the transition from stable equilibrium dynamics to self-sustained oscillatory behavior. Below this curve, the model has one or more stable equilibrium points. Above the curve, we find a region of complex behavior delineated by the period-doubling curve PD^{1-2}. Along this curve, the first period-doubling of the continuous spiking behavior takes place. In the heart of the region delineated by PD^{1-2} we find an interesting squid-formed structure with arms of chaotic behavior (indicated black) stretching down towards the Hopf bifurcation curve. Each of the arms of the squid-formed structure separates a region of periodic bursting behavior with n spikes per burst from a region with regular $(n + 1)$ spikes per burst behavior. Each arm has a period-doubling cascade leading to chaos on one side and a saddle-node bifurcation on the other. It is easy to see that the number of spikes per burst must become very large as k_S approaches zero.

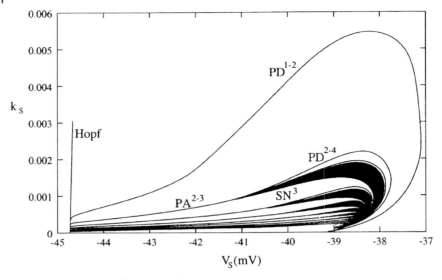

Fig. 2.9 Two-dimensional bifurcation diagram outlining the main bifurcation structure for the β-cell model in the (V_S, k_S) parameter plane. Notice the squid-shaped black region with chaotic dynamics. V_S characterizes the voltage dependence of the gating variable for the calcium conductance, and k_S is a measure of the ratio of the fast and the slow time scales.

2.5
Conclusions

We have presented three simple examples of the use of modeling and simulation techniques to establish a quantitative description of a number of pharmacological and physiological phenomena related to insulin absorption from the skin and to the regulation of pancreatic insulin secretion at the cellular and the systemic levels. In the final analysis, the important challenge to the pharmaceutical industry may not be the rising costs of drug development, but the complexity of living systems. Modeling is the "natural" way to build an integrated understanding of biomedical interactions. And we believe that *Biosimulation* as presented in this book will prove an essential tool in the establishment of a more rational drug development process, i.e. a process where empirical results are accumulated from test to test and new information continuously interpreted *vis á vis* existing knowledge.

As the first example, we considered the 2-h to 3-h pulsatile dynamics in insulin secretion rates reported for type II diabetics and age-matched normal persons by Polonsky et al. [18]. A main purpose of that part of our discussion was to emphasize some of the problems involved in defining the structure of the system, i.e. the involved physiological processes. Biological systems have feedback regulations at many different levels and time scales, and the most difficult part of the model formulation is generally to identify the relevant mechanisms. This often involves a significant number of experiments, but experiments of a type that the industry does not usually perform. The information required to produce a mechanism-based model is different from the information obtained in usual drug tests. Hence, the modeling approach requires a dedicated collaboration with university departments, and the industry must either establish its own basic research groups or base part of its modeling work on data from the open literature.

To establish a model of pulsatile insulin secretion, answers had to be found to questions such as:

- Are the oscillations associated with conditions characteristic for type II diabetic patients, e.g. insulin resistance or reduced pancreatic activity?
- Are the oscillatory behavior an expression of a normal physiological mechanism?
- Are the oscillations associated with signals from the central nervous system, or are they generated by a pancreatic pacemaker?
- If the oscillations derive from an instability in the insulin–glucose feedback regulation, where do we find the delay that can produce such an instability?
- And, finally, do the oscillations present an advantage to the physiological system, e.g. in the form of increased cellular glucose uptake?

Formulation of a model represents an effective way to think through all of these questions, and to define the conditions for the necessary critical experiments.

The second example considered the absorption of soluble insulin from subcutus. The problem here was to establish a set of consistent hypotheses that could explain the observed volume and concentration effects. At the time when the model was formulated there was no notion of the possible role of polymerization in the absorption process for insulin. Most experiments were performed at normal pharmacological concentrations (40 IU/ml) and injection volumes (0.3 ml), and the work was oriented towards elucidating the importance of exercise and skin temperature at the absorption site. Such experiments are obviously important, since variations in skin temperature may pose a problem in the control of labile type I diabetes. Analyses of a single set of data, obtained partly at micro-dose levels, allowed us to identify processes in the skin that were not amenable to direct experimentation.

The third example of a bursting pancreatic β-cell served to discuss some of the complicated dynamic phenomena that can be observed at the cellular level. Biological feedback regulation is not necessarily stable. By virtue of the constant supply of energy (food), living systems operate under far-from-equilibrium conditions, and many regulatory mechanisms are unstable and produce complicated dynamic phenomena. The discussion of nephron pressure and flow regulation in Chapter 12 of this book represents a relatively detailed example. Similar instabilities are likely to be involved in the formation of biological structures and forms (morphogenesis), and there is a clear analogy between the complex biological structures and the complicated temporal variations they display.

Simple oscillatory phenomena typically arise in feedback systems with delays. Such oscillations may serve as pacemakers or biological clocks to organize different processes that have to follow one another in a specific order. Opening and closing of the various ion channels in a cell, for instance, are typically to occur in a relatively strict sequence. The individual cell may display several rhythmic processes with different periods and associated both with electrophysiological processes and with different metabolic processes. This opens up for a variety of new dynamic phenomena, including quasiperiodicity, various forms of synchronization, and deterministic chaos. We suppose that transitions between these different states can represent an aspect of state of normal physiological regulation, and that the cells actively make use of the complex dynamics in their mutual communication. In other cases, transition to a different dynamic state may cause a disease, such as Parkinson's disease [38]. The development of *Nonlinear Dynamics* and *Complex Systems Theory* over the past few decades has provided us with a new and effective set of concepts and tools for the description of living systems.

Even models as simple as the models we describe in the present chapter can be extremely useful in unraveling the complex interactions of a particular biomedical problem. However, within a few years we imagine a situation where individual pharmaceutical companies accumulate information from experiment after experiment in order to gradually build up models that represent the most recent account of their medical and pharmacological knowledge in relation to a specific product line, disease, or patient group. Such models cannot replace databanks that hold all the underlying empirical data. However, it is likely that they will be significantly more useful in practice, because they represent the interpretation of large amounts of data in a consistent and prediction-oriented framework.

Acknowledgments

We would like to acknowledge many years of fruitful collaboration with Jeppe Sturis and Kenneth Polonsky in the area of insulin–glucose feedback regulation. Christian Binder provided the clinical results applied in our analysis of insulin absorption from subcutus. Brian Lading, Yuri Maistrenko, and Sergiy Yanchuk contributed to our study of bifurcation and synchronization phenomena in pancreatic β-cells.

References

1 J. J. Perez-Ruixo, F. De Ridder, H. Kimko, M. Samtani, E. Cox, S. Mohanty, and A. Vermeulen: *Simulation in Clinical Drug Development*. Chapter 1 in this book.

2 J. Andrew McCammon and Stephen C. Harvey: *Dynamics of proteins and nucleic acids*. Cambridge University Press, Cambridge, **1987**.

3 P. Baldi and S. Brunak: *Bioinformatics: the machine learning approach*. MIT Press, Cambridge, **1998**.

4 R. Frank and R. Hargreaves: Clinical biomarkers in drug discovery and development. *Nat. Rev. Drug Discov.* **2003**, 2:566–580.

5 H. Shoji, T. Shimizu, K. Shinohara, S. Oguchi, S. Shiga, and Y. Yamashiro: Suppressive effects of breast milk on oxidative DNA damage in very low birthweight infants. *Arch. Dis. Childhood Fetal Neonatal* **2004**, 89:F136–F142.

6 U. Gundert-Remy, S. G. Dahl, A. Boobis, P. Kremers, A. Kopp-Schneider,

A. Oberemm, A. Renwick, and O. Pelkonen: Molecular approaches to the identification of biomarkers of exposure and effect – report of an expert meeting organized by COST Action B15. *Toxicol. Lett.* **2005**, 156:227–240.

7 Schaefer et al.: *Biosimulation in Chemical Drug Development.* Chapter 17 in this book.

8 C. Tanaka, R. Kawai, and M. Rowland: Physiologically based pharmacokinetics of cyclosporine A: Reevaluation of dose-nonlinear kinetics in rats. *J. Pharmacokin. Biopharm.* **1999**, 27:597–623.

9 J. Sturis, K.S. Polonsky, E. Mosekilde, and E. Van Cauter: The mechanisms underlying ultradian oscillations of insulin and glucose: a computer simulation approach. *Am. J. Physiol.* **1991**, 260:E801–E809.

10 J. Sturis, C. Knudsen, N. M. O'Meara, J. S. Thomsen, E. Mosekilde, E. Van Cauter, and K. S. Polonsky: Phase-locking regions in a forced model of slow insulin glucose oscillations. *Chaos* **1995**, 5:193–199.

11 E. Mosekilde, K. S. Jensen, C. Binder, S. Pramming, and B. Thorsteinsson: Modeling absorption kinetics of subcutaneous injected soluble insulin. *J. Pharmacokin. Biopharm.* **1989**, 17:67–87.

12 E. Mosekilde, B. Lading, S. Yanchuk, and Y. Maistrenko: Bifurcation structure of a model of bursting pancreatic cells. *BioSystems* **2001**, 63:3–13.

13 J. M. T. Thompson and H. B. Stewart: *Nonlinear dynamics and chaos.* Wiley and Sons, Chichester, **1986**.

14 J. Guckenheimer and P. Holmes: *Nonlinear oscillations, dynamical systems and bifurcations of vector fields.* Springer-Verlag, Berlin, **1983**.

15 O. V. Sosnovtseva, E. Mosekilde, and N.-H. Holstein-Rathlou: *Modeling Kidney Pressure and Flow Regulation.* Chapter 12 in this book.

16 G. Leng: *Pulsatility in neuroendocrine systems.* CRC Press, Boca Raton, **1988**.

17 N. S. Bassett and P. D. Gluckman: Pulsatile growth hormone secretion in the ovine fetus and neonatal lamb. *J. Endocr.* **1986**, 109:307–312.

18 K. S. Polonsky, B. D. Given, and E. Van Cauter: Twenty-four-hour profiles of pulsatile patterns of insulin secretion in normal and obese subjects. *J. Clin. Invest.* **1988**, 81:442–448.

19 P. Bergsten and B. Hellmann: Glucose induced cycles of insulin release can be resolved into distinct periods of secretory actvity. *Biochem. Biophys. Res. Comm.* **1993**, 192:1182–1188.

20 L. Glass and M. C. Mackey: *From clocks to chaos. The rhythms of life.* Princeton University Press, New Jersey, **1988**.

21 F. Kronenberg, L. J. Cote, D. M. Linkie, I. Dyrenfurth, and J. A. Downey: Menopausal hot flashes: thermoregulatory, cardiovascular, and circulating catecholamine and LH changes. *Maturitas* **1984**, 6:31–43.

22 R. D. Hesch: *Endokrinologie, Teil A.* Urban and Schwarzenberg, München, **1989**.

23 G. Nicolis and I. Prigogine: *Self-organization in nonequilibrium systems.* Wiley, New York, **1977**.

24 M. J. Feigenbaum: Universal behavior in nonlinear systems. *Physica* **1983**, 7D:16–39

25 Y. Pomeau and P. Manneville: Intermittent transition to turbulence in dissipative dynamical systems. *Comm. Math. Phys.* **1980**, 74:189–197.

26 M.-C. Mormont and F. Lèvi: Cancer chronotherapy: principles, applications and perspectives. *Cancer* **2003**, 97:155–169.

27 A. Goldbeter et al.: *Optimising Temporal Patterns of Anticancer Drug Delivery by Simulations of a Cell Cycle Automaton.* Chapter 10 in this book.

28 A. Pikovsky, M. Rosenblum, and J. Kurths: *Synchronization – a universal concept in nonlinear sciences.* Cambridge University Press, Cambridge, **2001**.

29 E. Mosekilde, Yu. Maistrenko, and D. Postnov: *Chaotic synchronization – applications to living systems.* World Scientific, Singapore, **2002**.

30 I. Atwater, C. M. Dawson, A. M. Scott, G. Eddlestone, and E. Rojas: The nature of the oscillating behavior in electrical activity from pancreatic β-cell. *Horm. Metab. Res.* **1980**, 10:100–107.

31 D. L. Curry, L. L. Bennett, and G. M. Grodsky: Dynamics of insulin secretion by the perfused rat pancreas. *Endocrinology* **1968**, 83:572–584.

32 G. de Vries, A. Sherman and H.-R. Zhu: Diffusively coupled bursters, effect of cell heterogeneity. *Bull. Math. Biol.* **1998**, 60:1167–1200.

33 T. R. Chay: Effects of extracellular calcium on electrical bursting and intracellular and luminal calcium oscillations in insulin secreting pancreatic β-cells. *Biophys. J.* **1997**, 73:1673–1688.

34 R. M. Santos, L. M. Rosario, A. Nadal, J. Garcia-Sancho, B. Soria, and M. Valdeolmillos: Widespread synchronous $[Ca^{2+}]$ oscillations due to bursting electrical activity in single pancreatic islets. *Pflügers Arch. Eur. J. Physiol.* **1991**, 418:417–422.

35 E. Gylfe, E. Grapengiesser, and B. Hellman: Propagation of cytoplasmic Ca^{2+} oscillations in clusters of pancreatic β-cells exposed to glucose. *Cell Calcium* **1991**, 12:229–240.

36 M. Hoth and R. Penner: Calcium release-activated calcium current in rat mast cells. *J. Physiol.* **1993**, 465:359–386.

37 H. Peng, V. Matchkov, A. Ivasen, C. Aalkjaer, and H. Nilsson: Hypothesis for the initiation of vasomotion. *Circ. Res.* **2001**, 88:810–815.

38 P. A. Tass: Desynchronizing double-pulse phase resetting and application to deep brain stimulation. *Biol. Cybern.* **2001**, 85:343–344.

39 C. Simon, G. Brandenberger, and M. Follenius: Ultradian oscillations of plasma glucose, insulin and C-peptide in man. *J. Clin. Endocrinol. Metab.* **1987**, 64:669–675.

40 E. T. Shapiro, H. Till, K. S. Polonsky, V. S. Fang, A. H. Rubenstein, and E. Van Cauter: Oscillations in insulin secretion during constant glucose infusion in normal man: relationship to changes in phasma glucose. *J. Clin. Endocrinol. Metab.* **1988**, 67:307–314.

41 N. M. O'Meara, J. Sturis, J. D. Blackman, M. M. Byrne, J. B. Jaspan, D.C. Roland, J. R. Thistlethwaite and K. S. Polonsky: Oscillatory insulin secretion after pancreas transplant. *Diabetes* **1993**, 42:855–861.

42 E. Mosekilde: *Topics in nonlinear dynamics – applications to physics, biology and economic systems.* World Scientific, Singapore, **1996**.

43 I. M. Tolić, E. Mosekilde, and J. Sturis: Modeling the insulin-glucose feedback system: the significance of pulsatile

insulin secretion. *J. Theor. Biol.* **2000**, 207:361–375.

44 C. Binder: Absorption of injected insulin. *Acta Pharmacol. Toxicol.* **1969**, 27[Suppl. 2]:1–84 (1969).

45 O. Lund, J. Hansen, E. Mosekilde, J. O. Nielsen, and J.-E. Stig Hansen: A model of enhancement and inhibition of HIV infection of monocytes by antibodies against HIV. *J. Biol. Phys.* **1993**, 19:133–145.

46 O. Lund, O. S. Lund, G. Gram, S. D. Nielsen, K. Schønning, J. O. Nielsen, J.-E. Stig Hansen, and E. Mosekilde: Gene theraphy of T helper cells in HIV infection: mathematical model of the criteria for clinical effect. *Bull. Math. Biol.* **1997**, 59:725–745.

47 J. S. Thomsen, Li. Mosekilde, and E. Mosekilde: Quantification of remodeling parameter sensitivity – assessed by a computer simulation model, *Bone* **1996**, 19:505–511.

48 I. Atwater and P. M. Beigelman: Dynamic characteristics of electrical activity in pancreatic β-cells. *J. Physiol. (Paris)* **1976**, 72:769–786.

49 H. P. Meissner and M. Preissler: Ionic mechanisms of the glucose-induced membrane potential changes in β-cells. *Horm. Metab. Res.* **1980**, 10[Suppl.]:91–99.

50 S. Ozawa and O. Sand: Electrophysiology of endocrine cells. *Physiol. Rev.* **1986**, 66:887–952.

51 R. M. Miura and M. Pernarowski: Correlation of rates of insulin release from islets and plateau functions for β-cells. *Bull. Math. Biol.* **1995**, 57:229–246.

52 L. S. Satin and D. L. Cook: Calcium current inactivation in insulin-secreting cells is mediated by calcium influx and membrane depolarization. *Pflügers Arch.* **1989**, 414:1–10.

53 A. Sherman, J. Rinzel, and J. Keizer: Emergence of organized bursting in clusters of pancreatic β-cells by channel sharing. *Biophys. J.* **1988**, 54:411–425.

54 A. Sherman and J. Rinzel: Model for synchronization of pancreatic β-cells by gap junction coupling. *Biophys. J.* **1991**, 59:547–559.

55 T. R. Chay and H. S. Kang: Role of single-channel stochastic noise on bursting clusters of pancreatic β-cells. *Biophys. J.* **1988**, 54:427–435.

56 P. Smolen, J. Rinzel, and A. Sherman: Why pancreatic islets burst but single β-cells do not. *Biophys. J.* **1993**, 64:1668–1680.

57 T. R. Chay: Chaos in a three-variable model of an excitable cell. *Physica D* **1985**, 16:233–242.

58 A. Sherman: Anti-phase, asymmetric and aperiodic oscillations in excitable cells-I. Coupled bursters. *Bull. Math. Biol.* **1994**, 56:811–835.

59 Y.-S. Fan and T. R. Chay: Generation of periodic and chaotic bursting in an excitable cell model. *Biol. Cybern.* **1994**, 71:417–431.

60 X.-J.Wang: Genesis of bursting oscillations in the Hindmarsh-Rose model and homoclinicity to a chaotic saddle. *Physica D* **1993**, 62:263–274.

61 V. N. Belykh, I. V. Belykh, M. Colding-Jøgensen, and E. Mosekilde: Homoclinic bifurcations leading to the emergence of bursting oscillations in cell models. *Eur. J. Phys.* **2000**, E3:205–219.

62 B. Lading, E. Mosekilde, S. Yanchuk, and Yu. Maistrenko: Chaotic synchronization between coupled pancreatic b-cells. *Prog. Theor. Phys.* **2000**, 139[Suppl.]:164–177.

63 S. Yanchuk, Yu. Maistrenko, B. Lading, and E. Mosekilde: Chaotic synchronization in time-continuous systems.

Int. J. Bifurcation Chaos **2000**, 10:2629–2648.

64 V. Belykh, I. Belykh, and E. Mosekilde: Hyperbolic Plykin attractor can exist in neuron models. *Int. J. Bifurcation and Chaos* **2005**, 15:3567–3578.

65 R. Bertram, M. Gram Pedersen, D. S. Luciani, and A. Sherman: A simplified model for mitochondrial ATP production. *J. Theor. Biol.* **2006**, 243:575–586.

66 M. Gram Pedersen and M. P. Sørensen: The effect of noise on β-cell burst period. *SIAM J. Appl. Math.* (in press).

67 R. E. Pant and M. Kim: Mathematical descriptions of a bursting pacemaker neuron by a modification of the Hodgkin-Huxley equations. *Biophys. J.* **1976**, 16:227–244.

68 C. Morris and H. Lecar: Voltage oscillations in the barnacle giant muscle fiber. *Biophys. J.* **1981**, 35:193–213.

69 H. A. Braun, H. Bade, and H. Hensel: Static and dynamic discharge patterns of bursting cold fibers related to hypothetical receptor mechanisms. *Pflügers Arch.* **2000**, 386:1–9.

70 H. A. Braun, H. Wissing, K. Schäfer, and M. C. Hirsch: Oscillation and noise determine signal transduction in shark multimodal sensory cells. *Nature* **1994**, 367:270–273.

3
Biosimulation of Drug Metabolism

Martin Bertau, Lutz Brusch, and Ursula Kummer

Abstract

Computationally predicting the metabolic fates of drugs is a very complex task. This complexity is not only due to the huge and diverse biochemical network in the living cell, but also due to the fact that the majority of *in vivo* transformations occur by the action of hepatocytes and gastrointestinal microflora, thus not by a single cell type or even organism. However, the prediction of metabolic fates is definitely a problem worth solving since it would allow facilitating the development of drugs and rely less on animal testing. As a first step in this direction, *PharmBiosim* is being developed as a biosimulation tool which is based on massive data reduction and on attributing metabolic fates of drug molecules to functional groups and substituents. Initial work is done with yeast as a model organism and is restricted to drugs that are mainly transformed by central metabolism, especially sugar metabolism. The reason for the latter is that the qualitative functioning of the involved biochemistry is very similar in diverse cell types involved in drug metabolism. Results from the model are compared and validated with data from mammalian systems.

3.1
Introduction

Prior to approval for use in humans, a drug must undergo extensive studies to establish its efficacy and safety. An important factor in the evaluation of safety and efficacy of any drug is the knowledge of how the drug is metabolized, i.e. how a drug is structurally modified by enzymatic systems. This is important, because many secondary or tertiary occurring intermediates display pharmaceutical or toxicological action themselves. One particular important aspect is the high stereo-selectivity of enzymatic reactions. This affects the stereo-geometry of the reaction intermediates and products, which in turn determines the effect of a drug, as the thalidomide (Contergan) tragedy showed in the 1950s. Furthermore, besides the classic metabolism of a drug in the liver, perorally administered pharmaceutically

Biosimulation in Drug Development. Edited by Martin Bertau, Erik Mosekilde, and Hans V. Westerhoff
Copyright © 2008 WILEY-VCH Verlag GmbH & Co. KGaA, Weinheim
ISBN: 978-3-527-31699-1

active compounds (PAC) encounter the biotransformatory activity of the intestinal microflora in treated subjects. Therefore, drug action potential is greatly dependent on both first pass metabolism and the gastro-intestinal microbial barrier, and the understanding of drug metabolism plays an important role in the development of new drug entities [1].

Mammalian metabolism itself involves biotransformations of (generally hydrophobic) drugs in liver cells, kidney, and other organs to more polar, hydrophilic derivatives in order to allow excretion from the body. This process involves two types of reactions, classified as Phase I (functionalization) and Phase II (conjugation). This has been intensively reviewed and is covered by standard textbooks in pharmacology and pharmacy.

Phase I reactions involve oxidation, reduction, and hydrolysis. Hydrolysis reactions are generally viewed as being catalyzed by esterases or other hydrolytic enzymes. Latest results however demonstrate that hydrolases hardly contribute, while the major part of drug hydrolysis is due to the presence of glutathione, a stress response element of living cells [2]. Reduction reactions involve the reduction by $NAD(P)H$-dependent dehydrogenases of a keto compound to an alcohol. $FADH_2$-dependent pathways reduce olefins. Oxidation reactions involve reactions catalyzed by monoamine oxidase, flavine or cytochrome P450 monooxygenases.

Phase II reactions are synthetic reactions involving the conjugation of the drug or Phase I metabolites with endogenous substances such as glucuronic acid, glycine, acetate, methyl groups or inorganic acids, generally contributing to a further increase of hydrophilicity and facilitating the renal excretion of the drug and its metabolites. The eventual result of biotransformations of a complex drug molecule through sequential or parallel pathways is the formation of multiple metabolites.

Traditionally, drug metabolism studies rely on the use of model systems to predict the intermediates and products of drug metabolism in humans. For these purposes whole animal systems are in use, especially small laboratory animal models (e.g. rat, dog, cat, guinea pig, rabbit). *In vitro* studies are generally used to complement and specify the data obtained using perfused organs, tissue or cell cultures, and microsomal preparations. As discussed in more detail later, microorganisms can be used as model systems as well.

As described below, all of these methods suffer from a number of limitations, such as ethical concerns. In addition, the evaluation of drug metabolism suffers from a lack of insight into intracellular processes and the enzyme network and its regulation during the biotransformation of a drug. The use of a mathematical model of general principles of xenobiotic metabolism in living cells constitutes an alternative to the use of animal models, as it can ultimately mimic the mammalian metabolism and can give invaluable information about the metabolic fate of the drug. Hence, biosimulation is an indispensable tool in modern and future drug synthesis.

In the following, we give a short introduction to experimental and computational approaches to deciphering metabolic fates of drugs, as employed right now. Afterwards, we introduce the experimental basis and computational work of our own approach before concluding with an example application.

3.2
Experimental Approaches

3.2.1
Animal Test Models

Whole animal studies are very important in exploring xenobiotic metabolism and assessing the efficacy and safety of drugs. Metabolic information is retrieved from analyzing animal plasma and urine for the presence and identification of metabolites. Sophisticated modern approaches use transgenic animals bearing a luciferase gene, that allows the detection of PAC metabolites by biophotonic methods [3–6]. However, the toxicity of drugs limits the amount that can be administered, and therefore only small quantities of metabolites can be isolated (frequently just a few nanogrammes for minor ones). Despite the advances in highly sensitive analytical techniques, the structural identification of highly diluted metabolites remains difficult. The same applies to the identification of stereoisomeric structures [7].

Furthermore, although it is recognized that the use of animals in experiments is indispensable, it raises deep ethical concerns. Therefore, current approaches strive e.g. towards more efficient animal test models, not only to provide more detailed information with respect to potential drug metabolism and pharmacokinetics in man, but also to decrease the number of experiments necessary [8].

In addition to whole animal models, *in vitro* studies are generally used to complement and specify the data obtained using perfused organs, tissue or cell cultures, and microsomal preparations. The use of these models allows a more detailed or partial study of the effects of PAC and can decrease the ethical concerns, but it has disadvantages as well. For instance, the cells are often not easy to culture and maintain, which in turn results in diluted samples.

Unfortunately, microbial models are scarcely discussed as an alternative, although they can produce highly reliable results (*vide supra*). Animal models are indispensable, but microbiological ones can be a good alternative, or a starting point.

3.2.2
Microbial Models

The metabolic activity of microorganisms towards xenobiotics has been exploited since the early 1920s [9]. In the 1970s, attempts were intensified to explore the physiological activities and the metabolism of PAC in microorganisms [10–13]. Since then, several bacterial as well as fungal strains have been investigated, showing that microorganisms are well suited to reflect mammalian metabolism of drugs/PAC [14–17].

The hydrolytic and reductive capabilities of microorganisms, especially fungi, are extremely useful in evaluating the metabolic fate of drugs [18–21]. For instance, many pathways of oxidative reactions parallel the same ones in mammals. In addition, fungi have been shown to possess mono-oxygenase enzyme systems mechanistically similar to mammalian hepatic mono-oxygenases. The results of a systematic examination of microbial hydroxylations on a variety of model organic compounds strongly suggest that microbial transformation systems could closely mimic most of the mammalian phase I transformations [22–24].

The biotransformation reactions involved in the study of drug metabolism in microorganisms predominantly comprise those reactions of hydrophobic substrates that produce more polar metabolites. The use of microorganisms for simulating the mammalian metabolism of many molecules of pharmacological importance is well documented [23–27, 29, 30].

Fungi (e.g. *Cunninghamella* sp., *Aspergillus* sp., *Saccharomyces cerevisiae*) are eukaryotic organisms, like mammals, and are the most commonly utilized microorganisms in biotransformation studies [31, 32]. The use of bacteria (prokaryotes) is limited mostly to actinomycetes that seem to contain an enzyme system very similar to that of fungi. Other bacteria (e.g. *Pseudomonas*, *Escherichia coli*) are used occasionally, but their usefulness is limited [26, 33].

There are a number of practical advantages in the use of microbial systems as models for drug metabolism [26]:

1. Simple culture media are easily prepared at low cost.

2. Screening for the ability of a strain to metabolize a drug is simply done.

3. Because of the high rates of microbial metabolism and the often high culture densities, the drug concentrations that can be used (range 0.2–10.0 g l^{-1}) are much higher than those employed in other cell or tissue models. Consequently, the concentration of metabolites formed is typically in the range 20–200 mg l^{-1}, allowing easier detection, structural identification and comparison with mammalian metabolites.

4. In many cases micro-metabolites are not detectable except by microbial conversions. For instance Bertau and co-workers contributed to elucidating mechanisms of xenobiotic metabolism on a molecular level [2, 34, 35], delivering a multitude of novel insights into the micro-metabolism of xenobiotics, i.e. drugs, even pointing towards possible side-effects of drug administration.

5. The model system can be scaled up easily for the preparation of metabolites for pharmacological and toxicological studies.

6. The model system can be useful in cases where regio- and stereospecificity is required.

7. The maintenance of stock cultures of microorganisms is less resource-consuming and is ethically more favorable than the maintenance of cell or tissue cultures or laboratory animals.

3.3
The Biosimulation Approach

As an alternative to the above-described experimental approaches, the computational simulation of drug fates is a current topic of research and discussion. In the process of drug development, molecular modeling approaches are already a standard tool [36]. Here, computation of the protein structure of specific enzymes and their binding characteristics help to identify the most promising ligands/drugs [37]. In addition, computational modeling of pharmacokinetics, the time and distribution of a drug within the organism, has been employed for quite some time [38]. However, computational approaches taking intracellular biochemistry into account are still pretty rare. Usually, these are focused on individual compounds [39].

The theoretical analysis of the metabolic fate of PAC or xenobiotics in general demands a systems approach since individual components of the metabolic networks are closely linked via common substrates or cofactors as well as allosteric enzyme regulation [40, 41]. As shown in a later section, drug catabolites especially affect cell physiology and thereby greatly impinge on cellular drug action.

Currently, there are several projects that introduce a stronger development on the computational biochemistry side. Thus, within HepatoSys (the German systems biology program) there is a project dealing with the modeling of the detoxification reactions of the hepatocyte (http://www.systemsbiology.de). Another project focuses on drug metabolism in skin cells [42]. There are several more.

In contrast to all of these, our approach, *PharmBiosim*, which is described below in more detail, focuses on relatively simple systems (i.e. central metabolism of yeast) initially to study how abstractions can help to predict the fate of drugs that are mainly transformed by reactions of the central metabolism, both in microorganisms in the gut and in mammalian cells. This simple system will later be expanded to become more cell-specific and encompass more reactions.

3.4
Ethical Issues

As mentioned above, animal experiments are a generally accepted method to assess drug safety, irrespective of the burden for the experimental animals. Criticism is mainly directed at developing substances and products which merely serve to improve the life quality of (healthy) man, for instance cosmetics.

Nevertheless, it is generally agreed that all measures that may contribute to refine, reduce and to replace animal tests are desirable (3R principle) [43, 44]. The biosimulation of drug metabolism is a powerful means to achieve this goal. The combination of microbial models of drug metabolism with mathematical models of cell metabolism allows – within certain limits – a prediction of the outcome of parameter changes on drug metabolism in the liver as well as in the intestine [45, 46, 48]. Wherever in the course of drug development and approval the involvement of animals appears inevitable, biosimulation studies aim to drastically reduce the number of required experiments.

The same aim applies to clinical trials, where drug interactions are examined in human beings. Biosimulation methods allow one to check, for instance, the interference of meal uptake with drug ingestion. As this methodology is capable of predicting how metabolite formation and drug biotransformation product distribution depend on dose and sugar supply, drug effects on human physiology can be assessed much more efficiently, and ideally fewer trials will be required.

It is of utmost importance that the scientific community acts according to ethical standards when clinical trials and animal tests are unavoidable. Implementation of biosimulation into the drug approval process is a great step towards this objective.

3.5
PharmBiosim – a Computer Model of Drug Metabolism in Yeast

3.5.1
General Concept

In the human body, PAC, since it is mostly administered perorally, faces the gastro-intestinal microbial barrier by which the drug compound is actively metabolized prior to resorption by the gut mucosa. The fraction of the active pharmaceutical ingredient (API) which is metabolized during the gastro-intestinal passage is mostly unknown. Likewise, the intermediates and products formed are often unknown, since there is a vast number of possible enzymatic actions, depending on the structure of the drug. Yet, these gastro-intestinally produced metabolites enter the hepatic blood stream, where they are subject to the biotransformatory activity of the liver. Of these reactions, phase I–phase II metabolism is the best studied system. Unfortunately, current approaches lack information on the micro-metabolites which themselves may be physiologically very active, but are undetectable in the high dilution found in mammalian organisms.

To reduce the abundance of possible enzymatic actions on a whole variety of drugs to the most general principles, our approach *PharmBiosim* implies two critical layers of abstraction which are described below.

3.5.1.1 Chemical Abstraction
As enzymes mostly act on strategic positions in the target molecule, a holistic approach is pursued which aims at massive data reduction and generalization by attributing metabolic fates of drug molecules to functional groups and substituents, i.e. by making use of the comprehensive experience in whole-cell biotransformations of simple organic substrates.

This abstraction is based on the key–lock principle introduced by Emil Fischer in 1894. According to this principle, the interaction of a biological entity with a molecule is best when both species behave like a key and a lock. If we now extend this principle to pharmaceuticals, then the drug molecule is the key. Certain

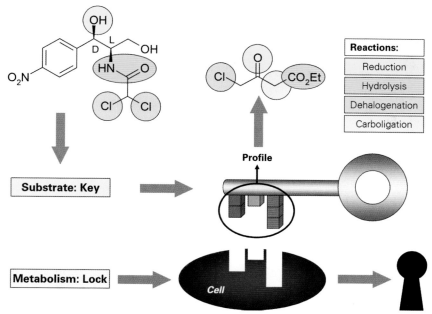

Fig. 3.1 Abstraction of substrate properties to common chemical principles is one of the basic concepts of the biosimulation approach (chemical abstraction). From the xenobiotic substrate molecule strategic positions for enzymatic attack are identified, since it is these that represent the reaction profile of a chemical compound in biological systems. The reaction profile determines regio-, chemo-, and stereoselectivity of enzymatic conversions. For these reasons model substrates representing the functionalities prone to enzymatic attack are studied instead of an unmanageable wealth of individual, complex, pharmaceuticals. This concept is outlined with the example of chloramphenicol, for which ethyl 4-chloro acetoacetate serves as representative model substrate. In this way the molecule's functional diversity is reduced to three functionalities. On this basis cellular metabolism is elucidated on a molecular level, and the obtained data can be implemented into the mathematical model.

functionalities at defined positions in the molecule are required in order to interact optimally with the target structure, the lock. If we regard the biochemical degradation of PAC, again certain functionalities at defined positions in the molecule are required for certain enzymatic actions to take place. In other words, chemical functionalities serve as targets for enzymes and determine the metabolic profile of a molecule. Consequently, since for instance esters are cleaved hydrolytically or ketones are converted reductively, it is the type of chemical reaction that can occur to a chemical entity in a biochemical system rather than the overall structure that determines its metabolic fate. Therefore, PAC from now on are regarded as the sum of chemical functionalities prone to enzymatic attack. In other words, the metabolic fate of a xenobiotic is primarily given by its chemical functionalities – the teeth of the key (Fig. 3.1).

In this view the multitude of chemical entities is reduced to what determines their chemical behavior in a biological system, and the wealth of data is drastically reduced to a manageable extent.

3.5.1.2 Biological Abstraction

The majority of *in vivo* transformations of PAC occurs by the action of hepatocytes and the gastro-intestinal micro-flora, the latter being composed of bacteria (prokaryotes) and fungi (eukaryotes). Biotransformations of drug compounds usually take place in the cytosol of the cell, and the most important cytosolic metabolic pathways affected by the presence of a xenobiotic compound in the cell are glycolysis on the one hand and fatty acid anabolism on the other [47]. Additional metabolic action in mitochondria or the nucleus has not yet been reported to be significantly involved. Glycolysis is the more active pathway in the cell, compared with fatty acid synthesis. This allows the introduction of biological abstraction in our approach. All cell types faced by the xenobiotic until its excretion are generally regarded as cytosols in which only glycolysis is affected. Thus, xenobiotic metabolism is less dependent on cell type at this level of abstraction. Therefore, a well studied organism, *Saccharomyces cerevisiae* was chosen as a model system to reflect the metabolism

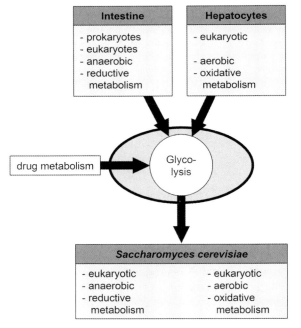

Fig. 3.2 Biological abstraction. Yeast cells reflect anaerobic, reductive metabolism (intestine) as well as aerobic, oxidative metabolism (liver), if glycolysis is regarded as the most active pathway. Therefore, the yeast *Saccharomyces cerevisiae* is a good model organism for studies of xenobiotic metabolism.

of model compounds irrespective of whether enzymatic degradation occurs in the intestine (anaerobic) or aerobically in the liver (Fig. 3.2).

A microbial test system such as *S. cerevisiae* offers an ideal starting point for the biosimulation approach. The reasons are: (1) easy culturing, allowing a fast practical approach, (2) possibility to detect micro-metabolites due to a higher applicable metabolite concentration and cell density, (3) possibility to measure cell stress responses linked to xenobiotic metabolism, (4) measurability of periodic phenomena in glycolysis, (5) ethical aspects. Moreover, this microorganism is one of the best understood organisms and is therefore the ideal model organism for promoting the understanding of physiological actions of a drug as well as predicting its metabolic fate. Results from this model organism can easily be compared with and validated with data from mammalian systems. In particular, glycolysis of *S. cerevisiae* and the subsequent fermention of glucose to ethanol is one of the best studied metabolic pathways, both experimentally and by modeling [48–54].

3.5.2
Initial Steps – Experimental Results

The metabolic fates of small mono- to tri-substituted organic compounds are investigated [48]. Literature sources as well as experimental work allow us to attribute cellular responses, i.e. enzymatic activities, to the presence of a functional group in an organic substrate (the chemical abstraction). Next, PharmBiosim follows the approach of systems biology by numerically analyzing the model system, the eukaryote *S. cerevisiae* (the biological abstraction). The results are extracted for generalized principles of xenobiotic metabolism which feed the computational tool with the aim of predicting the metabolic fates of pharmaceuticals in eukaryotic cells.

In this section, for the reasons outlined above (Fig. 3.1) and with the purpose of exemplification, three model xenobiotics are considered: ethyl acetoacetate (compound **1**), ethyl 4-chloro-acetoacetate (**2**), and ethyl 4,4,4-trifluoro-acetoacetate (**3**), the structures of which are given in Fig. 3.3.

The β-keto esters are reduced to the respective chiral β-hydroxy esters by at least two alternative enzymes one of which is D-directing the other one is L-directing (Fig. 3.4). A product mixture results that contains both enantiomeric forms, D and L,[1] of the carbinol (β-hydroxy ester) in varying degrees. In the case of ethyl acetoacetate (**1**) preferably the L-form of ethyl 3-hydroxybutyrate (L-**4**) is produced which is then secreted from the cell [55–58]. The L-directing enzyme is methyl butyraldehyde reductase (MBAR; EC 1.1.1.265), and the D-enantiomer is formed by the action of β-ketoacyl reductase (KAR; EC 1.1.1.100) which is a constituent of fatty acid anabolism (Fig. 3.4) [49, 59]. Alcohol dehydrogenase (ADH; EC 1.1.1.1) L-directing activity is classically attributed to was shown to be inactive – moreover the enzyme is even inhibited by the substrate [2, 34, 37].

---

1) The D, L nomenclature is chosen in this chapter in order to unmistakably distinguish homochiral stereoisomers. According to the specifications of the Cahn-Ingold-Prelog nomenclature **L-1** would equal (*S*)-1, while **L-2** equals (*R*)-2.

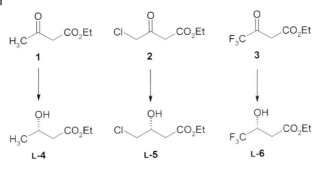

Fig. 3.3 Representative biotransformations serve as a basis for the development of a biosimulation-based approach. Model substrates for studying drug metabolism in the model eukaryote *Saccharomyces cerevisiae*: ethyl acetoacetate (**1**), ethyl 4-chloro-acetoacetate (**2**) and ethyl 4,4,4-trifluoro-acetoacetate (**3**).

Both competing reductions consume the cofactor nicotinamide adenine dinucleotide (NADH) and thereby interfere with the redox balance of the cell and feedback on glycolysis where NADH is regenerated on the one hand, while on the other hand NAD$^+$ is required to keep the glycolytic pathway running. The nonlinear dynamical model combines the network of glycolysis and the additional pathways of the xenobiotics to predict the asymmetric yield (enantiomeric excess, *ee*) of L-versus D-carbinol for different environmental conditions (Fig. 3.4). Here, the enantiomeric excess of fluxes v_L and v_D is defined as

$$ee = (v_L - v_D)/(v_L + v_D) \tag{1}$$

Carbonyl conversions provide insight into metabolic pathways activated in the metabolism of a xenobiotic, as the involved dehydrogenases belong to different metabolic pathways each. The stereoselectivities of carbonyl and alcohol bioconversions are monitored by classic chemical analysis of stereoisomer formation. Therefore, precisely defined reaction conditions are a strict prerequisite for establishing general principles of xenobiotic metabolism which are required for the highly challenging transfer to mammalian cells.

After enzymes and/or metabolic pathways have been identified and quantified in their contribution to the biotransformation of a drug (xenobiotic), its impact on cellular metabolism is investigated by monitoring periodic changes in cellular NADH concentration [49]. Periodic changes or oscillations of NADH concentrations in yeast are a long-known phenomenon. Back in the 1950s and 1960s, scientists discovered that all concentrations of metabolites participating in glycolysis oscillate in the living cell under certain circumstances and that cells in cell cultures are able to synchronize these oscillations [60]. Oscillations in glycolysis have been also observed for mammalian cells [61]. Studying the oscillatory state of glycolysis compared to a steady state offers several advantages. This methodology profits from the circumstance that dehydrogenases as well as dehalogenating enzymes, carbon backbone-modifying enzymes, and stress proteins use NAD(P)H/H$^+$ as a cofactor.

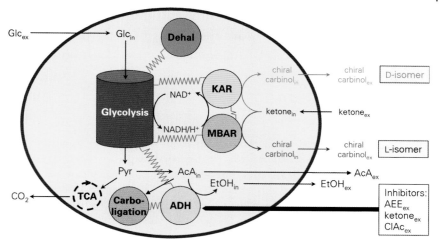

Fig. 3.4 The glycolytic pathway produces NADH which under regular conditions is oxidized to NAD⁺ while reducing acetaldehyde (ACA) to ethanol (EtOH), thereby in turn reducing NAD+ in order to keep hexose catabolism running. The actual cytosolic NADH concentration is determined by the respective conversion rates of the enzymes involved in the oxidation and regeneration of the compound. If these enzymes convert additional non-natural substrates (xenobiotics, i.e. drugs), the conversion rate changes. As a consequence, the cytosolic NADH concentration differs from the natural condition. Furthermore, if a xenobiotic acts as an enzyme inhibitor, e.g. for ADH, then NAD⁺ regeneration is substantially affected, which eventually results in altered cytosolic NADH concentration. Therefore the presence of a xenobiotic in the cell is conceivably a perturbation factor. Under the conditions where glycolytic oscillations become effective, the perturbation caused by the xenobiotic can be measured easily and with high accuracy. By means of classic chemical analysis the stereoisomer distribution can be determined which is the direct outcome of the individual contributions of MBAR and KAR. These data, together with ketone and glucose concentration provide all necessary information to fully describe the intracellular processes mathematically. The great advantage of this concept is that it is strictly linked to measured values. This allows to steadily adapt model and experiment until consistent predictions are obtained. Abbreviations: AcA – acetaldehyde, ADH – alcohol dehydrogenase, AEE – ethyl acetoacetate, ClAc – chloroacetone, Dehal – dehalogenase, Glc – glucose, KAR – bβ-ketoacyl reductase, MBAR – 2-methyl butyraldehyde reductase, Pyr – pyruvate, TCA – tricarboxylic acid cycle.

The upshot on the oscillation is a direct measure for the extent of perturbation on the metabolic network upon the uptake of a PAC. Glycolytic oscillations that are systematically perturbed by altered environmental conditions, i.e. exposure to the xenobiotic, constitute a direct and easily accessible measure of the intracellular behavior since the frequency and amplitude of oscillating metabolite concentrations and fluxes depend on both the perturbation and on most intracellular processes due to the coupled energy (ATP) and redox (NADH) balances (Fig. 3.4).

As demonstrated below, nonlinear model analysis can be used to predict the dependence of the oscillatory behavior of glycolysis on e.g. exposure of the cell to a xenobiotic. Experimentally, the oscillatory behavior is monitored via NADH fluo-

Table 3.1 Reactions and pathways involved in xenobiotic metabolism of model substrates **1–3**.

Substrate	Pathways	Reactions
Ethyl acetoacetate (**1**)	7	31
Ethyl 4-chloroacetoacetate (**2**)	25	118
Ethyl 4,4,4-trifluoroacetoacetate (**3**)	7	39

rescence. Oscillatory and steady behavior are easily distinguishable as individual cells synchronize their oscillations [49]. The temporal frequency of the oscillations can be measured with high accuracy. The modeling of cell metabolism and the observed changes in frequency and amplitude of the oscillations thus yields further insight into the intracellular mechanisms and will guide subsequent experiments.

Yet the xenobiotic metabolism even of such small compounds is highly complex, and many of the degrading pathways are branched, forming nonlinear graphs (Table 3.1).

Figure 3.5 gives a schematic overview of the major pathways that are involved in the degradation of substrate **2**. To cut a long story short, it is the cytosolic glutathione-dependent pathways II–VI which predominantly compete with reductive pathway I [35, 62].

Due to their significance in affecting glycolytic oscillations, these pathways are discussed shortly.

3.5.2.1 Dehalogenation (Pathways II and III)
Release of the chloride substituent in **2** produces ethyl acetoacetate (**1**) which is an inhibitor of alcohol dehydrogenase (ADH, EC 1.1.1.1) and thereby affects energy metabolism. Above all, NAD^+ regeneration is impaired. As a consequence acetaldehyde becomes an overflow metabolite which is excreted into the medium or reacts with activated positions in the molecule (Fig. 3.5).

3.5.2.2 *Retro*-Claisen Condensation (Pathway IV)
Cleavage of the carbon backbone is reported to be mediated by hydrolases [63]. However, there are no indications that this side-reaction proceeds enzymatically. Instead formation of *retro*-Claisen products **12** and **13** proceeds via GSH addition to the carbonyl center in **2**. The active role of GSH was demonstrated with GSH-depleted cells. After treatment with *N*-ethyl maleimide (NEM) [64] no *retro*-Claisen product was detectable. The same applied for NEM-treated cell liquor.

The *retro*-Claisen product **12** is further converted into acetic acid which together with **13** exerts acid stress what results in an increased energy demand in order to excrete protons from the cell and in this way concomitantly affects NADH and ATP concentration in the cell [65].

3.5.2.3 Ester Hydrolysis (Pathway VI)
Saponification of the ester group was hitherto assigned to the action of hydrolases [66, 67]. In fact, the occurrence of ester hydrolysis was found to be another exam-

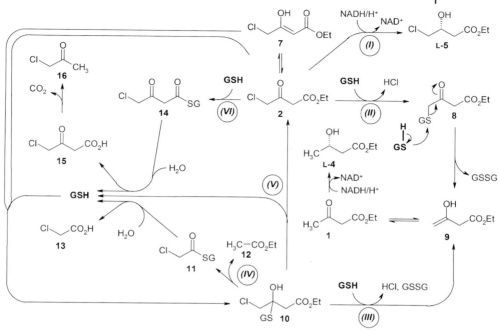

Fig. 3.5 Metabolism of **2** furnishes 118 metabolites via 25 different pathways of which those are depicted in this scheme that are gluthatione (GSH) dependent. From the exemplary biotransformation of **2** can be inferred that besides stereoselective reduction there exist competing pathways in *S. cerevisiae* as a result of xenobiotic cell stress: (*I*) stereoselective reduction, (*II*) dehalogenation after attack of GSH at C-4, (*III*) dehalogenation after attack of GSH at C-3, (*IV*) *retro*-Claisen condensation, (*V*) de-glutathionylation shifting the keto-enol equilibrium in favour of the enol form, and (*VI*) ester hydrolysis. The susceptibility of the Michael system for nucleophilic 1,4-addition renders also the enol a substrate for dehalogenation.

ple for the implications of GSH on the metabolic fate of xenobiotics. This GSH-catalyzed reaction is a sequence of thiolysis and subsequent hydrolysis of the unstable thioester (Fig. 3.5).

Whole cells which were depleted from GSH by addition of NEM did hydrolyze the substrate as the result of residual background esterase activity, to which ~30% of the hydrolytic activity can be attributed.

The ester hydrolysis final product **16** is dehalogenated to give acetone, which again is an inhibitor of ADH (EC 1.1.1.1).

These findings are of great importance in evaluating xenobiotic metabolism and are the first ever to demonstrate the extensive involvement of glutathione dependent processes. Moreover, glutathionylation obviously becomes effective already at low concentration. GSH-mediated side-reactions must generally be considered in assessing metabolic fates of drug compounds [2, 68].

Implementation of cytosolic glutathione concentration ([GSH]) into the mathematical model is therefore a strict requirement in order to reproduce correct results. Furthermore, this example demonstrates that biosimulation is a highly inter-disciplinary field in which modeling cannot exist without experimentation and *vice versa*.

3.5.2.4 Competing Pathways and Stereoselectivity

Ketone **2** and its metabolite chloro-acetone (**16**) are alkylating agents. This property is of special importance, when the substrate is being reduced stereoselectively by reductases with an essential sulfhydryl function. Because the sulfhydryl groups are inhibited irreversibly by alkylation [69], stereoisomer distribution in the product is considerably affected.

This was nicely outlined with the observed unusually high enantiopurity of the dehalogenation consecutive product ethyl L-3-hydroxybutyrate (L-**4**; Pathways II, III). Ester hydrolysis product **16** acts as an inhibitor of D-directing β-ketoacyl re-ductase of the fatty acid synthase (FAS) complex (EC 1.1.1.100) [69, 70], whereupon the fraction of L-**4** increased from 97% *ee* to >99.5% *ee*.

3.6
Computational Modeling

3.6.1
Selection of the Modeling Software

The modeling software used for our purposes should allow an easy model exchange with other software and scientists within BioSim. Therefore, this evaluation is re-stricted to software allowing the import and export of standard model files like SBML [71] and CellML [72]. Moreover, the software should be freely available to academic users in order to allow international free access to the respective models and simulation results.

Computational minimal requirements for modeling and simulation software to compute metabolic fates of xenobiotics in cells are the possibilities to:

- Model any biochemical reaction system.
- Simulate it by integrating ordinary differential equations (ODE).

Doubtlessly, there are also other computational analyses to be performed in the future, such as stochastic simulations, stoichiometric network analysis, sensitivity analysis, etc. However, for this initial survey, we only take this absolute minimum into account, while listing additional features of the respective software tools.

With these boundary conditions in mind, one can distinguish between SBML- and CellML-compatible software.

3.6.2
SBML-compatible Software

A list of SBML-compatible software is available at http://sbml.org/index.psp. With the minimal requirements mentioned above, we consider the following packages best suited: Cellware, Copasi, Ecell, SBW, JigCell, JSim, XPPAUT, Virtual Cell (Vcell).

3.6.2.1 Cellware
Cellware (http://www.bii.a-star.edu.sg/research/sbg/cellware/index.asp) offers network statistics, pathway homology, automated pathway layout, TauLeap algorithm for simulation, parameter estimation, plotting, stochastic algorithms (Gillespie, Gibson, Stochsim), and ODE solvers (Euler, Runge Kutta). It is platform-independent and offers a user-friendly graphical user interface (GUI).

All in all, it should be sufficient for many biochemical problems; however it offers only very bad integration routines, given that most biochemical systems are very stiff and stiff solvers are asked for, but not provided.

3.6.2.2 Copasi
Copasi (http://www.copasi.org) offers stochastic and deterministic time course simulation (the latter with LSODA, a stiff solver), steady state analysis (including stability), metabolic control analysis, elementary mode analysis, mass conservation analysis, parameter scans, and sliders for interactive parameter changes. It is platform-independent and offers a user-friendly GUI.

It should be sufficient for most biochemical problems.

3.6.2.3 Ecell
Ecell (http://ecell.sourceforge.net/) offers stochastic and deterministic time course simulation (including stiff solvers). It only runs on Windows and offers a user-friendly GUI.

It should be sufficient for the minimal tasks of building a model and running it.

3.6.2.4 JigCell
JigCell (http://jigcell.biol.vt.edu/) offers no simulation engine on its own, but relies on other sources, such as XPPAUT, XPP or BioPack. It allows parameter estimation and step functions. The GUI is not very user-friendly, since different packages have to be called and there is no central GUI, even though there are not many features in the software. It is platform-independent.

However, for the minimal tasks asked for in this study, it is sufficient.

3.6.2.5 JSim
JSim (http://nsr.bioeng.washington.edu/PLN/Members/butterw/JSIMDOC1.6/JSim_Home.stx/view) offers deterministic time course simulation with stiff solvers and is platform-independent. The GUI is easy to understand.

Again, the software is sufficient for the tasks asked for here.

3.6.2.6 **Systems Biology Workbench**

Systems Biology Workbench (SBW; http://sbw.sourceforge.net/) includes *Jarnac*, a biochemical simulation package for Windows, *JDesigner*, a visual biochemical network layout tool, *MetaToolSBW*, a network analysis tool, *Pasadena Twain*, a simple interactive ODE solver, and *GillespieService*, a module implementing a stochastic simulator, bifurcation analysis module, and optimization module running on Windows and Linux with a nice GUI.

It is certainly sufficient for the tasks asked for here.

3.6.2.7 **Virtual Cell**

Virtual Cell (Vcell; http://www.nrcam.uchc.edu/vcellR3/login/login.jsp) offers deterministic spatio-temporal simulations with stiff solvers. It runs web-based and has a nice GUI. In addition, it offers sensitivity analysis.

The software is sufficient for the tasks asked for here.

3.6.2.8 **XPPAUT**

XPPAUT (http://www.math.pitt.edu/~bard/xpp/xpp.html) offers deterministic simulations with a set of very good stiff solvers. It also offers fitting, stability analysis, nonlinear systems analysis, and time-series analysis, like histograms. The GUI is simple. It is mainly available under Linux, but also runs on Windows.

The software is sufficient for the above described tasks.

3.6.3
CellML-compatible Software

There are not as many tools fulfilling the above-mentioned requirements that are CellML-compatible. Basically, the only tool mentioned and described above is VCell.

However, CellML models can be converted to SBML via a freely available tool (http://sbml.org/software/cellml2sbml/) and thus be available for the above described software.

In one case (Cellware) there are limitations with regard to the quality of the numerical routines. All of the others should be suitable and it is more a matter of taste which tool is preferred by the individual user in BioSim. The compatibility with the biochemical file exchange standards ensures a simple and flexible sharing of models.

If more functionality is asked for, only a few software packages work as listed above.

For most purposes Copasi and XPPAUT come into use.

3.6.4
Kinetic Model

3.6.4.1 **Methods**

The influence of any metabolite is described by a single time-dependent variable, the concentration. The temporal evolution of the variables is then computed from ordinary differential equations (ODEs) that predict the immediate change of a variable according to the size of the fluxes that produce or degrade the particular metabolite. In turn, the size of each flux depends on a well defined set of variables, i.e. the instantaneous concentrations of substrates, products, cofactors and effectors, and a set of parameters. The state of the model, i.e. the values of all metabolite concentrations, evolves by updating the variables over subsequent time intervals which are repeated for different external conditions to yield theoretical time courses of concentrations and fluxes.

Since metabolic networks, e.g. glycolysis in *S. cerevisiae*, constitute connected sets of reactions, the corresponding kinetic models form large systems of coupled, nonlinear ODEs. Bifurcation analysis [73, 74] and direct numerical simulation [75] have proven to be essential for the analysis of complex cellular networks [76, 77]. For simulations we employed the software Copasi [78]. To numerically determine the unknown parameter values, repeated simulations serve to follow the model's behavior for many combinations of the parameter values. The parameters are updated subsequently depending on the model's performance for previous choices. This "optimization" strategy is automated by Copasi and allows a much faster convergence to a reasonable set of values than would be possible by random or dense sampling of the high-dimensional parameter space.

As the focus lies on steady and oscillatory states, it is most convenient to combine the computation of the system's state along "branches" in order to efficiently scan the parameter space, for which tasks the software packages XPP and Auto were used [79–81]. Therein, the solutions are represented as the root of an extended system of algebraic equations derived by discretising the ODE. Any new solution (the root) is then computed iteratively by the Newton algorithm. Specifically, the continuation algorithm exploits the knowledge of previously computed solutions to systematically compute similar solutions for other values of the parameters. Since the continuation monitors the solution over an entire interval of the parameter space, it also detects critical values of the parameters where the solution changes drastically, so called bifurcations. Bifurcations delimit domains in the parameter space with typical behavior, e.g. the same number of coexisting steady states or steady versus oscillatory states. They also represent constraints on the model's behavior and can themselves be followed through the parameter space by continuation.

Both Copasi and XPP comply with the SBML standard for exchange of biochemical models. The SBML-based PharmBiosim approach combines the above simulation-based strategy with the application of algorithms based on continuation and bifurcation analysis from applied mathematics.

3.6.4.2 Model Derivation

Glycolysis in yeast has been intensively studied especially under anaerobic conditions. Here, one of these previous kinetic studies is augmented by the core reactions of the xenobiotic ketone. The used model of glycolysis was devised by Hynne et al. and contains 22 variables for concentrations of involved metabolites and 24 reactions [53]. This model quantitatively accounts for most known details of enzyme regulation in order to precisely describe the supercritical onset of oscillations as observed experimentally [49, 82]. The following extensions have been introduced to the literature model (Fig. 3.4):

1. The extracellular concentration of ketone $[Ket]_x$ is used as a control parameter and the intracellular concentration of ketone $[Ket]$ is calculated according to new equations. The membrane transport $v_{Ket}^{in} = k_{Ket}^{in} \cdot ([Ket]x - [Ket])$ is assumed linear with $k_{Ket}^{in} = v_{ACA}^{in} = 27.7$ min^{-1} in analogy to the intermediate acetaldehyde in the full-scale model [53].

2. The nonlinear kinetics of the L-directed reduction of the ketone by the enzyme MBAR (EC 1.1.1.265) is modeled by Michaelis–Menten kinetics:

$$v_{Ket}^{L} = \frac{V_{Ket}^{L} \cdot [Ket] \cdot [NADH]}{(K_{NADH} + [NADH]) \cdot (K_{Ket}^{L} + [Ket])} \tag{2}$$

with K_M constant $K_{Ket}^{L} = 0.7$ mM that was experimentally determined and $V_{Ket}^{L} = 89.8$ mM min^{-1} as well as $K_{NADH} = 0.1$ mM chosen in analogy to the kinetics of ADH (EC 1.1.1.1).

3. The non-linear kinetics of the D-directed reduction of the ketone by means of the reductase KAR (EC 1.1.1.100) is modeled as:

$$v_{Ke}^{D} = \frac{V_{Ke}^{D} \cdot [Ket] \cdot [NADH]}{(K_{NADH} + [NADH]) \cdot (K_{Ke}^{D} + [Ket])} \tag{3}$$

with $K_{NADH} = 0.1$ mM assumed identical to ADH. In order to estimate V_{Ket}^{D}, the observation of $ee = 97\%$ was considered with an excess of ketone **1**. Inserting Eqs. (2) and (3) into Eq. (1) yields:

$$\frac{v_D}{v_L} = \frac{V_{Ket}^{D}}{V_{Ket}^{L}} \tag{4}$$

$$V_{Ket}^{D} = V_{Ket}^{L} \cdot \frac{1 - ee}{1 + ee} \tag{5}$$

and hence $V_{Ket}^{D} = 1.34$ mM·min^{-1}. The K_M constant $K_{Ket}^{D} = 0.4$ mM of KAR complies with what has been determined for several organisms [83–85].

Table 3.2

Ketone	K_i	ee
Ethyl acetocetate (**1**)	58.5 mmol l^{-1}	97% L
Ethyl 4-chloro-acetoacetate (**2**)	–	54% L
Ethyl 4,4,4-trifluoro-acetoacetate (**3**)	52.1 mmol l^{-1}	62% L

4. Experimentally, a competitive inhibition of ADH by ketones was observed and quantified as $K_{1,Ket}^{ADH}$ = 58.5 mM for ketone **1** (Table 3.2). The rate equation of ADH was extended accordingly with respect to the full-scale model:

$$v_{ACA}^{ADH} = \frac{V_{ADH} \cdot [ACA] \cdot [NADH]}{\left(K_{NADH} + [NADH]\right) \cdot \left(\left(1 + \frac{[Ket]}{g_{1,Ket}^{ADH}}\right) \cdot K_{ACA}^{ADH} + [ACA]\right)} \tag{6}$$

All other parameters remain unchanged with respect to the full-scale model [53] and hence the original model behavior is maintained for [Ket]$_x$ = 0 mM and captures the oscillations as observed experimentally [49].

Table 3.2 experimentally determined inhibition of ADH. The inhibition constant $K_{1,Ket}^{ADH}$ and enantiomeric excess *ee* of carbinol formation for three ketones. Competitive inhibition of ADH was observed for **1** and **3**, whereas irreversible inhibition by alkylation occurred for **2**.

The enantiomeric excess (*ee*) of carbinol formation was determined for the three β-keto esters **1**–**3**. Their inhibition constants for ADH are given in Table 3.2.

3.6.4.3 Results

In this section we discuss the model predictions for the ketone ethyl acetoacetate (**1**). With the ketone absent ([Ket]$_x$ = 0 mM), the extended model reproduces all previous results with oscillations of all system variables above [Glc]$_{x0}$ > 18.5 mM [53]. Figure 3.6 shows the system's response to a fixed glucose concentration [Glc]$_{x0}$ at 30 mM and an increase of [Ket]$_x$ to 1 mM. The oscillations vanish at [Ket]$_x$ = 0.23 mM in a supercritical Hopf bifurcation and the steady state is stable for [Ket]$_x$ > 0.23 mM. Figure 3.6a shows the minimum and maximum concentrations of NADH as two thick curves, while in all other panels the time averages of the plotted variables are shown, not the minimum and maximum values. Since the addition of ketone provides an alternative mode of oxidation of NADH, the concentration of NADH is decreasing in Fig. 3.6a whereas the fluxes of carbinol production are increasing in Fig. 3.6b.

The enantiomeric excess (*ee*) does not change significantly (Fig. 3.6d). However, the carbon flux distribution through the backbone of glycolysis is changing. Ethanol (dot-dashed line) and glycerol (solid curve) production decrease due to the lack of cofactor NADH that is consumed in the competing reductions of ketone (Fig. 3.6c). Secretion of the intermediate acetaldehyde acts as a mode of overflow. Since the total glucose influx remains almost constant at 95 mM C$_3$ min^{-1} and

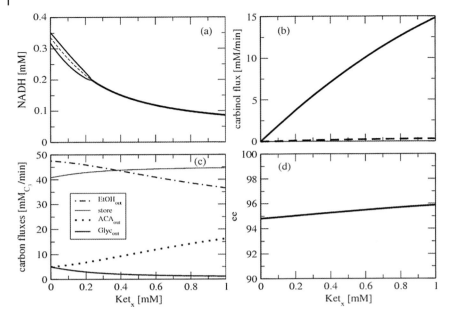

Fig. 3.6 Vanishing oscillations and flux re-routing for increasing ketone concentration. (a) NADH concentration oscillates between two solid curves, the unstable steady state is denoted by the thin dashed curve. (b) L-Carbinol (solid) and D-carbinol (dashed) fluxes. (c) C_3 carbon fluxes where time averages are shown in the oscillatory region. (d) Enantiomeric excess (*ee*), as defined in Eq. (1).

the glycerol pathway is gradually shut down, there is an increasing relative yield of ATP from glucose which in turn down-regulates phosphofructokinase activity and increases the amount of stored carbohydrates (thin solid curve).

3.6.5
Stoichiometric Model

3.6.5.1 Methods
The catabolism of **2** includes 118 reactions many of which are reversible (Fig. 3.5). Since kinetic data on micro-metabolites are difficult to determine experimentally, and in order to obtain an overall view of the xenobiotic metabolism, a stoichiometric model of the full network of degradation pathways of **2** was set up in addition to the network shown in Fig. 3.5. This network was then analyzed by means of elementary flux mode analysis [78].

3.6.5.2 Model Derivation
Using the modeling platform Copasi, the reaction network was encoded in the language SBML and the elementary flux modes were calculated, i.e. minimum sets

of reactions that can operate simultaneously at steady state. These were used for flux analysis.

3.6.5.3 **Results**

Four modes of degradation of **2** were discovered without the net consumption or production of redox equivalents. These four modes produce the final metabolites ethyl 3-ethoxy-4-chloro-2-butenoate, ethyl 3,3-bisethoxy-4-chlorobutanoate, oxol-2,4-dione, and 2,4-bischloromethyl-3,5-bisethoxycarbonyl-4-methyl pyridine. As a consequence, these modes do not depend on or disturb the glycolytic activity of the cell.

The analysis also revealed 20 other elementary modes of degradation of **2** that do consume NADH and therewith interfere with glycolysis. In order to facilitate these additional modes for a longer period of time, the cell has to ensure the necessary regeneration of NADH which under anaerobic conditions may proceed via the secretion of acetaldehyde or succinate. The analysis has shown that production of cytosolic intermediates along this large set of flux modes appears likely on the basis of the stoichiometric data.

The competing routes (I) through (VI) in Fig. 3.5 offer the possibility to direct the flux towards desired modes by inhibiting competing ones. As was shown for the enantiomeric excess in Fig. 3.6, it is now possible to predict the effect of inhibiting e.g. routes (II) and (III) in the reaction network (Fig. 3.5). Since both routes consume glutathione, cytosolic concentration of the latter increases and accelerates the flux along route (VI). With PharmBiosim the abundance of **7**, **12/13**, and **16** was predicted to increase by 36.1%, 5.4%, and 1.1%, respectively. Experimentally **(7)** was found to increase by 30.0%, **(12/13)** by 3.6% each, and **(16)** by 0.9%. From these deviations it becomes apparent that the current model reflects correct trends, but needs implementation of further metabolites. The results complement the findings depicted in Fig. 3.6, where trends are reproduced correctly, but with somewhat too low absolute values.

As was shown experimentally, micro-metabolites that are capable of acting as enzyme inhibitors may greatly affect product distribution. A biosimulation model can considerably assist drug approval by identifying physiological effects of drug metabolites too dilute in mammalian systems to be detected analytically. These results also show that the prediction of approximate product distribution requires implementation of a much more complex metabolic network, which is currently being done.

3.7
Application of the Model to Predict Drug Metabolism

In order to validate the abstraction concept (Fig. 3.1), chloramphenicol was submitted as a PAC to biotransformation by *S. cerevisiae*. Chloramphenicol metabolism in man and rat has been fully elucidated [86, 87], yet its complexity necessitated simplification:

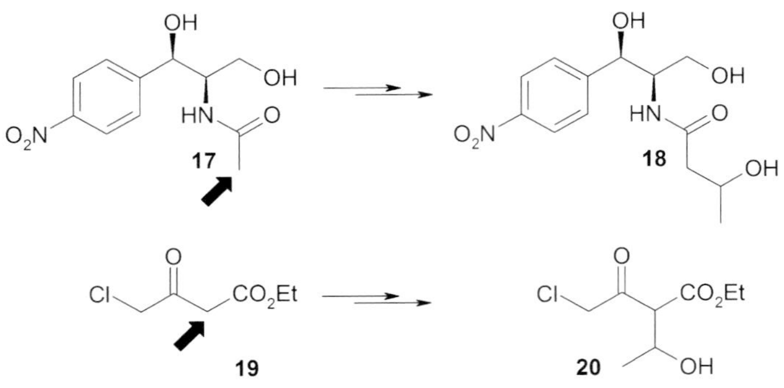

Fig. 3.7 Carboligation of bis-dechloro chloramphenicol (**17**). The doubly activated α-carboxylic position (arrow) reacts readily with acetaldehyde to give carbinols **18** and **20**, respectively. The adducts react further by dehydration followed by reduction.

1. Glucuronidation at C-1 was beyond the scope of these investigations.
2. Biotransformation of the nitro group will be implemented in a later stage of model development.
3. Whether or not a chloramphenicol stereoisomer is formed needs to be determined by investigating biochemical reduction of dehydrochloramphenicol.

Hence model substrate **2** was chosen, since stereoselectivity of carbonyl reduction could be studied directly [34].

In fact, biotransformation of **2** by *S. cerevisiae* fully represented the reaction pattern observed for the compound chloramphenicol in yeast and man. Most interestingly, carboligation occurred at the acetoamide residue of the fully dechlorinated chloramphenicol metabolite **17**. This reaction has not yet been described. The detection, after prediction by PharmBiosim, nicely illustrates the value of this approach (Fig. 3.7).

Further steps will be taken in order to implement bioconversion of the nitro group as well as biochemical glucuronidation and glutathionylation of the drug. Since hydrolysis of the amide bond was found to be caused by the action of glutathione, glutathione metabolism will be implemented into the metabolic network as well.

3.8
Conclusions

A living cell has a vast number of possibilities to degrade even simple xenobiotics. The metabolites arising from these reactions are in their majority physiologically active as well. For instance, chloro-acetone that acts as an enzyme inhibitor may greatly interfere with hepatic isomerization processes as they take place e.g. with ibuprofene. In other words, we have shown that it is indispensable to investigate pharmaceuticals for micro-metabolite production already in the intestine, since otherwise deleterious consequences for the patient may occur.

After the series of metabolic pathways had been elucidated for the three model compounds **1–3**, these data were implemented into the mathematical model PharmBiosim. The nonlinear system's response to varying ketone exposure was studied. The predicted vanishing of oscillatory behavior for increasing ketone concentration can be used to experimentally test the model assumptions in the reduction of the xenobiotic ketone. To generate such predictions, we employed as a convenient tool the continuation of the nonlinear system's behavior in the control parameters. This strategy is applicable to large systems of coupled, nonlinear, ordinary differential equations and shall together with direct numerical simulations be used to further extend PharmBiosim than was sketched here. This model already allows more detailed predictions of stereoisomer distribution in the products.

Acknowledgments

We gratefully acknowledge financial support by the European Union (NoE BioSim, Contract No. 005137), the Deutsche Bundesstiftung Umwelt, the Klaus Tschira Foundation, and the Fonds der Chemischen Industrie. Yeast was kindly provided by FALA Hefe GmbH, Kesselsdorf, Germany.

References

1 Ekins, S.: Predicting undesirable drug interactions with promicuous proteins *in silico. Drug Discov. Today* **2004**, 9:276–285.
2 Jörg, G., Hemery, T., Bertau, M.: Effects of cell-stress protectant glutathione on the whole-cell biotransformation of ethyl 2-chloro-acetoacetate with *Saccharomyces cerevisiae. Biocatal. Biotransform.* **2005**, 23:9–17.
3 Scatena, C.D., Hepner, M.A., Oei, Y.A., Dusich, J.M., Yu, S.F., Purchio, T., Contag, P.R., Jenkins, D.E.: Imaging of bioluminescent LNCaP-luc-M6 tumors: a new animal model for the study of metastatic human prostate cancer. *Prostate* **2004**, 59:292–303.
4 Jenkins, D.E., Yu, S.F., Hornig, Y.S., Purchio, T., Contag, P.R.: In vivo monitoring of tumor relapse and metastasis using bioluminescent PC-3M-luc-C6 cells in murine models of human prostate cancer. *Clin. Exp. Metastasis* **2003a**, 20:745–756.
5 Jenkins, D.E., Oei, Y., Hornig, Y., Yu, S.F., Dusich, J., Purchio, T., Contag, P.R.: Bio-

luminescent imaging (BLI) to improve and refine traditional murine models of tumor growth and metastasis. *Clin. Exp. Metastasis* **2003b**, 20:733–744.
6 Murray, L.J.: SU11248 inhibits tumor growth and CSF-1R-dependent osteolysis in an experimental breast cancer bone metastasis model. *Clin. Exp. Metastasis* **2003**, 20:757–766.
7 Azerad, R.: Microbial models for drug metabolism. *Adv. Biochem. Eng. Technol.* **1999**, 63:169–218.
8 Anzenbacherová, E., Anzenbacher, P., Svobodac, Z., Ulrichováa, J., Květinac, J., Zoulovác, J., Perlíkd, F., Martínková, J.: Minipig as a model for drug metabolism in man: comparison of in vitro and in vivo metabolism of propafenone. *Biomed. Pap.* **2003**, 147:155–159.
9 Neuberg, C., Hirsch, J.: Über ein Kohlenstoffketten knüpfendes Ferment (Carboligase). *Biochem. Z.* **1921**, 115:282–310.
10 Venisetty, R.K., Ciddi, V.: Application of microbial biotransformation for the new

drug discovery using natural drugs as substrates. *Curr. Pharm. Biotechnol.* **2003**, 4:153–167.

11 Briand, J., Blehaut, H., Calvayrac, R., Laval-Martin, D.: Use of a microbial model for the determination of drug effects on cell metabolism and energetics: study of citrulline-malate. *Biopharm. Drug Dispos.* **1992**, 13:1–22.

12 Romeo, J., Scheraga, M., Umbreit, W.W.: Stimulation of the growth and respiration of a methylotrophic bacterium by morphine. *Appl. Env. Microbiol.* **1977**, 34:611–614.

13 Smith, R.V., Rosazza, P.: Microbial systems for study of the biotransformation of drugs. *Biotechnol. Bioeng.* 17:785–814.

14 Cha, C.-J., Doerge, D.R., Cerniglia, C.E.: Biotransformation of malachite green by the fungus *Cunnighamella elegans*. *Appl. Env. Microbiol.* **2001**, 67:4358–4360.

15 Chatterjee, P., Kouzi, S.A., Pezzuto, J.M., Hamann, M.T.: *Appl. Env. Microbiol.* **2000**, 66:3850–3855.

16 Zhang, D., Freeman, J.P., Sutherland, J.B., Walker, A.E., Yang, Y., Cerniglia, C.E.: Biotransformation of chlorpromazine and methdilazine by *Cunninghamella elegans*. *Appl. Env. Microbiol.* **1996**, 62:798–803.

17 Griffiths, D.A., Best, D.J., Jezequel, S.G.: The screening of selected microorganisms for use as models of mammalian drug metabolism. *Appl. Microbiol. Biotechnol.* **1991**, 35:373–381.

18 Randez-Gil, F., Aguilera, J., Codon, A., Rincon, A.M., Estruch, F., Prieto, J.A.: Baker's yeast: challenges and future prospects. In: de Winde, J.H. (ed.), *Functional genetics of industrial yeasts* (*Topics in current genetics, vol. 2*), Springer-Verlag, Heidelberg, **2003**, pp 57–97.

19 De Souza Pereira, R.: The use of baker's yeast in the generation of asymmetric centers to produce chrial drugs and other compounds. *Crit. Rev. Biotechnol.* **1998**, 18:25–64.

20 Servi, S.: Baker's yeast as a reagent in organic synthesis. *Synthesis* 1990, **1990**:1–25.

21 Sih, C.J., Chen, C.S.: Microbial asymmetric catalysis – enantioselection reduction of ketones. *Angew. Chem. Int. Ed. Engl.* **1984**, 23:570–578.

22 Smith, R.V., Rosazza, J.P.: Microbial models of mammalian metabolism, aromatic hydroxylation. *Arch. Biochem. Biophys.* **1974**, 161:551–558.

23 Smith, R.V., Rosazza, J.P.: Microbial models of mammalian metabolism. *J. Pharm. Sci.* **1975**, 64:1737–1758.

24 Smith, R.V., Rosazza, J.P.: Microbial models of mammalian metabolism. *J. Nat. Prod.* **1983**, 46:79–91.

25 Abourashed, E.A., Clark, A.M., Hufford, C.D.: Microbial models of mammalian metabolism of xenobiotics: an updated review. *Curr. Med. Chem.* **1999**, 6:359–374.

26 Azerad, R.: Microbial models for drug metabolism. *Adv. Biochem. Eng. Biotechnol.* **1999**, 63:169–218.

27 Clark, A.M., Hufford, C.D.: Use of micro-organisms for the study of drug metabolism – an update. *Med. Res. Rev.* **1991**, 11:473–501.

28 Clark, A.M., McChesney, J.D., Hufford, C.D.: The use of microorganisms for the study of drug metabolism. *Med. Res. Rev.* **1985**, 5:231–253.

29 Rosazza, J.P., Smith, R.V.: Microbial models for drug metabolism. *Adv. Appl. Microbiol.* **1979**, 25:169–208.

30 Beukers, R., Marx, A.F., Zuidweg, M.H.J.: Microbial conversion as a tool in the preparation of drugs. In: Ariens, E.J. (ed.) *Drug design, vol. 3*. Academic Press, New York, **1972**, p. 1.

31 Lobastova, T.G., Sukhodolskaya, G.V., Nikolayeva, V.M., Baskunov, B.P., Turchin, K.F., Donova, M.V.: Hydroxylation of carbazoles by *Aspergillus flavus* VKM F-1024. *FEMS Microbiol. Lett.* **2004**, 235:51–56.

32 Hezari, M., Davies, P.J.: Microbial models of mammalian metabolism. Furosemide glucoside formation using the fungus *Cunninghamella elegans*. *Drug Metab. Disp.* **1993**, 21:259–267.

33 Edelson, J., McMullen, J.P.: *O*-Demethylation of *p*-nitroanisole by *Escherichia coli*. Stimulation by phenobarbital. *Drug Metab. Disp.* **1977**, 5:185–190.

34 Bertau, M.: How cell physiology affects enantioselectivity of the biotransformation of ethyl 4-chloro-acetoacetate with *Saccharomyces cerevisiae*. *Biocatal. Biotransform.* **2002**, 20:363–367.

35 Jörg, G., Bertau, M.: Fungal aerobic reductive dechlorination of ethyl 2-chloro-acetoacetate by *Saccharomyces cerevisiae*: mechanism of a novel type of microbial

dehalogenation. *ChemBioChem* **2004**, 5:87–92.

36 Williams, K. L., Zhang, Y., Shkriabai, N., Karki, R. G., Nicklaus, M. C., Kotrikadze, N., Hess, S., Le Grice, S. F., Craigie, R., Pathak, V. K., Kvaratskhelia, M.: Mass spectrometric analysis of the HIV-1 integrase-pyridoxal 5'-phosphate complex reveals a new binding site for a nucleotide inhibitor. *J. Biol. Chem.* **2005**, 280:7949–7955.

37 Geldenhuys, W. J., Gaasch, K. E., Watson, M., Allen, D. D., Van der Schyf, C. J.: Optimizing the use of open-source software applications in drug discovery. *Drug. Discov. Today* **2006**, 11:127–132.

38 Roy, A., Ette, E. I.: A pragmatic approach to the design of population pharmacokinetic studies. *AAPS J.* **2005**, 7:E408–E420.

39 Riley, R. J., Kenna, J. G.: Cellular models for ADMET predictions and evaluation of drug–drug interactions. *Curr. Opin. Drug Discov. Dev.* **2004**, 7:86–89.

40 Oliver, S. G.: From DNA sequence to biological function. *Nature* **1996**, 379:597–600.

41 Kitano, H.: *Foundations of systems biology*, MIT Press, Cambridge, **2001**.

42 Dimitrov, S. D., Low, L. K., Patlewicz, G. Y., Kern, P. S., Dimitrova, G. D., Comber, M. H., Phillips, R. D., Niemela, J., Bailey, P. T., Mekenyan, O. G.: Skin sensitization: modeling based on skin metabolism simulation and formation of protein conjugates. *Int. J. Toxicol.* **2005**, 24:189–204.

43 Zutphen, L. F. M. van: Use of animals in research: a science–society controversy? The European perspective. In: Gruber, F.P., Brune, K. (eds), *ALTEX-Buch 2002*, Spektrum Akademischer Verlag, Heidelberg, **2002**, pp 33–40.

44 Russell, W. M. S., Burch, R. L.: *The principles of humane experimental technique*, Methuen, London, **1959**.

45 Griffiths, D. A., Best, D. J., Jezequel, S. G.: The screening of selected microorganisms for use as models of mammalian drug metabolism. *Appl. Microbiol. Biotechnol.* **1991**, 35:373–381.

46 Zhang, D., Yang, Y., Castlebury, L. A., Cerniglia, C. E.: A method for the large scale isolation of high transformation efficiency fungal genomic DNA. *FEMS Microbiol. Lett.* **1996**, 145:261–266.

47 Sybesma, W. F. H., Straathof, A. J. J., Jongejan, J. A., Pronk, J. T., Heijnen, J. J.: Reductions of 3-oxo esters by baker's yeast: current status. *Biocatal. Biotransform.* **1998**, 16:95–134.

48 Rizzi, M., Baltes, M., Theobald, U., Reuss, M.: In vivo analysis of metabolic dynamics in *Saccharomyces cerevisiae*: II. Mathematical model. *Biotechnol. Bioeng.* **1997**, 55:592–608.

49 Danø, S., Sørensen, P. G., Hynne, F.: Sustained oscillations in living cells. *Nature* **1999**, 402:320–322.

50 Wolf, J., Heinrich R.: Effect of cellular interaction on glycolytic oscillations. *Biochem. J.* **2000**, 345:321–334

51 Wolf, J., Passarge, J., Somsen, O. J. G., Snoep, J. L., Heinrich, R., Westerhoff, H. V.: Transduction of intracellular and intercellular dynamics in yeast glycolytic oscillations. *Biophys. J.* **2000**, 78:1145–1153.

52 Teusink, B., Passarge, J., Reijenga, C. A., Esgalhado, E., van der Weijden, C. C., Schepper, M., Walsh, M. C., Bakker, B. M., van Dam, K., Westerhoff, H. V., Snoep, J. L.: Can yeast glycolysis be understood in terms of in vitro kinetics of the constituent enzymes? Testing biochemistry. *Eur. J. Biochem.* **2000**, 267:5313–5329.

53 Hynne, F., Danø, S., Sørensen, P. G.: Full-scale model of glycolysis in *Saccharomyces cerevisiae*. *Biophys. Chem.* **2001**, 94:121–163. The Silicon Cell version of this model is available at http://www.jjj.bio.vu.nl/database/ hynne.

54 Reijenga, K. A., Westerhoff, H. V., Kholodenko, B. N., Snoep, J. L.: Control analysis for autonomously oscillating biochemical networks. *Biophys. J.* **2002**, 82:99–108.

55 Beaudoin, F., Gable, K., Sayanova, O., Dunn, T., Napier, J. A.: A *Saccharomyces cerevisiae* gene required for heterologous fatty acid elongase activity encodes a microsomal beta-keto-reductase. *J. Biol. Chem.* **2002**, 277:11481–11488.

56 Gonzalez, E., Fernandez, M. R., Larroy, C., Sola, L., Pericas, M. A., Pares, X., Biosca, J. A.: Characterization of a $(2R,3R)$-2,3-butanediol dehydrogenase as the *Saccha-*

romyces cerevisiae YAL060W gene product. Disruption and induction of the gene. *J. Biol. Chem.* **2000**, 275:35876–35885.

57 Dahl, A. C., Madsen, J. O.: Baker's yeast: production of D-and L-3-hydroxy esters, *Tetrahedron: Asymmetry* **1998**, 9:4395–4417.

58 Kometani, T., Kitatsuji, E., Matsuno, R.: Bioreduction of ketones mediated by baker's yeast with acetate as ultimate reducing agent. *Agric. Biol. Chem.* **1991**, 55:867–868.

59 Ford, G., Ellis E. M.: Characterization of Ypr1p from *Saccharomyces cerevisiae* as a 2-methylbutyraldehyde reductase. *Yeast* **2002**, 19:1087–1096.

60 Richard, P.: The rhythm of yeast. *FEMS Microbiol. Rev.* **2003**, 27:547–557.

61 Prentki, M.: New insights into pancreatic betacell metabolic signaling in insulin secretion. *Eur. J. Endocrinol.* **1996**, 134:272–286.

62 Penninckx, M.: A short review on the role of glutathione in the response of yeasts to nutritional, environmental, and oxidative stresses. *Enzyme Microb. Technol.* **2000**, 26:737–742.

63 Grogan, G., Graf, J., Jones, A., Parsons, S., Turner, N. J. and Flitsch, S. L.: An asymmetric enzyme-catalyzed *retro*-Claisen reaction for the desymmetrization of cyclic diketones. *Angew. Chem.* **2001**, 113:1145–1148 [*Angew. Chem. Int. Ed.* 40:1111–1114].

64 Datta, J. and Samanta, T. B.: Characterization of a novel microsomal glutathione *S*-transferase produced by *Aspergillus ochraceus* TS. *Mol. Cell. Biochem.* **1992**, 118:31–38.

65 Stegemann, C.: Diploma thesis, Dresden University of Technology, **2005**.

66 Chin-Joe, I., Nelisse, P. M., Straathof, A. J. J., Jongejan, J. A., Pronk, J. T. and Heijnen, J. J.: Hydrolytic activity in baker's yeast limits the yield of asymmetric 3-oxo ester reduction. *Biotechnol. Bioeng.* **2000**, 69:370–376.

67 Breeuwer, P., Drocourt, J. L., Bunschoten, N., Zwietering, M. H., Rombouts, F. M. and Abee, T.: Characterization of uptake and hydrolysis of fluorescein diacetate and carboxyfluorescein diacetate by intracellular esterases in *Saccharomyces cerevisiae*, which result in accumulation of fluores-

cent product. *Appl. Environ. Microbiol.* **1995**, 61:1614–1619.

68 Cha, C. J., Coles, B. F., Cerniglia, C. E.: Purification and characterization of a glutathione *S*-transferase from the fungus *Cunninghamella elegans*. *FEMS Microbiol. Lett.* **2001**, 203:257–261.

69 Ushio, K., Ebara, K. and Yamashita, T.: Selective inhibition of *R*-enzymes by simple organic acids in yeast-catalysed reduction of ethyl 3-oxobutanoate. *Enzyme Microb. Technol.* **1991**, 13:834–839.

70 Wakil, S. J., Stoops, J. K. and Joshi, V. C.: Fatty acid synthesis and its regulation. *Annu. Rev. Biochem.* **1983**, 52:537–579.

71 Hucka, M., Finney, A., Sauro, H. M., Bolouri, H., Doyle, J. C., Kitano, H., Arkin, A. P., Bornstein, B. J., Bray, D., Cornish-Bowden, A., Cuellar, A. A., Dronov, S., Gilles, E. D., Ginkel, M., Gor, V., Goryanin, I. I., Hedley, W. J., Hodgman, I., Hofmeyr, J.-H., Hunter, P. J., Juty, N. S., Kasberger, J. L., Kremling, A., Kummer, U., Le Novere, N., Loew, L. M., Lucio, D., Mendes, P., Minch, E., Mjolsness, E. D., Nakayama, Y., Nelson, M. R., Nielsen, P. F., Sakurada, T., Schaff, J. C., Shapiro, B. E., Shimizu, T. S., Spence, H. D., Stelling, J., Takahashi, K., Tomita, M., Wagner, J., Wang, J.: The systems biology markup language (SBML): a medium for representation and exchange of biochemical network models. *Bioinformatics* **2003**, 19(4):524–531.

72 Cuellar, A. A., Nielsen, P. F., Bullivant, D. F., Hunter, P. J.: CellML 1.1 for the definition and exchange of biological models. *Conf. Proc. IFAC Symp. Model. Control Biomed. Syst.* **2003**, 2003:451–456.

73 Strogatz, S. H.: *Nonlinear dynamics and chaos*, Addison–Wesley, Reading, Mass., **1994**.

74 Kaplan D., Glass, L.: *Understanding nonlinear dynamics*, Springer, New York, **1995**.

75 Mendes, P.: Biochemistry by numbers: simulation of biochemical pathways with Gepasi 3. *Trends Biochem. Sci.* **1997**, 22:361–363.

76 Tyson, J. J., Chen K., Novak, B.: Network dynamics and cell physiology. *Nat. Rev. Mol. Cell Biol.* **2001**, 2:908–916.

77 Fall, C. P., Marland, E., Wagner J. M., Tyson, J. J.: *Computational cell biology*, Springer, NewYork, **2002**.

78 http://www.copasi.org

79 Ermentrout, B.: *Simulating, analyzing, and animating dynamical systems: a guide to XPPAUT for researchers and students*, SIAM, Philadelphia, **2002**, available at http://www.math.pitt.edu/bard/xpp/xpp.html.

80 Doedel, E., Keller, H., Kernevez, J.: Numerical analysis and control of bifurcation problems: (I) Bifurcation in finite dimensions. *Int. J. Bif. Chaos* **1991**, 1:493–520.

81 Doedel, E., Keller, H., Kernevez, J.: Numerical analysis and control of bifurcation problems: (II) Bifurcation in infinite dimensions. *Int. J. Bif. Chaos* **1991**, 1:745–772.

82 A public domain "SiliconCell" version of the model can be used to simulate online at http://www.jjj.bio.vu.nl, including a graphical overview and mathematical expressions.

83 Shimakata, T., Stumpf, P.K.: Purification and characterizations of β-ketoacyl-[acyl-carrier-protein] reductase, β-hydroxyacyl-[acylcarrier-protein] dehydrase, and enoyl-[acyl-carrier-protein] reductase from *Spinacia oleracea* leaves. *Arch. Biochem. Biophys.* **1982**, 218:77–91.

84 Sheldon, P.S., Kekwick, R.G.O., Smith, C.G., Sidebottom, C., Slabas, A.R.: 3-Oxoacyl-[ACP] reductase from oil seed rape (*Brassica napus*). *Biochim. Biophys. Acta* **1992**, 1130:151–159.

85 Sheldon, P.S., Kekwick, R.G.O., Sidebottom, C., Smith, C.G., Slabas, A.R.: 3-Oxoacyl-(acyl-carrierprotein) reductase from avocado (*Persea americana*) fruit mesocarp. *Biochem. J.* **1990**, 271:713–720.

86 Wal, J.M., Peleran J.-C., Bories, G.F.: Identification of chloramphenicol oxamic acid as a new major metabolite of chloramphenicol in rats. *FEBS Lett.* **1980**, 119:38–42.

87 Kunin, C.M., Glazko, A.J., Finland, M.: Persistance of antibiotics in blood of patients with acute renal failure. II. Chloramphenicol and its metabolic products in the blood of patients with severe renal disease or hepatic cirrhosis. *J. Clin. Invest.* **1959**, 38:1498–1508.

Part II
Simulating Cells and Tissues

4
Correlation Between *In Vitro*, *In Situ*, and *In Vivo* Models

Isabel González-Álvarez, Vicente Casabó, Virginia Merino, and Marival Bermejo

Abstract

To obtain a drug product with a good oral bioavailability is one of the main objectives during drug development, as the oral route is the most convenient and used. In order to optimize the absorbability of any new drug molecule, it is necessary to increase our knowledge about the factors influencing drug absorption and membrane permeation. Mathematical modeling of transport processes is a useful tool for this later purpose. Predicting the oral fraction absorbed is a multifactorial problem that depends upon physicochemical, physiological, and formulation factors. The drug must be released from the dosage form and get into solution, which is the first essential step before its absorption. All these processes are amenable for modeling and simulation with adequate input from different experimental systems. In this chapter, some biophysical absorption models are described as examples of how to use the physicochemical properties of the drug to obtain estimations about its absorbability and to elucidate the potential influence of some excipients. Second, the modeling approaches to predict oral fraction absorbed from permeability and solubility data are illustrated; and finally some examples are given of modeling methods to characterize active and passive transport parameters and the correlation between *in vitro* and *in situ* results.

4.1
Introduction

The oral route for drug administration is the most convenient and preferred by patients. So, in general, the main objective in the process of drug development is to obtain a drug product with a good oral bioavailability. Different approaches have been developed to predict intestinal permeability and oral fraction absorbed as the essential steps for determining bioavailability (a complicated and time-consuming process). The terms absorption and bioavailability are often misused. Win L. Chiou defined absorption as the movement of drug across the outer mucosal membranes of the gastro-intestinal tract (GI), while bioavailability is defined as availability of

Biosimulation in Drug Development. Edited by Martin Bertau, Erik Mosekilde, and Hans V. Westerhoff
Copyright © 2008 WILEY-VCH Verlag GmbH & Co. KGaA, Weinheim
ISBN: 978-3-527-31699-1

drug to the general circulation or site of pharmacological actions [1]. Correlations between experimental *in situ* and *in vitro* parameters and the oral bioavailability of drugs must include a term to account for first-pass effects to differentiate the oral fraction absorbed from the systemic appearance before establishing any correlation. In order to optimize the absorbability of any new drug molecule it is necessary to increase our knowledge about the factors influencing drug absorption and membrane permeation. Mathematical modeling of transport processes is a useful tool for this later purpose.

Predicting oral fraction absorbed is a multifactorial problem that depends upon three levels of factors, from the physicochemical characteristics of the drug (drug solubility, pK_a, lipophilicity, particle size, surface area, crystalline form, stability), the formulation (dosage form, disintegration, release mechanism and rate) and finally, the physiological variables (gastrointestinal pH, gastric emptying, intestinal motility and transit time, intestinal secretions, intestinal blood flow, membrane permeability). The drug must be released from the dosage form and get into solution, which is the first essential step before its absorption. For this reason, solubility (and dissolution rate) and permeability across the intestinal membrane can be identified as key parameters for a new chemical entity in order for it to become a lead compound. Different approaches have been developed to predict intestinal permeability and oral fraction absorbed for the drug itself and for the drug in the dosage form incorporating the dissolution and release processes in the models.

Among the wide range of assays available for estimating intestinal permeability, the main difference between the *in vitro* and the *in situ* or *in vivo* models are the throughput ability and the physiological variables that the model is able to reproduce. The simplest *in vitro* models (as pampa membranes) have a high-throughput capacity but lack many of the relevant characteristics as transporter expression or paracellular route. *In vitro* models, such as Caco-2 cultures, include transporter and paracellular route but there is a quantitative difference with the *in vivo* situation that must be modeled and considered to scale up the results. This is the reason for developing mathematical models that allow to scale up from the *in vitro* data and to predict human results. As shown here, all these *in vitro* systems and the corresponding models perform well for estimating the absorbability of compounds absorbed by passive diffusion but more research is necessary in order to improve the predictability of *in vitro* systems with regard to carrier-mediated processes; and in this case the development of mechanistic models incorporating the relevant physiological variables could be a helpful method.

However, the sensibility of the Caco-2 monolayers to the different standard operation procedures used for its culture increases the inter-laboratory variability and complicates the combination of data coming from different laboratories and the interpretation of results. Animal *in vivo* models reproduce the real situation better, as those models include all the physiological variables but, of course, have a low throughput capacity and are not adequate for screening a high number of compounds.

This chapter deals with the description of the biophysical absorption models that have been applied to study intestinal drug absorption and the influence of surfactants on intestinal permeability. Second, the modeling approaches to predict oral fraction absorbed are described; and finally some examples of modeling methods to characterize active and passive transport parameters and the correlation between *in vitro* and *in situ* results are shown.

4.2
Biophysical Models of Intestinal Absorption

Biophysical models are useful tools for studying drug absorption and the relationships between drug intestinal permeability and molecular descriptors. The methodology that many research groups apply consists of establishing correlations among intestinal permeability values and physicochemical indexes such as lipophilicity, and molecular weight. These relationships are different, depending on the absorption site in the gastrointestinal system, and those models explain the differences based of the particular physiological features of each segment. Figure 4.1 represents a scheme of the biophysical models describing the process of drug absorption in the gastric mucosa, small intestine, and colon. The rationale for these models is described in this section.

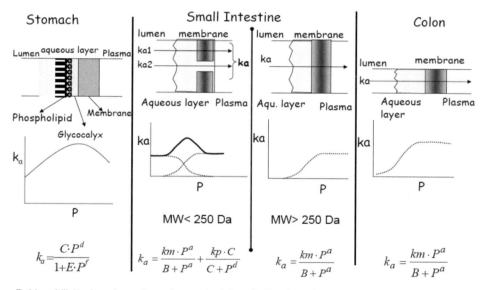

P: Lipophilicity, **ka**: absorption rate constant, **km**: limiting ka value
Other symbols: fitting parameters

Fig. 4.1 Summary of the biophysical models applied to predict drug absorption from lipophilicity in the gastrointestinal tract.

4.2.1
Colon

To describe the relationship between colonic absorption rate constants and the lipophilicity of xenobiotics, the hyperbolic equation proposed by Wagner applies [2].

$$k_a = \frac{k_m \cdot P^a}{B + P^a} \tag{1}$$

where a and B are the constants or parameters depending on the experimental conditions, P represents any lipophilicity index, and k_m is a parameter that represents the asymptote or maximum absorption rate constant in that condition for the series. Therefore, the absorption rate constant tends to increase as lipophilicity does, but the existence of a limiting step for diffusion stabilizes the k_a values for highly liphophilic members of the series, which can permeate this layer at a limiting constant rate, k_m. The limiting step for highly liphophilic compounds is the permeation across the boundary aqueous layer in the luminal side [3, 4].

In colon only the transcellular permeation route is considered to be relevant as the transcellular route is restricted due to the higher tightness of the intercellular junctions in this area.

4.2.2
Small Intestine

A double hyperbolic equation [5, 6] describes the absorption through the intestinal epithelia for compounds with a molecular size below 250 Da.

$$k_a = k_1 + k_2 = \frac{k_m \cdot P^a}{B + P^a} + \frac{k_p \cdot B'}{B' + P^{a'}} \tag{2}$$

These compounds are able to diffuse through both the membrane and aqueous pores and/or tight junctions. The global absorption rate constant is described by the sum of two absorption rate values which represent two different pathways; the penetration into the lipoidal membrane, and across the aqueous pores. B, B', a, and a' are constants that depend on the experimental technique used to obtain the absorption rate constants, P is the lipophilicity parameter, and k_m and k_p are the asymptotic values for the membrane and the paracellular way, respectively. For compounds with a molecular weight higher than 250 Da the contribution of the aqueous pathway is negligible, collapsing the equation to the one hyperbola model already described for colon.

4.2.3
Stomach

A bilinear equation applies in this special situation where phospholipid and glycocalyx layers preceed the lipoidal membrane. As in colon and small intestine there is an aqueous boundary layer close to the phospholipid layer. This idea seems in agreement with the approaches of Hills and coworkers [7] who described a lipid lining of natural amphiphiles that protects the gastric mucosae. This phospholipid layer is bonded to the glycocalyx by hydrogen bonds and van der Waal's forces. This structure behaves as a heterogeneous system in terms of xenobiotic diffusion: a stagnant aqueous layer (hydrophilic), a layer of phospholipids (liphophilic), the glycocalyx (hydrophilic), the lipoidal membrane, and the aqueous plasma sink. That heterogeneous path produces the bilinear correlation [8].

$$k_a = \frac{C \cdot P^d}{1 + E \cdot P^f} \tag{3}$$

where C, d, E, and f are the fitting parameters that depend on the technique, and P is the lipophilicity parameter.

4.3
Influence of Surfactants on Intestinal Permeability

Surfactants are included as excipients in many drug formulations with the objective of improving dissolution rate and increasing drug solubility. These aims are based on the ability of surfactants to reduce the interfacial tension and contact angle between solid particles and aqueous media, thus improving drug wettability and increasing the surface available for drug dissolution. Their influence on the overall solubility of the drug relates to the inclusion of the poorly soluble compounds in the non-polar core of the micelles.

The effects of surfactant on the solubility and dissolution rate of poorly soluble drugs are well characterized. In general, the surfactant increases both solubility and dissolution rate even if the increment in dissolution rate is less pronounced due to the low diffusivity of the drug loaded into the micelle [9–16].

These pharmaceutical additives have been also characterized as potential absorption enhancers. Nevertheless, their effects over intestinal membrane, (i.e. over drug permeability) are more complex and less well defined. It has been shown that most surfactants interact with the absorbing membranes enhancing permeability and facilitating intestinal absorption of some dissolved drugs. This effect becomes particularly apparent when the surfactant is below or just at its critical micelle concentration (CMC) in the drug solution, but it can be decreased and even reversed when CMC is surpassed [17, 18]. The role of lipophilicity in intrinsic drug absorption from surfactant solutions at CMC and at supramicellar concentration has been discussed [19–21]. This section summarizes how the previously described biophysical models are used to explain the permeability changes appearing in the presence of surfactants.

In order to explore the effects of surfactants on the gastrointestinal tract, absorption experiments (with families of homologous or structurally related compounds) were performed in absence and in the presence of surfactants and with the xenobiotic under study in solution. The surfactants were used at two different concentrations: a concentration below the CMC, and another one above it (supramicellar concentration, SMC). This experimental design allowed differentiation of the influence of surfactant monomers from the effect of micelles. The next step was the establishment of absorption–partition relationships in the absence and the presence of the additive to check the changes exerted by the surfactant.

The experiments were performed in stomach, small intestine, and colon. In order to establish a general hypothesis suitable for different drugs and surfactants, the research was performed including anionic, cationic, and neutral surfactants as well as drugs with acidic, alkaline, neutral or amphoteric character [22–25].

4.3.1
Absorption Experiments in Presence of Surfactants

4.3.1.1 Colon

Below or at the CMC In the experiments with synthetic surfactants at CMC, the best fit describing the relationship between absorption rate constant and lipophilicity corresponds to a potential equation [26]:

$$k_0 = C \cdot P^d \tag{4}$$

where d and C are the regression parameters.

Figure 4.2 shows the results obtained in rat colon with sodium lauryl sulfate and phenyl-alkyl-carboxylic acids. While correlations between k_a and P in the absence of surfactants are hyperbolic, correlations in the presence of CMC of synthetic surfactants are always potential with a rather low slope. Below or at the CMC [27, 28] the

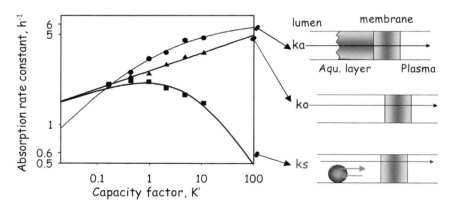

Fig. 4.2 Absorption–partition correlation obtained in rat colon for a family of acid compounds under different conditions: k_a and k_o are absorption rate constant values calculated in the absence and in the presence of sodium lauryl sulfate below its CMC, respectively, and k_s at a supramicellar concentration. (Adapted from Garrigues et al. [20]).

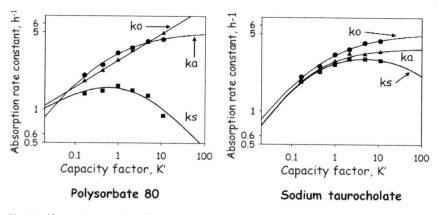

Fig. 4.3 Absorption–partition relationships obtained in rat colon in the presence of polysorbate 80 and sodium taurocholate at their CMC and SMC concentrations. (Adapted from Bermejo et al. [21]).

limiting effect on k_a values exerted by the boundary aqueous layer is lost and hence k_m cancels out. Besides, a second effect consisting of an increased membrane polarity by the surfactant also takes place. The overall result over the absorption rate constant of each compound depends on its lipophilicity and on the surfactants. As can be seen in Fig. 4.2, the change in the slope from the hyperbola to the potential equation can lead to an increase of the absorption rate constant but also to a decrease in this parameter.

Natural surfactants such as sodium taurocholate and sodium glycocholate are not able to eliminate the limiting character of the aqueous diffusion layer adjacent to the luminal side of the membrane [29, 30]. For that reason in the presence of these natural surfactants at their CMC, the absorption lipophilicity correlations are as hyperbolic as the ones obtained without additives. Figure 4.3 displays the difference between the correlations obtained in the presence of polysorbate 80 and sodium taurocholate [27, 28].

At the SMC The change in membrane polarity as well as removal of the aqueous layer limiting step are effects which, undoubtedly, must also be produced by the surfactant molecules. These effects are apparently masked by micelle solubilization governed by micelle–aqueous phase partitioning (P_a). According to the Collander equation [31], two partition coefficients found in two related systems can be expressed as follows:

$$\log P_a = f \cdot \log P + y \tag{5}$$

$$P_a = P^f \cdot 10^y = P^f \cdot E \tag{6}$$

The micelle-solubilized fraction is not available for absorption. The bilinear equation is in this case applicable to correlate absorption rate constant values and lipophilicity parameters:

$$k_s = \frac{k_0}{1 + P_a} = \frac{k_0}{1 + E \cdot P^f} \tag{7}$$

where k_0 had to be replaced by a suitable expression, that is Eq. (1), if bile acid salts are involved, or Eq. (4) when synthetic surfactants are employed, E and f are constants which depend on the technique employed. These correlations can be observed in Figs. 4.2 and 4.3 (k_s values).

4.3.1.2 Intestine

Below or at the CMC In this case it was assumed that the paracellular absorption rate constant, k_{02}, was the same as k_2 in Eq. (2), whereas the membrane absorption rate constant, k_{01}, was calculated as was k_1 in colon tests from Eq. (4). Accordingly, the global k_a value should be:

$$k_a = k_{01} + k_{02} = k_{01} + k_2 = C \cdot P^d + \frac{k_p \cdot B'}{B' + P^{a'}} \tag{8}$$

At SMC The reasoning already used for colonic pathway was used with the following modifications: (a) the membrane absorption rate constants, k_{s1}, were assumed to be governed by the bilinear-type equation deduced as:

$$k_{s1} = \frac{C \cdot P^d}{1 + E \cdot P^f} \tag{9}$$

and (b) the pore absorption rate constant, k_{s2}, was also calculated assuming the same micelle–aqueous phase partitioning:

$$k_{s2} = \frac{k_p \cdot B'}{(1 + E \cdot P^f) \cdot (B' \cdot P^{a'})} \tag{10}$$

and thus the global k_a value would be:

$$k_a = k_{s1} + k_{s2} = \frac{C \cdot P^d}{1 + E \cdot P^f} + \frac{k_p \cdot B'}{(1 + E \cdot P^f) \cdot (B' + P^{a'})} \tag{11}$$

An example of the correlations obtained in small intestine in the absence and in the presence of tetradecyltrimethylamonium bromide (TTAB) at 0.0125% (<CMC) as well as at 1% (>CMC) is represented in Fig. 4.4 [26]. As can be seen, the effect of each surfactant concentration is quite significant. But, a natural surfactant such as sodium taurocholate did not produce any significant change in the absorption rate constants of a series of phenyl–alkyl carboxylic acids at its CMC and it exerted an almost negligible solubilization effect at a supramicellar concentration [23].

4.3.1.3 Stomach
In the presence of surfactants below its CMC the correlations obtained in stomach become potential (synthetic surfactants) or hyperbolic (natural surfactants). Both

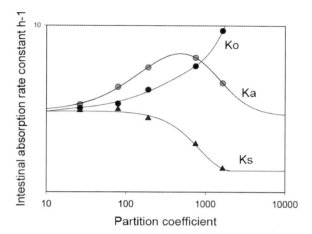

Fig. 4.4 Absorption–partition correlations obtained in rat small intesine for a family of aromatic amines in absence (k_a), and in presence of tetradecyltrimethylamonium below (k_o) and above (k_s) its CMC. (Adapted from Garrigues et al. [26]).

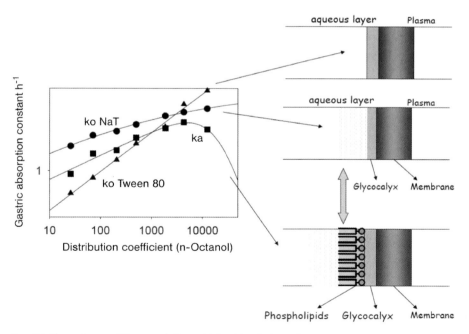

Fig. 4.5 Absorption–partition correlations obtained in rat stomach for a series of acid xenobiotics in absence of any additive (ka) and in presence of sodium taurocholate (NaT) and Tween 80 at a concentration below its CMC. (Adapted from Garrigues et al. [8]).

correlations are represented in Fig. 4.5 along with the correlation obtained in the absence of additives. In general, we observed an overall and significant increase in all the absorption rate coefficients. This fact suggests the possibility that the surfactants disrupt the hydrophobic lining, leaving the gastric mucosa exposed to the acidic environment. Once the phospholipid layer is removed, the lipid membrane is the main resistance to diffusion in the presence of synthetic surfactants, as they effectively reduce aqueous resistance, whereas they lead to a potential relationship in the presence of natural surfactants. The aqueous layer remains unaffected, and the correlations are hyperbolic. This difference can be clearly observed in Fig. 4.5 [6]. The ability to eliminate the phospholipid layer could explain the ulcerogenic activity of sodium taurocholate and other bile salts and synthetic surfactants.

All these results can be interpreted as follows:

1. Pharmaceutical surfactants seem to nullify the limiting effect on solute diffusion of the stagnant aqueous layer.
2. Pharmaceutical surfactants increase the polarity of the membrane, rendering it more permeable for highly hydrophilic substances. This effect is responsible for the lower slope of the potential relationships.
3. When the synthetic surfactant is added at SMC concentration, the above effects become almost completely masked by the micellar solubilization of the xenobiotics. Correlations become bilinear as a result of the partitioning process of the solutes between the micellar cores and the aqueous solution.
4. Natural surfactants show no ability of nullifying the aqueous layer resistance.
5. The solubilization potential of natural surfactant is lower than that observed for pharmaceutical surfactants.

The possible reasons for the different behavior of natural surfactants could include the following. Natural surfactants lead to surface tension values higher than those corresponding to the same concentration of a synthetic surfactant. The ability to nullify the aqueous layer resistance could be related with the surface tension values. However, the micelles of bile salts are smaller and more rigid than the micelles of synthetic surfactants. The solubilization potential of bile salts is increased in the presence of lecithins and fatty acids. For instance, the absorption rate constants obtained in the presence of sodium taurocholate and glycocholate mixed-micelles with lecithin for a series of acids were significantly lower than those obtained in the presence of simple micelles of the same bile salts [29, 30].

The reported effects of surfactants on membrane permeability have been observed by other authors using a different approach. Rege et al. [32] studied the inhibition activity of non-ionic surfactants on P-glycoprotein (P-gp) efflux and the relationship between inhibition and membrane fluidity. Tween 80 and Cremophor inhibited P-gp. These inhibition effects could be related to their effects on membrane fluidity, as these surfactants fluidized cell lipid bilayers. This fluidification mechanism could be related to the increase in membrane polarity that we observed with synthetic surfactants in our experiments; and it has been reported by other authors as a possible explanation for the inhibition effects of surfactants on P-gp efflux in the blood–brain barrier [33].

Apart from the effects already discussed, the fundamental conclusion of our work is that synthetic and natural surfactants behave in different ways. When these kinds of additives are used in a drug formulation their possible influence on drug permeability and absorption must be considered.

4.4
Modeling and Predicting Fraction Absorbed from Permeability Values

Drug solubility (and dissolution rate) and permeability across the intestinal membrane can be identified as key parameters for a new chemical entity in order for it to become a lead compound. These factors constitute the fundamental aspects of the *Biopharmaceutic Classification System* that has evolved into a modern tool to speed the drug development process. Different approaches have been developed to predict intestinal permeability and oral fraction absorbed for the drug itself and for the drug in the dosage form incorporating the dissolution and release processes in the models. This section focuses on the modeling approaches used to predict the oral fraction absorbed from permeability values obtained using different experimental systems. In this regard the predicted fraction absorbed corresponds to the maximum potential value in the absence of any other limiting factor, e.g. solubility or dissolution time.

4.4.1
Mass Balance, Time-independent Models

The simplest model consists of considering small intestine as a tube with area:

$$S = 2 \cdot \pi \cdot R \cdot L$$

where R is the radius and L the length of the segment.

Considering the mass balance in the segment:

$$-\frac{dM}{dt} = \phi \cdot (C_0 - C_f) = 2\pi \cdot R \cdot P_{eff} \cdot \int_0^L C \cdot dz \qquad (12)$$

Where M is the amount of drug absorbed, ϕ is the volumetric flow, C_o and C_f are the concentration at the beginning and at the end of the segment respectively, L is the length of the segment, R is the radius, P_{eff} is the drug permeability, and z is the axial distance.

At steady state the fraction absorbed is:

$$fa = 1 - \frac{C_f}{C_o} \qquad (13)$$

$$fa = 2 \cdot \frac{\pi \cdot R \cdot L \cdot P_{eff}}{\phi} \cdot \int_0^1 C^* \cdot dz^* \tag{14}$$

$$fa = 2 \cdot An \cdot \int_0^1 C^* \cdot dz^* \tag{15}$$

where C^* and dz^* are adimensional variables that corresponds to C_f/C_o and z/L. An is the absorption number, defined as the ratio between the transit time in the segment and the absorption time (R/P_{eff}). Table 4.1 summarizes the solutions for Eq. (15).

This mass balance approach in general renders good predictions for high solubility drugs; but for low solubility compounds there are other more accurate approaches based on microscopic mass balance that provide better estimations [34].

A second approach to obtain Eq. (16) is to consider the small intestine as a compartment from which the drug disappears following a first-order process. This process corresponds to the disappearance by absorption of the remaining amounts of a tested drug or xenobiotic from the perfusion fluid along the *in situ* absorption experiment. As A_r is the amount of the tested compound in the luminal fluid at any time, t, and A_0 is the initial amount (i.e. the total dose perfused), one can write:

$$A_r = A_0 \cdot e^{-k_a \cdot t} \tag{19}$$

which, if we subtract each member from A_0, gives:

$$A_0 - A_r = A_0 - A_0 \cdot e^{-k_a \cdot t} \tag{20}$$

Table 4.1 Solutions to Eq. (15). C_s is drug solubility. C_o and C_f are the concentration at the beginning and at the end of the intestinal segment respectively. An – Absorption number, Do – dose number. See text for explanation.

Comment	Conditions	Integral	
Highly soluble drugs. Permeability is the main parameter determining fa	$C_o \leqslant C_s$ and $C_f \leqslant C_s$	$Fa = 1 - e^{-2 \cdot An}$	(16)
Drug in solid form (suspension). If dissolution is faster than absorption then concentration is C_s	$C_o > C_s$ and $C_f > C_s$	$Fa = \dfrac{2 \cdot An}{Do}$	(17)
	$C_o > C_s$ and $C_{sf} \leqslant C_s$	$Fa = 1 - \dfrac{1}{Do} \cdot e^{-2 \cdot An + Do - 1}$	(18)

If each term is divided by A_0, we have:

$$1 - \frac{A_r}{A_0} = 1 - e^{-k_a \cdot t} \tag{21}$$

Where A_r/A_0 is the fraction of the initial dose remaining in the luminal fluid. $1 - A_r/A_0$ represents the fraction of the compound absorbed, provided that no presystemic losses exist, i.e. F. If we make t equivalent to the mean absorbing time (that is, the total time along which the absorption of the compound takes place), Eq. (16) is obtained.

4.4.2
Prediction of the Fraction of Dose Absorbed from *In Vitro* and *In Situ* Data

Absorption rate coefficients and permeability values obtained through different systems, i.e. *in vitro* or *in situ*, finally are determined with the objective in mind of predicting the fraction of a dose that will be absorbed through the intestine when the drug is orally administered. In order to use the data obtained in those systems it is essential to validate them properly. The present section reviews the capability of the parameters representatives of the absorption determining with different techniques to predict *in vivo* drug absorption in absence of any limited step (as dissolution or solubility) by means of the previously described model and Eq. (16). In particular, the absorption rate constant obtained with an *in situ* close loop assay, permeability through Caco-2 cell lines, and PAMPA models have been examined.

4.4.3
Prediction from *In Situ* Absorption Rate Constant Determined with Closed Loop Techniques

As explained elsewhere, *in situ* absorption rate constants (k_a) can be easily obtained after fitting a first-order equation to the data corresponding to the concentration of drug remaining in the intestine with the time determined using a closed loop technique. We demonstrated, using a series of seven fluoroquinolones, that from k_a it is possible to predict the fraction of dose that would be absorbed when administering those drugs orally [35].

The equation used to predict the fraction of dose absorbed *in vivo* (from a solution), F_a, of a drug or drug candidate belonging to a series, by means of a simple determination of its *in situ* intrinsic intestinal absorption rate constant (k_a) in anesthetized rats, is as follows:

$$F_a = 1 - e^{-k_a T} \tag{22}$$

in which T is the absorbing time, calculated by non-linear regression between the absolute bioavailability determined from the *in vivo* plasma level curves, obtained after oral administration of the tested compounds in solution, and k_a of the same compounds.

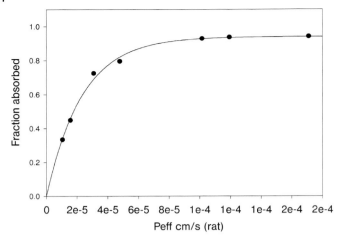

Fig. 4.6 Correlation between *in vivo* bioavailability in rat, *F*, based on oral/i.v. AUC ratios, and intestinal permeabilities, P_{eff} obtained in situ in rats. (Adapted from [35]).

It has to be emphasized that this expression is only applicable to compounds that are absorbed by diffusion and do not suffer a first-pass effect. The compounds that are actively absorbed (at least in part) along the enterocyte membrane, as well as those that are actively and substantially excreted to luminal fluid from the membrane or even the cytoplasm of the absorbing cells (i.e. P-gp substrates), should not be used for such approaches. For the compounds tested, the fraction of dose absorbed *in vivo* coincides with oral bioavailability (*F*), since they are passively absorbed and do not suffer first-pass metabolism.

Fitting the determined *F* and k_a values for the seven quinolones leads to an estimation of the absorption time $T = 0.93 \pm 0.06$ h. The correlation coefficient ($r = 0.989$) is highly significant. The precision of the parameter is also excellent (coefficient of variation $\sim$6.5%). Figure 4.6 graphically shows the correlation.

Table 4.2 shows the *F* values predicted after the "leave one out" procedure using Eq. (12). For all the compounds assayed, the prediction is fairly good. The predicted *F* values correlate very well with those determined *in vivo* from plasma level data.

These considerations are valid for the rat as experimental model. Because the majority of the intestinal absorption processes in rats and in humans seem similar, it may be feasible to use, with minor modifications (i.e. *T* value), the *in situ* k_a data to make human absorption predictions.

4.4.4
Prediction from Permeabilities Through Caco-2 Cell Lines

Caco-2 cell lines have been proposed by different authors as good substrates to study the permeability of drug candidates, especially if they can be actively effluxed to the intestine by MDR transporters. From our point of view this cellular line can

Table 4.2 K_a and F values determined in references [35–38].

Tested quinolone	k_a in situ (h^{-1})	P_{ab} Caco-2 $\times 10^{-5}$ $(cm\ s^{-1})$	F obtained *in vivo*	F predicted (leave one out) with Eq. (1)	with Eq. (2)
Norfloxacin	0.42±0.07	0.182±0.010	0.334	0.334	0.136
Pefloxacin	1.92±0.18	–	0.795	0.844	–
4N'-Propyl-norfloxacin	4.07±0.40	3.376±0.025	0.928	0.978	0.946
Ciprofloxacin	0.63±0.08	0.332±0.033[a]	0.448	0.442	0.226
4N'-Propyl-ciprofloxacin	4.80±0.52	2.750±0.192	0.936	0.989	0.905
3'-Methyl-ciprofloxacin	1.24±0.07	2.205±0.156	0.725	0.663	0.884
Flumequin	6.87±1.09	2.670±0.112	0.940	0.998	0.892
Grepafloxacin		2.30±0.156[a]			0.898
Sarafloxacin		1.03±0.110[b]			0.585

a) From reference [39].
b) From reference [40].

be used not only to identify the transporters implied in the absorption of a drug, but also to predict, in a general way, the potential absorption of a drug candidate.

The first step in this direction is to correlate the permeability values obtained in the Caco-2 cell lines (apical-basolateral direction, P_{ab}) with the fraction of dose absorbed *in vivo* in rat. For this proposal eight fluoroquinolones were assayed and the results found with Caco-2 cell lines were compared with those obtained *in vivo* in rat. In the Caco-2 cell lines the permeability of the quinolones was evaluated at different initial concentration, in order to test for no linearities in the absorption process. For some of them, it was observed that a secretion system worked in the opposite direction to passive diffusion; for this reason the permeability value used for the correlations, in the case of secretion, was the one obtained at the highest concentration of the quinolone, which corresponds to saturation of the secretion process.

The equation fitted to the data was similar to the one used for the *in situ* data:

$$F_a = 1 - e^{-P_{ab}\lambda} \tag{23}$$

Where λ is a parameter of the fitting that has no equivalence with any physiological parameter, since the solutions are exposed to the membrane at infinite dose during the permeation assays.

The results are outlined in Fig. 4.7. As can be seen, the correlation is significant. Nevertheless, the bioavailability of the more hydrophilic compounds (norfloxacin, ciprofloxacin) is underestimated by their permeabilities through the Caco-2 cell line. The mean prediction error of this model is higher than the one obtained from *in situ* data (Table 4.2).

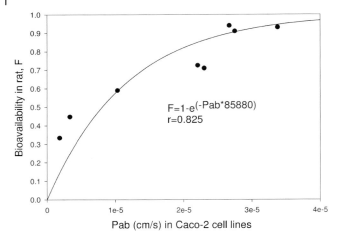

Fig. 4.7 Correlation between *in vivo* bioavailability in rat, *F*, based on oral/i.v. AUC ratios, and apical–basolateral permeability obtained in Caco-2 cell lines.

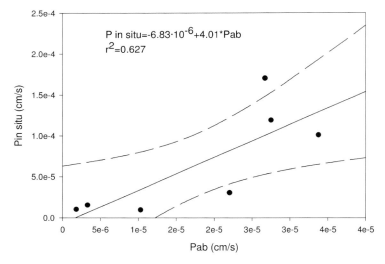

Fig. 4.8 *In situ* permeability values versus *in vitro* apical to basolateral permeabilities.

4.4.5
Prediction from the PAMPA *In Vitro* System

To predict *in vivo* permeabilities, the parallel artificial membrane permeability assay (PAMPA) was introduced in 1998. Since then, it has gathered considerable interest in the pharmaceutical industry [41–47]. This method uses a phospholipid-coated filter separating two aqueous compartments to mimic the passive transport of small molecules. It readily provides information about passive-transport permeability, not complicated by other mechanisms. Because of its speed, low cost, and versatility, it is a particularly helpful complement to cellular permeability models, such as Caco-2.

The artificial membranes are prepared by dispersing a 20% (w/v) dodecane solution of a lecithin mixture (PN 110669; pION, USA) on a microfilter disc (125 μm thick, 0.45 μm pores). The system mounted is adequately stirred to properly simulate *in vivo* situations, that is, to obtain the expected unstirred water layer (UWL) of 30–100 μm thickness [48].

The suitability of this system to predict absorption was explored with 17 fluoroquinolone derivatives.

For ionizable molecules, the membrane permeability, P_m, determined from the PAMPA system, may be derived from the effective permeability, P_e, when the effects of the UWL are taken into account [49]. The maximum possible value of P_m is realized at the pH where the solute is in its uncharged form. This limiting P_m is designated P_o, the (intrinsic) permeability of the uncharged species. The relationship between P_m and P_o may be stated as:

$$\frac{1}{P_e} = \frac{1}{P_u} + \frac{1}{P_m} = \frac{1}{P_u} + \frac{1 + 10^{+\text{pH}-\log K_1} + 10^{-\text{pH}+\log K_2}}{P_o} \tag{24}$$

Accordingly, the parameter used for the correlation was P_o, permeability of the uncharged species. The intrinsic permeability results obtained in the PAMPA system (P_o) were compared with more traditional *in vitro* methods, Caco-2 cell lines, as well as an *in situ* rat-based model which has already demonstrated a good ability for predicting *in vivo* oral absorption, as previously mentioned [50].

For the correlations, the absorption rate coefficients, k_a, obtained *in situ* were transformed into permeability values as mentioned above.

For the Caco-2 cell lines experiments, the effective permeability calculated (P_{app}; cm s^{-1}) was determined as previously described (this values corresponds with the permeability in the apical to basolateral direction, P_{ab}).

The results obtained, outlined in Fig. 4.9, demonstrated that PAMPA measurements can successfully predict *in vivo* data, based on the comparisons between rat *in situ*, Caco-2, and PAMPA permeability assays.

In Fig. 4.9a, the rat–PAMPA linear fit (solid line) is good, with regression equations being: log P_{app} (rat) = $-2.33 + 0.438$ log P_{mo} (PAMPA), with $r^2 = 0.87$, $s = 0.14$, $n = 17$, $F = 103$. It is notably better than that based on Caco-2 data: log P_{app} (rat) = $-0.66 + 0.752$ log P_o (Caco-2), with $r^2 = 0.63$, $s = 0.23$, $n = 17$, $F = 26$ (not shown in Fig. 4.9. The correlation between Caco-2 and PAMPA was similar, with $r^2 = 0.66$ (dashed line in Fig. 4.9a). Both correlations shown in Fig. 4.9 are statistically significant.

A close inspection of the Caco-2 data in Fig. 4.9a suggests a hyperbolic relationship, with log P_{app} values leveling off at about -5. Similar log P_{app} curve shapes have been observed elsewhere [41], and are thought to arise as a consequence of the unstirred water layer phenomenon.

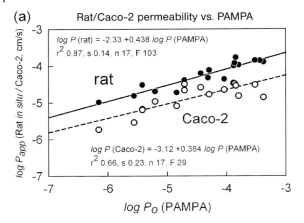

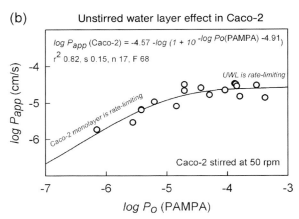

Fig. 4.9 (a) Logarithm of the apparent permeability coefficient, for perfusion experiments in rats (filled symbols) and Caco-2 (empty symbols) versus logarithm of the intrinsic permeability coefficients based on PAMPA. (b) The unstirred water layer effect in the Caco-2 data.

Hypothesizing this, a proposed equation was used to fit the Caco-2 data, as shown in Fig. 4.9b. It is remarkable that r^2 rises from 0.66 to 0.82, when the Caco-2 data are corrected for the effect of the unstirred water layer.

When the intrinsic PAMPA data (P_o) are corrected for the appropriate unstirred water layer permeability (P_u^{bio}), the unified relationship between the biological barrier data (P_{app}^{bio}) and the artificial membrane data is:

$$\log P_{app}^{bio} = -0.362 + 0.886 \left[\log P_o - \log \left(1 + \frac{P_o}{P_u^{bio}} \right) \right] \tag{25}$$

with $r^2 = 0.88$, $s = 0.18$, $n = 31$, $F = 209$.

As can be seen, PAMPA measurements can predict fluoroquinolone *in vivo* data remarkably well, as well as the standard benchmark, Caco-2. Accordingly, they can be complementary tools, quick in approaching the potential of absorption of a potential molecule in the early stages of a discovery project.

4.5
Characterization of Active Transport Parameters

When a saturable transporter is involved in the permeation process, the permeability is no longer a constant value but is dependent on the concentration of the substrate. In that case it is necessary to characterize the parameters of the carrier-mediated process, K_m, the Michaelis–Menten constant related with the affinity by the substrate and V_{max}, the maximal velocity of transport. If a passive diffusion process occurs simultaneously to the active transport pathway then it is necessary to evaluate the contribution of each transport mechanism. An example of how to characterize the parameters in two experimental systems and how to correlate them are described in the next section.

4.5.1
In Situ Parameter Estimation

The permeability values of CNV97100, a fluoroquinolone derivative, were obtained in different segments of rat small intestine. The compound was in solution and the pH of the perfusion fluids in each segment were the same. The permeability was lower in the terminal segment of the small intestine in accordance with the higher expression level of P-gp. In order to confirm that a saturable carrier was present, experiments at different concentrations of the quinolone were performed in the ileum and also in the whole small intestine. In both cases a non-linear correlation between permeability and concentration was found where, at the higher concentrations, the permeability values were higher thanks to saturation of the secretion transporter.

The permeability values obtained in the different intestinal segments and at different initial concentrations are shown in Fig. 4.10.

Data from whole small intestine, duodenum, and ileum were fitted simultaneously. This simultaneous analysis invoked the following assumptions:

1. Passive diffusional permeability was the same along the entire gastrointestinal tract, as the experiments were performed at the same luminal pH value.

2. The affinity for the transporter (K_m) was the same in all intestinal segments.

3. The maximal efflux velocity (V_m) depended on the P-gp secretion transporter expression level. A baseline value was taken to be the expression level in ileum (V_{mI}). The value of maximal velocity in duodenum (V_{mD}) and jejunum (V_{mJ}) was computed from the baseline value and a correction factor (E_f), based on

quantitative mRNA-MDR1 expression obtained in Wistar rats by Takara et al. [51]. From [51] the expression level in the proximal segment (proximal 10 cm) is 3.25 times lower than in ileum (last 40 cm) and 1.44 times lower in jejunum versus ileum. Hence, E_f in duodenum was 1/3.25, and E_f in ileum was 1/1.44.

The kinetic model is the following system of differential equations (Eqs. (26) and (27)):

Whole intestine

$$\frac{dC_i}{dt} = -\left(\frac{2}{R_a} \cdot P_{diff}\right) \cdot C_i + \frac{2}{R_a} \cdot \frac{V_{m\,Total} \cdot C_i}{K_m + C_i} \tag{26}$$

Segments

$$\frac{dC_i}{dt} = -\left(\frac{2}{R_x} \cdot P_{diff}\right) \cdot C_i + \frac{2}{R_x} \cdot \frac{V_{mX} \cdot C_i}{K_m + C_i} \tag{27}$$

where C_i represents the initial CNV97100 concentration, P_{diff} the passive diffusion permeability, and K_m is the Michaelis–Menten constant. R represents the effective radius and accounts for the volume/surface ratio. The subscript x refers to the segment in question: V_{mD} for duodenum, V_{mJ} for jejunum, and V_{mI} for ileum. The maximal velocity was estimated to be the product of the baseline value (in ileum) and the expression factor (E_f) for the segment.

$$V_{mD} = Ef_D \cdot V_{mI} \tag{28}$$

$$V_{mJ} = Ef_J \cdot V_{mI} \tag{29}$$

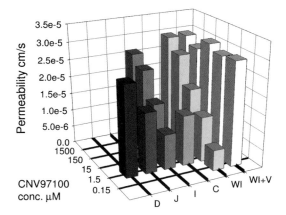

Fig. 4.10 CNV97100 rat perfusion permeability values in the different intestinal segments at different initial concentrations and in the presence of verapamil. Abbreviations: D = duodenum, J = jejunum, I = ileum, C = colon, WI = whole intestine, WI+V = whole intestine with verapamil.

Table 4.3 CNV97100 transport parameter estimated in whole small intestine, duodenum and ileum. Fits were performed simultaneously using Eqs. (26)–(29). No inhibitor was present. S.E. – standard error, CV % – coefficient of variation. V_{mTotal} maximal velocity in whole small intestine, V_{mD} maximal velocity in duodenum, V_{mJ} maximal velocity in jejunum, V_{mI} maximal velocity in ileum. V_{mD}, V_{mJ}, and V_{mTotal} are secondary parameters computed from V_{mI} (CV % is the same).

Parameters	Value	S.E.	CV %
P_{diff} (cm s^{-1})	2.96×10^{-5}	5.88×10^{-7}	1.98
V_{mTotal} (nmol cm^{-2} s^{-1})	1.91×10^{-4}	3.63×10^{-5}	19.03
V_{mD} (nmol cm^{-2} s^{-1})	8.79×10^{-5}	1.67×10^{-5}	19.03
V_{mJ} (nmol cm^{-2} s^{-1})	1.98×10^{-4}	3.77×10^{-5}	19.03
V_{mI} (nmol cm^{-2} s^{-1})	2.86×10^{-4}	5.43×10^{-5}	19.03
K_M (μM)	12.74	2.43	20.8

For the whole small intestine, the maximal velocity was taken to be the weighted average across all segments.

$$V_{mTotal} = 0.1 \cdot V_{mD} + 0.45 \cdot V_{mJ} + 0.45 \cdot V_{mI} \tag{30}$$

In this approach, the primary kinetic parameter is V_{mI}, whereas V_{mD}, V_{mJ}, and V_{mTotal} are calculated as secondary parameters from V_{mI}.

The transport parameters are summarized in Table 4.3 along with the indexes of goodness of fit.

4.5.2
In Vitro–In Situ Correlation

The same quinolone CNV97100 was used to perform bidirectional experiments in Caco-2 cells to obtain the apical to basal P_{ab} and basal to apical P_{ba} permeabilities at different initial concentrations. The relationship between previously obtained *in situ* rat permeability values and *in vitro* Caco-2 of CNV97100 was examined. In each system the *in vitro* and *in situ* paracellular permeability of CNV97100 was considered to be negligible. The systems differed in effective surface area (S_f) and in efflux transporter expression level. Although the Caco-2 expression level of P-gp is similar to the expression levels in rat ileum, the model allows for system differences (E_f).

Rat permeability values at different initial CNV97100 concentrations from whole small intestine and from ileum were combined with Caco-2 permeability from bidirectional studies (i.e. apical to basal and basal to apical). This dataset was fit to the following set of equations:

$$\text{Apical to Basal} \quad P_{eff} = P_{diff} - \frac{E_f \cdot V_m}{K_m + C} \tag{31}$$

$$\text{Basal to Apical} \quad P_{eff} = P_{diff} + \frac{E_f \cdot V_m}{K_m + C} \tag{32}$$

$$\text{Rat} \quad P_{eff}^{Rat} = S_F \cdot (P_{diff} - \frac{V_m}{K_m + C}) \tag{33}$$

where P_{eff} represents the experimental permeability, P_{diff} the passive diffusion component, S_F the surface area correction factor, and E_f a correction factor for expression-level differences between Caco-2 cells and small intestine.

The simultaneous fit of the whole data set (*in vitro* and *in situ* permeabilities) is shown in Fig. 4.11. The fitted parameters of the *in vitro–in situ* correlation are summarized in Table 4.4.

The rationale for these equations is described below.

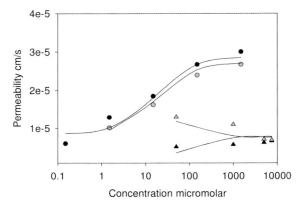

Fig. 4.11 Observed CNV97100 permeability values in whole small intestine (black circles) and ileum (gray circles) and in Caco-2 cells (A to B, black triangles) and (B to A, gray triangles). The lines are fitted values using Eqs. (31) and (33).

Table 4.4 Parameters of CNV97100 *in vitro–in situ* correlation.

Parameter	Estimate	S.E.	CV %
P_{diff} (cm s^{-1})	7.71×10^{-6}	5.98×10^{-7}	7.76
V_m (nmol cm^{-2} s^{-1})	9.06×10^{-5}	2.82×10^{-5}	31.15
K_M (μM)	16.53	5.50	33.29
S_f	3.53	0.30	8.57
E_f	3.02	1.13	37.49
r^2	0.965		
SSR	4.01×10^{-11}		
AIC	-394.98		

In cell monolayers the experimental observed permeability at each concentration, P_{ab} is calculated from the effective flux, i.e. amount of compound transported by unit time and unit area. The surface area used for the calculations correspond to the area of the insert. As the cells form a monolayer in the insert, this area value does not account for the increase in surface area due to the presence of microvilli.

However, in our experimental set up for rat *in situ* experiments, we measure the concentration versus time evolution reflecting the disappearance of the drug from the lumen because of the absorption process. From the absorption rate constants values obtained, permeabilities are calculated using the area/volume ratio. For this calculation we actually use the surface area of the geometrical cylinder (*A*) instead of the actual surface area available due to Kerkring's fold, villi and microvilli ($A \times S_f$, where *A* is the surface of the geometrical cylinder and S_f is the increase surface factor due to folds and villi). Accordingly, as we use an estimation of the intestinal surface area that is lower than the real value, the experimental permeability value in rat is overestimated (i.e. already includes S_f).

$$\frac{dC}{dt} = \frac{2}{R} \cdot S_f \cdot P_{eff} \cdot C = \frac{2}{R} \cdot P_{eff}^{rat} \cdot C$$

$$P_{eff}^{rat} = S_f \cdot P_{eff} \qquad (34)$$

This is one of the reasons for obtaining permeability values in rat (human) that are 5- to 10-fold higher than the corresponding values in Caco-2 cells.

First of all, we used this mathematical model to correlate the "*in vitro*" and "*in situ*" permeabilities of grepafloxacin and ciprofloxacin [39], and the area correction factor S_f obtained was around 4, in accord with results obtained by other authors [52]. This difference is explained by the differences in absorptive surface in the "*in situ*" versus the "*in vitro*" model, as the latter presents microvilli but not villi or folds. Now we have expanded the number of element of the correlation to all the quinolones included in Table 4.2, and the area correction factor does not suffer any variation (see Fig. 4.8, p. 104). Even if this model has been constructed using very simplistic assumptions, the results are promising and demonstrate that a good modeling approach helps to identify the system critical parameters and how the system behavior changes from the "*in vitro*" to the "*in situ*" level [39]. It is important to notice that with this linear correlation we make the assumption that the main difference between both systems is the actual effective area for transport. Nevertheless, since the plot is far from being perfect, it is probable that there are more differences in both experimental systems, such as different paracellular resistance or different expression levels of the transporter, that account for the deviation.

References

1 Chiou, W. L.: *J Pharmacokinet Biopharm* **2001**, 28:3–6.
2 Wagner, J. G., Sedman, A. J.: *J Pharmacokin Biopharm* **1973**, 1:23–50.
3 Ho, N. F. H., Po, J., Morozovich, W., Higuchi, W. I.: *Design of Biopharmaceutical Properties Through Pro-drugs and Analogs*, APHA, Washington, D.C., **1977**.
4 Higuchi, W. I., Ho, N. F. H., Park, J. Y., Komiya, I.: *Drug absorption*. MTP Press, Edinburgh, **1981**.
5 Pla-Delfina, J. M., Moreno, J.: *J Pharmacokinet Biopharm* **1981**, 9(2):191–215.
6 Martin-Villodre, A., Pla-Delfina, J. M., Moreno, J., Perez-Buendia, D., Miralles, J., Collado, E. F., Sanchez-Moyano, E., del Pozo, A.: *J Pharmacokinet Biopharm* **1986**, 14(6):615–633.
7 Hills, B. A., Butler, B. D., Lichtenberger, L. M.: *Am J Physiol* **1983**, 244(5):G561–568.
8 Garrigues, T. M., Climent, E., Bermejo, M. V., Martin-Villodre, A., Pla-Delfina, J. M.: *Int J Pharm* **1990**, 64:127–138.
9 Sun, W., Larive, C. K., Southard, M. Z.: *J Pharm Sci* **2003**, 92(2):424–435.
10 Rao, V. M., Lin, M., Larive, C. K., Southard, M. Z.: *J Pharm Sci* **1997**, 86(10):1132–1137.
11 Li, P., Zhao, L.: *J Pharm Sci* **2003**, 92(5):951–956.
12 Li, P., Patel, H., Tabibi, S. E., Vishnuvajjala, R., Yalkowsky, S. H.: *PDA J Pharm Sci Technol* **1999**, 53(3):137–140.
13 Jinno, J., Oh, D., Crison, J. R., Amidon, G. L.: *J Pharm Sci* **2000**, 89(2):268–274.
14 Crison, J. R., Weiner, N. D., Amidon, G. L.: *J Pharm Sci* **1997**, 86(3):384–388.
15 Crison, J. R., Shah, V. P., Skelly, J. P., Amidon, G. L.: *J Pharm Sci* **1996**, 85(9):1005–1011.
16 Alkhamis, K. A., Allaboun, H., Al-Momani, W. Y.: *J Pharm Sci* **2003**, 92(4):839–846.
17 Malik, S. N., Canaham, D. H., Gouda, M. W.: *J Pharm Sci* **1975**, 64(6):987–990.
18 Gibaldi, M., Feldman, S.: *J Pharm Sci* **1970**, 59(5):579–589.
19 Pla Delfina, J. M., Perez-Buendia M. D., Casabo V. G., Peris-Ribera J. E., Sanchez-Moyano E., Martin-Villodre A.: *Int J Pharm* **1987**, 37:49–64.

20 Garrigues, T. M., Bermejo, M. V., Martin-Villodre A.: *Int J Pharm* **1992**, 79:135–140.
21 Bermejo, M. V., Segura-Bono, M. J., Martin-Villodre, A., Pla-Delfina J. M., Garrigues, T. M.: *Int J Pharm* **1991**, 69:221–223.
22 Martinez-Coscolla, A., Miralles-Loyola, E., Garrigues, T. M., Sirvent, M. D., Salinas, E., Casabo, V. G.: *Arzneimittelforschung* **1993**, 43(6):699–705.
23 Fabra-Campos, S. C. E., Sanchis-Cortes A., Pla-Delfina J. M.: *Int J Pharm* **1994**, 107:197–207.
24 Fabra-Campos, S., Real, J. V., Gomez-Meseguer, V., Merino, M., Pla-Delfina, J. M.: *Eur J Drug Metab Pharmacokinet* **1991**, Spec No 3:32–42.
25 Collado, E. F., Fabra-Campos, S., Peris-Ribera, J. E., Casabo, V. G., Martin-Villodre, A., Pla-Delfina, J. M.: *Int J Pharm* **1988**, 44:187–196.
26 Garrigues, T. M., Fabra-Campos, S., Perez-Buendia, M. D., Martin-Villodre, A., Pla-Delfina, J. M.: *Int J Pharm* **1989**, 57:189–196.
27 Poelma, F. G., Tukker, J. J., Crommelin, D. J.: *J Pharm Sci* **1989**, 78(4):285–289.
28 Kimura, T., Inui, K., Sezaki, H.: *J Pharmacobiodyn* **1985**, 8(7):578–585.
29 Segura-Bono, M. J., Merino, V., Garrigues, T. M., Bermejo, M. V.: *Int J Pharm* **1994**, 107:159–166.
30 Garrigues, T. M., Bermejo, M. V., Merino, V., Pla-Delfina, J. M.: *Int J Pharm* **1994**, 107:209–217.
31 Collander, R.: *Acta Chem. Scand.* **1951**, 5:774–780.
32 Rege, B. D., Kao, J. P., Polli, J. E.: *Eur J Pharm Sci* **2002**, 16(4/5):237–246.
33 Batrakova, E. V., Li, S., Vinogradov, S. V., Alakhov, V. Y., Miller, D. W., Kabanov, A. V.: *J Pharmacol Exp Ther* **2001**, 299(2):483–493.
34 Oh, D. M., Curl, R. L., Amidon, G. L.: *Pharm Res* **1993**, 10(2):264–270.
35 Sanchez-Castaño, G., Ruiz-Garcia, A., Banon, N., Bermejo, M., Merino, V., Freixas, J., Garrigues, T. M., Pla-Delfina, J. M.: *J Pharm Sci* **2000**, 89(11):1395–1403.
36 Ruiz-Garcia, A., Bermejo, M., Merino, V., Sanchez-Castaño, G., Freixas, J.,

Garrigues, T. M.: *Eur J Pharm Biopharm* **1999**, 48(3):253–258.

37 Merino, V., Freixas, J., Bermejo, M. V., Garrigues, T. M., Moreno, J., Pla-Delfina, J. M.: *J Pharm Sci* **1995**, 84(6):777–782.

38 Bermejo, M., Merino, V., Garrigues, T. M., Pla Delfina, J. M., Mulet, A., Vizet, P., Trouiller, G., Mercier, C.: *J Pharm Sci* **1999**, 88(4):398–405.

39 Rodriguez-Ibanez, M., Sanchez-Castaño, G., Montalar-Montero, M., Garrigues, T. M., Bermejo, M., Merino, V.: *Int J Pharm* **2006**, 307(1):33–41.

40 Fernandez-Teruel, C., Gonzalez-Alvarez, I., Casabo, V. G., Ruiz-Garcia, A., Bermejo, M.: *J Drug Target* **2005**, 13(3):199–212.

41 Youdim, K. A., Avdeef, A., Abbott, N. J.: *Drug Discov Today* **2003**, 8(21):997–1003.

42 Kansy, M. F. H., Kratzat, K., Senner, F., Wagner, B., Parrilla, I.: High-throughput artificial membrane permeability studies in early lead discovery and development. In: Testa, B., Folkers, G., Guy, R. (eds) *Pharmacokinetics Optimization in Drug Research*, Verlag Helvetica Chimica Acta/Wiley-VCH, Zürich/Weinheim, **2001**.

43 Kansy, M., Senner, F., Gubernator, K.: *J Med Chem* **1998**, 41(7):1007–1010.

44 Faller B. W. F.: Physicochemical parameters as tools in drug discovery and lead optimization. In: Testa, B., Folkers, G., Guy, R. (eds) *Pharmacokinetics Optimization in Drug Research*, Verlag Helvetica Chimica Acta/Wiley-VCH, Zürich/Weinheim, **2001**.

45 Avdeef, A.: *Curr Top Med Chem* **2001**, 1(4):277–351.

46 Avdeef, A.: High-throughput measurements of permeability profiles. In: van der Waterbeemd, H. L. H., Artursson P. (eds) *Drug Bioavailability. Estimation of Solubility, Permeability, Absorption and Bioavailability*, Wiley-VCH, Weinheim, **2003**.

47 Avdeef, A.: *Absorption and Drug Development. Solubility, Permeability, Charge State.* Wiley–Interscience, New York, **2003**.

48 Lennernas, H.: *J Pharm Sci* **1998**, 87(4):403–410.

49 Ruell, J. A., Tsinman, K. L., Avdeef, A.: *Eur J Pharm Sci* **2003**, 20(4/5):393–402.

50 Bermejo, M., Avdeef, A., Ruiz, A., Nalda, R., Ruell, J. A., Tsinman, O., Gonzalez, I., Fernandez, C., Sanchez, G., Garrigues, T. M., Merino, V.: *Eur J Pharm Sci* **2004**, 21(4):429–441.

51 Takara, K., Ohnishi, N., Horibe, S., Yokoyama, T.: *Drug Metab Dispos* **2003**, 31(10):1235–1239.

52 Stewart, B. H., Chan, O. H., Lu, R. H., Reyner, E. L., Schmid, H. L., Hamilton, H. W., Steinbaugh, B. A., Taylor, M. D.: *Pharm Res* **1995**, 12(5):693–699.

5

Core-Box Modeling in the Biosimulation of Drug Action

Gunnar Cedersund, Peter Strålfors, and Mats Jirstrand

Abstract

Biosimulation has the potential to drastically improve future drug development: mathematical models may be used to obtain maximal information from experimental data, and can be used to pin-point potential drug targets, in a much more systematic manner than is the case today.

However, for this potential to be fulfilled, the developed models must themselves fulfill a number of requirements. For instance, the models must be trustworthy, and contain the necessary degree of mechanistic details. These two requirements are difficult to fulfill simultaneously, since a mechanistically detailed model typically becomes unidentifiable (i.e., over-parametrized with respect to existing data), which means that the model predictions are the result of arbitrary choices on a parameter manifold, and thus much less trustworthy.

In this chapter, we review a recently developed modeling framework – core-box modeling – which has been developed to be able to fulfill both of the above requirements. A core-box model is a combination of two models: a core model and a gray-box model. The gray-box model is a mechanistically detailed description of the relevant biochemical processes, while the core model is a simplified version of the same processes, where all aspects that cannot uniquely be estimated from the data have been eliminated. The core-box model consists of both these models, and a translation between them. The translation allows the two sub-models to be considered as two degrees of zooming of the same core-box model, where changes in, for instance, parameters are preserved upon the zooming. A path is suggested for the development of a core-box model, including identifiability analysis, model reduction, system identification, and back-translation. All of these steps are reviewed, and the steps and advantages of the modeling framework are demonstrated on a core-box model for insulin signaling.

Biosimulation in Drug Development. Edited by Martin Bertau, Erik Mosekilde, and Hans V. Westerhoff
Copyright © 2008 WILEY-VCH Verlag GmbH & Co. KGaA, Weinheim
ISBN: 978-3-527-31699-1

5.1
Introduction

The development of new drugs has become increasingly costly over the past years, and today the average cost to develop a single drug is about 1 billion Euro. With such an enormous budget it is of course highly relevant to search for ways to reduce the costs in different ways. One reason for the rising costs is increasing development times. A large part of the development is devoted to a search for possible drug targets, and hence a reduction in development time would be obtained if this search could be optimized. Further, if possible drug targets could be proposed more efficiently, fewer animal experiments would have to be carried out, and that would also be beneficial from an ethical point of view. It has often been proposed that biosimulation – the inclusion of mathematical modeling during drug developments – could contribute to such advances. It has also been pointed out that biosimulation-based drug developments would increase our understanding of the details of how drugs actually work – something which is relatively unknown for most current pharmaceutical agents. Such knowledge might also lead to other advantages, such as drugs with fewer side effects, and drugs that are more patient-specific. Thus, the potential of biosimulation in drug action is huge. However, in order to exploit this potential the mathematical models must fulfill a number of requirements.

One such requirement is that the developed models include the relevant degree of detail at which the drug acts. Usually, a drug acts at a specific protein, for instance a receptor, and the protein must be incorporated into the model. Further, the model must also include interactions between the drug target and its environment. This could mean that the signaling network of which the targeted protein is a part must also be included. Such detailed enzymatic network models are currently being developed in the scientific community using detailed mechanistic gray-box models [29, 35]. The ideal model would also include transfer of the actions of the enzymatic network to the effects on the whole body. Such a model requires the merging of whole-body, top-down models with existing detailed cellular models in multi-level hierarchical models. Whilst efforts are being made towards such hierarchical models, such models remain a long-term goal. In any case, many important benefits are available also before such complete hierarchical models are available.

Another requirement of the models is that they can make reliable predictions about the possible effects of a potential drug action. Such validations are partly obtained by comparing model predictions with validation data. However, even though a model might agree with the available validation data this is not a guarantee that all the model predictions are validated. This is especially true for most enzymatic models, as they are typically large models with many states and parameters, and where measurements are available for only a few of these states with a low time-resolution and a high noise level. In any case, it is essential to specify the degree of accuracy for each part of the model, or in other words to obtain a quality tag to the various model predictions. If that is established, it is possible to know how much faith should be placed in the prediction that a specific enzyme seems to be a promising drug target. If the prediction has a very poor quality tag, it can be discarded altogether; however, if it has a high-quality tag, it is a good argument to conduct further research on that specific enzyme.

In this chapter, we will review a new modeling framework – core-box modeling – that allows for such quality tags to be included in the large-scale gray-box models. One of the key methods for the establishment of quality tags is *practical identifiability analysis*, which specifies how well determined each parameter is from the given data. A typical gray-box model is so over-parametrized compared to available data, however, that most original parameters receive an infinite uncertainty from the direct application of these methods. Therefore, a reduction to a model with fewer states and parameters is first conducted. The quality tags are then calculated for this reduced model, denoted a *core model*. Finally, these results are back-translated to the original gray-box model. The result of this is a combination of a minimal core model with a detailed gray-box model – hence the name of the modeling framework.

The advantages of the framework can be demonstrated with two examples. The first example is a small hypothetical system, while the second example is a part of a model for insulin signaling in human fat cells. This latter system is instrumental in the development of drugs for type II diabetes mellitus, a disease which is becoming a severe problem worldwide. An overview is provided of the development of the core-box models for both of these systems. We show the advantages of the core-box model when analyzing the system compared to both a corresponding gray-box model, and compared to a corresponding minimal model. Finally, the potential of core-box modeling in the future development of whole-body hierarchical models is discussed.

5.2
Core-Box Modeling

The two modeling frameworks that lie closest to core-box modeling are full-scale mechanistic gray-box modeling and minimal modeling using hypothesis testing. Here, we will briefly review the essential parts of these frameworks, and highlight the shortcomings of both frameworks that core-box modeling intends to improve. We will then outline the major sub-steps in the core-box modeling framework, and finally provide a short review of the state-of-the-art methods for the novel sub-steps: reduction to an identifiable core model; system identification of the core model; and back-translation to a core-box model.

5.2.1
Shortcomings of Gray-Box and Minimal Modeling

5.2.1.1 Full-Scale Mechanistic Gray-Box Modeling
Full-scale mechanistic gray-box modeling is the most common type of modeling currently employed in the systems biology community. The attribute mechanistic is included to denote that the model structure is based on the assumptions of what

actually goes on in the system. The opposite to this is a *black-box model*, which only seeks to model the input–output relationship of the system, and where there is no established relationship between the states in the model and the entities in the real system. A mechanism-based model structure is often assumed automatically by the name gray-box model, but in some technical applications it is said that a gray-box model is obtained if one just forms more physically relevant input functions [21]. Finally, a model is referred to as full-scale if it includes all – or at least almost all – of the known variables and interactions that are known for the system.

In practice, a gray-box model is developed in steps. One early step is to decide which variables and interactions to include. This is often done by the sketching of an interaction-graph. It must then be decided if a variable should be a state or a dependent variable, and how the interactions should be formulated. In the case of metabolic reactions, the expression forms for the reactions have often been characterized in *in-vitro* experiments. If this has been done, there are also often *in-vitro* estimates of the kinetic parameters. For enzymatic networks, however, such *in-vitro* studies are much more rare, and it is hence typically less known which expression to choose for the reaction rates, and what a good estimate for the kinetic parameters is. In any case, the standard method of combining reaction rates, r_i, and an interaction graph into a set of differential equations is to use the stoichiometric coefficients, S_{ij}

$$\dot{x} = S \cdot r \tag{1}$$

where x is the n-dimensional vector of state variables (typically concentrations), S is the $n \times s$-dimensional matrix of stoichiometric coefficients, and r is the s-dimensional vector of reaction rates. Equation (1) is valid for systems where all variables are situated in compartments with identical and constant volumes, and where the temperature and pH are constant. This is the case for most biological systems. If these conditions are not fulfilled, then more complex expressions must be used.

In white-box modeling the process ends after the equations have been formed from the characterizations of the parts. This may be appropriate for small, non-complex systems, for example describing physical processes. However, for complex biological systems the combined errors of all the *in-vitro* characterizations of the parts become too large, and the resulting white-box model often displays a completely different behavior to that observed in the intact cell (see, e.g., Ref. [34] for an example of this). Note that this may occur even though the given model structure (i.e., the form of the differential equations) may give a fully acceptable agreement with the data after only small variations of the estimated parameters [18]. Since such small variations may very well lie within the parameter uncertainties, it is thus not possible to draw any conclusion from a disagreement between *in-vivo* data and a white-box model. This is why a gray-box modeling approach also involves optimization of the model parameters to make the agreement between the *in-vivo* data and the model as good as possible within the parameter uncertainties. One can then draw conclusions from the degree of success of this optimization. If the agreement between the gray-box model and the *in-vivo* data still is unacceptable, this means that the chosen model structure is insufficient to explain the data, and that one should reiterate some of the previous decisions in the modeling process. Likewise, if the agreement is acceptable, the conclusion can be drawn that the model structure is sufficient to explain the data. Either of these possible outcomes might be of high interest, depending on the biological question for which an answer is sought. Finally, the positive result of the optimization is especially interesting if the model also agrees with qualitatively different validation data, as this indicates that the model might be applicable for non-trivial predictions, for example in a biosimulation context.

Even if the gray-box model agrees with qualitatively different validation data, this is not a guarantee that the model as a whole has been validated, or that all predictions made by the model are trustworthy. This is because a typical gray-box model is not identifiable with respect to the existing *in-vivo* data. A model is identifiable with respect to data if all of its parameters may be uniquely estimated from the data with a reasonably small uncertainty. As this is not the case for most currently existing gray-box models of biological systems, there is an infinite number of parameter combinations that give a virtually identical agreement with the *in-vivo* data. These different parameter combinations might lead to quite different behaviors in the non-measured parts of the model. It is therefore not possible to draw any conclusions about the accuracy of the predictions from all parts of the model, even though some of these have been validated. On the other hand, there are parts and predictions of the model that are quite well-determined by the data. It would therefore be interesting to submit a quality-tag to the various parts and predictions of the model. Obtaining such quality tags, while retaining the strengths of the gray-box model, is the main objective of the core-box modeling framework.

5.2.1.2 Minimal Modeling Using Hypothesis Testing

The other framework that is close to core-box modeling is minimal modeling using hypothesis testing. A model is minimal if all smaller models are unacceptable, or at least are significantly worse. A model might be unacceptable either because the agreement with the available data is unacceptably poor, or because the model structure lacks a detail that is necessary to fulfill the purpose of the modeling. The most common way to obtain a minimal model is gradually to increase the complexity. One then starts with the most simple model that makes sense to consider. If this model has all the necessary details, and displays an acceptable agreement with the available data, it is kept and the modeling process is complete. If, on the other hand, the model is inacceptable, a more complex model is attempted. This model is treated in the same way as the first: if it is acceptably good it is retained and the modeling process is terminated if no more complex model is significantly better; otherwise, another complex model structure is tried. In this way the minimal model is found as the end of an iterative procedure testing increasingly complex model structures.

The agreement between the model and the data might be formalized using various statistical tests, one of the most common of which is the χ^2 test. The χ^2 test for differential equations is based on the difference between the *in-vitro* time-series given by

$$Z^N = \{y(t_i)\}_{i=1}^N \tag{2}$$

and the corresponding model output given by

$$\left\{\widehat{y}(t_i|\widehat{p})\right\}_{i=1}^N \tag{3}$$

where $\widehat{p}$ are the estimated values of the model's parameters. The sum of squares of the differences between the model output and the measured output normalized with the standard deviation of the measurement noise, σ, follows a χ^2 distribution if the true model has been found, if the only source of noise lies in the measurements, and if this noise is normally distributed. That means that one can form the following test entity:

$$\mathcal{T} = \sum_{i=1}^{n_y} \sum_{j=1}^{N} \left(\frac{y_i(t_j) - \widehat{y}_i(t_j|p)}{\sigma_i(t_j)} \right)^2 \tag{4}$$

where n_y is the number of measurement signals, and compare the value of this entity with an appropriate χ^2 distribution. Other common tests are the likelihood ratio test and the Akaike Information Criterion (AIC) [1, 2, 17, 24, 33]. These tests are slightly different because they are primarily comparison tests between two models. This is especially useful when distinguishing between two models that are both acceptable. By using these two tests one can therefore ensure that a more complex model than the chosen minimal model does not give a statistically improved agreement, as well as choose between models that are acceptable and of the same complexity.

There are some drawbacks with the minimal modeling approach. First, it does not necessarily give a unique minimal model. This might, in theory, be solved by the AIC test, which always gives a unique optimal model. However, if the likelihood ratio test cannot distinguish between two models, it is likely that the uniqueness provided by the AIC test is not meaningful in practice. Another similar problem is that there is no immediate way to choose the different model structures, which must be selected based on biochemical insight and various assumptions. The most important drawback of minimal modeling, however, is that it lacks much of the known mechanisms for the system. While a full-scale gray-box model is based on all such mechanistic characterizations, a minimal model is instead based on the available *in-vivo* data, and only includes so many details that a satisfactory agreement is possible. Those mechanisms that have been disregarded are sometimes important, for example when using the model to perform biosimulation of the action of a particular drug. Hence, this is an important drawback of a minimal model compared to a gray-box model. However, there are also advantages of the minimal model compared to the gray-box model. One may, for instance, draw many conclusions during model development in minimal modeling, as many models are rejected. A minimal model is also small and much more likely to have an easily established identifiability with respect to the *in-vivo* data. This means that the quality tags that are difficult to establish in a full-scale gray-box model are much more easily established for a minimal model. We will now examine the core-box modeling framework, which has been developed with the main purpose of combining the strengths of the gray-box model with those of the minimal model.

5.2.2
Outline of the Framework

The major steps in the core-box modeling framework are outlined in Fig. 5.1. Steps 1 and 2 in the framework are the construction of a gray-box model from characterizations of the parts. The first new step in the analysis is to detect all over-parametrizations in the model, using methods of identifiability analysis. These over-parametrized parts of the model are then replaced by simpler expressions, using different types of model reduction technique. These steps together are entitled *Model reduction to an identifiable core model*, and are represented by step 3 in Fig. 5.1. We refer to the resulting model as a *core model*. Note that an alternative way to obtain the core model is to formulate it directly from the same information that was used to obtain the gray-box model using the minimal modeling approach described above (step 3′ in Fig. 5.1).

Once the core model has been formulated, methods of system identification can be applied to estimate the parameters from the *in-vivo* data. This step is entitled *System identification of the core model*, and is depicted as step 4 in Fig. 5.1. The system identification framework provides methods that give an uncertainty of the estimated parameters. This uncertainty is then a combination of the original uncertainties from the *in-vitro* estimations of the parameters, with the uncertainty that remains after the additional information from the estimation to the *in-vivo* data. This model has therefore acquired the quality tags that were warranted above. However, all parts and processes in the core model have corresponding parts in the original gray-box model. The estimated features and these quality tags, can therefore be back-translated to the gray-box model, which is the next step. This step is entitled *Back-translation to the core-box model*, and is depicted as step 5 in Fig. 5.1.

5.2.3
Model Reduction to an Identifiable Core Model

Model reduction involves the identification and elimination of such parts of a model that are unrelated to some specific features of a model. The nature of such features might vary from situation to situation. In the core-box modeling framework, the feature in focus is identifiability (and agreement) with respect to the available data. We first review the state-of-the-art methods for identifiability, and then those for model reduction.

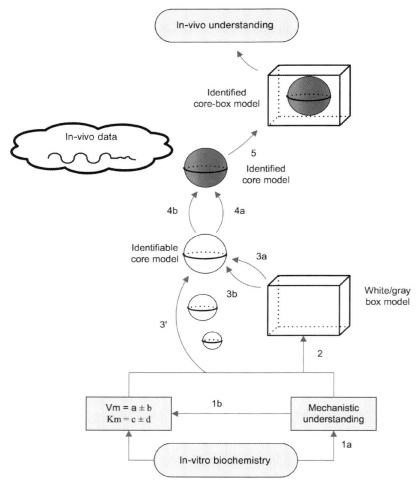

Fig. 5.1 A schematic drawing of the proposed modeling approach, leading from *in-vitro* biochemistry to an *in-vivo* understanding. In step 1, *in-vitro* experiments are carried out to determine the mechanisms of the involved reactions, and to determine estimates of the corresponding kinetic parameters and concentrations. In step 2, this information is collected into a mechanistic white-box model, which might also be adjusted to agree with the observed *in-vivo* behavior (gray-box modeling). In step 3, this model is simplified (step 3b) to a smaller model that is identifiable (step 3a) with respect to the *in-vivo* data, and has the same basic features as the original model. This is the *core model*, and an alternative way is to formulate it directly from the same information that was used to obtain the gray-box model, using the minimal modeling and hypothesis testing (step 3′). In step 4, the core model is estimated against one part of the *in-vivo* data (step 4a), and validated against another part (step 4b). This step also includes a determination of quality tags for the different parts of the model. In step 5, the well-determined parts of the core model are back-translated to the full-scale model, and the result is a gray-box model with an identified core (a core-box model). Such a model is proposed as the best candidate for obtaining an *in-vivo* understanding of the cellular processes, for example in the biosimulation of drug action.

5.2.3.1 Identifiability Analysis

Consider a parameter p_j in a model for which a particular data set, Z^N, has been collected. Loosely speaking, this parameter is *identifiable* with respect to the data if the data set contains sufficient information to uniquely determine the parameter. It is common to distinguish between two levels of identifiability: *structural* and *practical* [13].

Structural Identifiability If p_j is structurally identifiable it can, in principle, be estimated from the type of data present in Z^N, if the data were "perfect". One does therefore not take the practical limitations of the data into account. Such limitations include for instance the noise level, the type of excitation of the system, the actual number of samples, and the sample distance. Likewise, one does not take into account the practical limitations associated with the optimization. Structural identifiability is therefore a necessary, but not sufficient, requirement for practical identifiability. Finally, structural identifiability may be treated with differential algebra, and an implementation that performs the calculations in a reasonable time, although with the cost of probabilistic results, has been provided by Sedoglavic [31, 32].

Practical Identifiability If a parameter is practically identifiable it can really be estimated from the actual data. Estimation from a real data set is always associated with an uncertainty, and it is therefore also required that this uncertainty is acceptably low. The uncertainty may be quantified using bootstrapping [21]. This is a method that generates new time-series with the same statistical properties as the original time-series. The uncertainty is then approximated by the spread in estimated parameter values based on all of these different time-series. Another way of quantifying the uncertainty is to perform some type of sensitivity analysis. One such method is to calculate the gradient of the output with respect to the parameters, $\frac{\partial \widehat{y}(t)}{\partial p}$. These gradients may be calculated using the sensitivity equations, which are the derivative of the original state space equations. The sensitivity equations may for instance be generated and simulated in the Systems Biology Toolbox for Matlab [28]. Given these output sensitivities, one may then approximate the confidence regions for the i:th parameter estimated by the non-weighted least-squares cost function using the following formula [30]:

$$p_i^0 \in \left(\widehat{p} - t_{N-n_p}^{\alpha/2} s \sqrt{d_{ii}}, \ \widehat{p} + t_{N-n_p}^{\alpha/2} s \sqrt{d_{ii}} \right) \tag{5}$$

where p^0 is the true parameter value, d_{ii} is the i:th diagonal element in the following $n_y \times n_y$ matrix

$$\left[\sum_{i=1}^{N} \left(\frac{\partial y(t_i)}{\partial p} \right)^T \left(\frac{\partial y(t_i)}{\partial p} \right) \right]^{-1}$$

where n_p is the number of (identifiable) parameters, and $N - n_p$ is the degrees of freedom in the Student distribution t. Finally, the symbol s is the estimation of the standard deviation of the measurement noise, and α is the confidence level. This means that the true parameters are to be found in the given region to a probability of $100(1 - \alpha)\%$, and that the cumulative distribution function (CDF) of $t_{N-n_p}^{\alpha/2}$ is $\alpha/2$ for the given t distribution. It should finally be pointed out that the derivation of Eq. (5) is based on a linearization of the real non-linear regression problem, and that there are cases where such local approximations of the confidence regions provide very misleading results [14].

5.2.3.2 Model Reduction

There are many different methods for model reduction of differential equations, and this reflects the fact that no single method is superior in all situations, but rather that the appropriateness depends on the type of complexity in the original model, and on the purpose of the reduction. The purpose of the reduction in the core-box modeling context is to achieve a maximal identifiability, while retaining the connection to the relevant biological mechanisms. There exists a method with this former purpose, denoted balanced truncation [19], and this is frequently used in technical applications [16, 20]. However, for biological systems it is often less appropriate as the reduced model does not have variables that may easily be interpreted biochemically. Features that are possible to utilize in the model reduction of biochemical systems are the widely different time-scales at which the different reactions occur. There are several methods that use an analysis of the time-scales to decompose the system in a slow and a fast manifold. Two such methods are the computational singular perturbation and the algebraic approximation of the inertial manifold [26]. However, these methods are also inapplicable in the present context, as they replace the fast states by algebraic auxiliaries, and therefore do not eliminate any kinetic parameters in the model (see Fig. 5.2). Methods are available with the purpose of increased understanding, for instance of the source of oscillations [6, 12], but these are also typically not applicable in the present context.

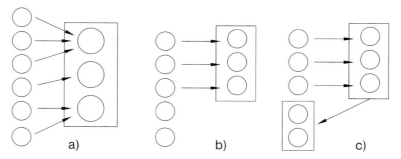

Fig. 5.2 Schematic representations of model reduction techniques. (a) Lumping pools together previous variables in disjoint groups. (b) Sensitivity analysis eliminates those entities that have a low impact on the output. (c) Time-scale based methods calculates the fast variables using algebraic relations.

There are, however, also reduction methods that are appropriate in the core-box modeling context. One such method is referred to as (discrete proper) *lumping*. Lumping is commonly used in biochemical modeling as it is often combined with the same kind of intuition that the modeling of the original gray-box model required. Nevertheless, lumping may also be based on a systematic analysis of the system, as for instance the identification of fast and slow reactions, or on a correlation analysis of the system [22]. In any case, the resulting model has states (and/or reactions) which are lumps of the states (and/or reactions) in the original model (see Fig. 5.2). Lumping thus leads to the exclusion of states and reactions. Other methods that lead to the exclusion of reactions are sensitivity analysis and term-based methods. Such methods remove the reactions or terms that have the least impact on the measured outputs [26]. Finally, in the case of metabolic models it is also often appropriate to simplify the original rate expressions in themselves [3, 6, 11].

Example 1. Let us consider an example that exemplifies step 3 in the core-box modeling framework. The system to be studied consists of one substance, A, with concentration $x = [A]$. There are two types of interaction that affect the concentration negatively: degradation and diffusion. Both processes are assumed to be irreversible and to follow simple mass action kinetics with rate constants p_1 and p_2, respectively. Further, there is a synthesis of A, which increases its concentration. This synthesis is assumed to be independent of x, and its rate is described by the constant parameter p_3. Finally, it is possible to measure x, and the measurement noise is denoted d. The system is thus given in state space form by the following equations:

$$\dot{x} = -p_1 x - p_2 x + p_3 \tag{6a}$$

$$y = x + d \tag{6b}$$

$$x(0) = x_0 \tag{6c}$$

We assume that *in-vitro* experiments have resulted in estimates $p_1 = 3 \pm 2$ and $p_3 = 0.02 \pm 0.01$, but that there is no estimate for p_2. We further assume that there is one experimental time-series, the one shown in Fig. 5.3. The time-series has been generated by simulations using the parameters

$$p = (p_1, p_2, p_3, x_0) = (2, 2.5, 0.015, 20)$$

and these parameter values are thus the "real" parameter values in this example system.

This example illustrates the problem of both structural and practical identifiability. The structural identifiability is present because p_1 and p_2 may never be distinguished from each other with only the present type of measurement possibilities. This is because the first two reactions may be written as $-(p_1 + p_2)x$, where one

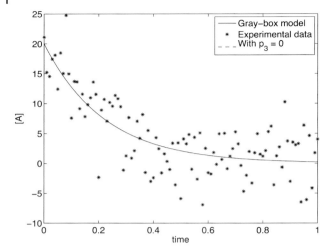

Fig. 5.3 The data from Examples 1, 2, and 3. Note that the solid and the dashed line coincides.

obviously cannot distinguish between p_1 and p_2. The reduction of this unidentifiability is obtained by a lumping of these two reactions. The lumped reaction is described by the new parameter $p' = p_1 + p_2$. The practical identifiability is present because the p_3 term is significantly smaller than the other two terms. That means that the effect of removing the p_3 term altogether is so small that it cannot be seen (see Fig. 5.3), and it is virtually impossible therefore to determine an accurate value for this parameter. Indeed, calculations show that the uncertainty of this parameter is large (much larger than the actual parameter value). The reduction of this practical unidentifiability is obtained by simply removing the p_3 term altogether.

5.2.4
System Identification of the Core Model

5.2.4.1 Parameter Estimation
The system identification step in the core-box modeling framework has two major sub-steps: parameter estimation and model quality analysis. The parameter estimation step is usually solved as an optimization problem that minimizes a cost function that depends on the model's parameters. One choice of cost function is the sum of squares of the residuals, $\varepsilon_i(t|p) = y_i(t) - \widehat{y}_i(t|p)$. However, one usually needs to put different weights, $w_i(t)$, on the different samples, and additional information that is not part of the time-series is often added as extra terms $k(p)$. These extra terms are large if the extra information is violated by the model, and small otherwise. A general least-squares cost function, $V_N(p)$, is thus of the form

$$V_N(p) = \sum_{i=1}^{n_y} \sum_{j=0}^{N} w_i(t_j)\Big(y_i(t_j) - \widehat{y}_i(t_j|p)\Big)^2 + k(p) \tag{7}$$

Note that the cost function is equal to the test function in Eq. (4) above, if no additional information is added, and if the weights are chosen as the inverse of the variance of the measurement noise. Note also that additional information may be added in other ways than by extra terms, for instance when the information concerns bifurcations [7, 9]. To use Eq. (7), where all $\widehat{y}(t|p)$ are given by a single simulation, is probably the most common choice of cost function. There are, however, important alternatives. Some of these alternatives concern the way the model output, $\widehat{y}(t|\widehat{p})$, is calculated. In the standard cost function, all model outputs are simulated in a single simulation starting at $x(0) = x_0$, where x_0 are the parameters specifying the initial condition. However, if p is not close enough to the correct parameters, p^0, the simulated output will lie far away from the measured output for the majority of the time-series. This complicates the search, as this causes the cost function to become large, and the direction in which the true parameters lie to be relatively unrelated to the gradient of the cost function. To improve this problem, one might restart the simulation in the known output a number of times during the time-series. Such methods are common in the prediction-error framework in time-discrete system identification [21], but the same principle is used also in the method of multiple shooting [4, 5, 27]. The latter method involves more parameters than does a classical output error approach, but the advantage of a more well-behaved cost function is sometimes worth this. The method is most applicable if all (or at least most) of the states in the model can be directly measured. However, if all states can be directly measured with a good data quality, many other interpolation-based methods suddenly also become available [15, 36], and they often imply an even more tractable estimation problem.

After the cost function has been specified, a unique model is – in principle – determined by the parameters that minimize it:

$$\widehat{p} = \arg\min_{p} V_N(p) \tag{8}$$

However, determination of the global minimum is often a highly difficult problem, and often one has to be satisfied with a local minimum. If the cost function is smooth and one has a good initial guess for the parameters, it might be possible to use a local gradient-based optimization method. If this is the case, it is usually advantageous, as gradient-based optimization methods have high convergence rates close to an optimum. Common choices are Gauss–Newton methods, such as the Levenberg–Marquardt method, and augmented Lagrangian methods [25]. However, the cost function is often full of local minima, and a sufficiently good initial guess is often not available. Therefore, one must often resort to global methods. A benchmark example has indicated that a method denoted sRES is sometimes advantageous [23], but our experience from other modeling examples is that a simulated annealing method combined with a nonlinear simplex method is often superior. In any case, both methods can handle constraints on the parameters, and are available in a C-implementation in the Systems Biology Toolbox for Matlab [28].

5.2.4.2 Model Quality Analysis

The model quality analysis consists of two major sub-steps: the first is concerned with the question of whether the obtained model is acceptable; and the second is concerned with a quality estimation of the various parts of the model. The question of whether the model is acceptable may be posed using the statistical tests mentioned above, for instance the χ^2 test. Such tests are particularly appropriate if the amount of data is limited, and of the same qualitative behavior. If there exists enough data to single out a part that is qualitatively different from the rest, one can also use the method of cross-validation. This method uses one part of the data for estimation, and the other part for validation. If the model can make predictions about the outcome of situations that are qualitatively different than the data that it has been optimized to agree with, this is a good argument that the model has captured some non-trivial information about the system, and that it is good enough to keep. Finally, the methods of cross-validation and statistical testing are non-exclusive and, since they provide different views of the same question, they may also be used in combination.

With regards to the analysis of the quality of the various parts of the model, one may use the same methods as are used for practical identifiability analysis. Since the same methods are used, albeit with different objectives, one sometimes refers to this model quality analysis as *a posteriori* identifiability (and the previous analysis as *a priori* identifiability). Now, however, one is also interested in how the parametric uncertainty translates to an uncertainty in the various model predictions. For instance, it might be so that even though two individual parameters have a high uncertainty, they are correlated in such a manner that their effect on a specific (non-measured) model output is always the same. Such a translation may be obtained by simulations of the model using parameters within the determined confidence ellipsoids. A global alternative to this is to consider the outputs for all parameters that correspond to a cost function that is below a certain threshold, for example 2% above the found minimum.

Example 2. We are now faced with the problem of estimating the following model structure:

$$\dot{x} = -p'x \tag{9a}$$

$$y = x + d \tag{9b}$$

$$x(0) = x_0 \tag{9c}$$

with respect to the data in Fig. 5.3. This problem is easily solved using the non-linear simplex method mentioned above, with the cost function in Eq. (7), where k was put to zero, and all weights w_i to one. There is no validation data, but a χ^2 test gives a test value of 15, which is way below the 95% rejection border at 125, indicating that the model may be approved. Finally, the p' parameter is estimated as $p' = 4.48 \pm 0.3$.

5.2.5
Back-Translation to a Core-Box Model

The central question in the core-box modeling framework is: What information has been drawn from the *in-vivo* data, and what information is solely based on guessing, literature values, and *in-vitro* characterizations? This question is answered increasingly well the more steps that are taken in the framework. This is especially true if the core model is obtained through reduction of the gray-box model. Then, an interpretation of the parameters, states, and quality tags in the core model is immediate, as one has determined the relationship Φ between the entities in the gray-box model and the core model during the reduction.

However, if the core model has been obtained through hypothesis testing, this information is not directly available. It can of course be obtained through biochemical reasoning, but in the case of an identifiable core model a quantitative mapping for the parameters may be obtained in a more automatic manner. The parameters in the gray-box model are denoted p, and they may be used to simulate a unique data set $Z^N(p)$. Further, since the parameters in the reduced model, denoted p', are identifiable with respect to a data set that is less informative than $Z^N(p)$, p' are also identifiable with respect to $Z^N(p)$. Therefore, each parameter vector p is mapped to a unique parameter vector p' through a mapping which is the combination of generating a data set $Z^N(p)$ and using it to estimate p' (see Fig. 5.4). Further information about the properties of this mapping may be deduced in terms of a Taylor expansion using numerical perturbation [6, 11]. In this way, the mapping Φ can also be deduced when the core model has been obtained without reduction of the gray-box model.

The determination of Φ has many advantages. First, it is a way to compare *in-vivo* and *in-vitro* parameters for the same system [11]. Second, it gives an interpretation of the results obtained in the system identification step of the core model. Nevertheless, the final step in the core-box modeling framework is not achieved until the gray-box model has obtained all the estimated features of the estimated core model,

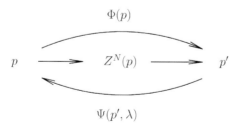

Fig. 5.4 The translation from gray-box model's parameters p to the core model's parameters p' is given by Φ. This mapping is given either by the reduction or, for instance if the core model is obtained through minimal modeling, by the intermediate generation of a simulated data series $Z^N(p)$ which is used to estimate the parameters p. The back-translation from p' to p is given by Ψ. The back-translation is made unique by the choice of one or several design variables λ.

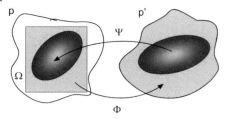

Fig. 5.5 The mappings involved in the translations of the uncertainties between the core and the gray-box model. The spaces for the parameters in the gray-box model are depicted to the left, and those for the core model to the right. The outer space to the left is denoted Ω, and refers to the space of allowed parameter values according to some general considerations (for instance that all parameters are positive). This space is restricted by the characterizations of the parts, which is often done in *in-vitro* experiments or by literature searches for values of other similar parameters; this space is depicted as the blue square. This space is mapped to the blue space in the parameter space of the core model's parameters. Inside this space an estimated parameter is obtained, and the uncertainties in this estimation are visualized by the red confidence ellipsoid. The final combined uncertainty is obtained by the determination of all parameters p that are mapped to the ellipsoid in the p' space, and that also lie within the uncertainties required by the characterizations of the parts.

particularly its key predictions and quality tags. This procedure is referred to as the back-translation step, and it results in the final core-box model. The mapping from p' to p is denoted Ψ. This mapping is in that form not unique as there are more parameters in p than in p'. Therefore, additional design parameters λ are part of the mapping, and with those fixed the mapping is unique. The design parameters may be chosen as any set that fulfills

$$p' = \Phi(\Psi(p', \lambda)) \quad \text{for all } \lambda \in \Lambda \tag{10}$$

where Λ is the space where the given translations are valid. The design variables λ are often used as a way to save information about the gray-box model to allow for a back-translation where all parts that have not been estimated by the data are the same as before. One can therefore often write $\lambda = \lambda(p)$. If this is the case, it would also be required that:

$$p = \Psi(\Phi(p), \lambda(p)) \quad \text{for all } \lambda \in \Lambda \tag{11}$$

If both Eqs. (10) and (11), are fulfilled one can almost consider the core model and the gray-box model as two versions of the same model – something which is advantageous for instance in hierarchical modeling. Regarding the choice of design variables, it is also advantageous if the λs have a clear biochemical interpretation. In any case, when Φ and Ψ have been determined the uncertainties between the gray-box model and the core model may be calculated. This procedure is visualized in Fig. 5.5, and the determinations of Φ and Ψ are exemplified in Example 3 and in Section 5.3.

By varying the design parameters one obtains a local approximation of the manifolds along which there is no information in the *in-vivo* data. For the general case, a tangent to this manifold might also be obtained from the eigenvectors corresponding to small eigenvalues of the Hessian to the cost function for the gray-box model. However, for some systems it is possible to achieve improved results, giving compact analytical solutions with a more global applicability. This is the case for structural unidentifiabilities. Then, the design parameters λ could be chosen as symmetry parameters, ω, in symmetry families. These families are mappings from p back to itself that leave the model outputs invariant

$$p \rightarrow p(\omega) \quad \text{so that} \quad \frac{\partial}{\partial \omega_i} \widehat{y}(t|p(\omega)) = 0 \quad \text{for all } \omega_i \tag{12}$$

There is a rich literature for the determination of these symmetry families [31], and this might be of use in the back-translation problem. In any case, for structural unidentifiability caused by conserved moieties the back-translation may be formulated by simple linear combinations [3, 6]. Another special case where improved solutions of the back-translation problem are possible is term elimination and lumping [6]; examples of both these cases in the model for insulin signaling are provided in the next section. First, however, we must consider the back-translation problem in the example introduced above.

Example 3. In Example 1, the core model was obtained using model reduction, and the mapping Φ is thus already given

$$p = (p_1, p_2, p_3) \rightarrow p' = \Phi(p) = p_1 + p_2 \tag{13}$$

Since p_3 is not part of the image in Φ, the p_3 value in the back-translation will not depend on the value of $\widehat{p}'$. We therefore choose p_3 always to be mapped to its previous *in-vitro* estimate, denoted by λ_1. The back-translation of p_1 and p_2 may be obtained by choosing the value of one of them (here p_1) as the other design variable, λ_2, and the other as $p' - \lambda_2$. This gives the following analytical translation formulas:

$$p \rightarrow \lambda(p) = (\lambda_1(p), \lambda_2(p)) = (p_1, p_3) \tag{14}$$

$$p' \rightarrow p = (p_1, p_2, p_3) = \Psi(p', \lambda) = (\lambda_2, p' - \lambda_2, \lambda_1) \tag{15}$$

The back-translation of the parameter values are given directly by Ψ. The back-translation of the parameter uncertainties are also given by Ψ using some simple calculations. Simply vary p' within its *in-vivo* uncertainties, and for each p' value vary λ within the regions allowed by the *in-vitro* uncertainties. The back-translated values of p are the back-translated uncertainties. Note that these uncertainties are a combination of the *in-vivo* uncertainties with the *in-vitro* uncertainties, but that the uncertainties in p' are more purely based in the *in-vivo* data alone.

In this example we only need to check the maximum and minimum values of p', and the only *in-vitro* estimates that restricts the variations of λ are those for p_1, which had been estimated to lie between 2.8 and 3.2. The *in-vivo* estimation of p' was 4.48 ± 0.3, and the minimum value, 4.18, gives the minimum p_2 value of $4.18 - 3.2 = 0.98$. Likewise, the maximum p_2 value is given by the maximum value of p' minus the minimum value of p_1: $4.78 - 2.8 = 1.98$. The uncertainties in p_3 are unchanged, since λ_2 and p' are in separate sub-systems in Eq. (15). We thus have the following back-translated parameter values with uncertainties

$$\Psi(\widehat{p'}) = (3 \pm 0.2, 1.48 \pm 0.5, 0.02 \pm 0.01) \tag{16}$$

Note that p_2 has been given an estimated value with an uncertainty, and that this is made possible because of the combination of the uncertainty of the other *in-vitro* estimations, and the *in-vivo* estimation of the core model parameter p'.

Finally, assume that we have a potential drug that may reduce p_2 by 50%. Let us see how the three available models (the gray-box, the core, and the core-box model) relate to that information, and what they can do to analyze the possible effects of such a drug. The gray-box model does not have any uncertainties related to the various parts of the model. It can thus make a prediction of the effect, but there is no way to know how to relate to that prediction – that is, to know whether it is trustworthy, or not. The core model has determined uncertainties, but not to the parameter p_2, since the core model is not detailed enough to include the flux from p_2. In order to use that model one would have to know how to translate the 50% reduction in p_2 into a reduction in p'. That translation has been characterized in the core-box model. By using that model, one can therefore say that the proposed drug would change the characteristic decay time of the whole system (which is given by p') by between 10% and 24%. This example clearly shows the advantage of the core-box model for biosimulation of drug action compared to both a corresponding gray-box model, and a corresponding core model.

5.3
A Core-Box Model for Insulin Receptor Phosphorylation and Internalization in Adipocytes

The adipocyte insulin response is one of the key systems in the regulation of glucose homeostasis, and is one of the first systems to malfunction in type II diabetes mellitus, a rapidly increasing disease of worldwide dimensions. There exists a core-box model developed for the parts of this response leading to the MAPK-kinase cascade [8]. Here, we will review the part of the model for the first step in this process, namely the response of the insulin receptor (IR). Upon binding of insulin (ins), the receptor autophosphorylates. Phosphorylation is the addition of an inorganic phosphate group to a molecule, and in the case of IR it is equivalent to an activation of its enzymatic activity towards other downstream signal transduction proteins, such as the IR substrate 1 and 2. However, upon phosphorylation another process is also started, namely the departure of a small segment of the membrane surrounding the IR into the cytosol. This process, which is referred to as *receptor internalization*, occurs commonly among tyrosine-phosphorylated receptors. Although the exact role of internalization is not known, several suggestions have been proposed. One is that the internalization acts to down-regulate further signaling by routing the internalized receptor towards lysosomal degradation. Another hypothesis is that internalization turns off the signal by providing access to dephosphorylating/deactivating protein phosphatases. A third hypothesis states that internalization provides for efficient signaling or compartmentalized signaling by providing efficient access to downstream signaling intermediates. Of course, any combination of these processes is also possible.

In an attempt to better understand the role of receptor internalization, a minimal model has been developed using hypothesis testing [10]. The model is based on experimental data on autophosphorylation of the IR. Upon addition of insulin to intact adipocytes, the IR rapidly autophosphorylates with an overshoot peak before $t = 0.9$ min, and then slowly declines to a quasi-steady state at around 15 min. The hypothesis testing shows that a satisfactory agreement with the data cannot be obtained using only reactions at the cell membrane. Further, a model that includes the internalization and dephosphorylation of IR, but not the recycling of the dephosphorylated receptor back to the membrane, also does not provide any satisfactory agreement with the data. However, if recycling of the dephosphorylated internalized receptors is included in the model, it is sufficient to use a simple model including only three states and three parameters to describe the data in a satisfactory manner. This model is henceforth referred to as $\mathcal{M}_1$. The interaction graph for $\mathcal{M}_1$ is shown in the upper part of Fig. 5.6. More complex models do not give a statistically significant improvement of the agreement, and $\mathcal{M}_1$ is thus a minimal model from a mathematical point of view.

Nevertheless, the center model structure ($\mathcal{M}_2$) in Fig. 5.6 was chosen as the main model in Ref. [10] as it includes the differentiation between the internalized and membrane-bound phosphorylated receptors (denoted IR$_i$ P-ins, and IR P-ins, respectively). However, $\mathcal{M}_2$ has also left out many known mechanisms for the IR subsystem, such as the binding of a second insulin molecule to the receptor, and receptor synthesis and degradation. These processes, as well as the reversal of all reactions except for the intracellular dephosphorylation, are included in the lower model structure ($\mathcal{M}_3$) in Fig. 5.6. This model structure is an example of a possible full-scale gray-box model for the IR subsystem.

Since the minimal models ($\mathcal{M}_1$ or $\mathcal{M}_2$) were not developed using reduction of the gray-box model, the translations Φ and Ψ need to be generated using post-analyses. In the present case this analysis is quite straightforward, as all models are relatively small, and basically only two types of reduction are involved: sensitivity analysis leading to reaction elimination; and variable lumping. The rate of dephosphorylation at the membrane, the internalization of the unphosphorylated receptor, and the recycling of the phosphorylated receptor are all significantly lower than the rates of the corresponding reversed reactions. This is concluded either by sensitivity analysis of optimized models where both the forward and reversed reactions are

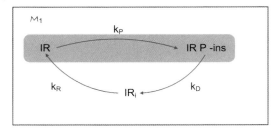

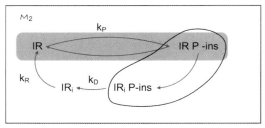

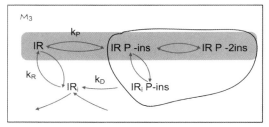

Fig. 5.6 The three model structures for the insulin receptor sub-system. The upper scheme shows a core-model, and the lower scheme a full-scale gray-box model. The center scheme may serve as either model, depending on the circumstances. In any case, translations between all three models are available, and they may be considered therefore as three versions of the same model, for example, for use in various hierarchical modeling situations. The gray horizontal shape represents the membrane, and the free-hand line indicates which variables that correspond to IR P-ins in the minimal core model.

included, and/or by previous data from other experiments. All these unidirectional reactions may therefore be omitted from the optimized gray-box models, without affecting its measured output in a statistically significant manner. Likewise, the rates of synthesis and degradation are so small that they can be omitted as they only marginally affect the model's behavior. These parameters will therefore all be of the same type in the translations as the p_3 parameter in Example 3; for all these parameters p_i we therefore have

$$p_i \rightarrow \lambda_i(p_i) \; ; \qquad \frac{\partial(\Phi(p))}{\partial p_i} = 0 \; ; \qquad \Psi_i(\widehat{p}', \lambda(p)) = \lambda_i(p_i) \pm \Delta_i \qquad (17)$$

where Δ_i is the uncertainty for the p_i parameter characterized prior to the back-translation. On the other hand, the kinetic parameter describing phosphorylation of the membrane-bound receptor (k_P), and the kinetic parameter describing recycling of the free internalized receptor (k_R) are present in all models, and they are mapped to each other directly. This means that

$$p_j = \Phi(p_j) = p'_j = \Psi(p'_j) \qquad \text{where } j = \{P, R\} \tag{18}$$

Finally, the parameter describing the dephosphorylation of the internalized receptor is mapped according to a standard lumping reduction [6]. This means that the reduced parameter p'_D in $\mathcal{M}_1$ is equal to the corresponding original parameter p_D times the fraction of the total amount of phosphorylated receptors that is made up of $IR_i \cdot$ P-ins

$$p'_D = \Phi(p_D) = p_D \frac{[IR_i \cdot \text{P-ins}]}{[IR \cdot P]_{tot}} \tag{19}$$

Lumping is based on the fast equilibration of the internal states in a variable pool, and the fraction is thus a relatively constant number over a simulation. Hence, the fraction is chosen as the final design variable, λ_f, for the back-translation

$$p_D = \Psi(p'_D, \lambda_f) = \frac{p'_D}{\lambda_f} \tag{20}$$

The variable λ_f depends on the other internal parameters describing the reactions inside the pool. The back-translation of the values of these internal parameters does not depend on p'_D, but as they affect λ_f they might achieve an improved uncertainty (similarly to the way the p_2 parameter achieved an improved uncertainty in Example 3). For these parameters p_l we thus have:

$$p_l \to \lambda_l(p_l); \qquad \frac{\partial \lambda_f(p)}{\partial p_l} \neq 0; \qquad \Psi_l(p') = \lambda_l(p_l) \tag{21}$$

A more detailed description of the complete core-box model, including the insulin signaling, is available in Refs. [6, 8, 10].

5.4
Discussion

The central questions in the core-box modeling framework are of the following form: What type of information can one extract from the available *in-vivo* data? How does this information relate to the details of the existing gray-box models for the system? What kind of predictions are most supported by the *in-vivo data*, and which parts of the gray-box models are merely based on *in-vitro* characterizations, or on even more vague experimental evidence?

For the example given in Section 5.2, the information available in the given data is primarily concerned with the lumped time-constant for the decay, p'. This time-constant may be accurately estimated from the data. The interpretation of the result in terms of the original parameters is given by the mapping Φ in Eq. (13). The back-translation mapping, Ψ, is given by the analytical formula in Eq. (15), and this formula allows for the back-translation of the obtained parameter values and uncertainties in a straightforward manner. The actual back-translation provides important novel information about the uncertainty of the p_2 parameter. It is shown in the example how this information allows the core-box model to be used in biosimulation experiments of a potential drug, that the other models are unable to perform. It is also clear that the p_3 parameter is not receiving any new information from the given *in-vivo* data.

For the insulin example in Section 5.3, the information in the available *in-vivo* data is concerned primarily with the importance and speed of the internalization. The minimal modeling shows that both the internalization, the subsequent dephosphorylation and the eventual recycling of the free receptor are essential to explain the given data. The time-constants for these processes may also be estimated with a relatively high accuracy. On the other hand, information about the reversal of these processes, and the details of the interconversions between the various forms of phosphorylated receptors, are virtually non-existent in the given data.

It should, however, be added that coarse information from the *in-vivo* data might also be obtained for the parameters that do not receive any input from the back-translation. For instance, the p_3 parameter in the example in Section 5.2 may not be estimated to any high accuracy measured as a percentage of the estimated value, but it may nevertheless be estimated with a high accuracy that the p_3 term is of a considerably lower magnitude than the other two terms combined. Similarly, one cannot estimate the detailed values for the parameters for the reversed reactions in $\mathcal{M}_2$ and $\mathcal{M}_3$, but it may be concluded with a high accuracy that the forward reaction rates are much higher than the backward reaction rates. These types of statement are most easily obtained by global searches in slightly over-parametrized models (such as the $\mathcal{M}_2$ model in Section 5.3).

The core-box modeling framework makes use of techniques and principles used in several other different research areas, in particular *a priori* and *a posteriori* identifiability analysis, parameter estimation, and model reduction. However, the final back-translation step to the complete core-box model seems to be virtually unstudied. The back-translation is associated with model reduction, but in the core-box modeling framework it appears in the more general case of two arbitrary models describing the same system. However, some techniques have now been developed, and these might be considered as the first in a new research area. These early initial results have also resulted in the important insight that the characterization of Φ and Ψ is, in general, a new form of modeling. The challenge remains, therefore, for future studies to characterize good model structures for these types of model, that for instance make use of the special structure in specific problem classes.

The obtaining of different models for the same system that may easily be exchanged for each other is important not only in the biosimulations of a single system, for instance in the core-box modeling framework, but also when developing models for systems of systems. Such models are often referred to as hierarchical models, because they are formulated at different levels of complexity. For an hierarchical model it is beneficial if changes in the internal variables of a sub-model are inherited by the new model, if the old model is being replaced. This property of inheritance is present in the current core-box models. For example, if one has replaced the detailed $\mathcal{M}_3$ model of Section 5.3 with the simplified $\mathcal{M}_1$ model for the same system using the mapping Φ, it would then also include the mappings to the design variables λ according to Eqs. (17) to (21). These design variables would then allow for all changes in the parameters p' in $\mathcal{M}_1$ to be saved when changing back to the $\mathcal{M}_1$ model, through the combined equations

$$p^{\mathcal{M}_1} = \Psi(p', \lambda(p)) \tag{22}$$

It is clear, therefore, that core-box models with the additional characteristics of Eq. (11) (such as the insulin model in Section 5.3) have major potential in the future developments of hierarchical models describing larger systems. In this way, the potential drug targets predicted by the core-box model may be judged, not only by their quality tag, but also through their translated importance on the whole-body behavior. Both of these possibilities are very important to achieve the full potential of biosimulation of potential drug targets.

5.5
Summary

This chapter has presented a review of a novel modeling framework, core-box modeling, and its possible advantages in the biosimulation of drug action. Core-box modeling attempts to combine the strengths of minimal modeling using hypothesis testing with the strengths of full-scale mechanistic gray-box modeling. The strengths of the minimal modeling framework are that the many model rejections provide reliable information about which features are essential to explain the given data, and that quality tags to the various parts and predictions are easy to establish. The strengths of the gray-box modeling framework is that much more mechanistic details may be included – details that are important to simulate drug action scenarios. In core-box modeling both a minimal model and a gray-box model have been developed, and translations Φ and Ψ between the models established. This is done either at the reduction step, if the core model is obtained using model reduction, or in a separate sub-step following the model developments. If the translations are successfully established, the two models may be easily interchanged one with another, and their strengths combined. This is advantageous both when simulating drug actions, evaluating the importance of the predictions, and when translating the predictions to larger contexts using hierarchical modeling. The advantages of the framework have been exemplified and discussed in relation to two examples: a minor hypothetical example with three reactions and one state; and a part of a real existing core-box model describing insulin signaling in human fat cells.

References

1 Akaike, H., A new look at the statistical model identification, *IEEE Trans. Autom. Control* **1974**, AC-19: 716–723.

2 Akaike, H., Modern development of statistical methods, *Trends and Progress in System Identification*, Pergamon Press, **1981**.

3 Anguelova, M., Cedersund, G., Johansson, M., Franzen, C.-J., Wennberg, B., Conserved moieties may lead to unidentifiable rate expressions in biochemical models, *IET Syst. Biol.* **2006**, 1(4): 230–237.

4 Bock, H. G., Modelling of chemical reaction systems, in: *Numerical Treatment of Inverse Problems in Chemical Reaction Kinetics*, Springer-Verlag, **1981**.

5 Bock, H. G., Recent advances in parameter identification techniques for O.D.E., in: *Numerical Treatment of Inverse Problems for Differential and Integral Equations*, Springer-Verlag, **1981**.

6 Cedersund, G., Core-box modeling – theoretical contributions and applications to glucose homeostasis related systems, PhD thesis, Chalmers Technical University, **2006**.

7 Cedersund, G., Elimination of the initial value parameters when estimating a system close to Hopf bifurcation, *IEE Proc. Syst. Biol.* **2006**, 153(6): 448–456.

8 Cedersund, G., Andersson, J., Roll, J., Knudsen, C., Danielsson, A., Strålfors, P., A core-box model for insulin signalling to MAPK-kinase response in human fat cells (in preparation).

9 Cedersund, G., Knudsen, C., Improved parameter estimation for systems with an experimentally determined Hopf bifurcation, *IEE Proc. Syst. Biol.* **2005**, 152(3): 161–168.

10 Cedersund, G., Ulfhielm, E., Roll, J., Tidefelt, H., Danielsson, A., Jirstrand, M., Strålfors, P., A model for insulin receptor signalling in human fat cells obtained using hypothesis testing, *FASEB J.* (submitted).

11 Danø, S., Cedersund, G., Madsen, M. F., Jirstrand, M., Fraenkel, D., Estimation of in vivo parameters in the phosphoglucoisomerase reaction (in preparation).

12 Danø, S., Madsen, M. F., Schmidt, H., Cedersund, G., Reduction of a biochemical model with preservation of its basic dynamical properties, *FEBS J.* **2006**, 273(21): 4862–4877.

13 Dochain, D, Vanrolleghem, P. A., *Dynamical Modelling and Estimation in Wastewater treatment Processes*, IWA Publishing, **2001**.

14 Donaldsson, J. R., Schnabel, R. B., Computational experience with confidence regions and confidence intervals for nonlinear least squares, *Technometrics* **1987**, 29: 67–82.

15 Gennemark, P., *Modeling and identification of biological systems with emphasis on osmoregulation in yeast*, Ph.D. Thesis. Chalmers, Gothenburg, Sweden, **2005**.

16 Hahn, J., Edgar, T. F., An improved method for nonlinear model reduction using balancing of empirical gramians, *Computers Chem. Eng.* **2002**, 26: 1379–1397.

17 Hall, P., Wilson, S., Two guidelines for bootstrap hypothesis testing, *Biometrics* **1991**, 47: 757–762.

18 Hynne, F., Danø, S., Sørensen, P. G., A full-scale model for yeast glycolysis in *Saccharomyces cerevisiae*, *Biophys. Chem.* **2001**, 94: 121–163.

19 Liebermeister, W., Baur, U., Klipp, E., Biochemical network models simplified by balanced truncation, *FEBS J.* **2005**, 272: 4034–4043.

20 Ljung, L., *Reglerteori – Moderna Analys- och Syntesmetoder*, Studentlitteratur. Swedish Publishing Company, **1981**.

21 Ljung, L., *System identification – theory for the user*, Prentice-Hall Inc., 2nd edition, **1999**.

22 Maertens, J., Donckels, B. M. R., Lequeux, G., Vanrolleghem, P. A., Metabolic model reduction by metabolite pooling on the basis of dynamic phase planes and metabolite correlation analysis, in: *Proceedings of the Conference on Modeling and Simulation in Biology, Medicine and Biomedical Engineering*, pp. 147–151, Linkøping, Sweden, **2005**.

23 Moles, C. G., Mendez, P., Banga, J. R., Parameter estimation in biochemical pathways: a comparison of global optimization methods, *Genome Res.* **2003**, 13: 2467–2474.

24 Müller, T. G., Faller, D., Timmer, J., Swameye, I., Sandra, O., Klingmüller, U., Tests for cycling in a signalling pathway, *Appl. Statist.* **2004**, 53: 557–568.

25 Nocedal, J., Wright, S. J., *Numerical Optimization*, Springer-Verlag, New York, **1999**.

26 Okino, M. S., Mavrovouniotis, M. L., Simplification of mathematical models of chemical reaction systems, *Chemical Rev.* **1998**, 98(2): 391–408.

27 Schittkowski, K., *Numerical Data Fitting in Dynamical Systems – A Practical Introduction with Applications and Software*, Kluwer Academic Publishers, **2002**.

28 Schmidt, H., Jirstrand, M., Systems biology toolbox for MATLAB: a computational platform for research in systems biology, *Bioinformatics* **2006**, 22: 514–515.

29 Schoeberl, B., Eichler-Jonsson, C., Gilles, E. D., Müller, G., Computational modeling of the dynamics of the MAP kinase cascade activated by surface and internalized EGF receptors, *Nat. Biotech.* **2002**, 20: 370–375.

30 Seber, G. A. F., Wild, C. J., *Nonlinear Regression*, John Wiley & Sons, **1989**.

31 Sedoglavic, A., A probabilistic algorithm to test algebraic observability in polynomial time, *J. Symbolic Computation* **2002**, 33: 735–755.

32 http://www.stix.polytechnique.fr/~sedoglav/.

33 Self, S. G., Liang, K.-Y., Asymptotic properties of maximum likelihood estimators and likelihood ratio tests under nonstandard conditions, *J. Am. Statist. Assoc.* **1987**, 82: 605–610.

34 Teusink, B., Passarge, J., Reijenga, C. A., Esgalhado, E., van der Weijden, C., Schepper, M., Walsch, M. C., Bakker, B. M., van Dam, K., Westerhof, H., Snoep, J. L., Can yeast glycolysis be understood in terms of *in-vitro* kinetics of the constituent enzymes? Testing biochemistry, *Eur. J. Biochem.* **2000**, 267: 5313–5329.

35 Iyengar, R., Bhalla, U. S., Ram, P. T., MAP kinase phosphatase as a locus of flexibility in a mitogen-activated protein kinase signaling network, *Science* **2002**, 297: 1018–1023.

36 Voit, E. O., *Computational Analysis of Biochemical Systems*, Cambridge University Press, UK, **2000**.

6
The Glucose–Insulin Control System

Christine Erikstrup Hallgreen, Thomas Vagn Korsgaard, René Normann Hansen, and Morten Colding-Jørgensen

Abstract

This chapter reviews the glucose–insulin control system. First, classic control theory is described briefly and compared with biological control. The following analysis of the control system falls into two parts: a glucose-sensing part and a glucose-controlling part. The complex metabolic pathways are divided into smaller pieces and analyzed via several small biosimulation models that describe events in beta cells, liver, muscle and adipose tissue etc. In the glucose-sensing part, the beta cell are shown to have some characteristics of a classic PID controller, but with nonlinear properties. Furthermore, the body has also glucose sensors in the intestine, the brain, the portal vein, and to some extent the liver, and they sense very different glucose concentrations. All sensors are incorporated in a dynamic network that is interconnected by both hormones and the nervous system. Regarding glucose control, the analysis shows that the system has many more facets than just keeping the glucose concentration within narrow limits. After glucose enters the cell and is phosphorylated to glucose-6-phosphate, the handling of glucose-6-phosphate is critical for glucose regulation. Also, this handling is influenced by insulin. Another facet is that the control system has to cope with the complex traffic of metabolites inside cells and between organs on time-scales from minutes to months or more. The analysis is evaluated using setups called "virtual experiments", i.e. biosimulation models describing common experimental scenarios like the fasting state, the fed state, glucose tolerance tests, and glucose clamps. The main finding is that the glucose–insulin control system does not work as an isolated control of the plasma glucose concentration. The system seems more designed to control the different nutrient and metabolic fluxes between storage, release, and oxidation, and between the different organs. In this control, both insulin and the nervous system are instrumental.

Biosimulation in Drug Development. Edited by Martin Bertau, Erik Mosekilde, and Hans V. Westerhoff
Copyright © 2008 WILEY-VCH Verlag GmbH & Co. KGaA, Weinheim
ISBN: 978-3-527-31699-1

6.1
Introduction

6.1.1
Glucose and Insulin

The food composition and the content of carbohydrate, fat, and protein in the food differs considerably from person to person and from culture to culture. The human metabolism must be able to cope with them all, sometimes immediately and sometimes after a period of adaptation. The nutrients are broken down in the intestine and appear in the blood mainly as glucose, amino acids, and fatty acids coupled to albumin or triglycerides stored in the lipoproteins. The blood transports the nutrients – or metabolites of the nutrients – from organ to organ, and the different organs collaborate in the utilization of the nutrients for energy needs, building or replacement of cells and tissues, and storage.

The storage option is important. Food intake is intermittent, with intervals of hours or even days. So in some periods the uptake from the intestine exceeds the organism's demands, and in other periods there is no uptake, so the organism must rely on its nutrient stores. Consequently there is traffic of nutrients and metabolites back and forth, and the regulation of this traffic is one of the major challenges of our metabolic control.

In the blood, the glucose concentration is held within narrow limits. The fasting plasma glucose is normally 4–5 mM, and even during large meals the increase is only a few millimolar in healthy persons and lasts only for a few hours. In contrast, the concentration of fatty acids is variable, from up to 1–2 mM during a long fast to 0.1 mM after a meal. Both the glucose and the fatty acid concentrations are mainly controlled by the hormone insulin, so it appears to be a fair conclusion that insulin is controlling the plasma glucose concentration and that the effect on fatty acids is secondary to the glucose control.

This view is supported by the findings that both low and high glucose concentrations are dangerous. The brain oxidizes almost exclusively glucose under normal circumstances. Too low a glucose concentration can therefore lead to loss of consciousness, and even be fatal. Too high a concentration can lead to metabolic disorders, as seen in diabetes, and the high concentration can in itself generate so-called Amadori products, a chemical binding between glucose and protein which leads to degeneration of cells and tissues.

This view has led to the concept that the insulin–glucose system acts as an isolated, classic control system to keep the plasma glucose constant. If the glucose concentration increases, the beta cells release more insulin that increases the glucose uptake mainly in muscle, and decreases the hepatic glucose output, so the glucose concentration is brought down again.

The reality is that there are numerous glucose sensors in the body, and that they are interconnected both by hormones and by the nervous system. Similarly, the fate of the glucose is complex, and control of the intracellular movements may be even more important than control of the uptake.

The control system may rather be regarded as a complex, self-organizing system that controls the detailed metabolic traffic, and where the movements of fatty acids play an integral role.

6.1.2
Diabetes Mellitus

Metabolic disorders are common, especially diabetes mellitus, a disorder of the glucose control. Most serious is type 1 diabetes, where the beta cells are destructed, typically by an autoimmune reaction, so the patient must be given insulin the rest of his/her life. Untreated it can lead to death within some months to a few years. It attacks mainly younger adults or children. The second, called type 2 diabetes, affects older people, typically in their 60s and typically obese. The disease is a combination of a decreased insulin production and an impaired glucose disposal. It evolves slowly and many patients can, at least in the beginning, be controlled with diet and exercise.

Fundamental in the disease is the elevated plasma glucose concentration. When it passes the renal threshold, 10–12 mM, large amounts of glucose are lost in the urine. This drags water osmotically, so the patient can produce 6–9 l of sweet urine per day. The name diabetes relates to the large urine volume, and the name mellitus, like honey, to the glucose content. The large water loss requires a similar water intake, so the patient is thirsty, and the glucose loss – up to 40–50% of the caloric intake – forces the patient to eat more than normal. Added to this, when glucose is not used for nutrition, the energy supply must come from protein and fat. This leads to a large production of ketone bodies, so the patient risks life-threatening ketoacidosis. The findings are most serious for patients with type 1 diabetes or patients with a severe type 2 diabetes.

With time, the increased glucose concentration leads to glycosylation of the proteins over Amadori products to advanced glycosylation products (AGEs) and cross-linking between proteins. This degenerative process leads to the so-called late diabetic complications (LDC) after some decades. The four main complications are:

1. Retinopathy that may lead to a slow loss of vision;
2. Nephropathy, giving albumin loss and risk of kidney destruction;
3. Neuropathy in the extremities with numbness, gangrene, and risk of amputations;
4. Macro-angiopathy, leading to increased risk of cardiovascular events.

The treatment aims at reducing the plasma glucose to minimize the risk of these serious complications: the lower the better. But on the other side of the coin is the risk of severe hypoglycaemia here and now. The result is a subtle balance, where an understanding of the mutual interaction between the different processes is *sine qua non*.

6.1.3
Biosimulation and Drug Development

The World Health Organization estimates that more than 300 million people in 2025 will suffer from diabetes, mainly type 2 diabetes, and primarily in industrialized countries. This places an economic burden on the health care system and the society. On one hand, the expenses to glucose control and patient education are large; but, on the other hand, the expenses for treating the complications are enormous.

Also, the development of new medicines is expensive, and the expenses are increasing year after year. Most of the expenses go to clinical trials that demonstrate that the medicine is both effective and safe; and with newer and more effective medicines, the safety issue becomes more and more acute. What helps one patient may be dangerous for another. The result is that the industry is beginning to discuss individualized medicine, where both the type of medicine and the treatment regime, dose and timing, are personalized.

Regarding diabetes, there now appears to be a spectrum of patients where the metabolism is altered in different aspects and in different combinations. This makes it much more difficult to develop a general drug and treatment regime for all patients. However, with a better understanding of the many diabetic variants, a much more effective and safe medicine can be developed for a particular patient type.

A characteristic problem with diabetes is the large span of time-scales involved. Metabolic experiments usually last a few hours and only address one particular pathway. The nutrient traffic involves time-scales of hours to days or weeks, but the development of late complications takes years or decades. It is therefore important to analyze the experimental results and place them in a greater context to see which effect a modification of the pathway in question will have on the overall progression of the disease.

To reveal this by conventional experiments or trials is almost impossible. However, a biosimulation model, or a set of models, can describe even very complex, interacting pathways and control systems quantitatively and give an estimate of the effect. Moreover, the models can indicate where and how a given metabolic disorder best can be corrected.

The biosimulation models need not to be large and comprehensive. Large models are difficult to validate, because small changes in the model structure can lead to different outcomes. The models in the following sections are all of the type which could be called "virtual experiments" because the setup resembles an experimental setup. The advantage is that the biosimulation models easily can analyze hundreds of "what if" situations within a second or two.

6.1.4
The Glucose–Insulin Control System

The present chapter has several aims. The first is to describe the glucose–insulin control system in a fashion that includes the most important physiological and bio-chemical processes governing the glucose metabolism. The second is to demonstrate how biosimulation models can be built and to show the quantitative arguments behind the setup of model equations. The third is to use these biosimulation models to reveal critical mechanisms in the glucose handling and to relate these mechanisms to experimental and clinical problems.

This chapter does not present a comprehensive, combined model of the entire glucose metabolism. Instead, it describes some important facets that may be poorly understood, and where a quantitative description of the mechanisms makes it easier to clarify the often very complex relations.

The advantages of this approach are manifold. The breakdown into small models makes it easier to validate each individual facet, to overview its role in the total metabolism, and to relate it to more complex frameworks. It also becomes possible to combine different facets to model new scenarios and to predict the outcome of new experimental setups. The approach is also valuable in demonstrating how a biosimulation model can be created via a quantitative argumentation based on experimental results.

As previously mentioned, the glucose–insulin control system is often regarded as a simple system to keep the plasma glucose concentration within narrow limits. In this context it has been compared to technical control systems and described by simple, often linear or linearized models. Section 6.2 of this chapter gives an outline of classic control and underlines some of the peculiarities of biological control systems.

Section 6.3 analyzes the glucose sensor concept, and it is shown how the different glucose sensors collaborate in a fashion that may resemble the technical control systems with elements mimicking proportional, integral, differential, and even predictive control. The glucose sensing appears to be dependent on the enzyme glucokinase that phosphorylates glucose into glucose-6-phosphate (G6P). Several autocatalytic effects contribute to give a glucose sensing with a threshold near the normal plasma glucose concentration.

Section 6.4 describes the glucose uptake. One of the main functions of insulin is to facilitate the glucose uptake over the cell membrane by stimulating the translocation of the glucose transporter, GLUT4, to the membrane. But insulin is also important for the phosphorylation of glucose and for the handling of the produced G6P. The main routes are: oxidation, glycogen synthesis, lactate formation, and lipogenesis. The different routes may take place in different organs, so the shuffling of glucose and glucose equivalents between organs is a central function of the control system.

Section 6.5 describes some of the concerted actions of the system. For example, the fasting plasma glucose does not only depend on hepatic glucose output (HGO) and glucose uptake, but also on the control of the free fatty acids in plasma. Also, futile cycles are demonstrated. One is the conversion of lactate to glucose in the liver and back again to lactate in the tissues. Another is the build up of fatty acids from glucose before it is oxidized as fat. The two cycles play a particular role in diabetes and overfeeding. Finally, some of the main features of the glucose clamp are considered; especially the interplay between the glucose infusion and the patient's energy expenditure.

The analysis reveals that the glucose–insulin control system functions as a regulator of the flux of nutrients – not only glucose – between the different organs in a smooth, controlled manner. In this control, the nervous system has an important role.

6.2
Biological Control Systems

6.2.1
Features of Biological Control

Biological control systems are often regarded as some sloppy variants of the more precise engineering control systems. Classic control theory considers linear, stable and stationary systems [1–3]. To this could be added: well defined. Biological systems are nonlinear, often unstable, and never stationary. They work with small feedback gains, typically less than 10 [4–6]; they are interwoven, so completely different systems share common routes (hormones, nerves, etc.); and their properties vary from person to person, even in healthy people.

A classic example is the control of arterioles in the skin [7]. The arterioles participate in blood pressure control, and in case of acute cardiac failure or blood loss, the arterioles contract in order to reserve the cardiac output for the vital organs. But the arterioles also participate in temperature regulation [7]. When they contract, heat loss from the body is minimized, so a patient in shock risks getting hyperthermia, because the blood pressure control overrides the temperature control. Moreover, the same arterioles are fundamental for the regulation of local nutritional flow to the skin and also to the smooth muscle in the vessel walls [7]. So even if skin nutrition is given a low priority compared with blood pressure, the smooth muscle has to get some nutrition to keep up the same blood pressure control. This example shows not only the complexity of biological control, but also how difficult it is to intervene when a system is malfunctioning.

Many parallels have been drawn between biological and technical control systems. Often the biological system is linearized. This can be justified if the variations are small, but it is typical for biological systems that they are not only nonlinear, but that the nonlinearities are important for proper functioning of the system. Many hormones, as for example insulin, are released in pulses. The amplitude and duration of the pulse and the interval between pulses can vary, so the signal is not just the mean concentration of the hormone, but the duration and the interval may be more important [8–10]. The pulsatility is typical for systems where hormones and nerve signals interact.

Added to this, the control must be able to work properly under extreme conditions (hard work, fever, injury, stress, etc.), which requires a very robust control.

The robustness arrives from the way the biological systems function and from coupling between different systems. If one system is malfunctioning the other systems can, at least partly, compensate for the malfunctioning, system, so the overall control is fairly maintained. An example is seen in respiratory control [7]. Normally, the ventilation frequency and volume are controlled by CO_2 receptors in the brain. This is an advantage, because the plasma CO_2 content varies much more than the O_2 content. But a patient with chronic lung obstruction lives with an increased and less varying plasma CO_2. As a consequence, the CO_2 system is suppressed, and the patient relies on the O_2 sensing receptors in the wall of the big arteries. A low plasma O_2 increases the ventilation until O_2 again is normalized. This works fairly well – although the O_2 supply to the tissues may be suboptimal. But if the patient is given pure oxygen – to increase the O_2 supply – the stimulus to the receptor vanishes [7] and the patient stops breathing, because now also the backup system is suppressed.

6.2.2
The Control System

Conventionally, a control system can be regarded as consisting of three main elements: sensor, effector, and controller, as depicted in Fig. 6.1. The controlled variable is here called CO. It may be a concentration, a movement, a process etc., so CO may be regarded as a vector or a set of very different entities. The actual magnitudes of its elements can be sensed by one or several sensors that can be placed anywhere in the body. Typically, the sensors act as transducers that convert the modality of CO to some other signal (CS), for example a hormone, nerve signal, metabolite, etc., that can be processed by the control system. The signal is picked up by the controller(s), processed and transmitted, often via some other signal type, to the effector(s), i.e. cells or organs that are able to influence the output, CO. In the figure, the different elements are depicted separately, but they can very well be located together, also in the same cell.

A crucial part of the system is the action of the controller. In technical control systems, the controller receives an input signal – called CI in Fig. 6.1 – that represents the value or trace that CO is supposed to follow. This is sometimes also the case in biological systems, and CI may be multidimensional like CO; but in many biological systems some entities appear to be strictly controlled, although the origin of the input or set point is either unknown or it is hidden in a mesh of series of interacting control systems.

In technical systems, the controller compares CI and CS and sends a signal (CC) to the effector, and the type of control is classified after how CC depends on CI and CS. It is often possible to identify similar classes of biological control systems as

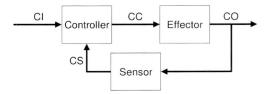

Fig. 6.1 Simplified structure illustrating the three main part of a classical control system: The controller, the effector and the sensor. CI is an input signal to the controller, CC is the signal from the controller to the effector, CO is the output from the effector and CS is the signal from the sensor to the controller.

demonstrated below, but in many cases it has not yet been possible to demonstrate the presence of a definite control system.

As the sensor mainly operates as a transducer that converts CO to CS to be compared with CI, the analysis in the following is simplified by assuming CS=CO. This corresponds, in principle, to characterizing CI in the same modalities as CO. Any time delay in the sensor is then shifted to be a part of the controller. An example of modality conversion was demonstrated by S.S. Stevens [11]. He showed that it is possible to compare the magnitude of different sensory modalities, such as light, sound, taste, length, etc., in a consistent fashion and that they are related via a power function with an exponent that is characteristic for the modalities in question.

6.2.3
Simple Control Types

6.2.3.1 Proportional Control

In technical systems, the best known control type is proportional control, where the control signal CC is proportional to the difference between the actual output (CO=CS) of the system and the desired value CI like:

$$CC = A \cdot (CI - CO) \tag{1}$$

where A is a constant. This type of control normally requires a high gain in the system, if the error CI–CO has to be small. Technical systems may have a gain factor of more than 100, giving an error less than 1%. The lower gain factors in biological systems may give a deviation from the desired value of 25–50%, which is exactly what could be called a sloppy control.

Many biological systems exhibit control types that mimic proportional control – or at least are described or modeled as proportional control – in particular homeostatic systems like regulation of temperature, pH, salt concentrations, etc. It is typical for these systems, however, that there are additional mechanisms that compensate for the sloppiness of the low gain.

6.2.3.2 **Integral Control**

An elegant way to overcome the effect of low gain is to use integral control. In this case, CC is not proportional to the error signal CI–CO but instead the rate of change dCC/dt is. This gives:

$$\frac{dCC}{dt} = A \cdot (CI - CO) \tag{2}$$

or

$$CC = A \cdot \int (CI - CO) \cdot dt \tag{3}$$

This means that CC will be regulated until CI=CO, so the system always ends up with perfect control no matter the gain. As shown later, many hormonal control systems and metabolic pathways utilize variants of integral control. Even some types of voluntary movements, for example length control via the muscle spindle, are dominated by integral control.

The main drawback of integral control is that it takes time to change CC. The rate of change is given by Eq. (2). A large difference in CI–CO gives a fast change, and as CO approaches CI, the change gets slower and slower. Consequently, a fast change requires a large difference between CI and CO, which however corresponds to a transient loss of control. To overcome this, the integral control is often combined with proportional control or, even better, differential or predictive control as described next.

Another shortcoming is that combinations of integral control often give rise to oscillations. Actually, many biological control systems may oscillate (blood glucose, blood pressure, many hormones, etc.), but usually the oscillations are limited.

6.2.3.3 **Differential Control**

A way to speed up a system is to add a type of differential control. Here the signal depends on the rate of change of CI–CO like:

$$CC = A \cdot \frac{d(CI - CO)}{dt} = A \cdot \left(\frac{dCI}{dt} - \frac{dCO}{dt} \right) \tag{4}$$

At steady state CC=0, so this system can not control the steady state. Alternatively, it can give a large signal, if the wanted value (CI) suddenly changes. In this way the system can to some degree overcome the sloppiness created by the integral system or other slow changes in the system. A minus is that this kind of control can create instabilities and amplify noise, so it can not stand alone.

An example of a strong differential regulation is seen in the control of eye movements [7]. When the head is suddenly turned, the acceleration is sensed by the vestibular apparatus in the inner ear. Among others, this signal controls the movements of the eyes, so when the turning is completed, the sight is rapidly adjusted to the new position.

6.2.3.4 **PID Control**

Many, more advanced, technical systems make use of a combination of all three types: *proportional, integral* and *differential* control. This is called the PID control [1–3]. The expression for CC becomes then:

$$CC = A_1 \cdot (CI - CO) + A_2 \cdot \int (CI - CO) \cdot dt + A_3 \cdot \frac{d(CI - CO)}{dt} \qquad (5)$$

where A_1, A_2, and A_3 are constants that determine the weight of the different control types on the total signal. The first part ensures a reasonably fast and simple action on a change. The second ensures that the system sooner or later comes to rest at CI=CO, or oscillate around CI=CO, and the third makes it possible to speed up the system and particularly compensate for the slow variation of the second term.

In Eq. (5) the three parts are added together, but in the biological system, CC can be multidimensional like CI and CO, so the three types of control may influence the parts of CO differently and in various combinations.

The PID control is relatively straightforward in a linear system, but biological systems are always nonlinear. A serious – and often neglected – nonlinearity is that production rates of hormones, nerve firing rates, etc. are always limited between some maximum and zero. With the complexity of biological systems this can make the system land in a dangerous state, where some variables are well controlled and other completely out of control.

6.2.3.5 **Predictive Control**

Even with the advanced PID control, regulation is never perfect. The system needs a difference between CI and CO to react; and with a small gain this difference can be large, so at least temporarily the control is partial. Moreover, an attempt to compensate for this, for example by increasing A_3 in Eq. (5), may lead to instabilities, oscillations, etc.

A simple way to circumvent the problem and improve the control is to use predictive control, often called feed forward. The idea is to use a control signal, CC_{FF}, which in some way reflects the expected time-course of CC that is needed to keep CO close to CI. In technical systems this control is frequently called model predictive control (MPC [12]), because CC_{FF} is typically derived from a mathematical model. In biological systems, the feed forward is mostly delivered by nerve signals, more rarely by hormones.

All exocrine and endocrine glands have a considerable nerve supply. Well known is Pavlov's conditioned reflex [13], where the digestive juices can be secreted via, for example, a sound. But also for many relatively fast reacting hormones – including insulin – the neural feed forward control plays a considerable role.

Another familiar example is the control of muscle movements [7, 14]. On top of complex sensory feedback systems in muscles and spinal cord, the cerebellum and basal ganglia deliver a feed forward signal that takes the person's reaction time into consideration and presets the muscles to a situation some 50–100 ms ahead. The PID control can then make small corrections. The precision of the movements can be improved by training that optimizes the feed forward control and thereby leaves little for the PID control to do.

The combination of feed forward and PID control can be described as:

$$CC = CC_{FF} + CC_{PID} \tag{6}$$

where CC_{PID} is given by Eq. (5). For hormonal systems, the CC_{FF} does not need to be very precise. It may be sufficient just to deliver a brief bolus that can increase the plasma hormone concentration rapidly and then let the PID control take over. However, it may play an important role in diseases, where the PID system is defective.

6.3
Glucose Sensing

Fundamental in the glucose–insulin control system is the glucose sensor concept. Traditionally, the idea of glucose sensing has been confined to the beta cells, but it now appears that there exists a widespread net of communicating glucose sensors in the brain, the intestine, the liver, etc. [15]. In principle, all cells that react to changing glucose concentration may be regarded as glucose sensors, but there appears to be some common features that characterize the network [16].

6.3.1
Glucokinase

6.3.1.1 Glucose Phosphorylation
Central in glucose sensing is the enzyme glucokinase (GK). In most glucose-utilizing cells, the incoming glucose is mainly transported over the cell membrane by GLUT1 and GLUT4 and phosphorylated by the enzyme hexokinase (HK), with a very high affinity for glucose. The Michaelis–Menten constant or half-saturation value K_M is only 0.1–0.2 mM, so the intracellular glucose concentration is kept low [17, 18]. In the glucose sensing cells, glucose is transported over the cell membrane via GLUT2 with a K_M in the order of 10–20 mM and phosphorylated by GK with a K_M in the order of 8 mM [19, 20]. The intracellular glucose concentration is therefore high in these cells.

GK is present in the cells in both an active (GK_A) and an inactive (GK_B) form. The inactive form is bound to intracellular membrane or to special regulating proteins, and the balance between the two forms is glucose sensitive [21–24]. The result is a varying activity of the amount of GK present in the cell, and a general finding is a relatively fast variation with time constants in the order of 20–60 min [21, 23, 24]. Also the total amount of GK may vary, but most investigators find that the total amount of glucokinase in the beta cell is relatively constant over the day [25, 26].

The glucose dependence of the GK activity is likely to be an essential part of glucose sensing, but as free glucose appears not to play a role inside the cell except for its own phosphorylation, glucose signaling in the cell must be secondary to glucose phosphorylation. The result is that it is not glucose but some metabolite, S (G6P or other), downstream of the glycolysis pathway that directly or indirectly controls the translocation. This implies that S controls its own formation through a positive feedback, a so-called autocatalytic mechanism, via GK translocation.

The setup is shown in Fig. 6.2a. Glucose (G) flows into the cell at a rate (J_{in}), and it is assumed that the glucose transport over the cell membrane is so fast in the glucose sensing cells that the intracellular glucose concentration equals the extracellular [19]. The inflowing glucose is phosphorylated to glucose-6-phosphate G6P, by the active GK_A. Some G6P may be dephosphorylated by G6P-phosphatase (GP), but this enzyme is only abundant in liver and kidney cells, so in other cells all or almost all the inflowing glucose must be oxidized, stored, or converted to something else. At steady state, the total removal must equal the inflow, and therefore also the phosphorylation rate.

The phosphorylation rate (v) of the isolated glucokinase enzyme is given by the Michaelis–Menten expression [27]:

$$v = \frac{v_{max}G^h}{K_{Hill}^h + G^h} = v_{max} f_{Hill}(G) \tag{7}$$

where v_{max} is the maximal rate, G the glucose concentration, $h = 1.7$ the Hill coefficient, and $K_{Hill} = 8$ mM [15]. The total rate, equal to J_{in}, is then:

$$J_{in} = v GK_A = v_{max} f_{Hill}(G) GK_A \tag{8}$$

where it is assumed, as mentioned, that the GLUT2 transport is not rate limiting, so the intracellular glucose concentration equals the plasma/interstitial concentration.

The removal of the inflowing glucose may be rather complex, as described later, but as a first approximation, it is assumed to be proportional to S, so:

$$J_{in} = v_{max} f_{Hill}(G) GK_A = bS \tag{9}$$

where b is a constant that also includes any G6P-phosphatase activity. Rearranging Eq. (9) gives:

$$\frac{S}{S_0} = X_A f_{Hill}(G) \tag{10}$$

where $S_0 = v_{max} GK_T/b$, $GK_T = GK_A + GK_B$, and $X_A = GK_A/GK_T$. The value of S thus scales with X_A. Figure 6.2b shows S/S_0 as a function of G for different values of X_A.

6.3.1.2 Translocation of Glucokinase

The balance between the active and the inactive form is described as a simple chemical reaction with two rate constants (k_{act}, k_2), as seen in Fig. 6.2a. It is assumed that

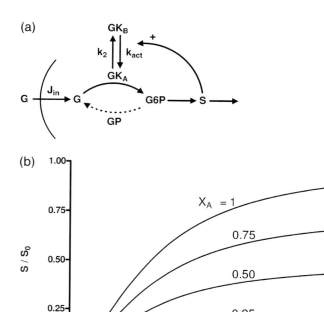

Fig. 6.2 (a) Glucose, G, flows into the cell at a rate, J_{in}, and is phosphorylated by the enzyme glucokinase, GK, to glucose-6-phosphate, G6P. G6P can in some cells be dephosphorylated by glucose-6-phosphatase, GP. The metabolite S downstream of the glycolysis is considered to be responsible for the removal of glucose. The balance between the active and inactive form of glucokinase is described as a chemical reaction with two rate constants k_{act} and k_2. k_{act} is assumed to depend on the metabolite S according to Eq. (11). (b) Steady state relation between the metabolite S normalized by $S_0 = v_{max}GK_T/b$ and the glucose concentration, G, for different values of X_A.

the active GK_A is bound with the rate constant k_2, and the bound GK_B is activated with a rate constant k_{act} that depends on S according to:

$$k_{act} = k_1 + k_3 S \tag{11}$$

With the active fraction $X_A = GK_A/GK_T$, the time course of the translocation is described by the equation:

$$\frac{dX_A}{dt} = k_{act}(1-X_A) - k_2 X_A = k_{act} - (k_{act} + k_2)X_A \tag{12}$$

where k_{act} depends on GK_A and G via S. At steady state, Eq. (12) gives:

$$X_A = \frac{k_{act}}{k_{act} + k_2} \tag{13}$$

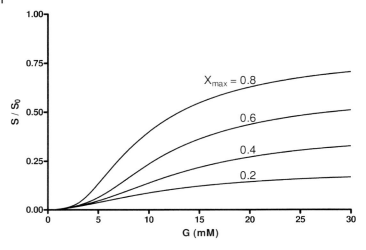

Fig. 6.3 Steady state relation between the metabolite S normalized by the value $S_0 = v_{max}GK_T/b$ and the glucose concentration, G, for different values of the maximal active fraction X_{max}. The smallest active fraction value, X_0, is set equal to 0.1. The main effect of the translocation of glucokinase is a right hand shift of the relation, resulting in a threshold value for the glucose sensor.

but as k_{act} depends on GK_A via S, the relation between G and GK_A is more complex. The smallest value, X_0, of X_A is found for $S=0$ and the largest, X_{max}, for $S=S_{max}=S_0X_{max}$, so:

$$X_0 = \frac{k_1}{k_1 + k_2} \tag{14}$$

$$X_{max} = \frac{k_1 + k_3 S_{max}}{k_1 + k_2 + k_3 S_{max}} \tag{15}$$

Figure 6.3 shows the relation between S/S_0 and G with $X_0=0.1$ and different values of X_{max}. From the figure follows that for higher values of X_{max}, the relation becomes more and more sigmoid. In this way the glucose dependent translocation of GK creates a threshold for the glucose sensor.

6.3.1.3 PI Control
From Eq. (10), it follows that a sudden increase in G gives a sudden increase in S. This corresponds to the "proportional control" (PI), although the relation between G and S is nonlinear. As time goes by, X_A increases as described by Eq. (12) which, again according to Eq. (10), gives a further, slow increase in S corresponding to the integral control. But again the nonlinearities make it a little more complex. The integral, for example, is not unlimited because S is always less than S_{max}.

The combination corresponds to a PI control as demonstrated in Fig. 6.4 for different values of the rate constant k_1. It starts with a rapid "proportional" response

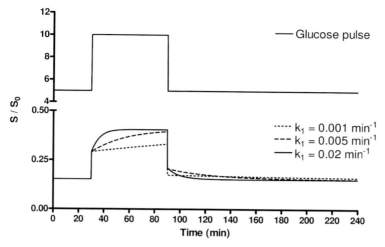

Fig. 6.4 Temporal behavior of the normalized signal S when a glucose pulse is applied, increasing the glucose from 5 mM to 10 mM and lasting for 60 min. The PI control features are clearly seen, starting with a rapid "proportional" response, followed by a slow "integral" control. Increasing the time constant k_1, responsible for the integral control, results in a faster saturation of the response. The smallest and maximal value of the active fraction is $X_0 = 0.2$ and $X_{max} = 0.8$ respectively.

followed by a slow "integral" increase as the G still is increased. For small rate constants or brief pulses, the integral appears to increase linearly, but for large rate constants or long pulses, the limitations of S are reached, and the signal saturates.

From the PI control, it follows that glucose sensing is not absolute in the sense that a given glucose concentration leads to a given signal (S) except at steady state. The design of the system seems rather to be directed towards processing the glucose influx from the intestine and directing it towards different fates (oxidation, storage, lipogenesis, etc.), depending on the state of the metabolism.

In this context, the different glucose sensing cell types have their particular role, so the tuning of the sensing mechanism and the response of the cells differ from cell type to cell type.

6.3.2
The Beta Cell

The beta cell has little or no glucose-6-phosphatase [28, 29], so removal of G6P by producing free glucose is not possible; and the beta cell lacks lactate dehydrogenase [30], so the beta cell cannot remove excess G6P by producing lactate. At the same time, there appears to be only little glycogen storage and lipogenesis in the beta cells [31], so all the glucose converted to G6P must be oxidized. This raises a problem. The influx of glucose depends on the glucose concentration and the GK activity, but the oxidation depends on the energy expenditure of the cell. As the energy expenditure generally is independent of the glucose supply, if sufficient, the glucose phosphorylation must be regulated in some way to keep phosphorylation equal to oxidation at steady state. This can be done in several ways. The most probable is an inhibition of GK, not by its product G6P as for hexokinase, but through a substance further down the glycolysis chain, for example the signal S described previously or some signal proportional to S [32].

Equation (9) can then be modified to:

$$J_{in} = v_{max}f_{Hill}(G)X_A GK_T \frac{K_S}{K_S + S} = G_{Ox} \tag{16}$$

where K_S is a constant and G_{Ox} the oxidation rate. Rearranging gives:

$$S = K_S(R_{ox}f_{Hill}(G)X_A - 1) \tag{17}$$

with $R_{ox} = v_{max}GK_T/G_{Ox}$ as ratio between the maximal possible GK phosphorylation rate and the actual glucose oxidation rate. Figure 6.5a shows the steady state relation between S/S_{max} and G with $X_0 = 0.2$ for different values of R_{ox} and with S_{max} as the value of S for $X_A = X_{max}$ and $G = \infty$. Normalization with S_{max} makes it easier to compare with the glucokinase curve (f_{Hill}). The main effect is an offset in G that is large if R_{ox} is small. Thus, there is a glucose threshold for S that depends on the total amount of GK and on the glucose oxidation rate of the beta cell. A large G_{Ox} and/or a small GK_T shifts the glucose dependence of the beta cell to the right.

Notice that below the threshold, the glucose phosphorylation is not sufficient to sustain the beta cells energy expenditures. This is particularly seen in the vertical jumps for small values of R_{ox}. However, in practice the threshold is smoother, because other nutrients like fatty acids and amino acids can take over [32].

6.3.2.1 First Phase of Insulin Secretion

It always takes some short time before the change in S can inhibit GK, so S may overshoot after a fast change. As the translocation process is assumed to be much

---

Fig. 6.5 (a) Steady state relation between the signal S normalized by the maximal value of S, S_{max}, and the glucose concentration, G, for different values of R_{ox}. The inhibition of the glucokinase by S, results in a threshold value for the beta cell glucose sensor. Below the threshold value, the glucose phosphorylation is not sufficient to sustain the beta cell energy expenditure, seen by the vertical jumps for small values of R_{ox}. $X_0 = 0.2$ and $X_{max} = 0.8$. (b) Relation between the normalized metabolite S and time, when the glucose concentration G is increased as a logistic function from 5 mM to 10 mM according to $G = G_0(\frac{1}{1+\exp(-\beta(t-t_0))})$, where G_0 denotes the initial glucose concentration, t_0 denotes the time where G is half the maximal value, and β determines the slope. Due to the delayed effect of S to inhibit GK, a fast change in glucose concentration results in an overshoot of S – a first phase. $X_0 = 0.2$, $X_{max} = 0.8$, $\tau = 10$ min, and $R_{ox} = 30$. (c) Relation between the normalized signal S and time, when the glucose concentration is increased as the logistic function given in (b) with a slope determined by $\beta = 1$ min^{-1}. The different control mechanisms of a proportional–integral–differential controller is indicated in the figure. $\tau = 5$ min. The other parameters are set as in (b).

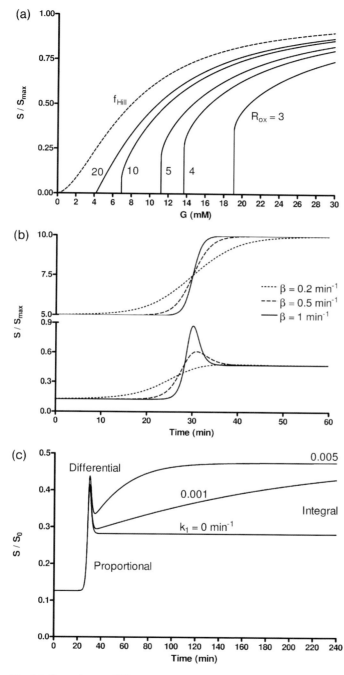

Fig. 6.5 (legend see p. 156)

slower, so X_A has no time to change substantially, the delayed value (S^*) is introduced instead of S in Eq. (17), and the time change is then given by:

$$\frac{dS^*}{dt} = (S - S^*)/\tau \tag{18}$$

where τ is a constant. Eliminating S^* gives:

$$S/K_S = R_{ox}X_A(f_{Hill}(G) + \tau f'_{Hill}(G)\frac{dG}{dt}) - 1 \tag{19}$$

This is a combination of a "proportional" control (first term of the parentheses) and a differential (second term). The beta cell is thus able to respond with a fast, but transient, response dependent on the rate of glucose change. This is demonstrated in Fig. 6.5b, where the glucose concentration increases from 5 mM to 10 mM as a logistic function with different slopes. The figure shows that there is an increasing overshoot for increasing slope. This type of differential control explains part of the so-called first phase of insulin release [33–35].

When the differential control is added to the effect of GK translocation, the beta cell acts as a PID controller [36], but with nonlinear characteristics and with saturable signals. This is shown in Fig. 6.5c. The mutual effect of the three parts depends on the actual parameters of the model. Notice that for fast changes ($k_1=0.005$ min^{-1}), the integrator tends to saturate [32].

6.3.3
The Liver Cell

The liver glucokinase undergoes both translocation and inhibition by the GK regulatory protein (GKRP [22]). Binding of GKRP to GK is competitive with glucose and reduces the activity of GK. At low plasma glucose concentrations, most of the liver GK is in an inactive form, but as increasing concentrations translocate and activate more and more, so also the liver exhibits a glucose threshold, as in Figs. 6.3 and 6.5a.

In the beta cells, glucose has only one dominating fate: oxidation. The situation in the liver is fundamentally different. G6P can be dephosphorylated to free glucose, it can be used for glycogen or fat synthesis, it can be oxidized, etc. At the same time, G6P can be produced via gluconeogenesis (GNG), stored as glycogen or converted to free glucose. The balance between these processes is discussed in Section 6.4, but it appears that the liver can not be regarded as a glucose sensor as such.

For the glucose producing cells in the liver and kidneys, there is a balance between the glucokinase and glucose-6-phosphatase GP, which can be called a push–push system. The net hepatic glucose output, J_{HGO}, can be described by:

$$J_{HGO} = J_{GP} - J_{GK} \tag{20}$$

where J_{GP} and J_{GK} are the fluxes through the corresponding enzymes. The net glucose uptake/production by the liver thus depends on the size of the two fluxes, and each of the fluxes depends on enzyme activity and substrate concentration. The conventional view is that if GK and G are high, as after a meal, the inward push is strong and the liver takes up glucose. If GP and G6P are high, as during a fast, the outward push dominates and the liver releases glucose into the blood.

However as demonstrated in Section 6.4, the picture is more complex. One of the liver's main roles in glucose metabolism is to build glucose from amino acids, glycerol, and lactate – the so-called gluconeogenetic precursors. In steady state, with constant glycogen stores, HGO only depends on GNG, which again only depends on the production rate of the precursors. Stimulating or inhibiting the liver only produces a transient change in HGO, until HGO again equals the precursor production rate.

The main function of the liver's glucose sensing activities is to provide a switch that directs GNG towards glycogen storage or towards glucose release. Only occasionally, J_{GK} becomes so large that there is a transient net glucose uptake [37, 38].

6.3.4
The Hepato-portal Sensor

The newest addition to the glucose sensors is the hepato-portal sensor [39]. The anatomical details of the sensor are still unknown. The sensor depends on the GLUT2 glucose transporter as most other glucose sensing cells and probably also on glucokinase, so its glucose sensing mechanism may resemble that of the beta cell [39]. The hormone GLP-1 is necessary for its proper function [40], which may explain why GLP-1 production is situated in the intestine. The output from the sensor is a nerve signal that goes to the central nervous system (CNS). The size of the output depends on the difference in glucose concentration in the portal and the systemic blood [39]. How this is done is still unknown.

The portal glucose concentration, C_{Po}, can be approximately

$$C_{Po} = C_a + \frac{J_{Po}}{Q_{Po}} \tag{21}$$

where C_a is the arterial glucose concentration, J_{Po} the absorption rate of glucose from the intestine, and Q_{Po} the portal flow rate. As the hepato-portal sensor reacts on the difference $C_{Po}-C_a$, it simply measures the glucose uptake rate from the intestine – at least with relatively constant portal flow. The concentration difference may be significant. With a J_{Po} in the order of 1–5 mmol min^{-1} and a Q_{Po} in the order of 0.8–1.5 l min^{-1} – depending on the glucose uptake in the erythrocytes – the concentration difference may reach 1–5 mM, i.e. up to a doubling of the arterial concentration. The sensor thus provides a robust measure of the actual glucose absorption rate.

The effect is a nerve-mediated increased insulin secretion, particularly increased first phase activity [41–43], but the effect on glucose uptake is more pronounced. It appears that the neural response can increase the translocation of GLUT4 much like the effect of exercise and without concomitant changes in insulin concentration [44].

The system works as a typical predictive control. After a meal, the glucose concentration is increased in the portal blood, so nerve signals go to the beta cells and muscles, etc., to prepare for the future glucose load. The system is so strong that a glucose load in the portal blood may create hypoglycaemia without increase in insulin release [45]. In this way, the system offers a glucose control system that is more or less independent of the beta cell sensor. In contrast, the system is apparently only active during the intestinal absorption of a meal, where the portal glucose concentration is higher than the systemic. When the absorption is finished, the sensor becomes inactive, even if the systemic glucose concentration is still increased [46–48].

The features of the hepato-portal system together with the previous mentioned integral control supports the concept that the glucose–insulin control system is not so much a control of plasma glucose concentration as a system that controls the nutrient fluxes after a meal.

6.3.5
The Intestine

Intestinal glucose sensors participate in the control of gastric emptying. Thus the L-cells produce GLP-1 in response to a glucose load [49]. GLP-1 is rapidly degraded, so one of its main effects may be on the hepato-portal sensor. Circulating GLP-1 can (although the concentration is small) increase the insulin release from beta cells – the so-called incretin effect. In both cases the effect is probably via an increased activity of glucokinase [50].

6.3.6
The CNS

Glucose sensitive neurons are present in the hypothalamus. They depend on glucokinase and act in many ways similar to beta cells [51]. They influence both insulin secretion and glucose uptake, and there appear to be neural connections between all glucose sensing cells [15]. The output may be part of the hepato-portal system, but also conditioned reflexes may participate in the neural stimulation of insulin release and glucose uptake [52]. In this sense the CNS also participates in predictive control.

6.3.7
Conclusion

From the previous sections, we conclude not only that there are many different glucose sensors in the body, but also that they sense very different glucose concentrations. The beta cell mainly senses the arterial concentration, because glucose uptake is small, and thus the gradient is small. The liver and the portal sensor sense a concentration that can be considerably higher. The same is the case in the intestine. However, the concentration in the cerebrospinal fluid is normally well below the plasma concentration. All in all, this supports the impression that glucose regulation is much more complex than the conventional view indicates, and that the different glucose sensors have their own regulatory function instead of participating in a common control of just the plasma concentration.

6.4
Glucose Handling

6.4.1
Glucose Intake

A standard, recommended diet is of the order of 2400 kcal day^{-1} (100 kcal h^{-1}), with a distribution of 50% carbohydrate, 35% fat, and 15% protein for a standard person (male, 25 years of age, 70 kg). This gives a daily intake of some 300 g carbohydrate, 100 g fat, and 90 g protein. Glycerol and the three-carbon end of odd-chained fatty acids are precursors for gluconeogenesis, so 100 g fat gives a production of some 12 g glucose (average estimate), and recycling of glycerol is estimated at 8 g, so fat metabolism results in a glucose production of 20 g. Similarly, 90 g protein results in a glucose production of 45 g day^{-1} [53, 54]. The total meal-related glucose load is thus about 365 g day^{-1} or 2 mol day^{-1}, corresponding to an average intake of 1.4 mmol min^{-1}.

When blood passes through the liver, some glucose may be absorbed. But as previously discussed, the main function of the liver is to produce glucose. So, except under very high glucose loads and high insulin, the liver adds glucose to the blood [37, 38], so the glucose concentration in the liver veins is higher than the portal concentration. Consequently, all the meal-related glucose load of 365 g passes the liver veins. To this comes a glucose production from lactate. The daily average lactate production is 1300–1500 mmol [55, 56], and 50% of this is converted to glucose [57], which amounts to some 55 g glucose. Hence, the total glucose flowing into the venous blood becomes 420 g day^{-1} or an average of 1.6 mmol min^{-1}.

6.4.1.1 Plasma Glucose

The dynamics of the plasma glucose concentration can be described by the simple equation:

$$V_G \frac{dG}{dt} = J_{abs} + HGO - J_{upt} \tag{22}$$

where V_G is the distribution volume for glucose [58], assuming a single large compartment, J_{abs} is the intestinal glucose absorption rate, HGO the hepatic (and renal) glucose output, and J_{upt} the removal of glucose from plasma by cellular uptake and, in case of high blood glucose concentration, also by renal elimination.

6.4.2
Glucose Uptake

Most of the literature focuses on glucose uptake, i.e. only the removal of glucose from blood into cells. This is, however, too simplified a view. The removal takes place mainly by facilitated diffusion via special transporters like GLUT1 and GLUT4 [59–61]. Inside the cell, glucose is phosphorylated to glucose-6-phosphate (G6P), which traps the glucose inside the cell. Only a few cell types, mainly liver and kidney, are able to dephosphorylate G6P back to glucose [62].

The fate of G6P is complex and differs from cell type to cell type. The following six routes represent the quantitatively most important fates of G6P (cf. Fig. 6.6a):

1. It can enter glycolysis and the Krebs TCA cycle to be oxidized into CO_2 and H_2O. This is the ultimate fate for glucose.

2. It can be stored as glycogen inside cells, mainly liver and muscle cells. The stores are limited to a total maximum near 1000 g glucose [63].

3. It can be converted to fat and stored inside cells. This is under normal conditions taking place in the adipocytes. The stores are practically unlimited. A year's glucose intake can be stored as 40 kg fat.

4. It can be converted to lactate and leave the cell again. This can take place in almost all cells in the body. Similarly, almost all cells can take up lactate, convert it to pyruvate, and let it re-enter one of the other routes.

5. It can be converted to fatty acids that leave the cell. The fatty acids are taken up by other cells and oxidized, mainly in muscle cells, or esterified to triglyceride (TG). In the liver the TG is released into the blood via the lipoprotein VLDL, and in the adipocytes it is stored.

6. Finally, as previously mentioned, it can be dephosphorylated to free glucose. This is the final step in gluconeogenesis that takes place in liver and kidneys. Other tissues have no or only little glucose-6-phosphatase, so they cannot form free glucose from G6P.

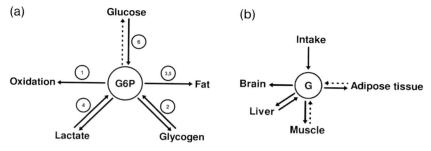

Fig. 6.6 (a) The most important pathways glucose can enter once it is phosphorylated to G6P inside the cell. The numbers refers to the different routes explained in the text. (b) Major organs participating in the use and/or reshuffling of glucose once it is absorbed into the plasma.

Options 4 and 5 are particularly interesting, because they represent a way that glucose equivalents can leave the cell again. This is an important part of glucose homeostasis, because, as discussed later, it appears to be the only way a cell can adjust the net influx of glucose to the cell's needs. Particularly, the lactate formation appears to play a decisive role in the distribution of carbohydrate between different cells and organs. As almost all cells can produce lactate and rebuild glucose from lactate, glucose equivalents can be reshuffled from cell to cell very easily.

The main routes for glucose movements are depicted in Fig. 6.6b. The glucose from intake and production ends up in the blood, from where it is supposed to be distributed to brain, liver, muscle, and adipose tissue. Its fate in these different organs will be described separately.

6.4.3
Brain

Under normal circumstances the brain oxidizes mainly glucose. As the energy expenditure of the brain is fairly constant during the day, the glucose consumption of the brain can be regarded as constant and in the order of 80 mg min^{-1} or 0.45 mmol min^{-1} [64]. However, the brain can take up large amounts of lactate [65] and ketone bodies, so in cases with high plasma concentration of lactate or ketone bodies, the glucose uptake goes down.

After a night's fast, the total hepatic glucose output is some 140 mg min^{-1} [66, 67], corresponding to a gluconeogenesis of 80 mg min^{-1} and breakdown of glycogen of 60 mg min^{-1} [67]. In this situation, the brain thus utilizes approximately 60% of the produced glucose.

The strong dependence of glucose supply makes the brain vulnerable during hypoglycaemia. Glucose transport over the blood–brain barrier depends on a glucose–sodium carrier. This gives an insulin independent glucose transport with an apparent K_M of 1 mM [62, 68]. The transport is thus some 83% saturated, at 5 mM glucose. If plasma glucose drops to 2 mM, the transport drops from 83% to 67%

and glucose uptake decreases by 20%. As the brain has virtually no energy stores, this decline is serious if the supply of lactate or ketone bodies is small.

With a constant energy need and a plasma glucose that normally is sufficient, there appears to be no need for any regulation of brain glucose uptake. There is evidence, however, of a slow regulation [69]. The transporters appear to be down regulated during hyperglycaemia and up regulated during hypoglycaemia. On one hand, this can enhance the effect of a sudden decrease in plasma glucose; on the other hand, it can help the subject through a long period with low plasma glucose.

6.4.4
Liver

6.4.4.1 Gluconeogenesis
The main routes in the liver are shown in Fig. 6.7a. Glucose is shuttled over the plasma membrane by the transporter GLUT2, and it is so effective that the intracellular glucose concentration (G_i) can be regarded as equal to the portal (G_{Po}) [70]. The glucose is phosphorylated by glucokinase as earlier described. Its activity depends on both glucose and insulin concentrations [71]. In liver and kidneys, G6P can be dephosphorylated by the enzyme glucose-6-phosphatase [62], so there is a balance between glucose, and G6P as described previously (Eq. (20)).

In fact, the liver must be regarded as a combination of a glucose producing organ and a glucose store. The daily production is in the order of 20+45+55=120 g day^{-1} or 80 mg min^{-1} [72]. The diurnal variations in the liver glycogen are about 50 g [73]. Taking a feeding period of 12 h and a fasting also of 12 h, the average glycogen build up or breakdown is 70 mg min^{-1}, so even in the feeding period, the net production is positive. A net glucose uptake only takes place transiently during

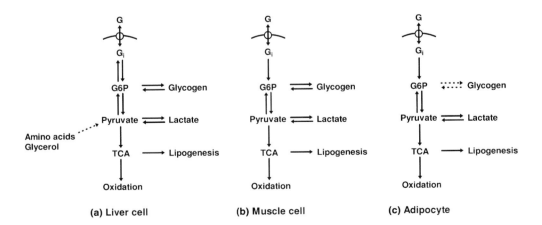

Fig. 6.7 The main routes of glucose in (a) liver cell, (b) muscle cell, and (c) adipocyte. For details about the different routes in the organs, see text.

very high portal glucose concentrations, a situation where the gluconeogenesis is probably also increased.

Each of the three main sources for gluconeogenesis (glycerol, amino acids, lactate) have a particular fate. Glycerol is released from lipolysis, and since the human adipocytes do not re-use glycerol [62], it must be converted to glucose in the liver. The adipocytes can then re-make glycerol from glucose [62]. As both lipolysis and esterification can occur at the same time, there is some recycling of glycerol/glucose. In contrast, all ingested fat that is oxidized liberates glycerol, so the net gluconeogenesis from glycerol to a large extent depends on fat oxidation. During a fast, the fat oxidation is high, so the release of glycerol during lipolysis from the adipocytes determines the gluconeogenesis. In the feeding state, adipocyte lipolysis is inhibited by insulin [74], but at the same time lipoprotein lipolysis is increased [74], so the feeding state also contributes to glycerol production. The overall effect is that gluconeogenesis from glycerol is fairly constant.

Regarding the amino acids, there is a continuous breakdown and re-synthesis of protein in the muscle cells [74]. To this come the amino acids come from ingested protein. The skeleton of most amino acids can enter the Krebs TCA cycle directly or indirectly [62], but to eliminate the amino group, mainly as urea, the amino acid must pass the liver or kidneys [74]. On average, half of the skeleton is converted to glucose and the other half enters the TCA cycle to provide energy for urea and glucose production [53, 54]. Glucose production may vary somewhat during the day, but most investigators assume a relatively constant urea production [53], so the gluconeogenesis from amino acids mainly depends on the protein oxidation that again depends on the amount of ingested protein.

The largest variation in gluconeogenesis comes from lactate. It is well known that the muscles produce large amounts of lactate during exercise, and it is generally believed that this lactate is converted to glucose in the liver – the Cori cycle [62, 75] – but the maximum conversion rate is far too little to be of substantial help for working muscles. A typical HGO during work is 1–2 mmol min^{-1} or about 180–360 mg min^{-1} [76, 77], which corresponds to 0.7–1.4 kcal min^{-1}, if it is completely oxidized. This is only a small fraction of the muscle energy expenditure, and if it is used anaerobically, the effect is vanishingly small.

Another source of lactate comes from meals. The ingested glucose is taken up by cells, but if the uptake rate is too high, the surplus is exported again as lactate [78]. The same is the case during other types of glucose/insulin load (clamps, IVGTT, insulin treatment, etc.) [56, 57, 79].

6.4.4.2 Hepatic Glucose Output and Gluconeogenesis

Taking all the gluconeogenetic precursors together, there is no typical pattern of variation in precursor production, and in all cases, production is taking place outside the liver. The liver's glucose production can therefore be regarded as a slave of precursor production [74]. With a constant precursor production, GNG should also be constant. Of course, it is possible to stimulate or inhibit liver GNG, but these variations are transient. For example, if GNG is inhibited, the concentration of pre-

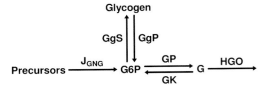

Fig. 6.8 Assuming a constant flux J_{GNG} from precursors to the pool of G6P, the hepatic glucose output HGO is more or less correlated to the changes in glycogen stores. The buildup of glycogen from G6P is controlled by the enzyme glycogen synthase, GgS, and the breakdown of glycogen is controlled by the enzyme glycogen phosphorylase, GgP.

cursors builds up and stimulates GNG, until it again equals precursor production. Similarly, a stimulation just decreases precursor concentration.

There is thus a more or less steady liver GNG. Variations in HGO are therefore more correlated with changes in the glycogen stores [80]. A simple model of the situation is depicted in Fig. 6.8. A constant inflow (J_{GNG}) to the pool of G6P is assumed – ignoring a possible lipogenesis directly from the precursors. The inflow can either go to the glycogen stores, symbolized and simplified by the enzyme glycogen synthase (GgS), or be dephosphorylated by glucose-6-phopsphatase (GP) to free glucose. At the same time, glucose is phosphorylated by glucokinase (GK) adding to the pool of G6P. There is thus a recycling between glucose and G6P. Finally, glycogen can be converted to G6P via the enzyme glycogen phosphorylase (GgP). This gives another recycling between glycogen and G6P. The two cycles are controlled by insulin and glucagon, or rather the ratio between insulin and glucagon [62]. At a high insulin/glucagon ratio, GK and GgS are stimulated and GP and GgP inhibited. This leads to glycogen storage. In contrast, a low ratio stimulates GP and GgP and inhibits GK and GgS, leading to glycogen breakdown and a high HGO [62]. Sooner or later, the glycogen store reaches a steady state, where the HGO equals J_{GNG}.

The system can, in a simple, linear form, be described by two coupled differential equations like

$$\frac{dGg}{dt} = k_{GgS} \cdot G6P - k_{GgP} \cdot Gg \tag{23}$$

and

$$\frac{dG6P}{dt} = J_{GNG} - k_{GgS} \cdot G6P + k_{GgP} \cdot Gg + k_{GK} \cdot G - k_{GP} \cdot G6P \tag{24}$$

and with:

$$HGO = k_{GP} \cdot G6P - k_{GK} \cdot G \tag{25}$$

where k_{GgS}, k_{GgP}, k_{GK}, and k_{GP} are the rate constants for the enzymes GgS, GgP, GK, and GP, respectively.

At steady state, HGO=J_{GNG} and the amount of glycogen is determined by the ratio k_{Ggs}/k_{GgP} and the concentration of G6P, while G6P is determined by the rate constants k_{GK} and k_{GP}, and the magnitudes of J_{GNG} and G.

The balance between the four rate constants determines the behavior of the system. After a night's fast, the different values can be estimated in the following way. To simplify, it is assumed that the distribution volume of glycogen, G6P, and glucose is 1 litre and the glucose concentration is 5 mM. This gives an amount of 900 mg. The glycogen content is taken to be 72 g or 72 000 mg [81] and the G6P concentration is taken to be 0.2 mM, corresponding to 36 mg. The degree of recycling is not very well determined [81], but to simplify, it is assumed that recycling is 20% of the fasting net flux in both cycles. The enzyme activities can then be calculated as rate constants. This gives:

$$k_{GK} = \frac{HGO}{G} \cdot 0.2 = \frac{140 \text{ mg min}^{-1}}{900 \text{ mg}} \cdot 0.2 = 0.03 \text{ min}^{-1}$$

$$k_{GP} = \frac{HGO}{G6P} \cdot 1.2 = \frac{140 \text{ mg min}^{-1}}{36 \text{ mg}} \cdot 1.2 = 4.67 \text{ min}^{-1}$$

$$k_{Ggs} = \frac{\text{Glycogen release}}{G6P} \cdot 0.2 = \frac{60 \text{ mg/min}^{-1}}{36 \text{ mg}} \cdot 0.2 = 0.33 \text{ min}^{-1}$$

$$k_{GgP} = \frac{\text{Glycogen release}}{\text{Glycogen}} \cdot 1.2 = \frac{60 \text{ mg min}^{-1}}{72000 \text{ mg}} \cdot 1.2 = 0.001 \text{ min}^{-1}$$

The fastest change is by far in the G6P concentration, because only tiny amounts of G6P are present, so changes in the fluxes can change the G6P concentration rapidly. The G6P therefore does not act so much as a buffer, but more like a switch that directs the glucose flux in the appropriate direction(s).

This is demonstrated in Fig. 6.9. Time zero corresponds to the overnight fast situation. After 60 min the glucose concentration is increased from 5 mM to 10 mM, and there is a rapid change in some of the enzyme activities for 180 min, whereafter the system is switched back to the initial values. Figure 6.9a shows the effect of increasing both k_{GK} and k_{Ggs} six times. The amount of G6P is about doubled and HGO is reduced to near zero. There is an initial negative spike in HGO that is mirrored at the end of the pulse. This corresponds to the transient phase, where G6P changes rapidly.

The balance between the different enzyme activities is important. This is shown in Fig. 6.9b. The mean values of G6P and HGO are determined during the 3-h test period and plotted against each other – all for changes in k_{GK} and k_{Ggs}, but in different proportions. It is seen that changing only k_{GK} gives a steep increase in G6P, but only a small decrease in HGO, and there is no net build up of glycogen. However, if only k_{Ggs} is increased, the G6P concentration drops and HGO goes down, because the drive on GP decreases. If k_{Ggs} increases five times faster than k_{GK}, the value of G6P is almost constant, although the HGO varies from high production to high uptake. The figure demonstrates clearly that the G6P concentration is a poor

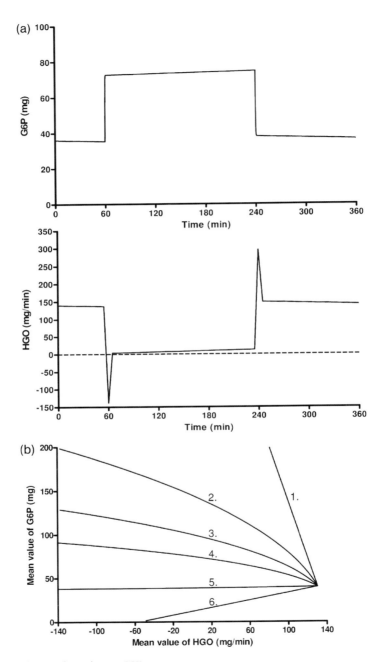

Fig. 6.9 (legend see p. 169)

indicator of HGO and glycogen synthesis. A similar picture is seen if the activities of k_{GP} and k_{GgP} are decreased, and a combination of both enhances the results seen in Fig. 6.9.

In the model, the effects of changing k_{GK} and changing G are similar, because only the product $k_{GK} \cdot G$ matters. In the liver cell the glucokinase has a Hill coefficient of 1.7, and GK translocation increases the k_{GK} activity in a nonlinear manner; but even if all this is taken into account, an isolated increase in glucose influx only has a moderate effect on HGO, because the increased G6P stimulates the efflux of glucose through GP. To block the HGO effectively by increasing the influx, it is necessary to dispose of the G6P, for example, by increasing the glycogen synthesis. Also the production of lactate and fatty acid (FA) can act as a sink for G6P and give effects similar to GgS action.

6.4.5
Muscle

6.4.5.1 Glucose Transport
In muscle, most glucose is transported into the cells via the two transporters GLUT1 and GLUT4 [59–61, 82]. The activity of GLUT1 is relatively constant and dominates glucose transport in the fasting state [83], where most GLUT4 is inactivated in intracellular stores [84]. Insulin is able to translocate GLUT4 to the cell membrane and in this way regulate the glucose transport capacity [59–61, 82]. Both transporters have a K_M for glucose in the order of 5 mM [74], so with a variation in glucose concentration of 5–10 mM, the variation in uptake rate is modest: only a 33% increase for a doubling in glucose concentration.

The main variation in the uptake is therefore taking place by varying the amount of active GLUT4 at the cell membrane [59–61, 82]. This is one of the important actions of insulin, but neural activity can also translocate GLUT4 independently of insulin [85]. This probably takes place during exercise [61, 82] and during stimulation from the portal sensor [44]. The glucose influx over the cell membrane can then be described like:

$$J_{influx} = \frac{V_{GLUT1}G}{K_{GLUT1} + G} + \frac{V_{GLUT4}G}{K_{GLUT4} + G} \tag{26}$$

Fig. 6.9 (a) Temporal changes in glucose-6-phosphate, G6P, concentration (upper figure) and hepatic glucose output, HGO (lower figure), when the glucose concentration is increased from 5 mM to 10 mM after 60 min, and held at 10 mM for 3 h, where after the glucose concentration is decreased to 5 mM again. During the 3 h both k_{GK} and k_{GgS} are increased sixfold. The two mirror spikes in HGO at the beginning and end of the test are a result of the fast change in G6P concentration in the two situations. (b) Relation between the mean value of glucose-6-phosphate, G6P, and the mean value of the hepatic glucose output, HGO, during the 3-h test period, for different balances between k_{GK} and k_{GgS}. (1) Only k_{GK} is changed. (2) k_{GK} is changed five times faster than k_{GgS}. (3) k_{GK} is changed two times faster than k_{GgS}. (4) k_{GK} and k_{GgS} are equally changed. (5) k_{GgS} is changed five times faster than k_{GK}. (6) Only k_{GgS} is changed.

with obvious constants, but as K_{GLUT1} and K_{GLUT4} are almost equal [74], Eq. (26) can be simplified to:

$$J_{influx} = \frac{V_{GLUT}G}{K_{GLUT} + G} \tag{27}$$

where V_{GLUT} corresponds to the total transporter activity.

It is generally believed that GLUT transporters are symmetric [61], so they can transport both ways with the same affinity. Thus, the efflux is given by:

$$J_{efflux} = \frac{V_{GLUT}G_i}{K_{GLUT} + G_i} \tag{28}$$

where G_i is the intracellular, free glucose concentration. The net uptake, J_{net}, then becomes:

$$J_{net} = \frac{V_{GLUT}G}{K_{GLUT} + G} - \frac{V_{GLUT}G_i}{K_{GLUT} + G_i}$$

$$= V_{GLUT} \left(\frac{G}{K_{GLUT} + G} - \frac{G_i}{K_{GLUT} + G_i} \right) \tag{29}$$

6.4.5.2 Glucose Phosphorylation

The inflowing glucose is phosphorylated by the enzyme hexokinase with a K_M in the order of 0.1–0.2 mM [17]. It is therefore able to keep G_i low, so the net influx in practice is independent of G_i. This is, however, a truth with modifications. The activity of hexokinase is inhibited by its product G6P, so if G6P is not removed fast enough, the activity of the hexokinase goes down.

The result is that, with low to moderate insulin concentrations, the net glucose uptake depends mainly on the inflow, so G_i is low. But at high concentrations, the inflow may be too large, so the phosphorylation becomes limiting and G_i increases [86]. There appears to be no report of systems that directly control G_i. Hence, if V_{GLUT} is large and the removal of G6P is too slow, the only way to control the influx is an increasing G_i [86], so the parenthesis in Eq. (29) decreases accordingly.

Figure 6.10 shows the inhibitory effect of rising G_i on the glucose influx. The figure shows that it requires a rather large G_i to get a strong inhibition of the influx. In contrast, a G_i of 1 mM reduces the flux with 33% of the maximum with $G = 5$ mM. Much higher G_i values have not been reported under normal circumstances, so either the hexokinase activity always suffices, or GLUT4 is not symmetric, or there exists some yet unknown system that decreases GLUT4 activity when G_i increases. The last option gains some support from the fact that GLUT4 may be translocated to the membrane without being activated [87, 88].

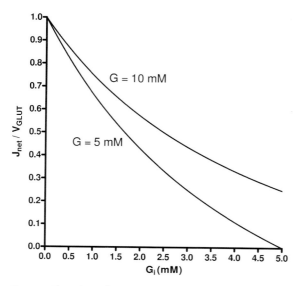

Fig. 6.10 The relation between the relative net glucose uptake, J_{net}/V_{GLUT}, and the intracellular free glucose, G_i, according to Eq. (29), for $G = 5$ mM and 10 mM.

The inhibition of hexokinase by G6P appears to be a so-called mixed inhibition [17], so the flux, J_{hex}, can be written as:

$$J_{hex} = \frac{V_{hex}G_i}{\left(1 + \dfrac{G6P}{K_{G6P}}\right) K_{hex} + \left(1 + \dfrac{G6P}{K^*_{G6P}}\right) G_i} \tag{30}$$

where the constants $K_{hex} = 0.188$ mM, $K_{G6P} = 0.068$ mM, and $K^*_{G6P} = 0.022$ mM are estimated from [17]. With a small K_{hex}, a substantial increase in G_i immediately saturates the hexokinase, so from then on, the phosphorylation activity only depends on G6P, and as also K^*_{G6P} is small, the flux approaches:

$$J_{hex} \approx \frac{V_{hex}K^*_{G6P}}{G6P} \tag{31}$$

The phosphorylation flux is in practice unidirectional. The muscle does not have G6P-phosphatase [62, 74], so when the free glucose has been phosphorylated, it is trapped inside the cell. The consequence is that the control of the glucose uptake becomes crucially dependent on the removal of the produced G6P. It is therefore not sufficient just to look at glucose transport as an effector of the glucose control system. The handling of G6P is in many cases much more important than the glucose transport.

This is illustrated in the following way. To simplify, it is assumed that the removal of G6P is proportional to the G6P concentration, so the flux, J_{G6P}, is given by:

$$J_{G6P} = R_{G6P} \cdot G6P \tag{32}$$

where R_{G6P} is a constant. At steady state the three fluxes (Eqs. (29), (30), and (32)) are equal, so:

$$J_{net} = J_{hex} = J_{G6P} \tag{33}$$

This gives a determination of G_i and G6P for different values of the constants, V_{GLUT}, V_{hex}, and R_{G6P}. Ignoring the absolute magnitude of the flux, the system

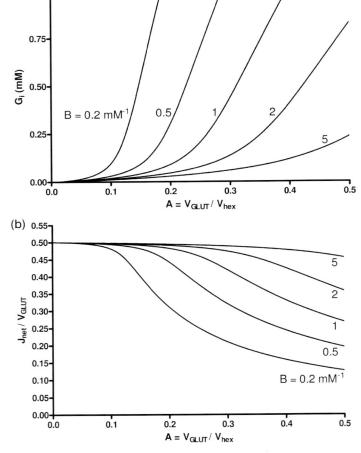

Fig. 6.11 (a) Relation between the intracellular glucose concentration, G_i, and the parameter A for different values of the parameter B.
(b) Inhibition of glucose uptake for different values of the parameter B.

can be simplified by introducing the ratios:

$$A = \frac{V_{\text{GLUT}}}{V_{\text{hex}}} \quad \text{and} \quad B = \frac{R_{\text{G6P}}}{V_{\text{hex}}} \tag{34}$$

so the effect of the flux balance on G_i and G6P can be analyzed.

The result is shown in Fig. 6.11a. Increasing A – high GLUT4 activity and/or low hexokinase activity – increases G_i. First moderately, but at a certain value (depending on B), G_i suddenly increases rapidly and approaches the outer glucose concentration (G). A high value of B can shift this transition to the right, so higher values of A are tolerated. The high G_i can diminish the uptake, as seen in Fig. 6.11b, but in glucose clamp experiments and with high glucose loads, there does not appear to be any substantial reduction in glucose uptake with time [89, 90], so G6P removal must in these cases be sufficient to keep G_i low.

It seems to be important to avoid high intracellular concentrations of free glucose. Most probably because it can activate the so-called sorbitol pathway and generate advanced glycosylation endpoints [91]. It is therefore helpful that not only glucose uptake, but also the phosphorylation and removal of G6P are stimulated by insulin [92].

The corresponding value of the G6P concentration is seen in Fig. 6.12a. The concentration increases with A until a certain level (depending on B) is reached, whereafter it is constant. From Eqs. (32) and (33) it then follows that the fluxes are also constant. This means that the net uptake according to Eq. (29) saturates, so the uptake becomes independent of changes in A, as seen in Fig. 6.12b. The consequence is that isolated changes in GLUT4 activity, for example by changing the insulin concentration, have no consequence for glucose uptake. The fate of the produced G6P is thus an important determinant of glucose uptake.

6.4.5.3 The Fate of Glucose-6-phosphate

The produced G6P can go into several different reactions, as depicted in Fig. 6.7b. During a fast, the resting muscle has a small glucose uptake that mainly goes to oxidation, and most of the energy comes from fat oxidation. After a meal, the increased insulin concentration stimulates all three fluxes of Eq. (33), but not necessarily to the same degree. The main G6P removal is glycogen synthesis, as in the liver, and insulin stimulates the glycogen synthase enzyme strongly [62]. However, the glycogen stores are limited. Maximum whole-body glycogen corresponds to some 1000 g [63] in a untrained person. The uptake rate declines more or less exponentially as the stores fill up; see Fig. 6.13a (redrawn from [93]). The half-time of the uptake depends on insulin stimulation, G6P, etc. The half-time in the figure is 36 h with an average insulin concentration of 120–140 pM. Also the steady state level may vary with stimulation, but to simplify, it is assumed to be a constant 1000 g.

Assuming that the stores are some 300 g after a night's fast, the glycogen build up is shown in Fig. 6.13b for different values of the half-time. The fastest (6 h) corresponds to a build up of 350 g in 6 h (0.97 g min^{-1}) or some 14 mg kg min^{-1} for a person of 70 kg. With the fast storage rate, the glycogen stores are filled rapidly,

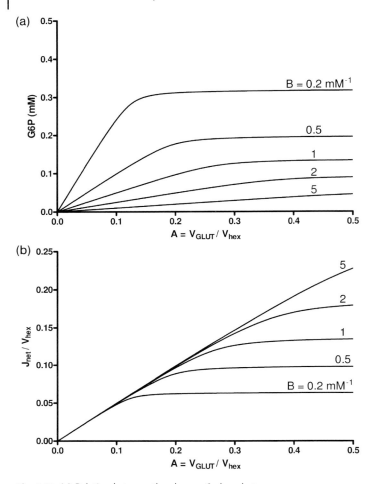

Fig. 6.12 (a) Relation between the glucose-6-phosphate concentration, G6P, and the parameter, A, for different values of the parameter, B. (b) Relation between the net uptake, J_{net}, normalized by V_{hex}, and the parameter A. Because G6P becomes constant at some value depending on B, according to (a), the net uptake, J_{net}, will be independent of A, from some value of A, depending on the value of B.

so even with a constant stimulation, the storage rate goes down. This is shown in Fig. 6.14a, where the initial storage rate, U_0, is given by:

$$U_0 = \frac{(Gg_{max} - Gg_0) \cdot \ln 2}{T_{1/2}} \tag{35}$$

where Gg_{max} is maximum glycogen (1000 g), and $Gg_0 = 300$ g is the initial value. It is seen that when the rate is high initially, it declines rapidly, while it is almost constant in cases with a small initial rate.

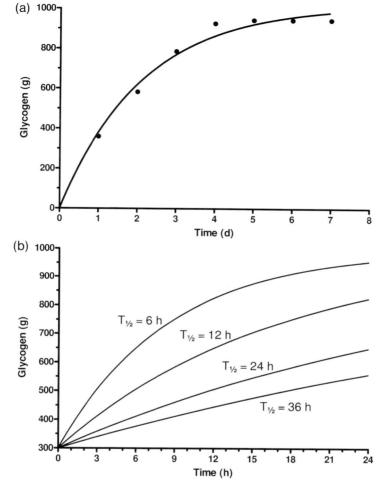

Fig. 6.13 (a) Relation between whole-body glycogen and time redrawn from [93]. The glycogen synthesis rate declines exponentially as the stores are filled up. (b) Relation demonstrating the exponentially decline in glycogen storage rate as the glycogen stores are filled up, for different half times, $T_{1/2}$.

A high glycogen storage rate is thus only possible for a limited time. If the glucose uptake continues at a high rate, the G6P must be disposed of elsewhere, to avoid a high free glucose. Most authors report that, during hyperinsulinaemic clamps, the glucose infusion rate does not appear to decline substantially with time [90], and during glucose loads, the plasma glucose concentration does not appear to increase [63]. Moreover, the energy expenditure of the resting muscle is low, so even if all muscle only oxidizes glucose, the oxidation flux is insufficient.

The muscle has then two options:

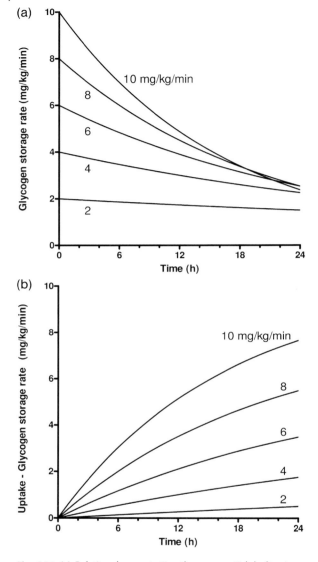

Fig. 6.14 (a) Relation demonstrating the exponential decline in glycogen storage rate when an increasing glucose load is applied. (b) Relation demonstrating the increase in lipogenesis, due to a decreasing glycogen storage rate when increasing glucose loads are applied.

1. Insufficient removal of G6P, leading to a high intracellular free glucose. This blocks the muscle uptake, but as the whole-body uptake continues, at least in healthy people, it requires a yet unknown and precisely timed signal from the muscle to other organs.

2. Activation of the glycolysis with lactate formation and/or lipogenesis, as shown in Fig. 6.7b.

It is reasonable to assume option 2, but option 1 may be possible in patients with type 2 diabetes, where glycogen synthesis appears to be decreased more than uptake and phosphorylation [94, 95].

The first choice is lactate formation. The lactate concentration is typically increased by a factor of two or three during high glucose uptake [90]. Lactate is not only an escape route for surplus glucose uptake, but as most cells can take up lactate, it can then be oxidized or stored elsewhere. The signal for lactate formation is unknown. A possible candidate is G6P, which increases as the glycogen stores fill up.

Also lactate has its limitations. The production of lactate can be very high, as seen during strenuous exercise, but a surplus of lactate cannot be stored as glycogen when the stores are full. The only way to store a surplus of carbohydrate is as triglyceride. This requires a *de novo* lipogenesis. The fat stores are virtually unlimited, so the only question is whether fatty acids can be synthesized fast enough to match the glucose uptake rate. Acheson et al. [63] found that, during overfeeding, at least 500 g glucose per day could be converted to fat at a relatively low insulin concentration. This corresponds to 5 mg kg^{-1} min^{-1}.

Muscle can synthesize fatty acids via *de novo* lipogenesis [96], but it is generally believed that adipocytes and liver are the major sites of lipogenesis. However, the muscle glucose uptake appears to continue as long as insulin is high [90] – although there is no firm evidence to confirm this for long periods of time – so the produced G6P appears to be removed fast enough to keep the free glucose low. This can be done by combining lipogenesis and lactate formation, but the control of the fluxes is unknown. Figure 6.14b shows that the lipogenesis in this case can be considerable.

6.4.6
Adipocytes

6.4.6.1 Triglyceride and Free Fatty Acid
The flux of glucose and its metabolites in adipocytes is depicted in Fig. 6.7c. The uptake and phosphorylation follow the same rules as for muscle [62], so Eqs. (29) and (30) are assumed valid for the adipocytes also. They can also build glycogen, but the amount is believed to be small [62]. The main part of the incoming glucose is therefore converted to lactate or used for lipogenesis. Only a small part is oxidized.

The destiny of ingested fat is shown in Fig. 6.15. The ingested triglyceride (TG) is broken down in the intestine to fatty acids (FAs) and glycerol or monoglycerides [74]. Inside the intestinal cell, the TG is rebuilt and released to the lymph as chylomicrons that later reaches the blood [74]. The TG in the chylomicrons is again broken down to FAs and glycerol by lipoprotein lipase (LPL) in the vessel wall [74]. The FAs are either released to plasma – the so-called spill over [97] – or transported into the adipocytes, rebuilt to TG and stored [74, 97].

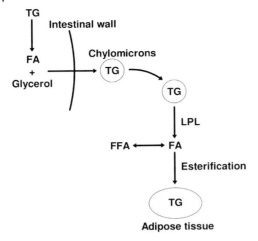

Fig. 6.15 The figure illustrates the complexity of fat metabolism,
where a sequence of breakdown of triacylglycerols, TG, to fatty acids,
FA, and glycerols, and rebuilding of TG from FA and glycerols takes
place, before TG enters the adipose tissue. For details see text.

FAs in plasma – the so-called free fatty acids (FFA) although they are bound to
albumin [74] – stem mainly from lipolysis inside the adipocytes and spill-over from
the LPL lipolysis [98]. FFA can be taken up and oxidized by most cells, particularly
muscle cells [74]. For example the heart lives mainly on fat oxidation [62, 74]. Also
the liver takes up FFAs. Some is oxidized, but a large fraction is rebuilt to TG and
released to blood as very low density lipoproteins (VLDLs) [74]. The VLDLs undergo
the same fate as chylomicrons, but their lipolysis rate is lower.

In the adipocytes there is a balance between TG formation from FA and glycerol
and the lipolysis of TG to FAs and glycerol, as shown in Fig. 6.16, but in contrast
to rodents, humans cannot re-use the glycerol liberated during lipolysis [62, 74].
The released glycerol goes to the liver to be re-build as glucose, and the glycerol
used in esterification comes from the inflowing glucose [62]. The storage of fat depends therefore on the inflow of glucose as a precursor for glycerol production [62].
Esterification is stimulated by insulin [74], so insulin both increases the presence
of glycerol and the TG formation rate. Locally, insulin also stimulates LPL activity
[74], so more FAs are released from the plasma lipoproteins. Added to this, insulin
inhibits lipolysis in the adipocytes [74]. All in all, insulin stimulates the storage of
fat as TG in the adipocytes.

It is interesting how the organs specialize their nutrient stores. Glucose is mainly
stored as glycogen in liver and muscle, and the stores are normally large enough to
contain the glucose intake of a day or two. The fat is mainly stored in the adipocytes,
where the stores are virtually unlimited. Ingested glucose and fat can then either
be oxidized or stored, but even with an eucaloric diet, the amount of glucose or fat
oxidized does not have to correspond to the amounts ingested on a daily basis, and
with over- or underfeeding the situation may be even more complex.

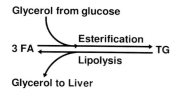

Fig. 6.16 Balance between formation of TG from FA and glycerols and breakdown of TG into FA and glycerols in the adipose tissue.

An important role of the insulin–glucose control system is therefore to shuffle the ingested nutrients between storing, releasing, and oxidation to smoothen the transitions between eating–fasting and exercise–rest, etc.

6.4.6.2 *De Novo* Lipogenesis

The formation of triglyceride from fatty acids and glycerol is often called lipogenesis. Another type of lipogenesis is the formation of fatty acids from glucose. To distinguish from the first, this is called *de novo* lipogenesis or DNL.

It is often stated that DNL is vanishingly small in normal humans – even during overfeeding [99–101]. This is perhaps due to the simple rule of thumb that ingested glucose is stored as glycogen in muscle and liver and ingested fat is stored directly in adipocytes. In the case of an eucaloric diet, the daily intake of nutrients corresponds to the daily energy expenditure, but, as mentioned above, the distribution between oxidation of carbohydrate and fat does not need to correspond to the distribution in the food on a daily basis. To this comes the many so-called futile cycles where the nutrients are converted back and forth between the different types, so the required food intake depends on the precise route of oxidation.

The experimental results are typically based on indirect calorimetry, i.e. measurements of RQ, the ratio between CO_2 release and O_2 uptake in the lungs [102]. Oxidation of carbohydrate gives a RQ of 1.0, while fat gives a RQ near 0.7. By measuring O_2 consumption and CO_2 production and subtracting the protein values, the oxidation rates of the two nutrients are calculated [102]. However, while fat cannot be converted to glucose in humans, glucose can easily be converted to fatty acids via DNL. When glucose is stored as glycogen, it has no influence on the RQ, but when some glucose is converted to FAs, the RQ goes up. From Frayn [102], the nonprotein RQ can be calculated as:

$$RQ = \frac{CO_2\text{-release}}{O_2\text{-uptake}} = \frac{a_1 c_0 + a_2 f_0 + a_3 c_f}{b_1 c_0 + b_2 f_0 + b_3 c_f} \tag{36}$$

where c_0 and f_0 are the oxidation rates of glucose and fat (g min^{-1}) and c_f is the DNL (g glucose min^{-1}). According to Frayn [102], the constants are $a_1 = 0.746$, $a_2 = 1.43$, $a_3 = 0.24$, $b_1 = 0.746$, $b_2 = 2.03$, and $b_3 = 0.028$ [all: l air (STPD)g^{-1}].

Assuming that the energy expenditure (EE) from glucose and fat oxidation is constant gives:

$$EE = q_G c_0 + q_F f_0 \tag{37}$$

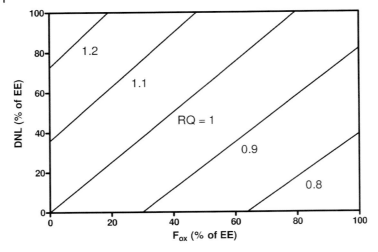

Fig. 6.17 Relation between *de novo* lipogenesis, DNL, and fat oxidation, F_{Ox}, according to Eq. (38), for different values of the respiratory quotient, RQ.

where it is assumed that glucose oxidation gives $q_G = 4$ kcal g^{-1} and fat oxidation $q_F = 9$ kcal g^{-1} [74]. Combining Eqs. (36) and (37) gives:

$$\frac{DNL}{EE} = \frac{b_1 RQ - a_1}{a_3 - b_3 RQ} + \frac{q_G(b_2 RQ - a_2) - q_F(b_1 RQ - a_1)}{a_3 - b_3 RQ} \frac{F_{Ox}}{EE} \tag{38}$$

where DNL and F_{Ox} are given in the same units as EE.

The result of Eq. (38) is shown in Fig. 6.17. Clearly, a substantial amount of DNL is possible with a RQ<1, if fat oxidation is also present. The fat content in a standard diet is 41% of EE, so if all this is oxidized, then even substantial overfeeding with carbohydrate – up to 50% of EE – gives RQ<1.

In other words, DNL only results in RQ>1 if DNL is large and F_{Ox} is small. This might happen briefly during a high glucose intake, but in the eucaloric case, where intake equals oxidation, it is not possible to get the average RQ above 1. Also with overfeeding with mixed diet, it is difficult to get RQ>1, because most of the excess energy is stored as fat from the diet, and because there will always be some fat oxidation by tissues, like the heart, that prefer fat as an energy source [62, 74]. Consequently, only a massive overload with carbohydrate and no or almost no fat can force RQ above 1 [101]. In all other cases there may be a substantial DNL together with a fat oxidation, but they take place in different tissues.

Under normal circumstances most DNL appears to happen in the adipose tissues and only a small fraction in the liver and muscle [101]. As most investigators have concentrated on the liver DNL, the importance of DNL has been underestimated [101]. In the adipocytes FAs can be built from the inflowing glucose or from lactate. The amount is not insignificant. As mentioned previously, Acheson et al. [63] found that at least 500 g glucose could be converted per day. The maximum DNL capacity is difficult to estimate. It may be as much as 1000 g day^{-1}, and assuming an energy loss of 25% by the conversion, this corresponds to a production of more than 300 g pure fat per day or 10 kg month^{-1}.

Tracer experiments have been used to evaluate the magnitude of transient DNL [103–106], but the tracers are difficult to follow in a transient situation. It therefore cannot be excluded that some glucose during high peaks is temporally stored as fat and later oxidized. This can help the removal of glucose in cases, where the glycogen synthesis is too slow.

It has been argued that, by first converting glucose to fat and later oxidizing the fat, there is an energy loss of some 20–25% [101], but there are many other futile cycles in the metabolism that releases energy without "work". Moreover, even muscles only work with an efficacy of 15–25% [7]. The rest of the energy goes to heat. The energy loss is only important if energy is sparse, but in such a case the glycogen stores are small and the ingested carbohydrate is stored as glycogen. Substantial DNL is therefore only seen, when a period – at least some days – of surplus food has restored the glycogen stores.

An interesting situation arises during long glucose loads. As the glycogen stores in muscle cells fill up, the DNL must take over. According to Acheson et al. [63] this takes place in a graded manner, so DNL increases while glycogen storage decreases. If the muscle glucose uptake continues, the produced G6P must be removed through lactate production or via DNL. However, with a net non-oxidative uptake in the order of, say, 6 mg kg min^{-1}, the lactate production should be 4.5 mmol min^{-1} leading to a very substantial, 1–2 mM, increase in plasma lactate. Experiments have shown a moderate increase [90] but not in that order of magnitude.

Consequently, the most of the excess G6P appears to be converted to fatty acids in the muscle, so in this case a substantial DNL is taking place in the muscle cells [96, 107]. The fate of the fatty acids is unknown. They may be oxidized or, at least partly, stored in the muscle cells, or transported to the adipose tissues and stored there. An interesting effect of fatty acid production is that they may inhibit GLUT4 [108, 109], leading to a kind of insulin resistance.

Again it appears that the metabolic control is able to anticipate the necessary changes in nutrient fluxes. In the beginning of a glucose load, glucose is directed to glycogen stores and oxidation is moderately increased. As the stores fill up with time, there is apparently no accumulation of glucose in plasma or inside cells. Instead, the glucose is directed out again as lactate and fatty acids and converted to triglyceride in the adipocytes, again with no piling up in the blood.

6.4.7
Conclusion

The analysis demonstrates that glucose uptake is very dynamic and complex. In the simple context, glucose is transported over the cell membrane, phosphorylated, and then oxidized, converted to lactate or fat, or stored as glycogen as shown in Fig. 6.6a. But these processes take place at very different rates in the different organs and they depend strongly on the metabolic state of the organism. The real challenge to the control system is therefore to coordinate all these fluxes, so there is no piling up of metabolites, and the nutrients can be passed on to the cells that need them.

6.5
The Control System at Large

As concluded in Sections 6.3 and 6.4, the function of the control system is to shift the nutrient fluxes between different end points and organs, so piling up of nutrients or metabolites is avoided. The control is not absolute in the sense that there are definite rules that govern the precise function of the control system. It rather appears that the system operates in a self-organizing manner that is able to cope with the very different possible nutrients regarding both composition, quantity, and timing.

The control system is typically investigated under more or less controlled circumstances, and to go from these to free living is not an easy task. In the following sections some of the typical setups are analyzed in the light of the different mechanisms described in the previous sections.

6.5.1
The Fasting State

The fasting state can be defined as the state where the stomach and intestine are empty, so there is no absorption of glucose. All glucose is then provided by the liver and kidneys, here again for simplicity combined as the hepatic glucose output (HGO). Assuming very slow changes in G, Eq. (22) gives:

$$V_G \frac{dG}{dt} = HGO - J_{upt} \approx 0 \tag{39}$$

or, using Eq. (26):

$$HGO = J_{upt} = J_{brain} + (V_{GLUT1} + V_{GLUT4}) \frac{G}{K_{GLUT} + G} \tag{40}$$

where J_{brain} is the – normally constant – brain uptake, and where it is assumed that G_i is small and that GLUT1 and GLUT4 have the same K_{GLUT}.

Equation (40) is interesting. First, the constancy of J_{brain} and to some extent V_{GLUT1} forces an enhanced variation in the other variables, when HGO varies. Second, the glucose dependence of the uptake is much weaker than normally assumed, and it is saturable. Third, the effect of insulin on V_{GLUT4} is crucial for the fasting glucose. This can be elucidated in the following way.

The fasting HGO (overnight fast) is as previously mentioned in the order of 140 mg min^{-1} (2 mg kg^{-1} min^{-1}). Of this the brain takes some 80 mg min^{-1}. The remainder, 60 mg min^{-1}, is the glucose dependent uptake. With $K_{GLUT} = 5$ mM and a fasting glucose also of 5 mM, the fasting value of:

$$V_{GLUT} = V_{GLUT1} + V_{GLUT4} \tag{41}$$

becomes 120 mg min^{-1}. The maximum uptake for that value of V_{GLUT} is thus 200 mg min^{-1}, corresponding to an increase in HGO of only 43% and with a value of G going towards infinity. Similarly, an increase in HGO to 170 mg min^{-1} (2.4 mg kg^{-1} min^{-1}) forces the value of G up to 15 mM. In contrast, HGO must be above 80 mg min^{-1} to meet the glucose requirement of the brain. Thus, there is a narrow interval for HGO, 80–200 mg min^{-1}, for $V_{GLUT} = 120$ mg min^{-1} and with G in principle going from zero to infinity.

The relation between G and HGO is shown in Fig. 6.18a for different values of V_{GLUT}, corresponding to different fasting insulin concentrations. It is seen that even moderate increases in HGO require an increase in V_{GLUT} to keep G low. The renal loss, shown in Fig. 6.18b, of course, always works as a safety valve that keeps the glucose within reasonable limits, so above a G of 10–15 mM there is a much weaker dependency on HGO.

6.5.1.1 Futile Cycles

Patients with type 2 diabetes have often an increased HGO [110–112]. As the glycogenolysis is decreased [111] and in all cases transient, the cause of augmented HGO is increased gluconeogenesis, mainly from lactate [111, 113]. The increased lactate production is interesting. To give a constant overproduction of lactate, glucose must be transported into cells but not oxidized, mainly because there is an increased level of FFA and therefore an increased fat oxidation [74]. The overproduction is thus caused by a mismatch between the fluxes of glucose and fat.

This gives a futile cycle, because the new glucose again is turned into lactate and then into glucose, etc., *ad infinitum*. The culprit is neither the liver, nor the muscles. Blocking gluconeogenesis in the liver gives lacto-acidosis; and increasing glucose uptake in the muscles just creates more lactate. Instead, the FFA concentration should be lowered, so more glucose can be oxidized. The FFAs stem mainly from adipose tissue and focus has been on the breakdown of TGs by the hormone sensitive lipase (HSL) which is strongly insulin dependent [74].

The idea is then that insulin resistance at the adipocytes gives a weaker inhibition of the HSL, so more FAs are released. This may very well be the case, but it cannot last for long, because then the fat stores would be emptied; and the typical type 2 patient is obese. Instead, a new theory is emerging. As previously mentioned, it seems that adipocytes can create considerable amounts of fat from glucose [63]. This fat can be stored or it can be oxidized. In the latter case, another so-called futile process is active. Instead of oxidizing the glucose directly, it is first converted into fat and then oxidized as fat.

The processes are called futile, because energy is lost. The lactate–glucose cycle costs some 10% of glucose energy, and lipogenesis costs some 25%. The futility is, however, questionable as previously mentioned. The processes only occur in situ-

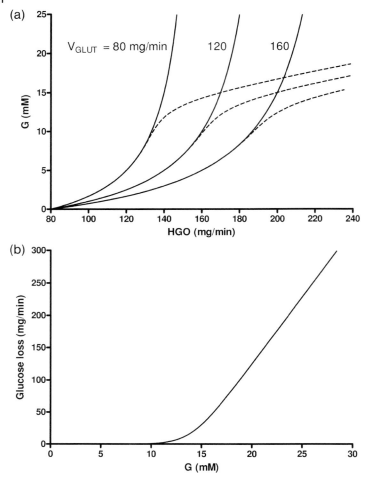

Fig. 6.18 (a) Relation between glucose concentration, G, and hepatic glucose output, HGO, according to Eq. (40), for different values of V_{GLUT}, corresponding to different fasting insulin concentrations. The dashed lines indicate the renal glucose loss. (b) Relation between renal glucose loss and glucose concentration, G, taken from [68].

ations of overfeeding, where carbohydrate and fat are competing for oxidation. In this case it is advantageous to increase the energy expenditure, and it appears that type 2 patients have a basal metabolic rate (BMR) some 5% higher than matched healthy persons [114]. This should give an energy increase near 100 kcal day^{-1}, corresponding to a total oxidation of 25 g carbohydrate.

6.5.2
Normal Meals

Equations (22) and (40) give – still assuming G_i is small:

$$V_G \frac{dG}{dt} = J_{abs} + HGO - J_{brain} - (V_{GLUT1} + V_{GLUT4})\frac{G}{K_{GLUT} + G} \qquad (42)$$

After the meal, J_{abs} increases to a maximum and then declines again to zero. The AUC for J_{abs} corresponds to the total amount of glucose absorbed, but the size of the maximum and the duration of absorption are very variable. First, liquid meals are absorbed much faster than solid [115, 116]. Second, the higher the fat content, the slower is the absorption [117, 118]. Actually, it appears that emptying of the stomach is controlled by the flow of energy through the duodenum [119].

The control system gets information about the meal in several ways. Hormones like GIP and GLP-1 and the intestinal nervous system informs about the events in the gut and intestine [120], while the hepato-portal sensor measures the actual glucose absorption rate. Finally, an increase in glucose concentration acts as a signal to the glucose sensors. This signal can, however, rather be regarded as a signal for the fine tuning of the flux rather than the main control.

The principle can be seen from Eq. (42). If the idea of the control system is to keep the glucose concentration constant, the equation can be used to find the conditions for this. Taking G as a constant and ignoring the variations in HGO, the equation gives:

$$V_{GLUT4} = (J_{abs} + HGO - J_{brain})\frac{K_{GLUT} + G}{G} - V_{GLUT1} = aJ_{abs} + b \qquad (43)$$

where a is a constant and b is the value of V_{GLUT4} before the meal. To keep G constant, the GLUT4 dependent uptake must vary proportionally to the glucose uptake. Unfortunately, the insulin system has a considerable reaction time. First, the glucose must increase; second, the insulin must be released; third, the insulin must reach the peripheral receptors; fourth, GLUT4 must be translocated. Then the increased uptake can begin.

Especially at the beginning it is important that the nervous system is able to sense the meal and start the GLUT4 translocation. Here the activity of the hepato-portal sensor is ideal, because it gives a signal that is precisely proportional to J_{abs}. It is typically seen that the plasma insulin concentration increases before any noticeable increase in glucose concentration [121].

6.5.3
Glucose Tolerance Tests

6.5.3.1 Intravenous Glucose Tolerance Test
The IVGTT, where a bolus of glucose is administered intravenously, is mostly used to test the first phase of insulin release and the insulin sensitivity of glucose uptake.

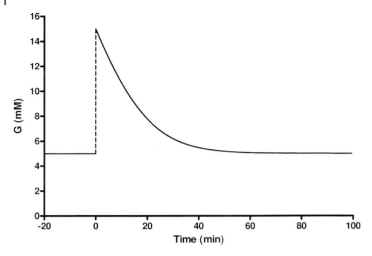

Fig. 6.19 Relation between glucose concentration, G, and time during an intravenous glucose tolerance test (IVGTT).

Several models of different complexity – e.g. the minimal model – describe the IVGTT [122–125]. Typically, beta cells respond with a large first phase, followed by a much smaller second phase as the glucose concentration rapidly declines.

One of the main problems with the IVGTT is its transient nature, because the time it takes to distribute the bolus evenly in the blood is considerable. The mean transit time for the circulation is approx. 1 min [64], but some blood returns fast, while other blood, for example to the legs, may take 10–15 min to return. The result is an initial distribution phase that mimics an uptake, so particularly the initial glucose dependent uptake is overestimated.

However, saturation of the transporters underestimates the transport capacity at normal glucose concentrations, particularly, because most models are linear [126] in G. The discrepancy is considerable. The peak plasma glucose is often some 15 mM, so the concentration is tripled, but with a K_{GLUT} in Eq. (42) of 5 mM this corresponds only to a 50% increase in transport. Simplifying Eq. (42) gives:

$$V_G \frac{dG}{dt} = V_{GLUT} \left(\frac{G_0}{K_{GLUT} + G_0} - \frac{G}{K_{GLUT} + G} \right) \tag{44}$$

where G_0 is the initial glucose concentration. This gives a decline which is more linear than exponential, particularly in the beginning as shown in Fig. 6.19, and as G usually undershoots G_0 after the initial increase [127, 128], it becomes difficult to determine the steady level that is necessary to estimate the rate constant of the decline. Another problem is the increase in GLUT4 during the IVGTT. It will bend the curve downwards and distort it even more from an exponential decline.

6.5.3.2 OGTT and MTT

In the oral glucose tolerance test (OGTT) and the meal tolerance test (MTT) the glucose is given orally [123–125]. This gives a slower increase in plasma glucose generally, but also a dependence on gastric emptying and intestinal motility. With the intestinal uptake, the hepato-portal sensor comes into play, so there is also an increased insulin independent glucose uptake. The IVGTT and OGTT therefore cannot be compared directly. Also the higher and more sustained portal concentration has a more lasting effect on the HGO.

For all tolerance tests, it is generally assumed that the glucose taken up is either oxidized or stored as glycogen [129], because the tests are normally performed on overnight fasted subjects, and because the glucose load is moderate. The tests may thus be valuable in predicting different experimental results, where the initial condition typically is a night's fast. They may, however, be less predictive for normal 24 h living conditions with varying over- and underfeeding.

6.5.4
The Glucose Clamp

6.5.4.1 Glucose Infusion Rate

One of the most used tests for the glucose–insulin control system and for new insulins is the glucose clamp. The idea is to keep the glucose concentration at a constant level by infusing glucose intravenously, replacing the glucose taken up by the cells [89, 90, 130]. Modifying Eq. (22) gives:

$$V_G \frac{dG}{dt} = \text{GIR} + \text{HGO} - J_{\text{upt}} \tag{45}$$

where GIR is the glucose infusion rate. Assuming $dG/dt = 0$ – which in practice is not an easy task – gives:

$$\text{GIR} = J_{\text{upt}} - \text{HGO} \tag{46}$$

This is the fundamental relation for the glucose clamp. The value of GIR depends then on the changes in J_{upt} and HGO after insulin stimulation and the clamped value of the glucose concentration. Introducing the uptake from Eq. (29) gives:

$$\text{GIR} = J_{\text{brain}} - \text{HGO} + V_{\text{GLUT}}\left(\frac{G}{K_{\text{GLUT}} + G} - \frac{G_i}{K_{\text{GLUT}} + G_i}\right) \tag{47}$$

where J_{brain} is the – normally constant – brain uptake, and where it is again assumed that GLUT1 and GLUT4 have the same K_{GLUT}. As G is constant, the main variations in the uptake come from variations in V_{GLUT4} and HGO.

In a typical setup, insulin is given subcutaneously and absorbed with a given plasma profile and a somewhat delayed interstitial profile. More rarely, insulin is infused to a constant insulin concentration. The interstitial insulin stimulates the

GLUT4 translocation that – with a further delay – gives a variation in V_{GLUT}. At the same time, the insulin inhibits HGO – at least for some time.

6.5.4.2 Nutrition During Clamp

If the subject is clamped at the fasting value of G, the GIR is zero at time zero. At this time, the energy expenditure is partly covered by the HGO – except for a small generation of lactate that recycles – and partly by fat oxidation and some small protein oxidation. As the subject is resting, the energy expenditure is in the order of 1.1–1.2 kcal min^{-1} for a healthy person of 70 kg [7]. If this is covered solely by glucose infusion, the GIR should be some 4 mg kg^{-1} min^{-1}. A GIR above that corresponds to overfeeding, and below that to underfeeding.

Typically, the clamp is initiated after an overnight fast, i.e. 10–12 h without food. During the clamp, the untreated group typically needs no or very little GIR [131], so with a 24 h clamp, the patient may have fasted for 36 h. At the other end of the scale, a large insulin concentration may require a GIR of 10 mg kg^{-1} min^{-1} or more. This is 2.5 times the energy expenditure and a very massive overfeeding. Continuing this for, say, 24 h has a dramatic influence on all the body's metabolic processes.

Between these extremes exists an interval where there is an interplay between the insulin given, the GIR and the fate of the infused glucose, and the movements and oxidation of the previously stored nutrients. Also, there is a yet unknown participation of the nervous part of the control system. The effect of all this is that each clamp must be viewed in its own context.

6.5.4.3 Clamp Level

In a trial with many patients it is desirable to have comparable setups, so the results can be pooled. The first choice is whether the clamp should be at a predefined value or at the patient's fasting blood glucose level [130]. Using the fasting G has the advantage that the patient is more or less at steady state. Continuing at that level mainly requires a GIR that can compensate for the reduction in glycogenolysis as the glycogen stores empties. Using another level requires some intervention: either a titration with glucose, if the level has to rise, or a titration with insulin, if it has to decrease. Sometimes, if the difference is small, it suffices just to wait, until the declining glycogenolysis brings the G down.

More crude methods have to be used when patients are to be clamped at a level much below their fasting glucose. The most used is simply to infuse insulin, until the wanted level is reached. The discussion is then whether the infusion has to be stopped when the test insulin is given, or has to be continued in order to give comparable results [131, 132]. From Eq. (40) it follows that, of a resting HGO of 2 mg kg^{-1} min^{-1}, some 1.14 mg kg^{-1} min^{-1} goes to the brain, and with $G = 5$ mM, the value of V_{GLUT} is 1.72 mg kg^{-1} min^{-1}.

A patient with a fasting G of, say, 15 mM has an increased HGO, or decreased V_{GLUT}, or both. With $G = 5$ mM the uptake is 0.86 mg kg^{-1} min^{-1}, and with $G = 15$ mM the uptake is 1.29 mg kg^{-1} min^{-1}. To achieve 15 mM, the HGO needs only to increase to 0.43 mg kg^{-1} min^{-1} or 22%, assuming normal V_{GLUT}. In contrast,

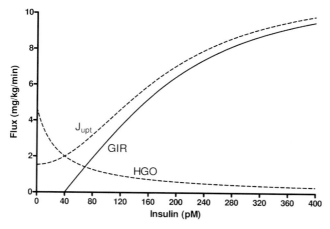

Fig. 6.20 Relation between the fluxes and insulin concentration during a glucose clamp. The glucose infusion rate, GIR, is calculated according to Eq. (46).

with a normal HGO, the V_{GLUT} has to decrease to 1.15 mg kg^{-1} min^{-1} or 33% to give the same effect.

Insulin infusion improves both. The HGO goes down and V_{GLUT} increases until the wanted glucose level is reached. If the wanted level is 7 mM and the distribution volume for glucose is 17 l, this requires the uptake of 24.5 g glucose, which can take several hours even with a noticeable insulin stimulation. On the other hand, to keep G at 7 mM requires only a small insulin stimulation, because of the saturation effect. This is seen in the following way. The insulin effect on V_{GLUT} can be described by [68]:

$$V_{GLUT4} \approx \frac{V_{max}I^2}{K_I^2 + I^2} \tag{48}$$

with V_{max} = 20 mg kg^{-1} min^{-1} and K_I = 180 pM. At 40 pM insulin and G = 5 mM and ignoring G_i this gives 0.47 mg kg^{-1} min^{-1}. Taking V_{GLUT1} to be 0.78 mg kg^{-1} min^{-1}, the insulin independent uptake becomes 0.39 mg kg^{-1} min^{-1}. Together with the brain uptake, the total uptake becomes 2 mg kg^{-1} min^{-1}, corresponding to the fasting HGO. The situation is depicted in Fig. 6.20.

The HGO is supposed to be inhibited (at least temporarily) by insulin according to:

$$HGO = \frac{HGO_{max} K_{HGO}}{K_{HGO} + I} \tag{49}$$

where HGO_{max} = 4.7 mg kg^{-1} min^{-1} is the value of HGO at I = 0, and K_{HGO} = 30 pM is a constant [68]. The figure demonstrates that the insulin effect on uptake is relatively small at low concentrations, while the effect on HGO is large. What is not shown is the insulin's effect on lipolysis. The most sensitive effect of insulin appears to be on lipolysis [74]. As FFA in plasma goes down, more glucose is oxidized and therefore there is less lactate to be rebuilt to glucose.

It is interesting that the combination of the two effects gives an almost linear relation between insulin concentration and GIR for small to moderate insulin concentrations. However, for long clamps, the uptake and HGO evolve completely differently, as described in the previous section.

The modest participation of the insulin dependent uptake at fast makes it relatively easy to compensate for variations in fasting G. At $G = 5$ mM and $I = 40$ pM, the contribution is 0.47 mg kg^{-1} min^{-1} or 23.5% of the total uptake. Decreasing V_{GLUT4} from 0.94 mg kg^{-1} min^{-1} to 0.37 mg kg^{-1} min^{-1} and keeping the other parameters constant, increases G from 5 mM to 15 mM. This can happen if K_I is increased by 60%. To reduce G back to 5 mM requires, under these circumstances, only an increase in insulin concentration from 40 pM to 48 pM, which is marginal.

Changing the clamp level at a given insulin concentration changes both the uptake rate and the HGO. The effect on uptake is dominant at large insulin concentrations, where the constant brain uptake has less weight. In this situation, the HGO is inhibited, so the effect on HGO is larger at small insulin concentrations.

6.6
Conclusions

6.6.1
Biosimulation

This chapter has presented a series of small biosimulation models that describe different facets of glucose metabolism. The description is far from complete, and especially the role of the nervous system in this control is still poorly understood.

In the past, the main focus has been on insulin's effect on glucose uptake and hepatic glucose production. It now appears that there are many more players. The fate of glucose – or G6P – after uptake appears to be even more important for glucose metabolism, both in healthy persons and in patients with diabetes. There is an intimate interplay between the nutrient stores and between the different organs. This interplay appears well controlled, but the control is dynamic, so the detailed setup of the control system varies with the amount and type of nutrients. In this way the system can cope with almost any eating habit.

Many results are presented as consequences of the biosimulation setups, and the experimental evidence for some of them may be questioned. This is not a weakness of the biosimulation. When a biosimulation model is constructed, it is based upon known experimental evidence, so the outcome of the model represents a series of conclusions based upon this evidence. The advantage is that the conclusions are quantitative and can point to new possible mechanisms that can be tested, to new experimental setups, and to new treatment targets.

The dynamics of the system makes it difficult to combine the many small models into a large, comprehensive model. Each facet is compared to some specific experimental setup, and it is typical that the variety of setups gives different, maybe even contradicting results. Merging the different facets may therefore lead to inconsistent results, particularly when the many effects of the nervous system are taken into account.

The present review should therefore be taken as an inspiration to construct new experiments that can elucidate glucose metabolism in the light of the described findings.

6.6.2
The Control System

The glucose–insulin system has been presented in a context of classic control theory as a combination of a PID control and predictive control. This context may be too simplified. With a network of sensors, a complex traffic of nutrients and metabolites, and day to day variations in food intake, it is difficult to assign a single controlled variable, like plasma glucose concentration, to the control system. Equally important variables could be intracellular glucose concentration, the flux of fatty acids and amino acids.

Central in the control is the hormone insulin. It influences the traffic of all nutrients and their fate. It has also been called "the hormone of plenty", because the insulin concentration is increased after a meal, and because its main effect is to guide nutrients to their destination: oxidation, conversion, and/or storage. Thereafter the insulin concentration declines slowly to lower values. As time passes and the insulin concentration becomes low, the traffic turns around, and nutrients are released from the stores to cover the body's needs.

6.6.3
Diabetes

The biosimulation models in this chapter describe mainly healthy persons. In the case of diabetes some of the relations, regarding both carbohydrate and fat metabolism, are altered. One main change is a decreased insulin release for a given glucose increase. For the type 1 diabetes patient the decrease is fast, months to a few years, and leads rapidly to incompetent insulin production. For the type 2 diabetes patient the progress is much slower and takes decades.

Another main change is in the insulin sensitivity at different tissues. For example, the effect of insulin on glucose uptake is decreased. Insulin has the same inhibitory effect on lipolysis and on hepatic glucose output, and inside the cells the removal of glucose-6-phosphate is decreased. The different changes in the metabolic traffic result in a spectrum of possible patient variants. Common for them all is an increased plasma glucose concentration and increased meal-related variations.

The increased plasma glucose can in itself lead to further destruction of beta cells, so the progression of the disease depends on the efficacy of the control system. A high plasma glucose, i.e. a defect control, leads to further deterioration of the beta cells, which leads to a further glucose increase and so on. Compared with the PID controller in Eq. (5), this corresponds to a decrease in the constants A_1, A_2, and A_3 as a function of the cumulative error, CI–CO.

The disease progression can be modeled as a slow change in one or more parameters. Different patients not only display different combinations and degrees of changes, but the evolution of the changes can follow different time-courses, depending on genetics, food intake, surroundings, treatment, and so forth. The path an individual patient follows is difficult to predict, but biosimulation models can help to describe a spectrum of possible scenarios and in this way indicate the best treatment.

6.6.4
Models and Medicines

The development of new medicines is both lengthy and expensive. Many experiments and trials are necessary before the medicine reaches its final form. Integrating biosimulation models in the development can speed up the development process considerably and save much expense. This is instrumental for the development of individualized medicines, because otherwise the sales would be too small to cover the development costs.

At present, few companies use biosimulation and only to a very limited extent. There is therefore a large, unmet need for good biosimulation models and modelers in all branches of the pharmaceutical industry. But to satisfy the need is not easy.

Biosimulation as a discipline has emerged from a combination of natural sciences with a strong representation of physics, chemistry, and mathematics. The modeling technique and the modelers still reflect this inheritance, with a focus on scientifically interesting results. But what is a good biosimulation model seen from the industry's view? Certainly, the model has to be mathematically correct and reflect the biology to the best of our knowledge, but there are several aspects that are crucial for the pharmaceutical industry and are rarely addressed with existing models.

The most important aspect is validation. Drug development goes forward in small steps, where the planning of each step builds on a comprehensive evaluation of the previous steps and on the knowledge of the pathophysiology of the disease in question. Each step may cost many millions of euros, and a suboptimal planning may jeopardize the whole project. A biosimulation model is indeed a valuable guidance, but all models are based upon some or many simplifications, unknowns, and so forth, so it is difficult for the decision makers – often without modeling experience – to know how much weight should be laid on the model results.

A real help for the industry and a possible breakthrough for biosimulation would be to find an easy way to validate each model and its results in a fashion that can compete with the high valuation of experimental and clinical results in the development process.

References

1 H. T. Milhorn, *The Application of Control Theory to Physiological Systems*, W. B. Saunders Company, Philadelphia, **1966**.

2 B. Thomas, *Identification, Decoupling and PID-Control of Industrial Processes*, Kompedenietryckeriet-Kållered, Sweden, **1990**.

3 G. F. Franklin, J. D. Powell, A. E.-Naeini, *Feedback Control of Dynamic Systems*, Addison-Wesley Publishing Company, USA, **1991**.

4 L. R. Young, L. Stark, Biological control systems – a critical review and evaluation. Developments in manual control, *NASA CR* **1965**, 1–121.

5 H. Hosomi, K. Sagawa, *Am. J. Physiol.* **1979**, 236(4):H607–H612.

6 E. Simon, F. K. Pierau, D. C. Taylor, *Physiol. Rev.* **1986**, 66(2):235–300.

7 A. C. Guyton, J. E. Hall, *Textbook of Medical Physiology*, W. B. Saunders Company, USA, **2000**.

8 L. Wildt, A. Hausler, G. Marshall, J. S. Hutchison, T. M. Plant, P. E. Belchetz, E. Knobil, *Endocrinology* **1981**, 109:376–385.

9 Yue-Xian L, A. Goldbeter, Frequency specificity in intercellular communication, *Biophys. J.* **1989**, 55:125–145.

10 E. Knobil, *News Physiol. Sci.* **1999**, 14:1–11.

11 S. S. Stevens, *Percept. Psychophys.* **1969**, 6:251–256.

12 J. A. Rossiter, *Model-Based Predictive Control*, CRC Press, USA, **2003**.

13 I. P. Pavlov, *Conditioned Reflexes: An Investigation of the Physiological Activity of the Cerebral Cortex*, Oxford University Press, London, **1927**.

14 R. Llinas, The nervous system. In: V. B. Brooks (Ed.) *Handbook of Physiology, Vol. II*, pp 831–976, American Physiology Society, Bethesda, Md., **1981**.

15 F. M. Matschinsky, M. A. Magnuson, D. Zelent, T. L. Jetton, N. Doliba, Y. Han, R. Taub, J. Grimsby, *Diabetes* **2006**, 55:1–12.

16 F. C. Schuit, P. Huypens, H. Heimberg, D.G. Pipeleers, *Diabetes* **2001**, 50:1–11.

17 C. J. Toews, *Biochem. J.* **1966**, 100:739–744.

18 T. Y. Kruszynska, T. P. Ciaraldi, R. R. Henry, Regulation of glucose metabolism in skeletal muscle. In: L. S. Jefferson, A. D. Cherrington (Eds.), *Handbook of Physiology, Sec. 7, Vol. II*, Oxford University Press, USA, **2001**.

19 S. Efrat, M. Tal, H. F. Lodish, *TIBS* **1994**, 19:535–538.

20 M. A. Magnuson, F. M. Matschinsky, Glucokinase as a glucose sensor: past, present and future. In: F. M. Matschinsky, M. A. Magnuson (Eds.), *Glucokinase and Glycemic Disease, from Basics to Novel Therapeutics, Vol. 16*, S. Karger, Basel, **2004**.

21 Y. Noma, S. Bonner-Weir, J. B. Latimer, A. M. Davalli, G. C. Weir, *Endocrinology* **1996**, 137:1485–1491.

22 L. Agius, S. Aiston, M. Mukhtar, N. de la Iglesia, GKRP/GK: control of metabolic fluxes in hepatocytes future. In: F. M. Matschinsky, M. A. Magnuson (Eds.), *Glucokinase and Glycemic Disease, from Basics to Novel Therapeutics, Vol. 16*, S. Karger, Basel, **2004**.

23 M. A. Rizzo, M. A. Magnuson, P. F. Drain, D. W. Piston, *J. Biol. Chem.* **2002**, 37:34168–34175.

24 I. Miwa, Y. Toyoda, S. Yoshie, Glucokinase in β-cell insulin-secretory granules. In: F. M. Matschinsky, M. A. Magnuson (Eds.), *Glucokinase and Glycemic Disease, from Basics to Novel Therapeutics, Vol. 16*, S. Karger, Basel, **2004**.

25 P. B. Iynedjian, P.-R. Pilot, T. Nouspikel, J. L. Milburn, C. Quaade, S. Hughes, C. Ucla, C. B. Newgard, *Proc. Natl Acad. Sci.* **1989**, 1989:7838–7842.

26 Y. Liang, H. Najafi, R. M. Smith, E. C. Zimmerman, M. A. Magnuson, M. Tal, F. M. Matschinsky, *Diabetes* **1992**, 41:792–806.

27 A. Cornish-Bowden, M. L. Cárdenas, Glucokinase: a monomeric enzyme with positive cooperativity. In: F. M. Matschinsky, M. A. Magnuson (Eds.), *Glucokinase and Glycemic Disease, from Basics to Novel Therapeutics, Vol. 16*, S. Karger, Basel, **2004**.

28 F. M. Matschinsky, *Diabetes* **1996**, 45:223–241.

29 M.-H. Giroix, A. Sener, W. J. Malaisse, *Mol. Cell. Endocrinol.* **1987**, 49:219–225.

30 N. Sekine, V. Cirulli, R. Regazzi, L. J. Brown, E. Gine, J. Tamarit-Rodriguez, M. Girotti, S. Marie, M. J. MacDonald,

C. B. Wollheim, G. A. Rutter, *J. Biol. Chem.* **1994**, 269:4895–4902.

31 C. B. Newgard, F. M. Matschinsky, Substrate control of insulin release. In: L. S. Jefferson, A. D. Cherrington (Eds.), *Handbook of Physiology, Sec. 7, Vol. II*, Oxford University Press, USA, **2001**.

32 T. V. Korsgaard, M. Colding-Jørgensen, Time-dependent mechanisms in beta cell glucose sensing, *J. Biol. Phys.* **2006**, 32:289–306.

33 G. Toffolo, E. Breda, M. K. Cavaghan, D. A. Ehrmann, K. S. Polonsky, C. Cobelli, *Am. J. Physiol. Endocrinol. Metab.* **2001**, 280:E2–E10.

34 E. Breda, M. K. Cavaghan, G. Toffolo, K. S. Polonsky, C. Cobelli, *Diabetes* **2001**, 50:150–158.

35 E. Breda, G. Toffolo, K. S. Polonsky, C. Cobelli, *Diabetes* **2002**, 51[Suppl. 1]:S227–S233.

36 A. E. Panteleon, M. Loutseiko, G. M. Steil, K. Rebrin, *Diabetes* **2006**, 55:1995–2000.

37 M. C. Moore, C. C. Connolly, A. D. Cherrington, *Eur. J. Endocrinol.* **1998**, 138:240–248.

38 M. Roden, K. F. Petersen, G. I. Shulman, *Recent Prog. Horm. Res.* **2001**, 56:219–238.

39 B. Thorens, The hepatoportal glucose sensor. In: F. M. Matschinsky, M. A. Magnuson (Eds.), *Glucokinase and Glycemic Disease, from Basics to Novel Therapeutics*, Vol. 16, S. Karger, Basel, **2004**.

40 R. Burcelin, A. Da Casta, D. Drucker, B. Thorens, *Diabetes* **2001**, 50:1720–1728.

41 B. Balkan, X. Li, *Am. J. Physiol. Reg. Integ. Comp. Physiol.* **2000**, 279:R1449–R1454.

42 F. Preitner, M. Ibberson, I. Franklin, C. Binnert, M. Pende, A. Gjinovci, T. Hansotia, D. J. Drucker, C. Wollheim, R. Burcelin, B. Thorens, *J. Clin. Invest.* **2004**, 113:635–645.

43 B. Thorens, *Horm. Metab. Res.* **2004**, 36:766–770.

44 R. Burcelin, V. Crivelli, C. Perrin, A. Da Costa, J. Mu, C. R. Kahn, P. Vollenweider, B. Thorens, *J. Clin. Invest.* **2003**, 111:1555–1562.

45 R. Burcelin, W. Dolci, B. Thorens, *Diabetes* **2000**, 49:1635–1642.

46 A. Gardemann, H. Strulik, K. Jungermann, *FEBS* **1986**, 202:255–259.

47 M. C. Moore, P.-S. Hsieh, D. W. Neal, A. D. Cherrington, *Am. J. Physiol. Endocrinol. Metab.* **2000**, 279:E1271–E1277.

48 D. Smith, A. Pernet, H. Reid, J. M. Rosenthal, I. A. Macdonald, A. M. Umpleby, S. A. Amiel, *Diabetologia* **2002**, 45:1416–1424.

49 T. J. Kieffer, J. F. Habener, *Endocrinol. Rev.* **1999**, 20(6):876–913.

50 C. Fernandez-Mejia, M. S. German, Regulation of Glucokinase by Vitamins and Hormones in *Glucokinase and Glycemic Disease, from Basics to Novel Therapeutics*, Vol. 16, F. M. Matschinsky, M. A. Magnuson (Eds.), S. Karger, Basel, **2004**.

51 B. E. Levin, V. H. Routh, L. Kang, N. M. Sanders, A. A. Dunn-Meynell, *Diabetes* **2004**, 53:2521–2528.

52 S. C. Woods, P. J. Kulkosky, *Psychosom. Med.* **1976**, 38:201–219.

53 R. L. Jungas, M. L. Halperin, J. T. Brosnan, *Physiol. Rev.* **1992**, 72(2):419–448.

54 M. C. Gannon, J. A. Nuttall, G. Damberg, V. Gupta, F. Q. Nuttall, *J. Clin. Endorinol. Metab.* **2001**, 86(3):1040–1047.

55 J. Levraut, C. Ichai, I. Petit, J.-P. Ciebiera, O. Perus, D. Grimaud, *Crit. Care Med.* **2003**, 31(3):705–710.

56 B. Phypers, J. M. Tom Pierce, *Cont. Educ. Anaest. Crit. Care Pain* **2006**, 6(3):128–132.

57 G. A. Brooks, *Biochem. Soc. Trans.* **2002**, 30(2):258–264.

58 J. R. Guyton, R. O. Foster, J. S. Soeldner, M. H. Tan, C. B. Kahn, L. Koncz, R. E. Gleason, *Diabetes* **1978**, 27:1027–1042.

59 M. Mueckler, *Eur. J. Biochem.* **1994**, 219:713–725.

60 A. Handberg, *Glucose transporters and insulin receptors in skeletal muscle: Physiology and Pathophysiology*, Danish Medical Bulletin, Denmark, **1995**.

61 A. Klip, A. Marette, Regulation of glucose transporters by insulin and exercise: cellular effects and implications for diabetes. In: L. S. Jefferson, A. D. Cherrington (Eds.), *Handbook of Physiology, Sec. 7, Vol. II*, Oxford University Press, USA, **2001**.

62 J. M. Berg, J. L. Tymoczko, L. Stryer, *Biochemistry*, W. H. Freeman and Company, USA, **2002**.

63 K. J. Acheson, Y. Schutz, T. Bessard, K. Anantharaman, J.-P. Flatt, E. Jéquier, *Am. J. Clin. Nutr.* **1988**, 48:240–247.

64 W. F. Ganong, *Review of Medical Physiology*, Lange Medical Publications, Canada, **1979**.

65 K. Ide, I. K. Schmalbruch, B. Quistorff, A. Horn, N. H. Secher, *J. Physiol.* **2000**, 522.1:159–164.

66 R. A. DeFronzo, R. Gunnarsson, O. Björkman, M. Olsson, J. Wahren, *J. Clin. Invest.* **1985**, 76:149–155.

67 K. F. Petersen, T. Price, G. W. Cline, D. L. Rothman, G. I. Shulman, *Am. J. Physiol.* **1996**, 1(Pt 1):E186–E191.

68 R. N. Hansen, *Glucose homeostasis – a biosimulation approach*, PhD thesis, Copenhagen, **2004**.

69 A. L. McCall, L. B. Fixman, N. Fleming, K. Tornheim, W. Chick, N. B. Ruderman, *Am. Physiol. Soc.* **1986**, 86:E442–E447.

70 F. Rencurel, G. Waeber, B. Antoine, F. Rocchiccioli, P. Maulard, J. Girard, A. Leturque, *Biochem. J.* **1996**, 314:903–909.

71 C. Postic, J.-F. Decaux, J. Girard, Regulation of hepatic glucokinase gene expression. In: F. M. Matschinsky, M. A. Magnuson (Eds.), *Glucokinase and Glycemic Disease, from Basics to Novel Therapeutics*, Vol. 16, S. Karger, Basel, **2004**.

72 H. F. Bowen, J. A. Moorhouse, *J. Clin. Invest.* **1973**, 52:3033–3045.

73 S. Wise, M. Nielsen, R. Rizza, *J. Clin. Endocrinol. Metab.* **1997**, 82(6):1828–1833.

74 K. N. Frayn, *Metabolic Regulation – A Human Perspective*, Blackwell Science Ltd., Oxford, **2003**.

75 A. Philp, A. L. Macdonald P. W. Watt, *J. Exp. Biol.* **2005**, 208:4561–4575.

76 J. Wahren, P. Felig, G. Ahlborg, L. Jorfeldt, *J. Clin. Invest.* **1971**, 50:1971–2725.

77 M. Kjær, K. Engfred, A. Fernandes, N. H. Secher, H. Galbo, *Am. Physiol. Soc.* **1993**, 93:E275–E283.

78 G. A. Brooks, *J. Physiol.* **2002**, 541.2:333–334.

79 M. Ellmerer, L. Schaupp, G. Sendlhofer, A. Wutte, G. A. Brunner, Z. Trajanoski, F. Skrabal, P. Wach, T. R. Pieber, *J. Clin. Endocrinol. Metab.* **1998**, 83 (12):4394–4401.

80 R. Taylor, G. I. Shulman, Regulation of hepatic glucose uptake. In: L. S. Jefferson, A. D. Cherrington (Eds.), *Handbook of Physiology, Sec. 7, Vol. II*, Oxford University Press, USA, **2001**.

81 A. D. Cherrington, Control of glucose production in vivo by insulin and glucagon. In: L. S. Jefferson, A. D. Cherrington (Eds.), *Handbook of Physiology, Sec. 7, Vol. II*, Oxford University Press, USA, **2001**.

82 A. Zorzano, C. Fandos, M. Palacín, *Biochem. J.* **2000**, 349:667–688.

83 F. Tremblay, M.-J. Dubois, A. Marette, *Front. Biosci.* **2003**, 8:1072–1084.

84 A. C. F. Coster, R. Govers, D. E. James, *Traffic* **2004**, 5:763–771.

85 J. S. Elmendorf, *J. Membr. Biol.* **2002**, 190:167–174.

86 B. Teusink, J. A. Diderich, H. A. Westerhoff, K. van Dam, M. C. Walsh, *J. Bacteriol.* **1998**, 180:556–562.

87 K. V. Kandror, *Sci. STKE* **2003**, 169:1–3.

88 M. Funaki, P. Randhawa, P. A. Janmey, *Mol. Cell. Biol.* **2004**, 24:7567–7577.

89 E. Bailhache, K. Ouguerram, C. Gayet, M. Krempf, B. Siliart, T. Magot, P. Nguyen, *J. Anim. Physiol. Anim. Nutr.* **2003**, 87:86–95.

90 M. Soop, J. Nygren, K. Brismar, A. Thorell, O. Ljungqvist, *Clin. Sci.* **2000**, 98:367–374.

91 M. Brownlee, *Diabetes* **2005**, 54:1615–1625.

92 C. Bouché, S. Serdy, C. R. Kahn, A. B. Goldfine, *Endocrine Rev.* **2004**, 25:807–830.

93 B. F. Hansen, S. Asp, B. Kiens, E. A. Richter, *Scandinav. J. Med. Sci. Sports* **1999**, 9(4):209–213.

94 A. W. Thorburn, B. Gumbiner, F. Bulacan, P. Wallace, R. R. Henry, *J. Clin. Invest.* **1990**, 85:522–529.

95 G. I. Shulman, *J. Clin. Invest.* **2000**, 106:171–176.

96 V. Aas, E. T. Kase, R. Solberg, J. Jensen, A. C. Rustan, *Diabetologia* **2004**, 47:1452–1461.

97 S. W. Coppack, R. D. Evans, R. M. Fisher, K. N. Frayn, G. F. Gibbons, S. M. Humphreys, M. L. Kirk, J. L. Potts, T. D. R. Hockaday, *Metabolism* **1992**, 41 (3):264–272.

98 B. A. Fielding, K. N. Frayn, *Br. J. Nutr.* **1998**, 80:495–502.

99 M. K. Hellerstein, *Eur. J. Clin. Nutr.* **1999**, 53[Suppl. 1]:S53–S65.

100 R. M. McDevitt, S. J. Bott, M. Harding, W. A. Coward, L. J. Bluck, A. M. Prentice, *Am. J. Clin. Nutr.* **2001**, 74:737–746.

101 Y. Schutz, *Int. J. Obesity* **2004**, 28:S3–S11.

102 K. N. Frayn, *J. Appl. Physiol.* **1983**, 55(2):628–634.

103 L. C. Hudgins, M. Hellerstein, C. Seidman, R. Neese, J. Diakun, J. Hirsch, *J. Clin. Invest.* **1996**, 97(9):2081–2091.

104 L. C. Hudgins, *P. S. E. B. M.* **2000**, 225:178–183.

105 I. Marques-Lopes, D. Ansorena, I. Astiasaran, L. Forga, J. A. Martinéz, *Am. J. Clin. Nutr.* **2001**, 73:253–261.

106 A. Strawford, F. Antelo, M. Christiansen, M. K. Hellerstein, *Am. J. Physiol. Endocrinol. Metab.* **2004**, 286:577–588.

107 A. G. Dullo, M. Gubler, J. P. Montani, J. Seydoux, G. Solinas, *Int. J. Obesity* **2004**, 28:S29–S37.

108 A. Dresner, D. Laurent, M. Marcucci, M. E. Griffin, S. Dufour, G. W. Cline, L. A. Slezak, D. K. Andersen, R. S. Hundal, D. L. Rothman, K. F. Petersen, G. I. Shulman, *J. Clin. Invest.* **1999**, 103:253–259.

109 G. Boden, *Endocrine Pract.* **2001**, 7(1):44–51.

110 G. M. Reaven, C. Hollenbeck, C.-Y. Jeng, M. S. Wu, Y.-D. I. Chen, *Diabetes* **1988**, 37:1020–1024.

111 I. Magnusson, D. L. Rothman, L. D. Katz, R. G. Shulman, G. I. Shulman, *J. Clin. Invest.* **1992**, 90:1323–1327.

112 A. Mitrakou, *Diabetes Obesity Metab.* **2002**, 4:249–254.

113 I. Toft, T. Jenssen, *Scand. J. Lab. Invest.* **2005**, 65:307–320.

114 C. Weyer, *Diabetes* **1999**, 48:1607–1614.

115 M. Camilleri, J.-R. Malagelada, M. L. Brown, G. Becker, A. R. Zinsmeister, *Am. J. Physiol.* **1985**, 249:G580–G585.

116 C. Tosetti, A. Paternicò, V. Stanghellini, G. Barbara, C. Corbelli, M. Marengo, M. Levorato, N. Monetti, R. Corinaldes, *Nuc. Med. Com.* **1998**, 19:581–586.

117 J. N. Hunt, M. T. Knox, *Am. J. Dig. Dis.* **1968**, 13:372–375.

118 M. B. Sidery, I . A. MacDonald, P. E. Blachshaw, *Gut* **1994**, 35:186–190.

119 J. N. Hunt, D. F. Stubbs, *J. Physiol.* **1975**, 245:209–225.

120 S. Efendic, N. Portwood, *Horm. Metab. Res.* **2004**, 36:742–746.

121 B. Ahrén, J. J. Holst, *Diabetes* **2001**, 50:1030–1038.

122 R. N. Bergman, Y. Z. Ider, C. H. Bowden, C. Cobelli, *Am. J. Physiol.* **1979**, 236:E667–E677.

123 A. Mari, *Curr. Opin. Nutr. Metab. Care* **2002**, 5:495–501.

124 B. Ahrén, G. Pacini, *Eur. J. Endocrinol.* **2004**, 150:97–104.

125 E. Ferrannini, A. Mari, *Diabetologia* **2004**, 47:943–956.

126 G. M. Steil, B. Clark, S. Kanderian, K. Rebrin, *Diabetes Tech. Ther.* **2005**, 7(1):94–108.

127 R. N. Bergman, R. Prager, A. Volund, J. M. Olefsky, *J. Clin. Invest.* **1987**, 79:790–800.

128 K. Tokuyama, Y. Higaki, J. Fujitani, A. Kiyonaga, H. Tanaka, M. Shindo, M. Fukushima, Y. Nakai, H. Imura, I. Nagata, A. Taniguchi, *Am. Physiol. Soc.* **1993**, 28:E298–E303.

129 L. J. C. Bluck, A. T. Clapperton, C. V. Kidney, W. A. Coward, *Clin. Sci.* **2004**, 106:645–652.

130 R. A. DeFronzo, J. D. Tobin, R. Andres, *Am. J. Physiol.* **1979**, 237:E214–E223.

131 J. C. Woodworth, D. C. Howey, R. R. Bowsher, R. L. Brunelle, *Diabetes Tech. Ther.* **2004**, 6(2):147–153.

132 C. Meyer, P. Saar, N. Soydan, M. Eckhard, R. G. Bretzel, J. Gerich, T. Linn, *J. Clin. Endocrinol. Metab.* **2005**, 90(11):6244–6250.

7

Biological Rhythms in Mental Disorders

Hans A. Braun, Svetlana Postnova, Bastian Wollweber, Horst Schneider, Marcus Belke, Karlheinz Voigt, Harald Murck, Ulrich Hemmeter, and Martin T. Huber

Abstract

Mental disorders are complex disturbances of the brain which lead to distinct emotional and cognitive alterations. These are manifested in an abnormal behavior of the patients as well as in biological markers such as nightly sleep interruptions and disturbed cortisol secretion, which suggests that the neuronal malfunctioning simultaneously affects different systems at different organizational levels and time-scales. A comparison of the functional characteristic of the different systems and their critical control parameters therefore may provide a promising link towards a better understanding of the origin of this multi-level and multi-scale disease. Here we compare actual computational approaches which have been developed to simulate: (1) the time-course of affective disorders, (2) the control of cortisol secretion and (3) the alteration of neuronal dynamics. With this approach we are able to demonstrate that the models' structures show similarities in principle and that their intrinsic rhythms can be related to the interaction of nonlinear positive and negative feedback loops which are operating at different activation levels and time-scales. Each control parameter which modifies these interactions can essentially change the system's dynamics and consequently may lead to inappropriate responses according to a disease state. Further approaches to elucidate the disease relevant interlinks between the different systems are discussed.

7.1
Introduction: Mental Disorders as Multi-scale and Multiple-system Diseases

The critical pathology of mental disorders concerns the emotional state of the person. The most frequently occurring type is "major depression" with a prevalence of about 10% in men and 20% in woman. The typical symptoms are: (1) depressed mood, (2) difficulties in concentrating, (3) loss of energy and interest and, as the most dangerous aspect, (4) thoughts and (not so rarely) commitment of suicide. Apart from these unipolar affective disorders there is a second main type with

Biosimulation in Drug Development. Edited by Martin Bertau, Erik Mosekilde, and Hans V. Westerhoff
Copyright © 2008 WILEY-VCH Verlag GmbH & Co. KGaA, Weinheim
ISBN: 978-3-527-31699-1

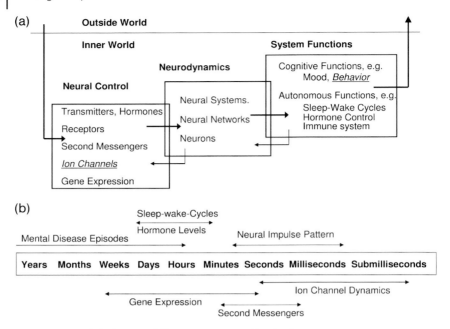

Fig. 7.1 (a) Interlinks between different organizational levels of biological functions – from the subcellular, molecular level to the cellular and network level, up to systemic functions. (b) The major time scales of biological dynamics and rhythms related to mental disorders.

still about 1% prevalence in both sexes. These are the so-called bipolar disorders where depressed states alternate with manic episodes. Mania is characterized by: (1) elevated mood, (2) grandiose ideas, (3) increased energy and expansiveness. The clinical data are summarized in detail by Kaplan et al. [1].

Mental disorders, also called affective disorders, are multi-level, multi-scale and multiple-system diseases (Fig. 7.1). Mental disturbances generally go along with disturbances of autonomous functions. These essentially are: (1) sleep disturbances, both sleep duration and sleep pattern, and (2) disturbances of the hypothalamic–pituitary–adrenal (HPA) axis, the so-called stress axis with elevated cortisol levels. It can be expected that disturbances of autonomous control systems as well as mood are caused by neuronal malfunctioning which may concern practically all neuronal levels: systemic interactions, neuronal network connections, single neuron dynamics, synaptic transmitters and/or receptors, ion channels, second messengers, and gene expression (Fig. 7.1a). Nevertheless, despite a manifold of data, there are only vague ideas so far about the differences in neuronal dynamics in the brain of a chronically depressed person compared with a person with a sensitive but balanced mood.

Most hypotheses about the cause of mental disorders assume a disturbance of neurotransmitter systems, which is supported by medications which enhance synaptic efficiency, for example, with monoamine oxidase inhibitors (MAOIs), tricyclic antidepressants (TCAs) or selective serotonin reuptake inhibitors (SSRIs). These drugs increase the availability (time and concentration) of the corresponding transmitters. However, the success of such treatments does not necessarily mean that the cause of the disturbance is repaired. Moreover, when reduced transmitter availability is the cause of the disease, re-uptake inhibition should have an immediate effect within hours. The drug effects, however, only occur with a long delay, mostly weeks, which means that secondary effects are involved. These long-lasting effects are assumed to be mediated via gene expression, e.g. for additional membrane receptors or synaptic sprouting, which should further increase the efficiency of synaptic transmitters. However, how these or other effects reinstall regular brain dynamics is completely unclear as it is also still unclear what brain dynamics determine whether a person is in a stable or vulnerable mental state.

A conceptual approach towards a better understanding of the transitions from stable to vulnerable states in the time course of manic-depressive disorders has recently been presented on the basis of computer modeling studies [2–7]. The model and its major outcomes will be reviewed in the first "Results" section. Although it is a rather formal, nonhomologous model of disease patterns without any concrete relations to neural dynamics, it can provide valuable insights into the principle system behavior with regard to vulnerability, kindling and autonomous progression as the most detrimental characteristics of mood disorders.

For further progress towards mechanisms based models, such phenomenological descriptions shall also be examined in context with disease-related disturbances of autonomous functions. This mainly concerns disturbances of sleep–wake cycles and cortisol release which are the most reliable biological markers of mental diseases, especially major depression, and can provide objective and quantifiable parameters (e.g. EEG frequency components, cortisol blood level) for the estimation of an otherwise mainly subjective and only behaviorally manifested illness. Moreover, there is a manifold of data which interlink the alterations of the autonomous system parameters (sleep states, cortisol release) with alterations of neural dynamics. Therefore, the most promising approach also to understand the interrelations between neural dynamics and affective disorders probably goes via the analysis of mood related disturbances of autonomous functions.

The autonomous functions exhibit circadian rhythms which are under the control of a neuronal pacemaker, located in the suprachiasmatic nuclei (SCN) of the hypothalamus. The circadian pacemaker arises from complex dynamics of gene expression and is synchronized to the external light. In the case of mental disorders, the regular rhythms of autonomous functions are obviously disturbed which led to the formulation of the "desynchronization hypotheses" [8, 9]. This does not necessarily contradict the "transmitter hypothesis". Transmitter imbalance, of course, also interferes with the inherent system dynamics and can change the endogenous rhythmicity, eventually with the result of desynchronization. A first computational approach which simulates the rhythmicity of the HPA axis and its alterations with scaling of transmitter mediated positive and negative feedback loops is briefly summarized in the second "Results" section.

The third "Results" section is referring to the neuronal rhythms considering single neurons as well as neural network dynamics. The transitions between different types of more or less rhythmic impulse patterns play an important role in many physiological and pathophysiological processes which especially holds true for the control of sleep–wake cycles and hormone secretion. Specifically, there are transitions from tonic, single-spike activity to rhythmic discharges of impulse groups, so-called "bursts" which are relevant for enhanced hormone secretion [10, 11] as well as for the transitions from wake to sleep states [12].

Moreover, tonic-bursting bifurcations can essentially facilitate neuronal synchronization [13–16] as is clearly the case in sleep states because the occurrence of EEG waves can only be expected when large brain areas exhibit rhythmically coordinated discharges. Complete synchronization, however, also means reduced sensitivity to environmental signals. This may be desired when a person is falling asleep but can have negative consequences in other situations, e.g. when fully synchronized discharges of larger brain areas lead to epileptic seizures. An inappropriate, i.e. less sensitive environmental responsiveness is the leading feature also of affective disorders and likewise may prevent autonomous systems to appropriately react on external signals. The decisive question therefore is how the neuronal dynamics can be kept in a stable but nevertheless sensitive state which allows them to successfully adapt their responsiveness to varying tasks and how this is disturbed in mental disorders. We assume that this is not only a question of the neuron's connectivity, e.g. synaptic weights or gap-junction coupling strengths, but is also a question of the neurons' internal dynamics and activity patterns.

These interactions are considered in our simulation studies which refer to physiological functions of different organizational levels (Fig. 7.1a) and also spread along quite different time scales (Fig. 7.1b). The major challenge is to combine them in a comprehensive approach which can explain stable and vulnerable moods as well as regular and disturbed autonomous functions on the basis of neuronal processes.

7.2
The Time Course of Recurrent Mood Disorders: Periodic, Noisy and Chaotic Disease Patterns

Most theories about the origin of mood disorders refer to the subcellular or molecular level, mainly emphasizing an imbalance between different transmitter systems [17] as a possible cause for unipolar depression as well as bipolar disorders. No concrete concepts exist to explain how and in which way such molecular disturbances modify the neural dynamics and lead to an increased mental vulnerability with progressive occurrence of disease episodes.

Additional information about the underlying dynamics comes from clinical observations which indicate that the occurrence of disease episodes in the course of disease progression follows general rules. Despite a large variability of individual time courses, there are typical patterns of disease episodes at different disease states (Fig. 7.2a). In the course of disease progression the patterns change from iso-

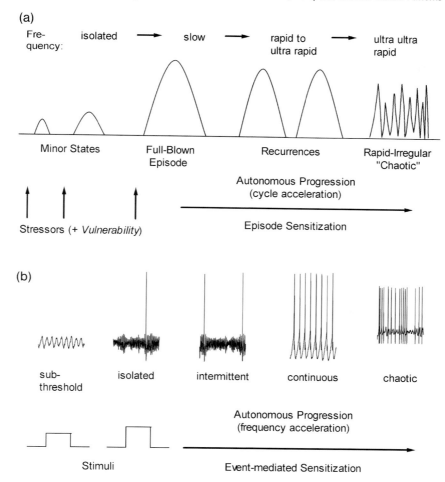

Fig. 7.2 (a) The prototypical time course of disease episodes in affective disorders (modified after [18]). Disease episodes are initially related to external stressors with increasing strength of the episodes (kindling) which then leads to autonomous progression (episode sensitization) with periodically occurring disease episode of increasing frequency up to ultra-rapid, chaotic mood fluctuations. (b) Analogous activity pattern of a neuronal model of subthreshold impulse generation [97, 98]. Initially it needs a sufficiently large stimulus for the generation of action potentials. When the neuron is sensitized with an additional current input the action potentials can spontaneously develop, exhibiting intermittent than continuous and finally chaotic patterns depending on the current amplitude. Figure modified after [2, 3, 6].

lated episodes at the beginning to intermittent and more rapid, rhythmically occurring episodes and finally end in ultrafast "chaotic" mood oscillations [18]. The first episodes can often be associated with psychosocial stressors while later episodes have an increasing tendency to occur autonomously.

These clinical observations led to the introduction of the so-called "kindling model" for the progression of mood disorders [18–21]. The "kindling model", which originally was developed in context with epileptic seizures, is a nonhomologous biological model which says that identical stimuli lead to continuously increasing stimulus responses with continuously decreasing stimulus thresholds until previously stimulus-related responses occur spontaneously. Moreover, as the probability for the occurrence of a disease episode increases with the number of previous episodes, the concept of episode sensitization was established, which assumes that each episode leaves behind a residue, i.e. implements a memory trace, as the result of so far unknown neuroplastic changes which progressively increase the vulnerability of neurobiological functions.

7.2.1
Transition Between Different Episode Patterns: The Conceptual Approach

Our conceptual approach for computer simulations is built on clinical observations about the time course of affective disorders and the characteristic changes of disease patterns (Fig. 7.2a). It is a nonhomologous modeling approach which first of all intends to understand the principal dynamics which may be able to generate such typical patterns. Such insights should then allow us to derive further approaches for more specific, goal-directed examinations. This is similar to the analysis of neuronal impulse patterns for the evaluation of membrane mechanisms when they are not directly measurable, as in peripheral sensory skin receptors where detailed pattern analysis led to a comprehensive oscillation theory [22–24]. Our "mood model" uses such an analogy to oscillating spike generation (Fig. 7.2b) and we also use biophysically based approaches of nonlinear data analysis which were essentially developed in context with neural dynamics.

The indications which justify such an approach are directly provided by the episode patterns. As described above, in specific parts of disease progression the disease episodes occur periodically, while they develop rather randomly before, with no exact bifurcation point in between. Accordingly, we do not expect the mechanism of episode generation to change completely when it goes from random to periodic disease episodes. We assume that this reflects disease-related alteration of the system dynamics and we expect that this also holds true for the transitions to chaotic disease patterns.

To account for such transitions we assume a process of subthreshold oscillations which trigger the disease episodes and which continuously increase in amplitude and frequency during disease progression. However, to explain the randomly occurring episodes at the beginning of the disease, we additionally have to include noise. Noise can reflect any kind of environmental influences on a person's mood. These can be positive and negative experiences. In vulnerable persons, strong psychosocial stressors may introduce disease episodes even when the internal oscillation are too small for intrinsic episode generation. To understand the transition from periodic to chaotic patterns no additional assumptions are needed. Chaos can be expected as an almost necessary consequence from the interactions between stronger and faster oscillation and more frequently triggered disease events. From a system-theoretical point of view chaos can be seen as the result of resonances between two coupled systems which operate at different levels and time scales.

7.2.2
A Computer Model of Disease Patterns in Affective Disorders

To simulate the disease episodes and subthreshold oscillations we again refer to our neuronal modeling approaches (Fig. 7.2b). The algorithms have been implemented with a simple but physiologically plausible approach, i.e. with two nonlinear feedback loops, one positive and one negative. Depending on the parameter setting, such a system can attain stable dynamics but also can develop oscillations. In our case, the episode generator is comparably stable, far away from spontaneous episode generation. The subthreshold mechanisms are "normally" also in a steady state but can be tuned with a disease variable into oscillatory limit cycles with increasing frequency and amplitude. Anyhow, the structure of the equations is the same for the two episodes and the two subthreshold variables and the similarities to the neuronal models immediately become evident from the equations. For a more detailed description see Huber's work [2–4, 7, 25].

The mood variable x is given by the sum of the feedback inputs F_i:

$$\tau_x \, dx/dt = -x - \sum F_i + \mathbf{gw} + S \tag{1}$$

S is the disease variable and $\mathbf{gw}$ accounts for Gaussian white noise.

The strength of the feedback and also whether it is positive or negative depends on the reference value x_i, or, more precisely, on its distance to the actual disease state x. Scaling parameters are given by the values w_i with corresponding activation variables a_i.

$$F_i = a_i w_i (x - x_i) \tag{2}$$

The feedback variables a_i go with first order time delays τ_i towards their final values $a_{i\infty}$. The variables of the subthreshold oscillations activate much slower than those for episode generation. In both subsystems, the positive feedback is faster than the negative feedback:

$$da_i/dt = (a_{i\infty} - a_i)/\tau_i \tag{3}$$

Nonlinearities are implemented with sigmoidal activation of $a_{i\infty}$ depending on the disease variable x with a half activation value x_{ih}:

$$a_{i\infty} = 1/\{1 + \exp[-s_i(x - x_{ih})]\} \tag{4}$$

With appropriate parameter scaling (see [2, 3]) the system is in a "healthy" steady state with $S = 0$. However, with increasing S the system goes through different dynamic states, including all the clinically described disease pattern. This is illus-

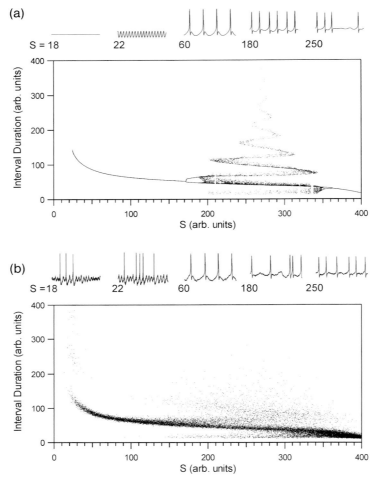

Fig. 7.3 Deterministic (a) and noisy (b) computer simulations of the time course of affective disorders showing the intervals between successive disease episodes (interval duration) as a function of a disease variable S and examples of episode generation from different disease states (figure modified after [2]). In deterministic simulations (a), there is a progression from steady state (S = 18) to subthreshold oscillations (S = 22) with immediate onset of periodic event generation at a certain value of S (slightly below S = 60). With further increase of S, the intervals between successive episodes are continuously shortened. At a critical value (at around S = 170), a period-doubling bifurcation (S = 180) and then a transition to chaos (S = 250) appears. At very high values of S, there is again periodic event generation (S = 250) of very high frequencies. Noisy simulations (b) show significant differences. In the deterministically steady state, addition of noise can induce subthreshold oscillations and even event generation (coherence resonance across two levels). The second major noise effect is a considerable extension of the range of irregular, chaotic dynamics.

trated by the bifurcation diagrams in Fig. 7.3 with examples of disease patterns from different disease states (in the traces above the bifurcation diagrams). The bifurcation diagrams show the intervals between successive disease episodes. Of course, no real person can have such a lot of episodes, as plotted here simply to make the relevant patterns easier to recognize.

The upper and lower diagrams in Fig. 7.3 compare the results of deterministic and noisy simulations, respectively. In deterministic simulations (Fig. 7.3a) the disease variable x goes from a steady state ($S = 18$) to subthreshold oscillations ($S = 22$) and than abruptly, at around $S = 24$, starts to generate regular rhythmic disease episodes (where the first dots in the bifurcation plot appear). With further increasing S the frequency of disease episodes continuously increases over a wide range. However, at around $S = 170$ the system is running into a homoclinic bifurcation with subsequent transitions to clearly chaotic dynamics before it again turns into periodic, high frequent episode generation.

It is evident that only the noisy simulations (Fig. 7.3b) can account for the sporadically occurring episodes which are typical for the beginning of the disease ($S = 18$; $S = 22$). Moreover, application of noise considerably expands the range of irregular patterns. Indeed, several more significant noise effects have been observed in such simulations which can be of relevance for the disease progression and eventually also for its treatments [3, 25, 26].

7.2.3
Computer Simulations of Episode Sensitization with Autonomous Disease Progression

While the above simulations describe how the disease pattern vary as a function of the disease state, the following simulations show that our model can also account for kindling phenomena and autonomous progression. This needs some model extensions which were made in reference to the above-mentioned assumption of episode sensitization which are assumed to be due to residues (memory traces) of previous disease episodes. For simplicity, and because the real mechanisms are unknown, we introduced an additional, positive feedback loop which is implemented exactly in the same way as the other feedback variables [4–7, 25]. The model now also includes a dynamic disease variable S_F (Fig. 7.4a). The specialities are that it only activates when a disease episode occurs (episode sensitization) and that it has long relaxation times (memory trace).

It depends on the parameter adjustments, i.e. on the actual vulnerability, whether the system runs into autonomous progression. This will never happen in a "healthy" steady state. Even a strong external stimulus can induce only one and only a transient "depressive" episode. The situation drastically changes when the system is tuned to a vulnerable state. Also, when the original dynamics are still in a steady state, the initiation of one disease episode can be sufficient to induce successive episodes with autonomous progression (Fig. 7.4b).

In these situations, again, cooperative effects of noise and nonlinear dynamics can play an essential role [7, 25] and even may allow to discriminate between

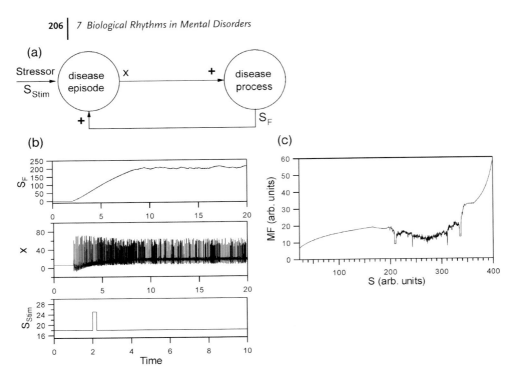

Fig. 7.4 (a) Implementation of a positive feedback loop (dynamic disease variable S_F) to account for sensitization and autonomous progression. (b) Autonomous progression is initiated with a single episode generating stimulus S_{Stim} (lowest trace). Occurrence of events can be recognized in the middle plot (disease variable x). The upper plot shows the increase of the sensitization variable S_F which stabilizes when the range of chaotic episode pattern is reached. (c) Stabilization in chaos can be explained by a reduction of the mean frequency of events (MF) at the transition from periodic to chaotic dynamics (at around $S = 250$) which also decreases the effects of the sensitization variable S_F. (Modified after [4, 25]).

healthy and vulnerable steady states [27]. However, chaotic dynamics also come prominently into play – specifically in the final states of disease progression. Most remarkably, in the course of progression the model does not run into complete depression but into chaotic dynamics – which exactly corresponds to the clinical observations. To understand this eventually counterintuitive behavior it is necessary to consider the implications of chaotic dynamics for the frequency of events. In contrast to the regular periodic situation with continuously increasing frequency, the transition to chaotic dynamics is associated with unpredictable alterations of episode intervals of short but also long duration. As a consequence, the frequency of episodes can decrease (Fig. 7.4c) and with this also the effect of the positive feedback loop. The system would have to go to extraordinarily high disease values for further progression which, however, will not be reached due to the inversion of the frequency curve with the occurrence of chaotic dynamics.

Such inversions may play an important role also for the stabilization of autonomous functions. Moreover, in neural transduction processes an inversion of stimulus sensitivity, i.e. frequency maximum curves, have repeatedly been seen in experimental recordings not only at the transitions to chaos but even more frequently as the result of noise effects [23, 24, 28], to which we will come back in Section 6.4, which examines neural dynamics.

7.3
Mood Related Disturbances of Circadian Rhythms: Sleep–Wake Cycles and HPA Axis

Autonomous functions, especially the circadian rhythms of sleep–wake cycles and cortisol release, are significantly disturbed during depression. However, it is still unclear whether these circadian alterations are reliably linked with psychopathology and whether they provide clues to the underlying mechanism, in particular with respect to the neurotransmitter models of depression [8] and the CRH-overdrive hypothesis [29].

The significant meshing of the neuronal control areas for sleep and hormone release and their connections to mood relevant brain areas suggest that functional interdependencies also exist and that these also become evident in the system disturbances. The block diagram in Fig. 7.5 mainly emphasizes on the parts which are of particular relevance for those autonomous parameters which are the most clearly accessible markers of mental disorders: the increased blood cortisol level and changes of the sleep EEG pattern.

7.3.1
The HPA Axis and its Disturbances

Increased cortisol levels belong among the most significant biological markers of depressive disorders. This was known more than 30 years ago, when increased cortisol secretion (predominantly in the afternoon and early morning hours) was described [30].

The secretion of cortisol is controlled by the HPA axis. As cortisol is often called the "stress hormone", so the HPA axis is also known as the "stress axis". The major control pathways are schematically illustrated on the left part of Fig. 7.5. The paraventricular nucleus (PVN) of the hypothalamus secretes the corticotropin releasing hormone (CRH) into the pituitary portal circulation. In the pituitary gland, CRH stimulates the secretion of the adrenocorticotropic hormone (ACTH) into the blood. In the adrenal cortex, ACTH leads to cortisol synthesis and release. Cortisol is the final product of the HPA system and acts on numerous target tissues. In contrast to this comparably simple-looking feedforward pathway there are several feedback loops of cortisol not only to the pituitary and the PVN but also to the limbic system and to the cortex which, in turn have projections back to the PVN. These can be negative as well as positive feedback loops, depending on the target cells, the type of cortisol receptors, and the prevailing homeostatic vulnerability of the involved systems.

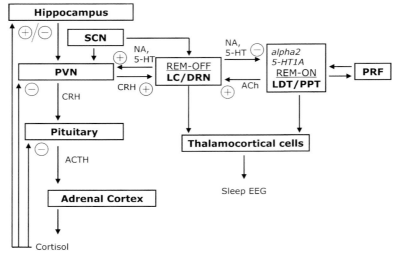

Fig. 7.5 Interrelations between functional structures of the brain and their transmitter and hormone systems which are involved in the control of cortisol secretion (HPA axis, left part) and sleep patterns (right part). Abbreviations of the diverse nuclei: SCN = suprachiasmatic nucleus, PVN = paraventricular nucleus, LC = locus coeruleus, DRN = dorsal raphe nuclei, LDT = laterodorsal tegmental nucleus, PPT = pedunculopontine tegmental nucleus, PRF = pontine reticular formation. and thalamocortical cells. Abbreviations of hormones and neurotransmitters: CRH= corticotropin releasing hormone, ACTH = adrenocorticotrope hormone, NA= noradrenaline, 5-HT = serotonin, Ach = acetylcholine.

Hypercortisolism in depression is the peripheral indicator of a general disturbance of the entire stress regulation. A primary reason for the enhanced cortisol release is obviously a central hypersecretion of CRH at the hypothalymic site [29, 31, 32]. The consequence of CRH hypersecretion is not only an increased cortisol level but also a "down"-regulation of glucocorticoid and mineralocorticoid receptors at the level of the pituitary and the hippocampus [33], i.e. a disturbance of the feedback system.

The central control areas of the HPA axis are intimately connected to mood-relevant brain areas (hippocampus, prefrontal cortex, etc.) and also to the neuronal systems monitoring circadian rhythms and sleep patterns. There is the common input from the circadian pacemaker in the suprachiasmatic nucleus (SCN), and there is a manifold of direct neuronal connections between the different nuclei (Fig. 7.5) as well as direct hormone effects on the sleep architecture, e.g. the early morning cortisol peaks as the trigger for awakening. Furthermore, drug-induced alterations in neurotransmitter systems which have significant mood effects generally also interfere with both sleep and cortisol release, which additionally underlines the strong interrelations.

7.3.2
Sleep EEG in Depression

Sleep disturbances are best characterized by the EEG. Most clear deviations in the EEG of depressed patients compared with control EEGs can be seen during the first part of the night. Depressed patients have reduced slow wave sleep (SWS) during the first sleep cycle, which is associated with significant alterations of rapid eye movement (REM) sleep. The REM latency is shortened and the amount of REM sleep is increased which, accordingly, increases the ratio of REM to nonREM sleep [34–36]. The most sensitive parameter for the discrimination of patients with major depression from patients with other psychiatric disorders and healthy subjects is the density of rapid eye movements during REM sleep (index of the number of eye movements during REM sleep), which is substantially increased only in depressed patients, independent of their age [35, 37].

Clinical improvement of depression is frequently accompanied by a normalization of sleep continuity disturbance. Difficulties in falling asleep, maintaining sleep, and problems with early morning awakening are going down [38], while disturbance of sleep architecture (reduced SWS and REM latency) may persist even in successfully treated patients [39, 40].

Some parameters of sleep architecture and REM sleep seem to be related to the response of the patients to therapeutic treatments and the long-term course of the disease [41–44]. Predominantly, the increase of delta ratio in the initial treatment phase seems to discriminate treatment responder from nonresponder [45]. Recent data from Hatzinger et al. [46] show that patients with a still increased REM density and reduced SWS after a successful antidepressant therapy are more likely to develop recurrence than patients who do not show these features.

It is not yet clear whether persistent alterations of sleep architecture like reduced SWS and REM latency and increased REM density reflect an increased vulnerability for the development of depression or a neurobiological scar acquired during the course of the disease [39, 46–48]. Nevertheless, it is evident that depression and sleep EEG alterations are closely linked and that the EEG can provide information about the intensity of the disease (state aspect) and about the vulnerability for developing further episodes, i.e. the stability of the long-term course (trait-like aspect).

7.3.3
Neurotransmitters and Hormones Controlling Sleep Pattern and Mood

An explanation for sleep EEG alterations in depression is provided by the well established monaminergic–cholinergic imbalance model [49] which includes monoaminergic dysfunction in depression. The result is a disturbance of the mutual interactions between the diverse nuclei which control the transitions between REM and nonREM states (see Fig. 7.5, right part). This concept was developed in close relation to the monamionergic–cholinergic imbalance model of depression [50] which is based on observations that inhibition of the aminergic system by reserpin as well as stimulation of the cholinergic system by cholinomimetics can induce depressive states. Accordingly, anticholinergic drugs such as scopolamine may exert antidepressant effects. This also is consistent with elevated mood which can be achieved by compounds which, like antidepressants, increase aminergic neurotransmission. Their main effect on sleep EEG parameters is a substantial suppression of REM sleep, resulting in a prolonged REM latency.

The measurement of hormonal secretion during night sleep and the parallel assessment of sleep EEG provide substantial evidence for a mutual relationship between sleep EEG pattern and nocturnal hormonal secretion. In this context, CRH plays a major role but other hormones and neuropeptides are also involved, e.g. growth hormone (GH), which is released from the pituitary under the control of the hypothalamic growth hormone-releasing hormone (GHRH).

GH secretion appears in a single peak during the first SWS episode [51]. In contrast, cortisol remains on a basal level during the first half of the night and increases in the second half of the night until awakening [51]. Experimental studies show that the application of cortisol increases SWS [52, 53], while the application of CRH decreases SWS and reduces sleep efficiency [54, 55]. Opposite effects have been observed with the application of GH, which reduces SWS [56], and GHRH which increases SWS and improves sleep continuity [57].

These findings are brought together in the extended two-process model of sleep regulation [54, 58] relating GHRH to sleep patterns and CRH to the circadian component. While GHRH may be active at the beginning of the night, inducing sleep, CRH exerts its activity in the second half of the night and thereby not only induces cortisol increase but also reduces SWS and facilitates awakening. These effects are strengthened under chronically increased cortisol levels and this is exactly what is seen in depression.

7.3.4
A Nonlinear Feedback Model of the HPA Axis with Circadian Cortisol Peaks

Several models of sleep regulation and the HPA axis already exist. The sleep models are rather formal descriptions, mostly based on the so-called two-process model [59–63] and do not provide an insight into the systems dynamics. Also the HPA axis models, sometimes biophysically very detailed [64–66], do not consider so far the highly nonlinear dynamic interactions of different positive and negative feedback loops, which seem to be the most vulnerable parts and therefore of particular importance in the context of depression. Our modeling studies of the autonomous functions are still at the very beginning. In the following we mainly describe a new conceptual approach for HPA axis simulation, which is specifically designed for the evaluation of the nonlinear system's dynamics with positive and negative feedback loops. The structure of the model allows one not only to systematically extend the model by sleep and mood components but also to endow the different parts with model components of realistic neural activity. The principle design strictly follows the above-described interdependencies between the different functional entities.

The hypothalamus releases CRH depending on the circadian pacemaker C under feedback control (F_H) from cortisol:

$$CRH = C \times F_H \tag{5}$$

The pituitary releases ACTH depending on CRH (following Michaelis–Menten kinetics). It also is under feedback control (F_P) from cortisol and degrades with a time constant k_{ACTH}:

$$d(ACTH)/dt = \frac{I_{max,ACTH} \times CRH}{(K_{M,ACTH} + CRH) \times F_P} - k_{ACTH} \times ACTH \tag{6}$$

The adrenal cortex releases Cortisol (CORT) depending on ACTH (also following Michaelis–Menten kinetics) and degrades with a time constant k_{CORT}:

$$d(CORT)/dt = \frac{I_{max,CORT} \times CORT}{K_{M,CORT} + CORT} - k_{CORT} \times CORT \tag{7}$$

The circadian process C is implemented by a simple cosine function with period t of 1 day (1 day = 1440 min), a phase shift ($t$phase) and an arbitrary amplitude (amp) and base line shift (base).

$$C = \cos[(t - t\text{phase}) \times 2\pi/\text{day}] \times \text{amp} + \text{base} \tag{8}$$

The feedback mechanisms are mediated via mineralocorticoid and glucocorticoid receptors (MR, GR), which means overlapping of nonlinear positive and negative feedback loops with nonlinear dependencies on cortisol (Michaelis–Menten kinetics, sigmoidal activation curves of corticoid receptors).

Feedback to the pituitary is modeled following Michaelis–Menten kinetics for an allosteric inhibited reaction which gives:

$$F_P = 1/(1 + CORT/k_{INH}) \tag{9}$$

Feedback to the hypothalamus and other brain areas goes via glucocorticoid receptors (GR) and mineralocorticoid receptors (MR) with inhibiting and stimulating effects, respectively on the CRH release. This constitutes a negative and a positive feedback loop with sigmoidal activation curves (Fig. 7.6) which are determined by the maximum values E_{max}, , the slope of activation S, and the half activation value EC50:

$$GR = E_{max,GR}/[1 + \exp(-S_{GR}/(CORT - EC50_{GR})] \tag{10}$$

$$MR = E_{max,MR}/[1 + \exp(-S_{MR}/(CORT - EC50_{MR})] \tag{11}$$

The GR receptors have low affinity (high EC50$_{GR}$) and high capacity (high $E_{max,GR}$). The MR receptors have high affinity (low EC50$_{MR}$) and low capacity (low $E_{max,MR}$).

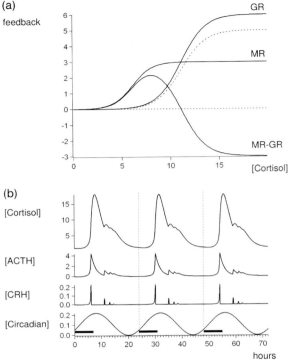

Fig. 7.6 Computer simulations of circadian cortisol release. (a) Activation curves of the positive (MR) and negative (GR) feedback loops of cortisol to diverse brain areas (see Fig. 6.5) via mineralocorticoid receptors (MR) and glucocorticoid receptors (GR), respectively. In combination (MR-GR), there is a transition from positive to negative feedback depending on the cortisol concentration. (b) Simulations of the time course of cortisol, ACTH (adrenocorticotrope hormone), CRH (corticotropin releasing hormone) in relation to a circadian pacemaker (from top to bottom). The bars indicate sleep states.

The full feedback to the hypothalamus and other brain areas above the pituitary is given by:

$$F_H = MR - GR \tag{12}$$

According to the different affinity and capacity of MR and GR (Fig. 7.6a) this leads to positive feedback at low cortisol concentrations which turns to negative feedback at high cortisol concentrations. These are, in combination with the multiplicative effects of the pacemaker, the relevant determinants for the system dynamics. As illustrated in Fig. 7.6b, the cortisol levels drastically increase in the morning, due to the strong pacemaker drive, followed by a plateau phase with some small peaks still in the afternoon. The pulsatile control of cortisol release is particularly well recognized in the CRH levels. These peaks arise when the negative feedback loop is transiently shut off by the decreasing cortisol levels and the circadian pacemaker can still exert its stimulating power on the positive feedback loop which is not the

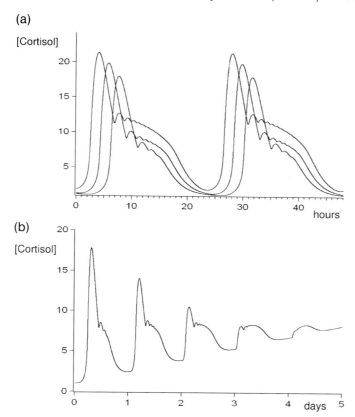

Fig. 7.7 (a) Alterations of cortisol secretion (superimposed recordings) when the negative feedback is impaired corresponding to reduced glucocorticoid receptor sensitivity (as assumed in the case of depression): the cortisol secretion is increased, lasts longer and occurs earlier. (b) Alteration of cortisol secretion due to increased cortisol blood concentrations (as a marker of depression). At higher cortisol level also negative feedback loop is constantly activated and finally dominate over the dynamics of the positive feedback loop. The result is a reduction of the cortisol peaks.

case at the lower pacemaker values during late afternoon, in the evening and during the first part of the night.

Although this model still does not represent all relevant dynamics of cortisol secretion it can already account for some specific features of mental disease states. When the negative feedback is impaired, as expected in depressive disorders, the cortisol peak increases, the cortisol level remains for a longer time on increased values and the onset of the morning peak occurs earlier (Fig. 7.7a). This corresponds quite well to the clinical observations of higher cortisol levels as major markers of depressive disorders. Also the early morning awakening of depressive patients might be associated with earlier cortisol peaks.

A different type of dynamics can be obtained with continuously increased cortisol levels which lead to attenuation of the cortisol peaks (Fig. 7.7b). This occurs because the increasing steady-state cortisol levels activate the negative feedback loop until it diminishes the dynamic sensitivity of the positive feedback loop. In pathophysiological situations of mental depression, the enhanced cortisol secretion may simultaneously reduce the efficiency of the feedback loops due to a reduction of receptor sensitivity. Hence, a combination of the effects from reduced sensitivity (Fig. 7.7a) and continuously enhanced cortisol secretion (Fig. 7.7b) may be still closer to the physiological reality. Such more complex interactions are currently being analysed. A major advantage of this approach is that it allows one to tune a diversity of disease-relevant parameters which have a mechanistic, physiological or pathophysiological correlation.

7.4
Neuronal Rhythms: Oscillations and Synchronization

In mental disorders, in contrast to other neural diseases like epilepsy or Parkinson's disease, not much is known about the relevant alteration of neural dynamics. It can be just assumed from the patient's inappropriate environmental responsiveness that neuronal adaptability and flexibility are disturbed, accompanied by a changed neuronal sensitivity on external stimuli. Fortunately, there is more information with regard to the neural control of sleep–wake cycles and cortisol release, which are very reliable biological markers of mental disorders and therefore may provide a link for a better understanding also of the neuronal dynamics which control mood.

The most significant alterations of neural activity in the control of autonomous functions are transitions between tonic-firing single-spike and grouped impulse discharges, so-called bursts. Transitions from tonic-firing to burst discharges have been seen in the neurosecretory hypothalamic nuclei [13, 67–70] where such transitions drastically increase hormone release [10, 11]. They can also be observed in the thalamocortical neurons where these neuronal transitions between different impulse patterns are correlated with states of vigilance and transitions from wake to sleep [12, 71–73].

Tonic-to-bursting transitions can facilitate synchronization in neuronal networks and between diverse nuclei [13–16, 74–77]. Synchronization obviously can spread across large brain areas, as indicated by the occurrence of slow potential waves in the sleep EEG, which can only be expected when a very large number of neurons rhythmically discharge in symphony.

Transitions to burst discharges with synchronization of large brain areas seem to be accompanied with a loss of sensitivity. Reduced sensitivity to environmental stimuli, of course, is not a problem but even required at the obviously well coordinated transition to sleep. However, it can become a problem when this happens while the neurons and network should be in a sensitive state for appropriate reactions on environmental information. Transitions to bursts and associated synchronization of complete nuclei can tune the system into an oscillatory and very stable internal state which makes it practically insensitive to external stimuli, e.g. during epileptic seizures, and overwhelms intentional motor control, e.g. in Parkinson's tremor [78, 79].

In physiological situations, synchronization should only be partial and transient, as it is described, for example, as a possible neural correlate of the "binding" of information from simultaneous stimulus inputs [80–82]. The neuronal systems seem to operate in a balanced state which allows specific neurons to synchronize under specific stimulus conditions but also give them the chance to desynchronize, to go together with other partners when the situation changes. This is a sensitive but, of course, also a very vulnerable state which can easily be disturbed when the neuronal dynamics and network interactions change. Disturbances may occur due to transmitter imbalance, changed neural connectivity, i.e. synaptic weights or gap-junction coupling strengths, and even from changed ionic compositions as well as from the ionic processes which determine the individual neuron's internal state and activity patterns. Morover, there are significant interdependencies. Synaptic transmitters, for example, mainly change ionic conductances and therefore also the individual neuron's internal states which, in turn, determine the firing pattern, i.e. the synaptic input to the follower neurons. Indeed, alterations of neuron connectivity are rather expected for the longer-term run and might occur, for example, during disease progression or, the other way round, during disease treatment. Actual information processing of external stimuli, also when synaptically mediated, has to be fast and probably goes first of all via the alteration of ionic conductances.

For a better understanding of the neuronal basis of flexible and adaptable brain dynamics and their reduced responsiveness in disease states, it is necessary not only to consider the disease-related alterations of neuronal connectivity but also the impact of ionic conductances on neuronal and network sensitivity. To bridge the gap between ionic membrane processes, neuronal impulse patterns and network dynamics, our model neurons are of the Hodgkin–Huxley type. For systematic analysis of the network dynamics we first chose electrotonic nearest neighbor gap-junction coupling. Our models generally also include a "noise" term which is not only for more realistically looking simulations but also because of the well known cooperative effects between noise and nonlinear dynamics which can have significant implications on neuronal impulse pattern and stimulus encoding [23, 75, 83–91]. We also consider the neurons' biological variability which, again, is not only because it is physiologically more realistic but also because heterogeneity could be an important feature for the neuronal network's flexibility [92–94].

7.4.1
The Model Neuron: Structure and Equations

All simulations described below are done with a Hodgkin–Huxley type single compartment model neuron [22, 95–98]. Compared with the original equations, this model is, on the one hand, simplified and, on the other hand, extended. The main simplification is the description of ion channel opening and closing with sigmoidal current–voltage relations which gives us an easier control of the dynamics and brings the model even closer to the experimental situations where mostly

I-V curves rather than ion channel kinetics are recorded. These simplifications do not significantly change the original action potential dynamics. We are anyhow not so much interested in the shape of the action potentials but in their temporal patterns. This is why the extensions were made which bring in the extraordinarily rich dynamics of action potential generation and allow one to reproduce all of the types of impulse patterns which, to our knowledge, have experimentally been recorded so far. This is achieved with two additional currents which activate on slower time scales but below the threshold of the spike generation (subthreshold currents). These currents may reflect an often described persistent sodium current and the sum of diverse slowly activating potassium (K) currents. The slow K currents are either implemented with direct or indirect voltage-dependencies, the latter reflecting Ca^{++}-dependent K^+ currents, which again are realized in a simplified form [96, 98]. Whether one or the other type of K^+ current is used depends both on the neurons which are actually under examination and on what the experimental data suggest in this respect [24, 99, 100]. The dynamics are not principally different. Here we only show data which are obtained with the Ca^{++}-dependent K^+ current.

The membrane equation is given by:

$$C_M dV/dt = -g_l(V - V_l) - I_{Na} - I_K - I_{Na,p} - I_{K,Ca}$$

$$-I_{noise} - I_{syn} - I_c \tag{13}$$

where V is the membrane voltage and C_M the membrane capacitance. Apart from the leakage current, given by $g_l (V - V_l)$, there are the fast Na^+ and K^+ currents for spike generation (I_{Na} and I_K) and the two slow currents for subthreshold oscillations, the persistent Na^+ current ($I_{Na,p}$) and the Ca^{++}-dependent K^+ current ($I_{K,Ca}$). I_{noise} accounts for stochastic fluctuations, e.g. from ion channel openings and closings or synaptic input. I_{syn} is for external current application to simulate a tonic synaptic input. I_c is for specific, nearest neighbor gap-junction coupling.

The voltage-dependent currents are calculated by the following equations (i = Na, K, Na,p, K,Ca):

$$I_i = g_i a_i(V - V_i) \tag{14}$$

$$a_{i\infty} = 1/\{1 + \exp[-s_i(V - V_{0i})]\} \tag{15}$$

$$da_i/dt = (a_{i\infty} - a_i)/\tau_i \tag{16}$$

with V_i the equilibrium (Nernst) potentials, g_i the conductances. V_{0i} and s_i are half-activation potentials and slopes of the steady-state activation variable $a_{i\infty}$ with τ_i as a voltage independent time constant for calculation of the actual value a_i.

Exceptions are: (a) instantaneous activation of the fast depolarizing current and (b) direct coupling of the slow repolarizing current to the slow depolarizing current according to:

(a) $a_{\text{Na}} = a_{\text{Na}\infty}$

(b) $da_{\text{K,Ca}}/dt = (-\eta I_{\text{Na,p}} - k a_{\text{K,Ca}})/\tau_{\text{K,Ca}}$ (17)

with η as a coupling constant and k as a relaxation factor.

Gaussian white noise is implemented according to the Box–Mueller algorithm [101]. I_c is the coupling current in network simulations which accounts for electrotonic gap-junction connections between the neurons. For bidirectional coupling of two neurons the coupling current I_c are of the form [94]:

$$I_c(1) = g_c(V_1 - V_2) \quad \text{and} \quad I_c(2) = g_c(V_2 - V_1)$$ (18)

For nearest neighbor coupling in bigger networks of, for example, ten by ten neurons the coupling current $I_c(i,j)$ of a neuron at position (i,j) is the sum of input currents from the neighboring neurons:

$$I_{c,i,j} = \sum_{n,m=-1}^{1} g_c(V_{ij} - V_{i+n,j+m}) \quad (i, j = 1, \ldots, 10)$$ (19)

7.4.2
Single Neuron Impulse Patterns and Tonic-to-Bursting Transitions

With appropriate adjustment of the membrane parameter values, the single neuron model can already be tuned to different dynamic states with a great diversity of impulse patterns [22, 96]. The map in Fig. 7.8 (left lower part) shows an example of different types of patterns which are passed with scaling of one internal and one external model parameter: the maximum conductance of the Ca^{++}-dependent K$^+$ current $g_{\text{K,Ca}}$ and the current input I_{syn}, respectively. Examples of voltage traces are given on the right. A bifurcation diagram of interspike intervals at $I_{\text{syn}} = 0$ with scaling of $g_{\text{K,Ca}}$ (according to the arrow in the map) is shown in the upper trace. Similar activity patterns as illustrated here can be obtained with tuning of other parameters as well.

The map exhibits a broad range of constant membrane potentials (steady state) when $g_{\text{K,Ca}}$ and/or I_{syn} is high, and another broad range of pacemaker-like tonic firing when $g_{\text{K,Ca}}$ and/or I_{syn} is low [Fig. 7.8, small panel (a)]. In between, there are subthreshold oscillations which generate bursts [Fig. 7.8, small panel (c)], whereby the number of spikes per burst can considerably change. A second type of tonic firing activity appears when the oscillations generate only one spike per oscillation cycle [Fig. 7.8, small panel (d)]. Towards the border to the steady state regime, a range of subthreshold oscillations without spike generation can be seen [Fig. 7.8, small panel (e)].

In deterministic simulations, these subthreshold oscillations may cover only a narrow regime. However, with the addition of noise a broad regime of functionally most interesting patterns can develop exactly in and around the area of subthreshold oscillations. Noise can introduce a random mixture of subthreshold oscillations

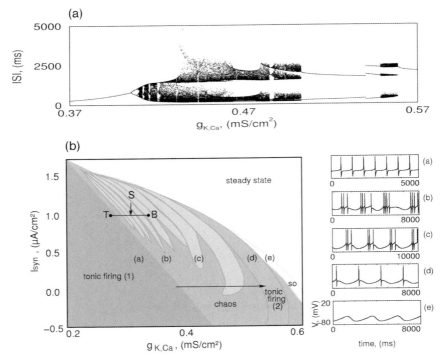

Fig. 7.8 (a) Bifurcation diagram of interspike intervals (ISI) generated by the neuronal model with subthreshold oscillation with scaling of the maximum conductance of the slow repolarizing Ca^{++}-dependent K$^+$ current-current $g_{Ca,K}$. (b) The map on the left shows a rough estimation of different activity patterns which the model can generate with tuning of the Ca^{++}-dependent K$^+$ current $g_{Ca,K}$ and a synaptic current I_{Syn}. Positive values are hyperpolarizing and negative values depolarizing currents. The activity patterns are additionally illustrated by the voltage traces on the right. The letters indicate the corresponding position in the map. The map shows big areas of steady state and tonic firing (1) activity in the right upper and left lower corner, respectively. In between there are burst discharges (numbers indicate the numbers of spikes per burst) and a tonic firing (2) activity range as well as a range of subthreshold oscillations (SO) at the border to the steady state. Moreover, there are large areas of chaotic dynamics especially in the lower part of the map, i.e. towards the range of depolarizing currents. T and B indicate tonic and bursting states which are used for the simulation in Fig. 6.10 with coupling of heterogeneous neurons which, in the completely synchronized state, meet at the position of the arrow (S). The voltage traces on the right illustrate that tonic firing (1) arises from pacemaker-like depolarization (a) while tonic firing (2) is generated by sub-threshold oscillations (d) which also can be seen without spike generation (e). The burst discharges are riding on membrane potential oscillations (b) which, depending on their duration, amplitude and base line, can generate bursts with a highly variable number of spikes. The numbers in the map indicate the number of spikes per burst for regimes up to five spikes. The chaotic impulse patterns (c) mostly look like irregular burst discharges. Even the slow membrane potential oscillations still can be seen. (Modified after [94]).

with and without spike generation whereby slight alterations of the oscillations amplitude, frequency or base line can lead to drastic changes of the spiking parameters, especially the spiking probability. The remarkable tuning properties of such dynamics, which obviously play an important role for information processing in diverse cortical areas [102–109] and for sensory transduction in peripheral neurons [23, 24, 28, 84, 95, 97, 98, 110], have experimentally and theoretically been demonstrated. In this paper, however, we focus on other types of impulse patterns of high physiological and pathophysiological relevance which occur at the bifurcations from pacemaker-like tonic firing to oscillatory burst discharges. As described above, these are the functionally relevant transitions in thalamic neurons which are associated with the control of sleep–wake cycles and in hypothalamic neurons which control hormone secretion.

It can easily be seen in the map of Fig. 7.8 that the oscillating, bursting regimes are embedded in a large area of chaotic dynamics which is particularly broad at lower values of I_{syn}. A broad range of chaotic dynamics can also be recognized in the bifurcation diagram of interspike intervals in the upper graph. The chaotic impulse sequences mostly look like irregular bursts as shown in the voltage traces in [Fig. 7.8, small panel (b)] and which is also indicated by the two clearly separated bands of short and long intervals in the bifurcation plot. These chaotic bursts are often hardly to distinguish from noisy bursts, especially in experimental recordings which are notoriously contaminated by noise. The question is whether the differences in the deterministic dynamics (chaotic or limit cycle dynamics) may have further implications, e.g. on the neurons' sensitivity or synchronization properties.

7.4.3
Network Synchronization in Tonic, Chaotic and Bursting Regimes

The implications of individual neuron dynamics on neuronal network synchronization is evident. In Fig. 7.9 (from Schneider et al., unpublished data) this is demonstrated with network simulations (10×10 neurons) of nearest neighbor gap-junction coupling. It is illustrated in quite a simple form which, in a similar way, can also be experimentally used: with the local mean field potential (LFP). In the simulations LFP simply is the mean potential value of all neurons. In the nonsynchronized state LFP shows tiny, random fluctuations. In the completely in-phase synchronized states the spikes should peak out to their full height.

When tonically firing neurons (Fig. 7.9a) are coupled it needs comparably high coupling strengths before the neurons begin to synchronize, as indicated by continuously increasing amplitudes of the LFP. In the chaotic regime (Fig. 7.9b) indications of synchronization, i.e. LFP oscillations, appear much earlier. However, beyond a certain coupling strength they do not further increase but rather become more irregular again. The network, even transiently, can switch back to an almost unsynchronized state. These switches can occur at unpredictable times and with unpredictable duration. When bursting neurons are coupled (Fig. 7.9c) synchronization seems initially not to increase significantly more than in chaotic situations. However, at certain, intermediate coupling strengths, where chaotic synchro-

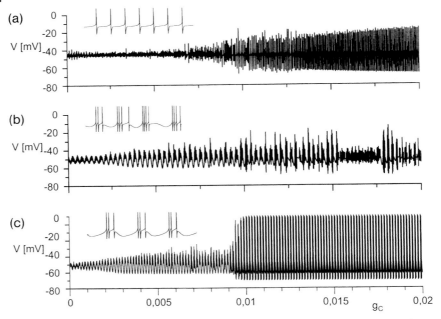

Fig. 7.9 Computer simulations of a neuronal network (10×10 neurons) with nearest neighbor gap-junction coupling (see equations). The diagrams show the local mean field potentials (LFP) which are calculated as the mean potential values V of all neurons during continuously increasing coupling strength g_C. Insets illustrate the different types of patterns in which the neurons originally operate (with randomly set initial values). (a) The network of tonic firing neurons needs a certain coupling strength until it develops continuously increasing LFP amplitudes. (b) Coupling of chaotic neurons leads to earlier occurring LFP waves which, however, can transiently disappear at certain, unpredictable coupling strengths. (c) Bursting neurons can very suddenly go from intermediate to complete in-phase synchronization as indicated by the sudden increase of the LFP amplitude to its maximum value.

nization becomes irregular, the bursting neurons abruptly go into full in-phase synchronization. Again, the exact coupling strength where this transition occurs is not exactly predictable and depends, in deterministic simulations, on the initial conditions. Altogether, there are significant differences in network synchronization depending on the dynamic states in which the individual neurons originally operate. This can easily be seen even when the impulse pattern looks quite similar, as in a comparison of regular and chaotic bursts. When the neurons are coupled, the differences are disclosed.

7.4.4
Synchronization Between Neurons at Different Dynamic States

For a more detailed analysis of the neuronal dynamics in the course of synchronization we reduced the network model to only two bidirectionally gap-junction

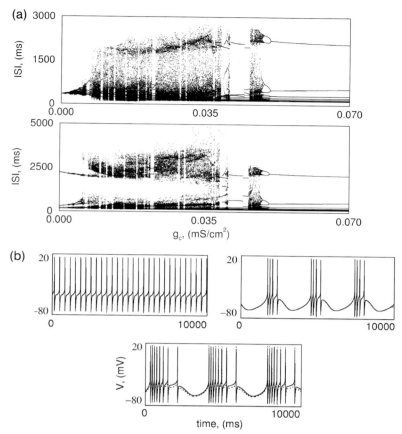

Fig. 7.10 Computer simulation of two gap-junction coupled neurons which originally are operating at different dynamic states, one in the tonic firing regime and the other one in the bursting regime (also indicated in Fig. 6.8b by the points T and B, respectively with the arrow S pointing on the completely synchronized state). (a) Bifurcation diagrams of interspike intervals (ISI) of the originally

tonic firing (upper trace) and originally bursting neuron (lower trace) as a function of continuously increasing coupling strength g_c. Both neurons go through a broad range of chaotic dynamics until they synchronize in a bursting regime. (b) Voltage traces of the uncoupled neurons (upper diagrams) and from a state of almost complete in-phase synchronization (lower trace).

coupled neurons. Moreover, we go one step further also to analyze the effects of gap-junction coupling when the neurons are originally in different dynamic states [94]. Such conditions can lead to very complex dynamics before the neurons finally, with sufficiently high coupling strengths, go into complete synchronization.

The bifurcation diagrams in Fig. 7.10a show the alteration of impulse patterns with increasing coupling strengths when one neuron is originally in the pacemaker-like tonic firing regime (upper diagram in Fig. 7.10a) and the other one is in the bursting regime of period 4 (lower diagram in Fig. 7.10a). The original, uncoupled voltage traces are shown in Fig. 7.10b (upper diagrams). The positions are also indicated in the map of Fig. 7.8, where T indicates the tonic firing and B the bursting dynamics. Complete synchronization ends in a bursting regime of period 8 (lower voltage trace in Fig. 7.10b). This is necessarily in the middle of the bifurcation diagram that goes from one neuron's starting point to the other. The position is indicated by the arrow S in the map of Fig. 7.8. Hence, in such a very idealized situation, it can principally be predicted which pattern the fully synchronized state will attain [94].

The clearly more complex dynamics appear at intermediate coupling strengths. At least one or, as in this example, both neurons go through chaotic dynamics over a broad range of coupling strengths. The patterns which thereby are passed are not necessarily the activity patterns found on the connection line between the two starting points. In the simulation shown here, there is only a narrow band of chaos between the starting points (see map of Fig. 7.8). Obviously, the neurons leave this line during the course of synchronization and only come back to it in the fully synchronized state. Further studies are required for a better understanding of these complex dynamics, especially to see whether the chaotic dynamics are a necessary consequence of the heterogeneity of the neurons and what the implications are for the neuronal network sensitivity.

7.5
Summary and Conclusions: The Fractal Dimensions of Function

We have shown computer models of biological rhythms in mood disorders and related systems which have been observed at different organizational levels at quite different time scales:

1. There are the very variable rhythms in the occurrence of disease episodes with considerable changes of disease intervals during the course of disease progression: from years to months to weeks and even to days and hours. Such global behavioral rhythms are manifestations of the mental disease while the mood of mentally stable, healthy persons rather randomly fluctuates mainly following the environmental influences with mostly stochastic up and downs.

2. We have emphasized the fact that the progression of mental diseases, especially major depression, is associated with significant disturbances in the circadian rhythms of autonomous functions, especially sleep–wake cycles and hormone release. These rhythms are naturally occurring but are distorted in affective disorders. The most clear distortions are often seen in the ultradian components in specific phases of the circadian cycle, e.g. in the morning peak of cortisol release or in the REM to nonREM transition during the sleep state.

3. We have considered that all these systems are under neuronal control. While the relation between neural activity and different mood states is unknown so far, it is well established that physiologically relevant transitions in autonomous functions like sleep–wake cycles and cortisol release are associated with signifi-

cant alterations of the neuronal rhythms and impulse patterns. This specifically concerns the tonic to bursting bifurcations in neuronal discharges which can drastically increase the secretion of neurohormones and also are relevant for the transition from wake to sleep.

We have shown computer simulations from all these different levels: First, we modeled the time course of affective disorders and showed that clinical observation can be mimicked in remarkable details with a combination of oscillatory dynamics and noise. Second, we presented an initial, still very basic, model of circadian cortisol release which nevertheless provided new insights also into eventual disease relevant alterations. Third, we showed single neuron and neuronal network simulations to elucidate the relevant interdependencies between ionic conductances and network interactions with regard to neuronal synchronization at different dynamic, also heterogeneous states.

In a comparison of the different simulation approaches it can easily be recognized that the model structures show principle similarities. Each model is composed of two potentially oscillating subsystems. One represents subthreshold mechanisms and the other one event-generating processes which, depending on the different systems, give rise to action potentials, cortisol peaks or mental disease episodes, respectively. Only in the model of the HPA axis is the subthreshold process not a dynamically oscillating system but so far explicitly implemented by simple sine wave equations. The other two subthreshold mechanisms have an identical structure which is the same also for all event-generating parts. In general, each subsystem consists of positive and negative feedback loops and each feedback loop includes nonlinearities and time delays. These subsystems can easily be tuned from steady state to oscillations of variable forms and frequencies with appropriate parameter scaling.

The complete system responses are determined by the interactions, i.e. resonances, between the subthreshold and event-generating mechanisms. Again, in the HPA axis model, these interactions, so far, are rather simple and unidirectional. The circadian pacemaker modulates the HPA feedback loops in a nonlinear multiplicative way but is itself not influenced by the dynamics of the cortisol releasing processes. In contrast, in the neuronal and psychiatric models, the two subsystems are interlinked by a common control variable, the membrane voltage and the disease variable, respectively. These interlinks are a major source of very complex, inclusively chaotic dynamics [96, 111, 112].

As a general rule, which also holds true for the HPA axis model, the subthreshold mechanisms determine the timing of events, but as soon as the event-generating mechanisms are activated they overtake the control. As long as the two subsystems are weakly coupled, i.e. have weakly overlapping activation ranges, this does not significantly disturb the subthreshold mechanisms. One of the reasons is that the events are relatively short compared with the time scale of the subthreshold oscillations. The more complex dynamics occur in parameter ranges where the events become significantly stronger and/or appear with significantly increased frequency, which allows them to essentially interfere with the subthreshold oscillations. Even-

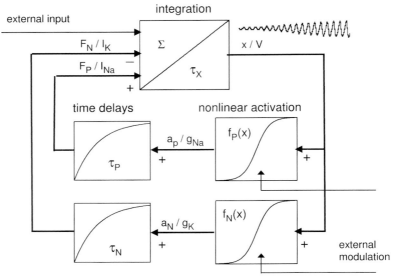

Fig. 7.11 The main common features of the different modeling approaches for computer simulations of mental disorders, hormone release and neural discharges. Despite significant differences in details and several additional components in the specific models, the principle dynamics originate from a combination of positive and negative feedback loops. The necessary ingredients are: (1) nonlinear activation functions $f(x)$ and (2) different time delays τ which (3) should be shorter for the positive feedback loop. The system's output x can go from steady state to oscillatory activity of different amplitude and frequency depending on the system parameters which also can be externally modulated. The output x may reflect a mood variable as well as cortisol levels or the neuronal membrane voltage V. As an example, in case of neurons the activation variables a_P and a_N correspond to the steady state ionic conductances g_{Na} and g_K and the feedback variables F_P and F_N are the ionic sodium and potassium currents I_{Na} and I_K. Please note that the full systems always consist of two of such subsystems which, apart from the actual HPA axis model, are coupled via the common output variable x. The complexity of the full system's dynamics depends on the overlapping of the subsystems activation range and time scales. Such principle approaches can continuously be extended, e.g. with the coupling of single neurons to neural networks. Moreover, the output of a lower lever system may constitute a part of the feedback loops in higher level systems which indicates a fractal dimension of physiological functions. The possible structure of these fractals is summarized in this block diagram.

tually, the event-generating mechanisms completely overtake the system's control. In neuronal discharges, this seems to be the case when the transition from periodic to chaotic bursts finally leads to tonic firing, for which the subthreshold oscillations are still required but no longer have the ability to determine the discharge rhythms. They even can no longer be recognized in the voltage traces [96]. Principally the same rules govern the transition to chaos and subsequent periodic events in the model of affective disorders (Fig. 7.11).

Hence, the "competition" between the two subsystems of slow oscillations and fast event generation seems to play a major role for physiologically and pathophysiologically relevant transitions. The more successful the subthreshold mechanisms are with the generation of more frequent and more stronger events, the higher is the probability that they themselves lose control of the timing of events. Chaotic patterns seem to occur in an intermediate range where neither the subthreshold nor the event-generating mechanisms have full control [96].

A new dimension is opened when such already complex systems again are coupled, which we have done with the neuronal model. The situation may appear not very complicated as long as identical neurons are electrotonically connected. Nevertheless, significant differences in the synchronization pattern can be seen not only depending on the coupling strength but even more on the individual neuron's dynamics (Fig. 7.9). Therefore, it is of particular interest to see the network effects when neurons of different activity patterns are connected [94]. Remarkably, even when only two neurons are connected and when their original pattern is regular, the coupled system exhibits a clearly chaotic pattern over a very broad range of coupling strengths (Fig. 7.10). It seems that it is again the competition between different time scales which almost necessarily leads to chaotic patterns.

Altogether, we have shown computer simulations of rhythmic, inclusively chaotic activity patterns from different systems which are operating at different organizational levels and time scales. All these systems are somehow involved in the control of mood and its disturbances. Therefore, the crucial point is to elucidate the relevant interlinks between the different functional levels, which only can be done with implementation of the neuronal dynamics approaches in simulations of higher order functions. This brings up a principal methodological problem. Whenever lower level dynamics are endowed in all details into models of higher orders, the full system dynamics soon become very complicated – not easy to overlook, to analyze and to understand. Therefore, it is of particular importance to understand which properties of the lower level systems have to be considered at the higher levels [113]. It is therefore of particular interest that comparison of the different models has elucidated significant similarities at all organizational levels, i.e. a fractal dimension of function (Fig. 7.11). Such principal similarities allow us to successfully use a neuron-based approach for simulating the time course of affective disorders and our model of the HPA axis is going in the same direction. These functional fractals let us expect that the development of more detailed and mechanism-based models of hormone secretion, sleep–wake cycles and mental disorders can go in parallel with physiologically justified dimension reduction. Dimension reduction which preserves the functionally relevant details appears the only promising approach for the elucidation of mood relevant control parameters at the neuronal as well as at the hormonal level, while also being accessible for pharmaceutical treatment.

Acknowledgment

This work was supported by the European Union through the Network of Excellence BioSim, Contract No. LSHB-CT-2004-005137.

References

1 Kaplan, H., Saddock, B., and Grebb, J.: *Synopsis of Psychiatry, 7th Edn*, Williams and Wilkins, Baltimore, **1994**.

2 Huber, M.T., Braun, H.A., and Krieg, J.C.:Consequences of deterministic and random dynamics for the course of affective disorders. *Biol Psychiatry* **1999**, 46:256–262.

3 Huber, M.T., Braun, H.A., and Krieg, J.C.: Effects of noise on different disease states of recurrent affective disorders. *Biol Psychiatry* **2000**, 47:634–642.

4 Huber, M.T., Braun, H.A., and Krieg, J.C.: On the impact of episode sensitization on the course of recurrent affective disorders. *J Psychiatr Res* **2001**, 35:49–57.

5 Huber, M.T., Braun, H.A., Voigt, K., and Krieg, J.C.: Some computational aspects of the kindling model for neuropsychiatric disorders. *Neurocomputing* **2001**, 38:1297–1306.

6 Huber, M.T., Braun, H.A., and Krieg, J.C.: Noise, nonlinear dynamics and the timecourse of affective disorders. *Chaos, Solitons and Fractals* **2000**, 11:1923–1928.

7 Huber, M.T., Braun, H.A., and Krieg, J.C.: Recurrent affective disorders: nonlinear and stochastic models of disease dynamics. *Int J Bifurc Chaos* **2004**, 14:635–652.

8 Wirz-Justice, A.: Chronobiology and mood disorders. *Dialog Clin Neurosci* **2003**, 5:315–325.

9 Wehr, T., and Goodwin, F.: Biological rhythms in manic-depressive illness. In: T. Wehr and F. Goodwin, eds., *Circadian rhythms in psychiatry*, The Boxwood Press, Pacific Gove, **1983**, pp 129–184.

10 Cazalis, M., Dayanithi, G., and Nordmann, J.J.: The role of patterned burst and interburst interval on the excitation-coupling mechanism in the isolated rat neural lobe. *J Physiol* **1985**, 369:45–60.

11 Dutton, A., and Dyball, R.E.: Phasic firing enhances vasopressin release from the rat neurohypophysis. *J Physiol* **1979**, 290:433–440.

12 McCormick, D.A.: Are thalamocortical rhythms the Rosetta Stone of a subset of neurological disorders? *Nat Med* **1999**, 5:1349–1351.

13 Gahwiler, B.H., Sandoz, P., and Dreifuss, J.J.: Neurones with synchronous bursting discharges in organ cultures of the hypothalamic supraoptic nucleus area. *Brain Res* **1978**, 151:245–253.

14 Bahar, S.: Burst enhanced synchronzation in an array of noisy coupled neurons. *Fluct Noise Lett* **2004**, 4:L87–L96.

15 Belykh, I., de Lange, E., and Hasler, M.: Synchronization of bursting neurons: what matters in the network topology. *Phys Rev Lett* **2005**, 94:88–101.

16 Wakerley, J.B., and Ingram, C.D.: Synchronization of Bursting in Hypothalamic Oxytocin Neurones:Possible Coordinating Mechanisms. *NIPS* **1993**, 8, 129–133.

17 Nestler, E.J., Barrot, M., DiLeone, R.J., Eisch, A.J., Gold, S.J., and Monteggia, L.M.: Neurobiology of depression. *Neuron* **2002**, 34:13–25.

18 Post, R.M., and Weiss, S.R.B.: The neurobiology of treatment resistant mood disorders. In: F.E. Bloom and D.J. Kupfer, eds., *Psychopharmacology: the fourth generation of progress*, Raven Press, New York, **1995**, pp 1155–1170.

19 Post, R.M.: Transduction of psychosocial stress into the neurobiology of recurrent affective disorder. *Am J Psychiatry* **1992**, 149:999–1010.

20 Kramlinger, K.G., and Post, R.M.: Ultra-rapid and ultradian cycling in bipolar affective illness. *Br J Psychiatry* **1996**, 168:314–323.

21 Weiss, S.R., and Post, R.M.: Kindling: separate vs. shared mechanisms in affective disorders and epilepsy. *Neuropsychobiology* **1998**, 38:167–180.

22 Braun, H.A., Schaefer, K., Voigt, K., and Huber, M.T.: Temperature encoding in peripheral cold receptors: oscillations, resonances, chaos and noise. *Nova Acta Leopoldina* **2003**, 88:293–318.

23 Braun, H.A., Wissing, H., Schafer, K., and Hirsch, M.C.: Oscillation and noise determine signal transduction in shark multimodal sensory cells. *Nature* **1994**, 367:270–273.

24 Braun, H.A., Bade, H., and Hensel, H.: Static and dynamic discharge patterns of bursting cold fibers related to hypothetical receptor mechanisms. *Pflugers Arch* **1980**, 386:1–9.

25 Huber, M.T., Braun, H.A., and Krieg, J.C.: On episode sensitization in recurrent affective disorders: the role of noise. *Neuropsychopharmacology* **2003**, 28[Suppl 1]:S13–S20.

26 Huber, M.T., Krieg, J.C., Braun, H.A., Pei, X., Neiman, A., and Moss, F.: Noisy precursors of bifurcations in a neurodynamical model for disease states of mood disorders. *Neurocomputing* **2000**, 32, 823–831.

27 Gottschalk, A., Bauer, M.S., and Whybrow, P.C.: Evidence of chaotic mood variation in bipolar disorder. *Arch Gen Psychiatry* **1995**, 52:947–959.

28 Braun, H.A., Schäfer, K., and Wissing, H.: Theories and models of temperature transduction. In: Bligh J.and Voigt K.H., eds., *Thermoreception and temperature regulation*, Springer, Heidelberg, **1990**, pp 19–29.

29 Holsboer, F.: Stress, hypercortisolism and corticosteroid receptors in depression: implications for therapy. *J Affect Disord* **2001**, 62:77–91.

30 Sachar, E.J.: Twenty-four-hour cortisol secretory patterns in depressed and manic patients. *Prog Brain Res* **1975**, 42:81–91.

31 Holsboer, F.: Psychiatric implications of altered limbic-hypothalamic-pituitary-adrenocortical activity. *Eur Arch Psychiatry Neurol Sci* **1989**, 238:302–322.

32 Holsboer, F.: The rationale for corticotropin-releasing hormone receptor (CRH-R) antagonists to treat depression and anxiety. *J Psychiatr Res* **1999**, 33:181–214.

33 Sapolsky, R.M., Krey, L.C., and McEwen, B.S.: Stress down-regulates corticosterone receptors in a site-specific manner in the brain. *Endocrinology* **1984**, 114:287–292.

34 Reynolds, C.F. 3rd, and Kupfer, D.J.: Sleep research in affective illness: state of the art circa 1987. *Sleep* **1987**, 10:199–215.

35 Benca, R.M., Obermeyer, W.H., Thisted, R.A., and Gillin, J.C.: Sleep and psychiatric disorders. A meta-analysis. *Arch Gen Psychiatry* **1992**, 49:651–668; discussion pp 669–670.

36 Riemann, D., Schnitzler, M., Hohagen, F., and Berger, M.: [Depression and sleep – the status of current research]. *Fortschr Neurol Psychiatr* **1994**, 62, 458–478.

37 Lauer, C.J., Riemann, D., Wiegand, M., and Berger, M.: From early to late adulthood. Changes in EEG sleep of depressed patients and healthy volunteers. *Biol Psychiatry* **1991**, 29:979–993.

38 Holsboer-Trachsler, E., Hemmeter, U., Hatzinger, M., Seifritz, E., Gerhard, U., and Hobi, V.: Sleep deprivation and bright light as potential augmenters of antidepressant drug treatment – neurobiological and psychometric assessment of course. *J Psychiatr Res* **1994**, 28:381–399.

39 Steiger, A., von Bardeleben, U., Herth, T., and Holsboer, F.: Sleep EEG and nocturnal secretion of cortisol and growth hormone in male patients with endogenous depression before treatment and after recovery. *J Affect Disord* **1989**, 16:189–195.

40 Linkowski, P., Mendlewicz, J., Kerkhofs, M., Leclercq, R., Golstein, J., Brasseur, M., Copinschi, G., and Van Cauter, E.: 24-hour profiles of adrenocorticotropin, cortisol, and growth hormone in major depressive illness: effect of antidepressant treatment. *J Clin Endocrinol Metab* **1987**, 65:141–152.

41 Kupfer, D.J., Spiker, D.G., Coble, P.A., Neil, J.F., Ulrich, R., and Shaw, D.H.: Depression, EEG sleep, and clinical response. *Comp Psychiatry* **1980**, 21:212–220.

42 Rush, A.J., Erman, M.K., Giles, D.E., Schlesser, M.A., Carpenter, G., Vasavada, N., and Roffwarg, H.P.: Polysomnographic findings in recently drug-free and clinically remitted depressed patients. *Arch Gen Psychiatry* **1986**, 43:878–884.

43 Giles, D.E., Jarrett, R.B., Roffwarg, H.P., and Rush, A.J.: Reduced rapid eye movement latency. A predictor of recurrence in depression. *Neuropsychopharmacology* **1987**, 1:33–39.

44 Reynolds, C.F. 3rd, Perel, J.M., Frank, E., Imber, S., and Kupfer, D.J.: Open-trial maintenance nortriptyline in geriatric depression: survival analysis and preliminary data on the use of REM latency as a predictor of recurrence. *Psychopharmacol Bull* **1989**, 25:129–132.

45 Kupfer, D.J., Frank, E., McEachran, A.B., and Grochocinski, V.J.: Delta sleep ratio. A biological correlate of early recurrence

in unipolar affective disorder. *Arch Gen Psychiatry* **1990**, 47:1100–1105.

46 Hatzinger, M., Hemmeter, U.M., Brand, S., Ising, M., and Holsboer-Trachsler, E.: Electroencephalographic sleep profiles in treatment course and long-term outcome of major depression: association with DEX/CRH-test response. *J Psychiatr Res* **2004**, 38:453–465.

47 Kupfer, D.J., Buysse, D.J., and Reynolds, C.F. 3rd: Antidepressants and sleep disorders in affective illness. *Clin Neuropharmacol* **1992**, 15[Suppl 1A]:360A–361A.

48 Thase, M.E., Simons, A.D., and Reynolds, C.F. 3rd: Abnormal electroencephalographic sleep profiles in major depression: association with response to cognitive behavior therapy. *Arch Gen Psychiatry* **1996**, 53:99–108.

49 McCarley, R.W., and Hobson, J.A.: Neuronal excitability modulation over the sleep cycle: a structural and mathematical model. *Science* **1975**, 189:58–60.

50 Janowsky, D.S., el-Yousef, M.K., Davis, J.M., and Sekerke, H.J.: A cholinergic–adrenergic hypothesis of mania and depression. *Lancet* **1972**, 2:632–635.

51 Steiger, A., Herth, T., and Holsboer, F.: Sleep-electroencephalography and the secretion of cortisol and growth hormone in normal controls. *Acta Endocrinol (Copenh)* **1987**, 116:36–42.

52 Born, J., Spath-Schwalbe, E., Schwakenhofer, H., Kern, W., and Fehm, H.L.: Influences of corticotropin-releasing hormone, adrenocorticotropin, and cortisol on sleep in normal man. *J Clin Endocrinol Metab* **1989**, 68:904–911.

53 Friess, E., U, V.B., Wiedemann, K., Lauer, C.J., and Holsboer, F.: Effects of pulsatile cortisol infusion on sleep-EEG and nocturnal growth hormone release in healthy men. *J Sleep Res* **1994**, 3:73–79.

54 Ehlers, C.L., Reed, T.K., and Henriksen, S.J.: Effects of corticotropin-releasing factor and growth hormone-releasing factor on sleep and activity in rats. *Neuroendocrinology* **1986**, 42:467–474.

55 Holsboer, F., von Bardeleben, U., and Steiger, A.: Effects of intravenous corticotropin-releasing hormone upon sleep-related growth hormone surge and

sleep EEG in man. *Neuroendocrinology* **1988**, 48:32–38.

56 Mendelson, W.B., Slater, S., Gold, P., and Gillin, J.C.: The effect of growth hormone administration on human sleep: a dose–response study. *Biol Psychiatry* **1980**, 15:613–618.

57 Steiger, A., von Bardeleben, U., Guldner, J., Lauer, C., Rothe, B., and Holsboer, F.: The sleep EEG and nocturnal hormonal secretion studies on changes during the course of depression and on effects of CNS-active drugs. *Prog Neuropsychopharmacol Biol Psychiatry* **1993**, 17:125–137.

58 Steiger, A.: Physiology and pathophysiology of sleep. *Schweiz Med Wochenschr* **1995**, 125:2338–2345.

59 Achermann, P., and Borbely, A.A.: Mathematical models of sleep regulation. *Front Biosci* **2003**, 8:s683–s693.

60 Borbely, A.A.: A two process model of sleep regulation. *Hum Neurobiol* **1982**, 1:195–204.

61 Achermann, P., and Borbely, A.A.: Combining different models of sleep regulation. *J Sleep Res* **1992**, 1:144–147.

62 Achermann, P.: The two-process model of sleep regulation revisited. *Aviat Space Environ Med* **2004**, 75:A37–A43.

63 Borbely, A.A.: The S-deficiency hypothesis of depression and the two-process model of sleep regulation. *Pharmacopsychiatry* **1987**, 20:23–29.

64 Liu, B.Z., and Deng, G.M.: An improved mathematical model of hormone secretion in the hypothalamo-pituitary-gonadal axis in man. *J Theor Biol* **1991**, 150:51–58.

65 Jelic, S., Cupic, Z., and Kolar-Anic, L.: Mathematical modeling of the hypothalamic-pituitary-adrenal system activity. *Math Biosci* **2005**, 197:173–187.

66 Dempsher, D.P., Gann, D.S., and Phair, R.D.: A mechanistic model of ACTH-stimulated cortisol secretion. *Am J Physiol* **1984**, 246:R587–R596.

67 Gahwiler, B.H., and Dreifuss, J.J.: Transition from random to phasic firing induced in neurons cultured from the hypothalamic supraoptic area. *Brain Res* **1980**, 193:415–425.

68 Dewald, M., Anthes, N., Vedder, H., Voigt, K., and Braun, H.A.: Phasic bursting activity of paraventricular neurons is modulated by temperature and angiotensin II. *J Therm Biol* **1999**, 24:339–345.

69 Braun, H.A., Hirsch, M.C., Dewald, M., and Voigt, K.: Temperature dependent burst discharges in magnocellular neurons of the paraventricular and supraoptic hypothalamic nuclei recorded in brain slice preparations of the rat. In: K. Pleschka and R. Gerstberger, eds., *Integrative and cellular aspects of autonomic functions: temperature and osmoregulation*, John Libbey Eurotext, Paris, **1994**, pp 67–75.

70 Andrew, R.D.: Isoperiodic bursting by magnocellular neuroendocrine cells in the rat hypothalamic slice. *J Physiol* **1987**, 384:467–477.

71 McCormick, D.A., and Huguenard, J.R.: A model of the electrophysiological properties of thalamocortical relay neurons. *J Neurophysiol* **1992**, 68:1384–1400.

72 Llinas, R.R., and Steriade, M.: Bursting of thalamic neurons and states of vigilance. *J Neurophysiol* **2006**, 95:3297–3308.

73 Wang, X.J.: Multiple dynamical modes of thalamic relay neurons: rhythmic bursting and intermittent phase-locking. *Neuroscience* **1994**, 59:21–31.

74 Jefferys, J.G., and Haas, H.L.: Synchronized bursting of CA1 hippocampal pyramidal cells in the absence of synaptic transmission. *Nature* **1982**, 300:448–450.

75 Sosnovtseva, O.V., Postnova, S., Mosekilde, E., and Braun, H.A.: Inter-pattern transition in a bursting cell. *Fluct Noise Lett* **2004**, 4:L521–L533.

76 Postnov, D.E., Sosnovtseva, O.V., Malova, S.Y., and Mosekilde, E.: Complex phase dynamics in coupled bursters. *Phys Rev E Stat Nonlin Soft Matter Phys* **2003**, 67:016215.

77 Sosnovtseva, O.V., Setsinsky, D., Fausboll, A., and Mosekilde, E.: Transitions between beta and gamma rhythms in neural systems. *Phys Rev E Stat Nonlin Soft Matter Phys* **2002**, 66:041901.

78 Jensen, M.S., and Yaari, Y.: Role of intrinsic burst firing, potassium accumulation, and electrical coupling in the elevated potassium model of hippocampal epilepsy. *J Neurophysiol* **1997**, 77:1224–1233.

79 Lopes da Silva, F., Blanes, W., Kalitzin, S.N., Parra, J., Suffczynski, P., and Velis, D.N.: Epilepsies as dynamical diseases of brain systems: basic models of the transition between normal and epileptic activity. *Epilepsia* **2003**, 44[Suppl 12]:72–83.

80 Singer, W., and Gray, C.M.: Visual feature integration and the temporal correlation hypothesis. *Annu Rev Neurosci* **1995**, 18:555–586.

81 Engel, A.K., and Singer, W.: Temporal binding and the neural correlates of sensory awareness. *Trends Cogn Sci* **2001**, 5:16–25.

82 Llinas, R.R., Leznik, E., and Urbano, F.J.: Temporal binding via cortical coincidence detection of specific and nonspecific thalamocortical inputs: a voltage-dependent dye-imaging study in mouse brain slices. *Proc Natl Acad Sci USA* **2002**, 99:449–454.

83 Zhou, C., and Kurths, J.: Noise-induced synchronization and coherence resonance of a Hodgkin-Huxley model of thermally sensitive neurons. *Chaos* **2003**, 13:401–409.

84 Braun, H.A., Huber, M.T., Anthes, N., Voigt, K., Neiman, A., Pei, X., and Moss, F.: Noise-induced impulse pattern modifications at different dynamical period-one situations in a computer model of temperature encoding. *Biosystems* **2001**, 62:99–112.

85 Sosnovtseva, O.V., Fomin, A.I., Postnov, D.E., and Anishchenko, V.S.: Clustering of noise-induced oscillations. *Phys Rev E Stat Nonlin Soft Matter Phys* **2001**, 64:026204.

86 Bulsara, A., and Gammeitoni, L.: Tuning in to noise. *Phys. Today* **1996**, 1996:39–45.

87 Longtin, A., and Hinzer, K.: Encoding with bursting, subthreshold oscillations, and noise in mammalian cold receptors. *Neural Comput* **1996**, 8:215–255.

88 Mosekilde, E., Sosnovtseva, O.V., Postnov, D., Braun, H.A., and Huber, M.T.: Noise-activated and noise-induced rhythms in neural systems. *Nonlin Stud* **2004**, 11:449–467.

89 White, J.A., Klink, R., Alonso, A., and Kay, A.R.: Noise from voltage-gated ion channels may influence neuronal dynamics in the entorhinal cortex. *J Neurophysiol* **1998**, 80:262–269.

90 Postnov, D.E., Sosnovtseva, O.V., Han, S.K., and Kim, W.S.: Noise-induced multimode behavior in excitable systems. *Phys Rev E Stat Nonlin Soft Matter Phys* **2002**, 66:016203.

91 Mosekilde, E., Sosnovtseva, O.V., Postnov, D., Braun, H.A., and Huber, M.T.: Noisy neural rhythm generators. *Appl Nonlin Dyn* **2004**, 11:95–110.

92 Zhou, C., and Kurths, J.: Hierarchical synchronization in complex networks with heterogeneous degrees. *Chaos* **2006**, 16:015104.

93 Motter, A.E., Zhou, C., and Kurths, J.: Network synchronization, diffusion, and the paradox of heterogeneity. *Phys Rev E Stat Nonlin Soft Matter Phys* **2005**, 71:016116.

94 Postnova, S., Wollweber, B., Voigt, K., and Braun, H.A.: Impulse-pattern in bidirectionally coupled model neurons of different dynamics. *BioSystems*, **2007** 89:135–172.

95 Huber, M.T., and Braun, H.A.: Stimulus–response curves of a neuronal model for noisy subthreshold oscillations and related spike generation. *Phys. Rev. E* **2006** 73:04129.

96 Braun, H.A., Voigt, K., and Huber, M.T.: Oscillations, resonances and noise: basis of flexible neuronal pattern generation. *Biosystems* **2003**, 71:39–50.

97 Huber, M.T., Krieg, J.C., Dewald, M., Voigt, K., and Braun, H.A.: Stimulus sensitivity and neuromodulatory properties of noisy intrinsic neuronal oscillators. *Biosystems* **1998**, 48:95–104.

98 Braun, H.A., Huber, M.T., Dewald, M., Schäfer, K., and Voigt, K.: The neuromodulatory properties of "noisy neuronal oscillators". In: J.B. Kadke, A. Bulsara, eds., *Applied nonlinear dynamics and stochastic systems near the millenium*, The American Institute of Physics, Washington, D.C., **1998**, pp 281–286.

99 Schafer, K., Braun, H.A., and Rempe, L.: Discharge pattern analysis suggests existence of a low-threshold calcium channel in cold receptors. *Experientia* **1991**, 47:47–50.

100 Schafer, K., Braun, H.A., and Isenberg, C.: Effect of menthol on cold receptor activity. Analysis of receptor processes. *J Gen Physiol* **1986**, 88:757–776.

101 Fox, R.F., Gatland, I.R., Roy, R., and Vemuri, G.: Fast, accurate algorithm for numerical simulation of exponentially correlated colored noise. *Phys Rev A* **1988**, 38:5938–5940.

102 Hutcheon, B., Miura, R.M., Yarom, Y., and Puil, E.: Low-threshold calcium current and resonance in thalamic neurons: a model of frequency preference. *J Neurophysiol* **1994**, 71:583–594.

103 Llinas, R.R., Grace, A.A., and Yarom, Y.: In vitro neurons in mammalian cortical layer 4 exhibit intrinsic oscillatory activity in the 10- to 50-Hz frequency range. *Proc Natl Acad Sci USA* **1991**, 88:897–901.

104 Llinas, R., and Yarom, Y.: Oscillatory properties of guinea-pig inferior olivary neurones and their pharmacological modulation: an in vitro study. *J Physiol* **1986**, 376:163–182.

105 Gutfreund, Y., Yarom, Y., and Segev, I.: Subthreshold oscillations and resonant frequency in guinea-pig cortical neurons: physiology and modelling. *J Physiol* **1995**, 483(3):621–640.

106 Lampl, I., and Yarom, Y.: Subthreshold oscillations of the membrane potential: a functional synchronizing and timing device. *J Neurophysiol* **1993**, 70:2181–2186.

107 Alonso, A., and Klink, R.: Differential electroresponsiveness of stellate and pyramidal-like cells of medial entorhinal cortex layer II. *J Neurophysiol* **1993**, 70:128–143.

108 Klink, R., and Alonso, A.: Ionic mechanisms for the subthreshold oscillations and differential electroresponsiveness of medial entorhinal cortex layer II neurons. *J Neurophysiol* **1993**, 70:144–157.

109 Klink, R., and Alonso, A.: Muscarinic modulation of the oscillatory and repetitive firing properties of entorhinal cortex layer II neurons. *J Neurophysiol* **1997**, 77:1813–1828.

110 Braun, H., Schäfer, K., Wissing, H., and Hensel, H.: Periodic transduction processes in thermosensitive receptors. In: W. Hamann and A. Iggo, eds, *Sensory receptor mechanisms*, World Scientific, Singapore, **1984**, pp 147–156.

111 Feudel, U., Neiman, A., Pei, X., Wojtenek, W., Braun, H., Huber, M., and

Moss, F.: Homoclinic bifurcation in a Hodgkin–Huxley model of thermally sensitive neurons. *Chaos* **2000**, 10:231–239.

112 Braun, W., Eckhardt, B., Braun, H.A., and Huber, M.: Phase–space structure of a thermoreceptor. *Phys Rev E* **2000**, 62:6352–6360.

113 Arhem, P., Braun, H.A., Huber, M.T., and Liljenström, H. (2005). Dynamic state transitions in the nervous system: from ion channels to neurons to networks. In: H. Liljenstrom and U. Svedin, eds, *Micro-meso-macro: addressing complex systems couplings*, World Scientific, London, **2005**, pp 37–72.

8

Energy Metabolism in Conformational Diseases

Judit Ovádi and Ferenc Orosz

Abstract

Increasing evidence indicates that accumulation of unfolded/misfolded proteins results in protofibril and protein aggregate formation, inducing mitochondrial dysfunction and failure of synaptic, transport and other crucial physiological processes. Our knowledge on molecular defects of the energy metabolism in the pathogenesis of neurodegenerative diseases is scarce. One problem that investigators face is distinguishing primary from secondary events: is the impairment of energy production a consequence of the development of neurodegeneration, or it is an active contributor? Genomic and proteomic data provide useful information for understanding the structural and functional perturbations during the development of neurodegeneration, but the metabolomic alterations are much less disclosed. This paper reviews the direct and indirect role of the damage of energy metabolism, focusing on the metabolism of glucose as major energy source of the brain. Special attention is paid to the functional consequences of the associations of glycolytic enzymes with unfolded proteins that specifically determine the nature of the so-called "conformational diseases". This review shows that the available data are largely inconclusive for the evaluation of the pathomechanism of energy production defects in human brains of patients of conformational diseases at system level, and the genuine need for development of rational models (biosimulation) of neuropathological conversion of glucose to ATP via glycolysis and oxidative phosphorylation.

8.1
What is the Major Energy Source of the Brain?

The brain is an organ with high energy demands. Glucose has been considered the almost exclusive energy source utilized by the brain. The glucose taken up by brain cells is metabolized by glycolytic enzymes, the mitochondrial respiratory chain and the oxidative phosphorylation system. In addition to glucose, neurons might rely on lactate to sustain their activity. Activation of brain areas is induced by the arrival

Biosimulation in Drug Development. Edited by Martin Bertau, Erik Mosekilde, and Hans V. Westerhoff
Copyright © 2008 WILEY-VCH Verlag GmbH & Co. KGaA, Weinheim
ISBN: 978-3-527-31699-1

of different impulses, which release glutamate into the synaptic cleft. Neurotransmitter molecules diffuse across the cleft and stimulate the postsynaptic cell, causing Na^+ channels to open (Fig. 8.1), resulting in membrane depolarization. This leads to the initiation of action potentials as a consequence of the movement of ions along their electrochemical gradients. Na^+ entry and K^+ release during electrical activity (action potential) initiate increased energy metabolism within neurons. The movement of ions activates the Na^+/K^+ ATPase pump which restores the polarization of the membrane supplying ATP. These processes imply rapid supply of energy produced by glycolysis (conversion of glucose to pyruvate) and by the tricarboxylic acid (TCA) cycle and mitochondrial oxidative phosphorylation. Activation of glycolysis lowers the glucose level, causing enhanced flux of glucose into neurons. According to the conventional hypothesis, neurons metabolize glucose nearly fully in an aerobic manner: The generated ATP is largely used to restore Na^+/K^+ balance via Na^+/K^+ ATPase. The rapid increase of glycolysis promotes $NADH/NAD^+$, as well as pyruvate production, leading to the accumulation of lactate that diminishes the utilization of lactate produced by astrocytes [1].

Astrocytes, sub-type of the glial cells in the brain, outnumber neurons by ten to one and play a number of active roles in the brain. Astrocytes have a critical role as the main energy stores and transporting materials and small molecules between blood and brain tissue as well as within the brain tissue. Even the supportive activity of astrocytes is driven by neuronal activity and interplay. Interactions between neurons and astrocytes are critical for signaling, energy metabolism, extracellular ion homeostasis, volume regulation, and neuroprotection in the central nervous system. In astrocytes, oxidative glucose metabolism is activated by uptake of glutamate and the basic activation process is similar to that in the neuron (cf. Fig. 8.1) [1]. Therefore, the conventional hypothesis asserts that glucose is the primary substrate for both neurons and astrocytes and that lactate is produced during neural activity and can be utilized by neurons as an energy source. However, the question is whether neurons do utilize glial-produced lactate as their principal energy substrate during activity, as proposed by the astrocyte–neuron lactate shuttle hypothesis (ANLSH).

The ANLSH challenged the classic view [2, 3]. It postulates compartmentalization of brain lactate metabolism between neurons and astrocytes: the activity-induced uptake of glucose takes place predominantly in astrocytes, which metabolize glucose anaerobically. Lactate produced from anaerobic glycolysis in astrocytes is then released from astrocytes and provides the primary metabolic fuel for neurons. The increased lactate in the neurons is converted to pyruvate via lactate dehydrogenase (LDH), which enters the TCA cycle, and increases ATP production in the neurons via oxidative phosphorylation (Fig. 8.1). This view is highly discussed, pro [4, 5]) and contra [1, 6].

A modeling approach was used to determine which mechanisms are appropriate to explain typical brain lactate kinetics observed upon activation [7]. The model takes into account the mechanisms that are known to determine extracellular lactate concentration and includes a systematic study of the effects of changing parameters, such as the effect of cellular production or consumption of lactate, regional

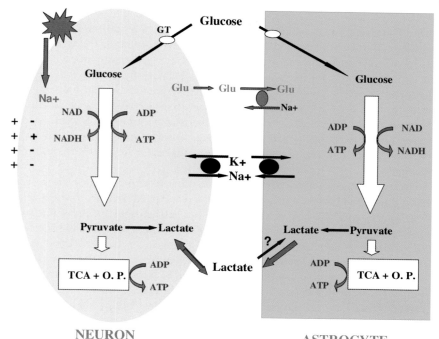

NEURON

ASTROCYTE

Fig. 8.1 Glucose metabolism in coupled neuron and astrocyte system. ATP is produced via oxidative energy metabolism (glycolysis, TCA cycle and oxidative phosphorylation) in neurons and in astrocytes. Na^+ entry during electrical activity initiates increased oxidative energy metabolism within neurons. The activation of neuronal Na^+-K^+ ATPase in the plasma membrane leads to reduced levels of ATP, which rapidly activates glycolysis. This process requires an elevated glucose level, which is transported via the neuronal glucose transporter (GT). The generated ATP can restore the Na^+/K^+ balance via Na^+-K^+ ATPase. The rapid increase of glycolysis results in increased $NADH$/NAD^+ and increased cytoplasmic pyruvate. In astrocytes, the uptake of glutamate via high-affinity glutamate transporters activates oxidative glucose metabolism, which provides ATP. The basic activation process is considered similar to that in the neuron. According to the astrocyte–neuron lactate shuttle hypothesis, the lactate is transported across the glial membrane and enters into the neuron via the extracellular space, where it is converted to pyruvate by LDH to produce ATP by the TCA cycle and oxidative phosphorylation (O.P.) [1]. It is hypothesized that, in the case of damaged glucose uptake or ATP production of astrocytes (e.g., due to inclusion formation), lactate could be transported from the extracellular space into the damaged astrocytes.

cerebral blood flow, exchanges through the blood–brain barrier, and extracellular pH variation. The modeling of the glucose flux was based on a similar model. The estimated contribution of lactate (versus glucose) to neuronal extra pyruvate supply was found to be between 30 and 60%. The model enabled a simulation of the early dip in extracellular lactate concentration and its subsequent increase ("overshoot") found experimentally by Hu and Wilson [8], thus the validity of ANLSH was suggested.

The ANLSH is based largely on *in vitro* experiments, and it postulates that lactate is derived from the uptake of synaptically released glutamate by astrocytes. However, the time course of changes in lactate, derived from *in vivo* experiments, appeared to be incompatible with the ANLSH. Neuronal activation leads to a delayed rise in lactate, followed by a slow decay which greatly outlasts the period of neuronal activation. This led to the proposition that the uptake of stimulated glutamate released from astrocytes, rather than that of synaptically released glutamate, is the source of lactate released following neuronal activation [6]. The increase of lactate occurs too late to provide energy for neuronal activity. Furthermore, it was considered that there is no evidence that lactate undergoes local oxidative phosphorylation.

Positron emission tomography (PET) and certain other techniques require long sampling times to obtain a sufficient signal and are therefore more likely to detect long-lasting signals of the glucose uptake. Thus, these images generated primarily reflect astrocytic activity rather than neuronal activity, even though the activation of both cell types is correlated in the majority of cases [9]. Similarly, results obtained by magnetic resonance spectroscopy (MRS), which monitors changes in glucose and lactate levels with low temporal resolution, are more likely to reveal the late component of the metabolic response associated with astrocytes [10, 11]. Thus, the increase observed in lactate concentration would correspond to enhanced glycolysis in astrocytes with associated sustained production of lactate. In contrast, use of MRS with higher temporal resolution [12] allows much faster sampling times enabling detection of the early lactate "dip". The detected "dip" is consistent with the rapid activation of oxidative phosphorylation followed by lactate oxidation via the TCA cycle in neurons, before lactate production by astrocytes fill up the extracellular lactate pool [9].

Recently, functional images using two-photon fluorescence signal of NADH suggested the existence of the astrocyte–neuron lactate shuttle [13]. (It is well known that the lactate/pyruvate ratio critically depends on the NADH/NAD ratio [14].) These data showed that initial consumption of NADH in dendrites caused by oxidative metabolism in mitochondria is followed by glycolytic (nonoxidative) metabolism in astrocytes. The latter process generates lactate, which is shuttled out of the astrocyte and into the neuron, where it can then be "burned" within the mitochondria (see Fig. 8.1). Therefore, experimental evidence suggested the coupling between oxidative metabolism in the dendritic shaft and glycolytic activity in the astrocyte to provide sustained neuronal energy in the form of ATP.

Astrocytes face the synapses and form an extensive network interconnected by gap junctions. They participate in synaptic signaling by providing metabolic support to the active neurons. Disturbances of these neuron–astrocyte interactions are likely to play an important role in neurologic disorders. Understanding the brain energetics might be critically important because several neurodegenerative diseases exhibit diverse alterations in energy metabolism.

There are virtually no data whether neurons use ambient glucose, and/or glial-derived lactate under neoropathological conditions. It has been proposed that glycolytic enzymes such as phosphofructokinase (PFK), and glyceraldehyde-3-

Table 8.1 Major characteristics of selected conformational diseases (selected data from [54]).

Disease	Diseased proteins	Major affected region	Pathology
AD	Aβ and hyper-phosphorylated tau	Cortex, hippocampus, basal forebrain	Neuritic plaques and neurofibrillary tangles
PD	α-Synuclein	Substantia nigra, cortex	Lewy bodies and Lewy neurites
HD	Huntingtin with poly-glutamin expansion	Striatum, basal ganglia, cortex	Intranuclear inclusions and cytoplasmic aggregates

phosphate dehydrogenase (GAPDH) are inhibited by interacting proteins, such as tau, β-amyloid, mutant huntingtin, α-synuclein, that play crucial role in the pathogenesis of conformational diseases (see Table 8.1). As consequences of these heteroassociations, glycolytic enzymes are impeded, resulting in less ATP and pyruvate. The reduced pyruvate production decreases the ATP production *via* the oxidative energy metabolism. The appearance of these aberrant protein–protein associations in the cytoplasm and/or the formation of insoluble aggregates due to the accumulation of these heteroassociates within inclusions of neurons or glia cells could impede glycolysis. ATP production could slow down, then lactate could be transferred between neurons and astrocytes depending on the request, and utilized as oxidative substrate during neuronal activity as suggested by the ANLS concept (Fig. 8.1). If inclusions are formed in glial cells, like in the case of multiplex system atrophy, and the ATP supply cannot be fulfilled by the damaged glia cells, the energy (ATP) can be fuelled by the extracellular lactate released from the healthy glia or neurons. Indeed, there may be no restriction for a glial cell to take up lactate and produce pyruvate by its LDH. One can hypothesize that the transfer of lactate from neurons to astrocytes might occur under pathological conditions. All these possibilities could be the target of biosimulation studies considering also the reduced activities of key glycolytic enzymes, hexokinase (HK), PFK, GAPDH or pyruvate kinase (PK) according to the experimental data measured in diseased tissues (see Section 8.2). However, the data referring to the brain samples including those from the well-studied Alzheimer's disease (AD) brain are so far inconsistent and controversial. The activities varied depending on the sampling techniques [15], on the difference of the brain sections [16], and on the origin of the disease, e.g. decrease of HK activity was found exclusively in the case of familial AD [17]. Nevertheless, a part of the differences of the findings was explained by technical reasons, the different conditions and the instability of the key glycolytic enzymes [18].

Another important difficulty in the evaluation of the correct activity values for enzymes involved in energy metabolism in neurons resides in the fact that opposite processes may occur simultaneously, like the reduction of neuronal metabolism and compensatory enhancement in glial activity. The altered activities of some glycolytic enzymes in brain homogenates represent a net-effect, and it is unclear how the energy state (ATP level) of the neurons is perturbed if the metabolic activation is restricted to astrocytes. This explains why the extensive search for well-established information on the involvement of glycolytic enzymes and energy metabolism in neurodegenerative processes gave limited amount of *in vitro* and *in vivo* data. Even the results obtained by determination of glycolytic activities in normal control and pathological brains of patients did not provide straightforward results [16].

8.2
Unfolded/Misfolded Proteins Impair Energy Metabolism

The major part of the intracellular proteins has well defined secondary and tertiary structures; however, a number of *in silico* and experimental data have been accumulated indicating that a part of the proteins does not bear well defined 3D structure [19]. These proteins denoted as *intrinsically unstructured* or *natively unfolded* proteins, are rather common in living cells and fulfil essential physiological functions [20]. These functions are linked with their structural states and interactions with other cellular targets. They are sometimes form fibrils or protein aggregates with distinct ultrastructures. Many of them, termed as "neurodegenerative proteins", are known to form intracellular inclusions, tightly linked to the development of different neurodegenerative diseases [21]. It has become apparent that unfolded and misfolded proteins are extensively involved in different unrelated neurodegenerative diseases such as AD, Parkinson's disease (PD), and Huntington's disease (HD). These diseases are frequently termed conformational diseases because the initiation of the diseases is likely caused by unstructured proteins, like α-synuclein, tau or mutant huntingtin protein (Table 8.1) and other unfolded/misfolded proteins, such as a recently isolated new brain-specific protein, termed TPPP/p25, which induces aberrant tubulin aggregates and is enriched in the inclusions characteristics for PD and other synucleinopathies (for a review, see [22]).

The development of neurodegeneration is a multistep process: the unfolded/misfolded proteins enter into aberrant protein-protein interactions that finally lead to the formation of inclusions in different area of the brain producing characteristic pathology. However, the exact molecular connections between these steps are unclear yet. "Precisely when in the disease process such interactions occur is unclear, as is any structural understanding of how altered protein conformations or aggregates trigger neuronal dysfunction" [23]. In addition, there is no clear distinction concerning the primary and secondary nature of the events [24], such as accumulation of unfolded or misfolded proteins, protofibril formation, different stress effects, failure of axonal and dendritic transport, mitochondrial dysfunction coupled with the damage of energy metabolism. A combination of these and other events that impair normal neuronal function can cause chronic neurodegenerative disorders such as the most common ones: AD, PD, and HD. The major characteristics of these conformational diseases are summarized in Table 8.1.

The impaired glucose turnover can cause not only ATP deficit in the cells involved but may induce other deleterious effects [25]. For example, a low glucose turnover can highly accelerate progress leading to serious dementia. Relevant data originate from the use of different imaging techniques for monitoring glucose consumption *in vivo* [26]. These data showed that the low glucose turnover resulted in cholinergic deficit as a consequence of decreased acetyl-coenzyme A (AcCoA) synthesis and reduced acetylation of choline to acetylcholine [27]. Pyruvate dehydrogenase multienzyme complex, which is responsible for the synthesis of AcCoA, was inhibited due to its phosphorylation by mitochondrial tau protein kinase I/glycogen synthase kinase-3beta activated by the amyloid β peptide (Aβ) [28, 29]. Hypometabolism of glucose can also lead to alterations in homeostasis of ions, to unfolding of proteins, altered protein synthesis, disturbance of the transport and degradation of proteins, loss of cell potentials, all of which could be relevant in the development of conformational diseases [30–33].

8.3
Interactions of Glycolytic Enzymes with "Neurodegenerative Proteins"

AD associated with progressive impairment of memory and cognition is characterized by three pathological lesions: senile plaques, neurofibrillary tangles, and loss of synapses. Plaques are mostly extracellular and consist of deposits of a fibrillous protein referred to as Aβ surrounded by dying neurites. Aβ is heterogeneous and produced from a precursor protein (amyloid precursor protein; APP; 130 kDa) by two sequential proteolytic cleavages that involve β- and γ-secretases. In fact, secretase represents a potentially attractive drug target since it dictates the solubility of the generated Aβ fragment by creating peptides of various lengths, namely Aβ (40) and Aβ (42). Tangles are intracellular and are abnormally altered deposits of tau, a microtubule-associated protein, whose normal function involves intracellular axonal transport. Senile plaques develop extracellularly with the main component being aggregated Aβ, whereas neurofibrillary tangles (NFTs) develop intracellularly with the main component being aggregated forms of phosphorylated tau. However, recently evidence has been presented that Aβ binds strongly to tau in solution that may be a precursor event to later self-aggregation of both molecules [34]. Since the insoluble Aβ and tau appear simultaneously in the same neurons, even if the dominant fraction of Aβ plaques are extracellular particles, it indicates that both molecules can be aggregated intracellularly at an early stage of the disease. This might represent an intermediate state when Aβ could be associated with glycolytic enzymes causing their inhibition.

APP functions as a neuronal receptor and stimulates dendritic and synaptic outgrowth. This transmembrane protein is split under non-physiological condition when the glucose turnover-generated and oxidative phosphorylation-generated ATP supply is insufficient [35]. A critical level of glucose and ATP turnover is crucial for the proper insertion of APP into the cellular or synaptic membranes. The accumulation of Aβ initiates an undesired cycle by binding to the glycolytic enzyme, PFK [36], that decreases glucose turnover by inhibition of the catalytic activity of PFK, resulting in reduced ATP production (Table 8.2). The decreased PFK activity in se-

Table 8.2 Involvement of glycolytic enzymes in impairing the energy metabolism of conformational diseases.

Enzyme	AD	PD	HD	TPI deficiency
HK			Activation [46, 116]	Activation [81]
PFK	Binding to Aβ: inactivation [36]			
Aldolase				Activation [81]
TPI	Nitration [32, 63, 66]	Inactivation? [75]		Binding to MT: inactivation [79, 86]
GAPDH	Binding to APP [37]; inactivation [39]; nitration [67]	Binding to α-synuclein [49, 52]	Binding to CAG repeat: inactivation [39, 45, 46]	Activation [81]
Enolase	Nitration: inactivation [32, 66]			
PGM	Inactivation [70]; nitration: decreased protein expression and activity [66]			
LDH	Activation [16]			Activation [81]
PK	Activation [16]		Activation [16]	Activation [81]

nile demented brain cortex resulting from PFK-Aβ interaction can decline glucose turnover that accelerates Aβ generation.

Association of APP to an other glycolytic enzyme, GAPDH has been also reported [37]. This enzyme, GAPDH displays multiple functions in eukaryotic cells. It binds to nucleic acids, translocates into the nucleus affecting transcription processes, exhibits pro-apoptotic function, binds to cytoskeleton filaments, tubulin and actin, and associates with other glycolytic enzymes ([38] and references therein). Functional role other than the glycolytic activity was identified for GAPDH in conformational diseases which is related to its specific binding to proteins such as APP and huntingtin implicated in neurodegeneration. The association of GAPDH with these proteins, similarly to the PFK-Aβ interaction, reduced the energy production in diseased cells and tissues. Mazzola and Shirover [38, 39] found that there was no measurable difference in GAPDH glycolytic activity in crude whole-cell sonicates of AD and HD fibroblasts as compared to normal controls. However, they could not exclude the disruption of macromolecular associations of GAPDH resulting in diminution of GAPDH glycolytic activity. In fact, studies on the glycolytic activity and subcellular distribution of GAPDH protein in a number of cells revealed the impairment of GAPDH glycolytic function in AD and HD subcellular fractions despite unchanged gene expression. In the nuclear fraction, deficits of 27% and 33% in GAPDH function were observed in AD and HD, respectively. This finding supports a functional role for GAPDH in neurodegenerative diseases; the reduced GAPDH activity in the cytoplasm of AD cells can cause a significant deficit in energy production. This would greatly contribute or possibly even initiate the disease process [40–42]. This idea was supported by *in vivo* data as well which demonstrated that glucose utilization is decreased by 20–40% in both AD and HD brain tissue (see Table 8.2) [43].

HD is an inherited neurodegenerative disorder caused by an insertion of multiple CAG repeats in huntingtin gene resulting in N-terminal polyglutamine (polyQ) expansion of the large huntingtin protein similarly to other polyQ-related conformational diseases. Huntingtin expression is ubiquitous, the greatest level is in neurons, occurring in membrane, cytosol, nucleus, mitochondrium. In mouse models the N-terminal fragments of mutant huntingtin caused neurodegeneration, suggesting the toxicity of the caspase-produced polyQ fragment [44]. The mutant huntingtin protein-induced neurodegenerative process is associated with increased intracellular calcium level that triggers the cleavage of huntingtin and the translocation of the polyQ into the nucleus.

GAPDH was found to bind extensively to the polyQ tract of huntingtin protein [45]. The direct effect of the polyQ tract on the catalytic function of the glycolytic enzymes are unclear yet, however, reduced catalytic activity of GAPDH was found in HD fibroblasts [46] and brain samples (see above). Since GAPDH plays an important role in the glucose metabolism of the brain, the inhibition of this enzyme causing the defective energy metabolism at system level, could contribute to the development of the neurodegenerative diseases [47]. The activity of HK in fibroblasts expressing mutant huntingtin was higher as compared to the control at 1–3 weeks after withholding fresh medium [46]. The finding that the activity of the key glycolytic enzyme, HK, is altered in fibroblasts expressing mutant huntingtin may support connection between the impairment of glycolysis and the disease as well.

PD is the most common neurodegenerative movement disorder caused by selective and progressive degeneration of dopaminerg neurons. One important feature of PD is the presence of cytoplasmic inclusions of fibrillar misfolded proteins, Lewy bodies, in the affected brain areas (see Table 8.1). The exact composition of Lewy bodies is unknown, but they contain ubiquinated α-synuclein, parkin, synphilin, neurofilaments, synaptic vesicle proteins (for a review, see [24]), and a recently identified protein, TPPP/p25 [48]. Immunohistochemical studies provided evidence for the co-localization of α-synuclein with GAPDH in Lewy bodies [49, 50]. A punctuate pattern of GAPDH distribution for other inclusions of PD nigral section has also been described [51]. There are *in vitro* binding data for the direct association of GAPDH with both α-synuclein [52] and TPPP/p25 protein [50], however, neither the counteraction of these associations nor their effect on the ATP production has been established. Interaction between α-synuclein and TPPP/p25, and TPPP/p25 promoted assembly of α-synuclein into filaments within Lewy bodies was revealed as well [53]. The data referring to the impairment of energy metabolism in PD are very limited; virtually only mitochondrial injury and altered cellular transport have been proposed [54].

Indeed, not very much data have been accumulated that effectively contribute to our understanding how the appearance/accumulation of the unfolded/misfolded proteins affects the energy metabolism of cells, whether the expression of these proteins directly or indirectly interferes with the glucose metabolism. One of these studies was performed on a stable SK-N-MC neuroblastoma cell line expressing TPPP/p25 (denoted as K4 cells), and the energy metabolism was studied at cellular and molecular levels in the transfected cells as compared with the controls [55].

Mitochondrial membrane potency, which regulates the production of high-energy phosphate, is an important parameter determining the fate of the cells [56]. For this reason, the hyperpolarization of mitochondrial membrane was determined by immunfluorescence microscopy monitoring the accumulation of a fluorescence marker dye. The comparison of the fluorescence images of the control (SK-N-MC) and transfected (K4) cells showed that the expression of TPPP/p25 did not damage, rather stimulated, the production of the intracellular high-energy phosphate as long as the cells preserved their morphology similar to the control. Quantitative data was obtained by determination of the cellular ATP level. The ATP concentration was found 1.5-fold higher in the extract of the K4 clone as compared with that of the control cells (26.6 ± 3.0 μM/g protein and 17.2 ± 1.8 μM/g protein, respectively). Analysis of the activities of the key glycolytic enzymes revealed that, except aldolase, the activities of other enzymes from the upper part of glycolysis were significantly higher in the case of K4 cells than in the control. The flux of glucose consumption, computed using a mathematical model with kinetic parameters determined experimentally and by well established rate equations of individual enzymatic reactions, showed that it was stimulated by the presence of TPPP/p25. The cellular, biochemical, and computational results suggested that the expression of the unfolded TPPP/p25 in K4 cells could be tolerated by responding with extensive energy production. This situation may mimic a very early stage of the accumulation of the undesired toxic proteins.

8.4
Post-translational Modifications of Glycolytic Enzymes

The impairment of glucose utilization could result from the modification of the glycolytic enzymes under oxidative stress effects. Oxidative stress is an important factor leading to the pathophysiologcal alterations in conformational diseases. Oxidative stress is manifested in protein oxidation, lipid peroxidation, DNA oxidation, and advanced glycation end-products, as well as reactive oxygen species (ROS), and reactive nitrogen species (RNS) formation. Either the oxidants or the products of oxidative stress could modify the proteins or activate other pathways that may lead to additional impairment of cellular functions and to neuronal loss [57, 58].

Nitration of some glycolytic enzymes has been reported in AD [32], PD [59], and other neurodegenerative diseases (see Table 8.2) [60, 61]. Nitric oxide (NO) is, indeed, implicated in the pathophysiology of a number of neurodegenerative diseases. NO reacts with the superoxide anion producing highly reactive peroxynitrite [62], which causes protein nitrotyrosination, a marker of cell damage reported in neurons and glial cells from AD brains [32]. Vascular amyloid deposits correlate with nitrotyrosination in brain vessels from AD patients [63]. Aβ fibrils act as a source of superoxide anion [64], which can react with the basal levels of NO, owing to the high affinity of NO for the superoxide anion [65], thereby triggering nitroty-rosination.

The redox proteomics approach utilized extensively for the identification of oxidized proteins has provided data consistent with the biochemical and pathological alterations in AD ([66] and references therein). Among other proteins, the glycolytic enzymes GAPDH, triosephosphate isomerase (TPI), phosphoglycerate mutase (PGM), and enolase were identified as the targets of nitration in AD hippocampus. Perturbation in energy metabolism and mitochondrial functions by specific protein nitration could be one of the mechanisms for the onset and progression of AD [67].

In AD, neuronal and synaptic loss occurs in a region-specific manner. Regions of the brain (for example the hippocampus) where oxidized forms of the glycolytic enzymes occurred showed an extensive deposition of Aβ, compared with the age-matched control brains. Oxidation of α-enolase and TPI was found not only in the hippocampus but also in the inferior parietal lobule, which implies a common mechanism operating in the two different regions of brain [66]. Why some brain regions are more sensitive in AD is unclear yet. The identification of common targets of oxidative damage would help us to understand the pathogenesis of the disease and could assist the development of more specific therapeutic strategies.

Enolase produces phosphoenolpyruvate, the second of the two high-energy intermediates that generate ATP in glycolysis. A proteomics analysis of AD brain showed that the protein level of the α-subunit of enolase is increased as compared to control brain [66, 68], and α-enolase as a nitrated protein was identified [66, 69]. Due to oxidation of the two isoforms of enolase existing in brain tissue (α and γ), the activity of this enzyme in AD brain was decreased [66, 70]. Enolase is also a target for carbonylation in AD hippocampus, which implies changes in cellular pH and bioenergetics that could impair enzyme activities leading to neurodegeneration.

PGM catalyzes the interconversion of 3-phosphoglycerate and 2-phosphoglycerate. A significant increase in oxidation of PGM and a decrease in protein expression were observed in AD hippocampus [66] that is consistent with the reported decreased expression and activity of PGM in AD brain, as compared with the age-matched controls [70, 71].

GAPDH is an excessively nitrated protein in AD hippocampus [67]. Aβ-induced oxidative disulfide bond formation of GAPDH, promoting the nuclear accumulation of the detergent-insoluble form of the enzyme in brain samples of AD patients and transgenic AD mice, was also observed [72]. Oxidation and subsequent loss of function of these glycolytic enzymes result in decreased ATP production, a finding consistent with the altered glucose tolerance and metabolism of patients suffering from neurodegenerative diseases [73, 74].

TPI, another glycolytic enzyme that catalyzes the interconversion of dihydroxy-acetone phosphate (DHAP) and D-glyceraldehyde-3-phosphate in glycolysis, was also oxidatively modified [32, 63, 66]. However, no change of TPI activity was observed [66], probably because the carbonyl groups added were localized away from the catalytic site of this enzyme. TPI was strongly inhibited by endogenous

β-carbolines due to their binding to TPI under *in vivo* conditions. β-Carbolines are known as potent inhibitors of mitochondrial respiration and cause neurodegeneration after injection into the substantia nigra [75]. Increased level of β-carboline derivatives was observed in the cerebrospinal fluid of patients with *de novo* PD [76]. Since it is known that TPI deficiency (originated by mutations of the isomerase) manifests itself in severe progressive extrapyramidal disorder and other neurological symptoms [77], it was thus suggested that inhibition of TPI by β-carbolines could contribute to neurodegeneration [75]. It is worth noting from this aspect that the structure of β-carbolines closely resembles that of the neurotoxin MPTP (1-methyl-4-phenyl-1,2,3,6-tetrahydropyridine) that is routinely used to mimic PD in animals.

8.5
Triosephosphate Isomerase Deficiency, a Unique Glycolytic Enzymopathy

The deficiency of the glycolytic enzyme, TPI, results not only in hematological symptoms, as in the case of the majority of the glycolytic enzymes, but in extrapyramidal neurodegeneration [78]. It is a rare disease with childhood death. It has been suggested that the disease is developed due to the mutation of TPI causing protein misfolding which leads to the partial inactivation of the isomerase [79]. Thus TPI deficiency could be considered as a "conformational disease". The extent of the loss of activity depends on the position of the mutation as well as on the types of blood cells ([78] and references therein). It is has to be noted that no brain samples have been available for research, even from *post mortem* brain tissues. However, the investigation of glycolytic pathway in the erythrocytes of patients rendered it possible to evaluate the effect of mutation on the energy metabolism.

The classic function of TPI is to adjust the rapid equilibrium between the two triosephosphates, glycerinealdehyde-3-phosphate and DHAP. Patients with TPI deficiency have unimpressive alterations in glucose utilization, ATP and lactate production. Modeling studies and experimental data suggested that the physiological ATP level was maintained due to the activation of enzymes involved in the pentosephosphate and glycolytic pathways (Fig. 8.2) [80, 81]. The interconnection of the two pathways with increased activities can compensate for the reduced TPI activity of deficient cells in TPI-deficient erythrocytes [80, 81].

Dramatic change in erythrocytes is characterized by the marked increase in DHAP and in a much lesser increase in fructose-1,6-bisphosphate levels [78, 81]. An inverse relationship between TPI activity and DHAP concentration was detected in all TPI-deficient patients. Computer models were evaluated to simulate this inverse relationship in normal and deficient red blood cells [82–84]. However, the low TPI activity-derived DHAP accumulation was not supported by the previous models unless a very low TPI activity was introduced which was not supported by experimental data [82, 84]. Recently, a realistic model of human erythrocyte glycolysis was elaborated on the basis of experimentally determined kinetic parameters [81]. It was shown that the mutation-derived activity

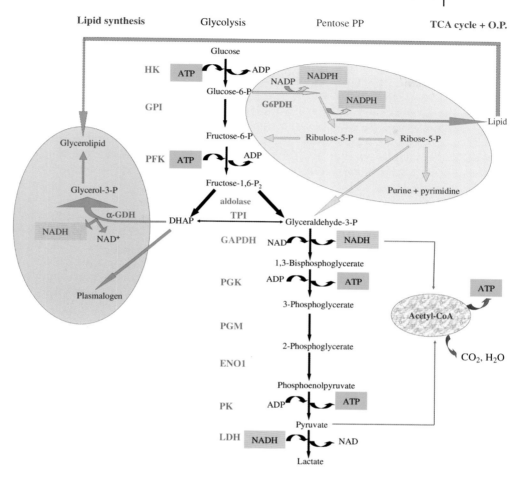

Fig. 8.2 Glycolysis and related pathways. Glycolysis is a central metabolic machinery in which one mole of glucose is catabolized to two moles of pyruvate, NADH, and ATP. Under aerobic conditions, pyruvate is further oxidized by mitochondrial system. In erythrocytes DHAP is a dead-end product; however, in brain it can be converted into direction of lipid synthesis. Glycolysis and the pentose phosphate pathway (pentosePP) are interconnected via fructose-6-P and glyceraldehyde-3-P. A high level of NADPH favors lipid synthesis via pentose phosphate shunt (pentosePP). At TPI inhibition (TPI deficiency), glyceraldehyde-3-P can be produced via G6PDH as well, to contribute to the glycolytic flux. α-GDH catalyzes the oxidation of NADH to NAD$^+$ and the reduction of DHAP to glycerol-3-P, which is a substrate of biosynthesis of glycerolipids (i.e., phospholipids, triacylglyceride). At low level of NADPH, DHAP can be transformed to plasmalogens. The reverse situation can occur at reduced TPI activity and neurological disorders [77]. Enzymes: HK = hexokinase, GPI = glucosephosphate isomerase, PFK = phosphofructokinase, TPI = triosephosphate isomerase, GAPDH = glyceraldehyde-3-phosphate dehydrogenase, PGK = phosphoglycerate kinase, PGM = phosphoglycerate mutase, ENO = enolase, PK = pyruvate kinase, LDH = lactate dehydrogenase, α-GDH = α-glycerolphosphate dehydrogenase, G6PDH = glucosephosphate dehydrogenase.

decrease of TPI was associated with an increase in the activity of the other glycolytic enzymes, including those (HK and PFK) which are known to control predominantly the glycolytic flux in erythrocytes. The mathematical modeling of the glycolytic flux required a consideration of changes in the kinetic parameters of the glycolytic enzymes, in addition to that of TPI, to obtain relevant information on the energy state of TPI-deficient erythrocytes. For the modeling of the whole glycolytic pathway (see Fig. 8.2), the V_{max} values of the glycolytic enzymes as well as the K_{equ} (= [DHAP]/[GAP]) and K_m values of the TPI reaction were determined experimentally in the hemolysates from both the control and the diseased cells (see http://www.BiochemJ.org/bj/392/bj3920675add.htm). Simulation data showed that the rate of glucose conversion into lactate was higher for the mutant cell, and the steady-state flux was increased by a 2.5-fold factor due to the enhanced activities of HK, PFK, and GAPDH. The mathematical model corresponded to the pathological criterion, namely, this realistic model was appropriate to mimic the higher DHAP level in the hemolysate of the patient. In fact, a new ratio of triosephosphate concentrations was adjusted for the steady-state flux of glycolysis in the diseased cell, due to the altered activity of the other enzymes of the glycolytic pathway. It was proposed that the TPI mutation-derived alterations of key glycolytic enzymes might represent a "repairing mechanism" [80] that could ensure a virtually normal energy state in erythrocytes [85].

A number of experimental and theoretical data has been accumulated related to the mutant TPI-connected metabolism in diseased cells, however, it is still a mystery which extent the TPI mutation is responsible for the development of neurodegenerative disorder. Recently, Bonnet and co-workers [75] suggested a qualitative model, in which they incorporated results published mostly by Hollán and her co-workers [77, 86]. The compensation of reduced activity of TPI by activation of the pentose phosphate shunt [80] leads to the accumulation of NADPH, which is mainly utilized for the synthesis of fatty acids in brain cells (Fig. 8.2). Thus, inhibition of TPI likely stimulates indirectly the biosynthesis of fatty acids. DHAP is a substrate of glycerol-3-phosphate dehydrogenase, an oxidoreductase which accepts both NADH and NADPH as co-substrates, depending on the supply. The product of the enzymatic reaction is glycerol-3-phosphate, a common substrate of various glycerolipids such as phospholipids and triacylglyceride. Furthermore, DHAP also serves as a precursor for the formation of etherlipids among them that of plasmalogens (see Fig. 8.2). High levels of NADPH shift the utilization of DHAP to glycerol-3-phosphate and cause a reduction in plasmalogens, specifically in phosphatidyl-ethanolamines [77]. These changes lead to an imbalance between bilayer- and nonbilayer-forming lipids in membranes, severely affecting the activity of several enzymes and the association of protein kinase C and G proteins to the membrane. Thus, the decrease in plasmalogens impairs the protection against oxidative stress with consecutive worsening of the neurodegenerative process.

The classic interpretation of the development of neurodegeneration in the case of TPI deficiency is based upon the extensive inhibition of TPI activity by mutation, which prevents the adjustment of the physiological ratio of triosephosphates. Indeed, in erythrocytes, it results in the accumulation of DHAP. However, this may not be valid for neurons or astrocytes. Even in the case of lymphocytes, which display some similarity to brain cells, only slight elevation of DHAP concentration was detected [79]. Therefore, further investigations are necessary to support or exclude the direct role of the mutation of TPI in the development of neurodegeneration. Recently, significant changes in mRNA levels of nitric oxide synthase and prolyl oligopeptidase were reported in the lymphocytes of a patient suffering TPI deficiency [81]. These two enzymes have been suggested to be indirectly involved in neurodegenerative diseases [87, 88]. The nitration of TPI has been documented in AD (see Table 8.2), and a 15-fold increase in the urinary 3-nitrotyrosine was found in a TPI-deficient patient as compared with that of the control [89].

8.6
Microcompartmentation in Energy Metabolism

Most of the proteins involved in neuronal homeostasis are synthesized within the body of the neuronal cell. Many of these proteins are required in the axon or at the nerve terminal far from the point of their synthesis. Most synapse-bound cargoes are transported directly to their destinations through the action of microtubules and actin-based motors [90]. The bulk of neuronally synthesized proteins destined for the axon is transported about 100 times slower than the vesicular traffic of fast axonal transport (100 mm/day). Two slow axonal transport components, SCa and SCb, have been identified [91]. The SCb is primarily made of actin, but it includes glycolytic enzymes with other known and not yet identified components. The binding of the slowly transported proteins to the motor or some intermediate could be in a "piggy-back" manner. Such piggy-back binding of sequential glycolytic enzymes along sarcomeres of *Drosophila* flight muscles has been demonstrated [92]. The co-localization of a set of six glycolytic enzymes, aldolase, glycerol-3-phosphate dehydrogenase, GAPDH, TPI, 3-phosphoglycerate kinase (PGK), and PGM, that catalyse consecutive reactions along the glycolytic pathway was revealed. It became clear that the presence of active enzymes in the cell is not sufficient for muscle function; a highly organized cellular system of the specific isoforms of enzymes is required ensuring mechanisms by which ATP is supplied to the filament ATPase [93].

In brain tissues, specific isoforms of glycolytic enzymes are also expressed; there are specific brain isoforms for PFK (PFK-C), fructose-1,6-bisphosphate aldolase (aldolase C), enolase (enolase γ), but not for GAPDH. The isoforms bear the same catalytic functions; however, they could be specialized to form different ultrastructural entities. For example, muscle PFK (a dissociable tetrameric form) binds to microtubules and bundle them [94, 95], however, the brain isoenzyme (stable tetramer) does not [96].

Glucose is not only the major energy source, but it is essential for maintaining normal neuronal function. Recent evidence suggests a direct correlation between glucose utilization and cognitive function ([97] and reference therein). Reduction of glucose levels results in pathophysiological state, however, this occurs long before significant alteration in tissue ATP levels is detected. Substitution of pyruvate for glucose does not support normal evoked neuronal activity, although tissue ATP level returns to normal. All these results suggest that glycolysis or glycolytic intermediate(s) are necessary for normal synaptic transmission independently of global cellular ATP levels.

Microcompartmentation of a metabolite is defined as a substrate having more than one distinct pool. It has been documented, for example, for glutamate metabolism at the multi- and single-cellular level in brain. Glycogen metabolism of astrocytes was investigated by exposing it to (U-^{13}C) glucose in the absence and presence of a glycogen phosphatase inhibitor [98]. The results indicated distinct compartments for lactate derived from glycogenolysis and that derived from glycolysis, and suggested a linkage between glycogen turnover and glycolytic and TCA cycle activity [99]. It has been proposed that the TCA cycle intermediates and TCA cycle-derived amino acids are preferentially generated from the glucose metabolized by glycolysis, whereas the glutamine mainly originated from glycogenolysis. Since glutamine is a precursor of the neurotransmitter glutamate, glycogen turnover could play a role in glutamatergic activity [99]. In addition, mitochondrial heterogeneity within single astrocytes was revealed.

Importance of glucose metabolism in maintaining neural activity was investigated by measuring the effect of lowering the concentration of glucose in perfusion medium on synaptic activity (population spikes, PS) and on the level of high energy phosphates in various regions of the hippocampus of pigs. Lowering the glucose concentration, ATP and creatine phosphate levels were well preserved [100]. However, significant differences were detected in the activity of PFK, in contrast to HK that showed no significant differences in the different regions of the hippocampus. The sensitivity in maintaining synaptic activity at lowered glucose concentrations varied parallel with the activity of PFK in the different regions of the hippocampus, indicating that the glycolytic metabolism regulated by PFK is crucial for the maintainance of synaptic activity.

The synaptic vesicles are capable of accumulating the excitatory neurotransmitter glutamate by harnessing ATP. It has been also demonstrated that GAPDH and its sequential glycolytic enzyme, PGK, are enriched in synaptic vesicles, forming a functional complex [101]. The vesicle-bound GAPDH/PGK can produce ATP at the expense of their substrates. The inhibition of mitochondrial ATP synthesis, however, has minimal effect on any of these parameters. This suggests that the coupled GAPDH/PGK system, which converts ADP to ATP, ensures maximal glutamate accumulation into presynaptic vesicles.

The maximum rate of some enzyme activities related to glycolysis, Krebs cycle, acetylcholine catabolism, and amino acid metabolism were determined in different types of synaptosomes obtained from rat hippocampus. Characterization of the enzyme was performed on two synaptosomal populations defined as "large" synaptosomes mainly containing the glutamatergic nerve endings and "small" synaptosomes, which typically contains the cholinergic nerve endings mainly arising from septohippocampal fiber synapses involved in cognitive processes [102]. In control animals, the activities of enzymes participating in energy metabolism were different in the two synaptosomal populations. It was proposed that this energetic micro-heterogeneity of the synaptosomes may underlay their different behavior during both physiopathological events and pharmacological treatment due to the different sensitivity of neurons confirming the existence of different metabolic machinery.

The postsynaptic density (PSD) organelle is a dense concentration of proteins attached to the cellular surface of the postsynaptic membrane in dendritic spine heads. It contains a large number of proteins; for example in the cerebral cortex more than 30 kinds of proteins were identified by characteristic 2D electrophoretic mobility, immunoblotting with specific antibodies, and N-terminal and peptide sequencing [103]. Recently, much more (>1000) protein constituents of the PSD were identified by proteomic analysis [104–110], of which 466 were validated by their detection in two or more studies [110]. These proteins include the receptors and ion channels for glutamate neurotransmission, proteins for maintenance and modulation of synaptic architecture, sorting and trafficking of membrane proteins, generation of anaerobic energy, scaffolding and signalling, local protein synthesis, and correct protein folding and breakdown of synaptic proteins.

PSD is enriched for spectrin, actin, tubulin and microtubule-associated protein II, myosin, glycolytic enzymes, creatine kinase, elongation factor 1 alpha, receptor, and neurofilament proteins. In fact, all glycolytic enzymes were found to be present in the PSD fraction. Among them GAPDH and PGK, directly involved in the generation of ATP, lactate dehydrogenase (LDH) used for the regeneration of NAD$^+$, were identified [111]. The association of GAPDH with synaptosomal particles but not with the synaptosomal plasma membranes was revealed as well [112]. A physical association of GAPDH with endogenous actin in synaptosomes and PSDs was demonstrated very early [111]. These data suggest a role of glycolysis in synaptosomes, as well as the involvement of GAPDH in nonglycolytic functions, such as actin binding and actin filament network organization.

The presence of GAPDH and fructose-bisphosphate aldolase was validated by their detection in all of the above-mentioned studies. It was postulated that these proteins provide immediate availability of glycolytic source of ATP to the synapse. This is especially important during acute increase in synaptic activity when the mitochondrial supply of energy does not meet the transient and highly localized increased demand in energy. Indeed, the glycolytic generation of ATP at the isolated PSD was demonstrated. In addition to the proteins involved in glycolysis, other proteins also involved in energy generation (i.e. creatine kinase-B) or in the phosphate transfer system (i.e. adenylate kinases) were detected. These proteins play important roles in providing the site-specific burst of high-energy phosphate. The ATP utilized by kinases (Ca^{2+}/calmodulin-dependent protein kinase, protein kinase A and C) was also found in the PSD. Therefore, it is likely that localized ATP is supplied not by the mitochondria, but by the glycolytic complex at the PSD [113].

Compartmentation of energy-producing systems in the central nervous system points to their crucial role in cell communication. ATP produced by mitochondria in the dendritic shaft may travel into the dendritic spine, the site of the synapse, because there are no mitochondria in the spine itself. However, there is another energy-producing system forming ATP in the spine itself, produced by glycolytic enzymes localized in the PSD at the postsynaptic membrane of the synapse [113]. The significance of these findings is that the ATP generated at this important synapse site can be used at ion channels, by various protein kinases involved in signal transduction, and also for protein synthesis in the spine. Thus, there could exist two energy-producing systems, one existing in the astrocyte and in the dendritic shaft in the production of energy for general "housekeeping" metabolic functions, and the other at the postsynaptic PSD site for the processing of nerve signal transduction. The communication requirement of the central nervous system almost invites a specialized, localized, structural site for its most important mission. These results show that the activity-dependent glycolytic and oxidative metabolic responses in the central nervous system are highly compartmentalized, on one hand, between astrocytes and neurons, on the other hand, in the neuron itself.

8.7
Concluding Remarks

Numerous studies have provided intriguing evidence for the role of energy production in the resting and activated function of the neuronal system. Much less is, however, known about the molecular mechanism for coupling between energy production and the development of conformational diseases that include neurodegenerative disorders, such as AD, PD, and HD, and TPI deficiency (a new candidate entering this family of diseases). This paper reviews the interrelations among glucose and lactate consumption, ATP production and utilization, physiological neuronal activation and aberrant protein–protein associations. It is clear that the interrelationships among glycolysis, ATP supply, and synaptic transmission is an important issue to understand the potential role of the damage of energy metabolism in the neurodegenerative processes. One of the major difficulties is, however, that two opposite processes may occur simultaneously: a reduction of neuronal metabolism and compensatory enhancement in glial activity [16]. The altered activities of some glycolytic enzymes in brain homogenates represent a net effect and it is unclear, if the metabolic activation is restricted to astrocytes, how the energy state (ATP level) of the neighboring neurons is being perturbed. Energy metabolism that uses lactate as an oxidizable energy substrate, produced by either neurons or astrocytes, is an important pathway for neuronal energy production under prolonged stimulation and may be under pathological conditions [114]. The appearance of clinical symptoms such as protein aggregates, fibres, inclusions, is frequently initiated by conformationally unfolded/misfolded proteins that enter aberrant protein–protein interactions. In these interactions well characterized "neurodegenerative proteins" (Aβ, α-synuclein, hyperphosphorylated tau, TPPP/p25) and metabolic enzymes involved in energy production are participating, resulting in nonphysiological ATP levels by slowing down the glucose consumption and lactate oxidation.

Microcompartmentation of metabolic enzymes has been shown to be a powerful way to control energy production in living cells. PSD is an excellent example for the ATP utilized by ion channels or protein kinases involved in signal transduction. Microcompartmentation can alter the flux of the whole pathway by producing high local metabolite concentration, resulting in different kinetic parameters as compared with that determined for individual enzymes. The formation of such organized enzyme structures (metabolon) [114] creates non-ideal behavior of the systems, which can be described in a phenomenological manner. However, for understanding the mechanism of energy metabolism in normal and pathological conditions, the evaluation of rational models is needed, including relevant data that determine metabolic flux and well established mathematical equations of the consecutive enzymatic reactions and transport processes. Coupling of more extended modeling data obtained for physiological and pathological situations is expected to improve the chances of early diagnosis as well as proposing novel drugs for the treatment of different stages of the diseases.

Acknowledgments

This work was supported by FP6-2003-LIFESCIHEALTH-I: Bio-Sim and Hungarian National Scientific Research Fund Grants OTKA T-046071 (to J.O.) and T-049247 (to F.O.), and NKFP-MediChem2 1/A/005/2004 (to J.O.).

References

1 Chih, C. P., Lipton, P., Roberts, E. L. J.: Do active cerebral neurons really use lactate rather than glucose? *Trends Neurosci.* **2001**, 24:573–578.

2 Pellerin, L., Magistretti, P. J.: Glutamate uptake into astrocytes stimulates aerobic glycolysis: a mechanism coupling neuronal activity to glucose utilization. *Proc. Natl Acad. Sci. USA* **1994**, 91:10625–10629.

3 Pellerin, L., Pellegri, G., Bittar, P. G., Charnay, Y., Bouras, C., Martin, J. L., Stella, N., Magistretti, P. J.: Evidence supporting the existence of an activity-dependent astrocyte-neuron lactate shuttle. *Dev. Neurosci.* **1998**, 20:291–299.

4 Gjedde, A., Marrett, S., Vafaee, M.: Oxidative and nonoxidative metabolism of excited neurons and astrocytes. *J. Cereb. Blood Flow Metab.* **2002**, 22:1–14.

5 Pellerin, L., Magistretti, P. J.: How to balance the brain energy budget while spending glucose differently. *J. Physiol.* **2003**, 546:325.

6 Fillenz, M.: The role of lactate in brain metabolism *Neurochem. Int.* **2005**, 47:413–417.

7 Aubert, A., Costalat, R., Magistretti, P. J., Pellerin, L.: Brain lactate kinetics: Modeling evidence for neuronal lactate uptake upon activation. *Proc. Natl. Acad. Sci. USA* **2005**, 102:16448–16453.

8 Hu, Y., Wilson, G. S.: A temporary local energy pool coupled to neuronal activity: fluctuations of extracellular lactate levels in rat brain monitored with rapid-response enzyme-based sensor. *J. Neurochem.* **1997**, 69:1484–1490.

9 Pellerin, L., Magistretti, P. J.: Let there be (NADH) light. *Science* **2004**, 305:50–52.

10 Prichard, J., Rothman, D., Novotny, E., Petroff, O., Kuwabara, T., Avison, M., Howseman, A., Hanstock, C., Shulman, R.: Lactate rise detected by 1H NMR in human visual cortex during physiologic stimulation. *Proc. Natl. Acad. Sci. USA* **1991**, 88:5829–5831.

11 Frahm, J., Kruger, G., Merboldt, K. D., Kleinschmidt, A.: Dynamic uncoupling and recoupling of perfusion and oxidative metabolism during focal brain activation in man. *Magn. Reson. Med.* **1996**, 35:143–148.

12 Mangia, S., Garreffa, G., Bianciardi, M., Giove, F., Di Salle, F., Maraviglia, B.: The aerobic brain: lactate decrease at the onset of neural activity. *Neuroscience* **2003**, 118:7–10.

13 Kasischke, K. A., Vishwasrao, H. D., Fisher, P. J., Zipfel, W. R.: Neural activity triggers neuronal oxidative metabolism followed by astrocytic glycolysis. *Science* **2004**, 305:99–103.

14 Berg, J. M., Tymoczko, J. L., Stryer, L.: *Biochemistry*, 5th edn. Freeman, New York, **2002**.

15 Sims, N. R., Blass, J. P., Murphy, C., Bowen, D. M., Neary, D.: Phosphofructokinase activity in the brain in Alzheimer's disease. *Ann Neurol.* **1987**, 21:509–510.

16 Bigl, M., Brückner, M. K., Arendt, T., Bigl, V., Eschrich, K.: Activities of key glycolytic enzymes in the brains of patients with Alzheimer's disease. *J. Neural Transm.* **1999**, 106:499–511.

17 Sorbi, S., Mortilla, M., Piacentini, S., Tonini, S., Amaducci, L.: Altered hexokinase activity in skin cultured fibroblasts and leukocytes from Alzheimer's disease patients. *Neurosci. Lett.* **1990**, 117:165–168.

18 Bigl, M., Bleyl, A. D., Zedlick, D., Arendt, T., Bigl, V., Eschrich, K.: Changes of activity and isozyme pattern of phosphofructokinase in the brains of patients with Alzheimer's disease. *J. Neurochem.* **1996**, 67:1164–1171.

19 Dunker, A. K., Brown, C. J., Lawson, J., Iakoucheva, L. M., Obradovic, Z.: Intrinsic disorder and protein function. *Biochemistry* **2002**, 41:6573–6582.

20 Wright, P. E., Dyson, H. J.: Intrinsically unstructured proteins: re-assessing the protein structure-function paradigm *J. Mol. Biol.* **1999**, 293:321–331.

21 Johnston, J. A., Ward, C. L., Kopito, R. R.: Aggresomes: a cellular response to misfolded proteins. *J. Cell Biol.* **1998**, 143:1883–1898.

22 Ovádi, J., Orosz, F., Lehotzky, A.: What is the biological significance of the brain-specific tubulin-polymerization promoting protein (TPPP/p25)? *IUBMB Life* **2005**, 57:765–768.

23 Muchowski, P., Wacker, J. L.: Modulation of neurodegeneration by molecular chaperones. *Nat. Rev. Neurosci.* **2005**, 6:11–22.

24 Bossy-Wetzel, E., Schwarzenbacher, R., Lipton, S. A.: Molecular pathways to neurodegeneration. *Nat. Med.* **2004**, 10[Suppl]S2–S9.

25 Vannucci, R. C., Vannucci, S. J.: Glucose metabolism in the developing brain. *Semin. Perinatol.* **2000**, 24:107–115.

26 Duara, R., Grady, C., Haxby, J., Sundaram, M., Cutler, N., Heston, L., Moore, A., Schlageter, N., Larson, S., Rapoport, S. I.: Positron emission tomography in Alzheimer's disease. *Neurology* **1986**, 36:879–887.

27 Tucek, S., Ricny, J., Dolezal, V.: Advances in the biology of cholinergic neurons. *Adv. Neurol.* **1990**, 51:109–115.

28 Hoshi, M., Takashima, A., Murayama, M., Yasutake, K., Yoshida, N., Ishiguro, K., Hoshino, T., Imahori, K.: Nontoxic amyloid beta peptide 1-42 suppresses acetylcholine synthesis. Possible role in cholinergic dysfunction in Alzheimer's disease. *J. Biol. Chem.* **1997**, 272:2038–2041.

29 Hoshi, M., Takashima, A., Noguchi, K., Murayama, M., Sato, M., Kondo, S., Saitoh, Y., Ishiguro, K., Hoshino, T., Imahori, K.: Regulation of mitochondrial pyruvate dehydrogenase activity by tau protein kinase I/glycogen synthase kinase 3beta in brain. *Proc. Natl. Acad. Sci. USA* **1996**, 93:2719–2723.

30 Erecinska, M., Silver, I. A.: ATP and brain function. *J. Cereb. Blood Flow Metab.* **1989**, 9:2–19.

31 Castegna, A., Aksenov, M., Aksenova, M., Thongboonkerd, V., Klein, J. B., Pierce, W. M., Booze, R., Markesbery, W. R., Butterfield, D. A.: Proteomic identification of oxidatively modified proteins in Alzheimer's disease brain. Part I: creatine kinase BB, glutamine synthase, and ubiquitin carboxy-terminal hydrolase L-1. *Free Radical Biol. Med.* **2002**, 33:562–571.

32 Castegna, A., Thongboonkerd, V., Klein, J. B., Lynn, B., Markesbery, W. R., Butterfield, D. A.: Proteomic identification of nitrated proteins in Alzheimer's disease brain. *J. Neurochem.* **2003**, 85:1394–1401.

33 Hoyer, S. Causes and consequences of disturbances of cerebral glucose metabolism in sporadic Alzheimer disease: therapeutic implications. *Adv. Exp. Med. Biol.* **2004**, 541:135–152.

34 Guo, J. P., Arai, T., Miklossy, J., McGeer, P. L.: $A\beta$ and tau form soluble complexes that may promote self aggregation of both into the insoluble forms observed in Alzheimer's disease. *Proc. Natl. Acad. Sci. USA* **2006**, 103:1953–1958.

35 Tabaton, M.: Research advances in the biology of Alzheimer's disease. *Clin. Geriatr. Med.* **1994**, 10:249–255.

36 Meier-Ruge, W. A., Bertoni-Freddari, C.: Pathogenesis of decreased glucose turnover and oxidative phosphorylation in ischemic and trauma-induced dementia of the Alzheimer type. *Ann. NY Acad. Sci.* **1997**, 826:229–241.

37 Schulze, H., Schuyler, A., Stuber, D., Dobeli, H., Langen, H., Huber, G.: Rat brain glyceraldehyde-3-phosphate dehydrogenase interacts with the recombinant cytoplasmic domain of Alzheimer's β-amyloid precursor protein. *J. Neurochem.* **1993**, 60:1915–1922.

38 Ovádi, J., Orosz, F., Hollán, S.: Functional aspects of cellular microcompartmentation in the development of neurodegeneration: mutation induced aberrant protein–protein associations *Mol. Cell. Biochem.* **2004**, 256/257:83–93.

39 Mazzola, J. L., Sirover, M. A.: Reduction of glyceraldehyde-3-phosphate dehydrogenase activity in Alzheimer's disease and in Huntington's disease fibroblasts. *J. Neurochem.* **2001**, 76:442–449.

40 Barinaga, M.: An intriguing new lead on Huntington's disease *Science* **1996**, 271:1233–1234.

41 Roses, A. D.: From genes to mechanisms to therapies: Lessons to be learned from neurological disorders. *Nat. Med.* **1996**, 2:267–269.

42 Chuang, D. M., Ishitani, R.: A role for GAPDH in apoptosis and neurodegeneration. *Nat. Med.* **1996**, 2:609–610.

43 Meier-Ruge, W., Bertoni-Freddari, C.: The significance of glucose turnover in the brain in the pathogenic mechanisms of Alzheimer's disease. *Rev. Neurosci.* **1996**, 7:1–19.

44 Mangiarini, L., Sathasivam, K., Seller, M., Cozens, B., Harper, A., Hetherington, C., Lawton, M., Trottier, Y., Lehrach, H., Davies, S. W., Bates, G. P.: Exon 1 of the HD gene with an expanded CAG repeat is sufficient to cause a progressive neurological phenotypes in transgenic mice. *Cell* **1996**, 87:493–506.

45 Burke, J. R., Enghild, J. J., Martin, M. E., Jou, Y.-S., Myers, R. M., Roses, A. D., Vance, J. M., Strittmatter, W. J.: Huntingtin and DRPLA proteins selectively interact with the enzyme GAPDH. *Nat. Med.* **1996**, 2:347–350.

46 Cooper, A. J., Sheu, K. F., Burke, J. R., Strittmatter, W. J., Blass, J. P.: Glyceraldehyde-3-phosphate dehydrogenase abnormality in metabolically stressed Huntington disease fibroblasts. *Dev. Neurosci.* **1998**, 20:462–468.

47 Sharp, A. H., Lowv, S. J., Li, S. H., Li, X. J., Bao, J., Waster, M. V., Kotzuk, J. A., Steiner, J. P., Lo, A., Hedreen, J., Sisodia, S., Snyder, S. H., Dawson, T. M., Ryugo, D. K., Ross, C. A.: Widespread expression of Huntington's disease gene (IT15) protein product. *Neuron* **1995**, 14:1065–1074.

48 Kovács, G. G., László, L., Kovács, J., Jensen, P. H., Lindersson, E., Botond, G., Molnár, T., Perczel, A., Hudecz, F., Mező, G., Erdei, A., Tirián, L., Lehotzky, A., Gelpi, E., Budka, H., Ovádi, J.: Natively unfolded tubulin polymerization promoting protein TPPP/p25 is a common marker of alpha-synucleinopathies. *Neurobiol. Dis.* **2004**, 17:155–162.

49 Tsuchiya, K., Tajima, H., Kuwae, T., Takeshima, T., Nakano, T., Tanaka, M., Sunaga, K., Fukuhara, Y., Nakashima, K., Ohama, E., Mochizuki, H., Mizuno, Y., Katsube, N., Ishitani, R.: Pro-apoptotic protein glyceraldehyde-3-phosphate dehydrogenase promotes the formation of Lewy body-like inclusions. *Eur. J. Neurosci.* **2005**, 21:317–326.

50 Oláh, J., Tőkési, N., Vincze, O., Horváth, I., Lehotzky, A., Erdei, A., Szájli, E., Medzihradszky, K. F., Orosz, F., Kovács, G. G., Ovádi, J.: Interaction of TPPP/p25 protein with glyceraldehyde-3-phosphate dehydrogenase and their co-localization in Lewy bodies. *FEBS Lett.* **2006**, 580:5807–5814.

51 Tatton, N. A.: Increased caspase 3 and Bax immunoreactivity accompany nuclear GAPDH translocation and neuronal apoptosis in Parkinson's disease. *Exp. Neurol.* **2000**, 166:29–43.

52 Lindersson, E., Beedholm, R., Hojrup P, Moos, T., Gai, W., Hendil, K. B., Jensen,

P. H.: Proteasomal inhibition by alpha-synuclein filaments and oligomers. *J. Biol. Chem.* **2004**, 279:12924–12934.

53 Lindersson, E., Lundvig, D., Petersen, C., Madsen, P., Nyengaard, J. R., Hojrup, P., Moos, T., Otzen, D., Gai, W. P., Blumbergs, P. C., Jensen, P. H.: p25alpha stimulates alpha-synuclein aggregation and is co-localized with aggregated alpha-synuclein in alpha-synucleinopathies. *J. Biol. Chem.* **2005**, 280:5703–5715.

54 Ross, C. A., Poirier, M. A.: Protein aggregation and neurodegenerative disease. *Nat. Med.* **2004**, 10:S10–S17.

55 Orosz, F., Kovács, G. G., Lehotzky, A., Oláh, J., Vincze, O., Ovádi, J.: TPPP/p25: from unfolded protein to misfolding disease: prediction and experiments. *Biol. Cell* **2004**, 96:701–711.

56 Iijima, T., Mishima, T., Akagawa, K., Iwao, Y.: Mitochondrial hyperpolarization after transient oxygen-glucose deprivation and subsequent apoptosis in cultured rat hippocampal neurons. *Brain Res.* **2003**, 993:140–145.

57 Butterfield, D. A., Castegna, A., Lauderback, C. M., J. Drake, J.: Evidence that amyloid beta-peptide-induced lipid peroxidation and its sequelae in Alzheimer's disease brain contribute to neuronal death. *Neurobiol. Aging.* **2002**, 23:655–664.

58 Lovell, M. A., Xie, C., Markesbery, W. R.: Acrolein is increased in Alzheimer's disease brain and is toxic to primary hippocampal cultures. *Neurobiol. Aging.* **2001**, 22:187–194.

59 Good, P. F., Hsu, A., Werner, P., Perl, D. P., Olanow, C. W.: Protein nitration in Parkinson's disease. *J. Neuropathol. Exp. Neurol.* **1998**, 57:338–342.

60 Cookson, M. R., Shaw, P. J.: Oxidative stress and motor neurone disease. *Brain Pathol.* **1999**, 9:165–186.

61 Walker, L. M., York, J. L., Imam, S. Z., Ali, S. F., Muldrew, K. L., Mayeux, P. R.: Oxidative stress and reactive nitrogen species generation during renal ischemia. *Toxicol. Sci.* **2001**, 63:143–148.

62 Radi, R.: Nitric oxide, oxidants, and protein tyrosine nitration. *Proc. Natl. Acad. Sci. USA* **2004**, 101:4003–4008.

63 Coma, M., Guix, F. X., Uribesalgo, I., Espuna, G., Solé, M., Andreu, D. F., Munoz, J.: Lack of oestrogen protection in amyloid-mediated endothelial damage due to protein nitrotyrosination. *Brain Res.* **2005**, 128:1613–1621.

64 Butterfield, D. A., Bush, A. I.: Alzheimer's amyloid beta-peptide (1-42): involvement of methionine residue 35 in the oxidative stress and neurotoxicity properties of this peptide. *Neurobiol. Aging* **2004**, 25:563–568.

65 Huie, R., Padmaja, S.: The reaction of NO with superoxide. *Free Radic. Res. Commun.* **1993**, 18:195–199.

66 Sultana, R., Boyd-Kimball, D., Poon, H. F., Cai, J., Pierce, W. M., Klein, J. B., Merchant, M., Markesbery, W. R., Butterfield, D. A.: Redox proteomics identification of oxidized proteins in Alzheimer's disease hippocampus and cerebellum: An approach to understand pathological and biochemical alterations in AD. *Neurobiol. Aging* **2006**, 27:1564–1576.

67 Sultana, R., Poon, H. F., Cai, J., Pierce, W. M., Merchant, M., Klein, J. B., Markesbery, W. R., Butterfield, D. A.: Identification of nitrated proteins in Alzheimer's disease brain using a redox proteomics approach. *Neurobiol. Dis.* **2006**, 22:76–87.

68 Castegna, A., Aksenov, M., Thongboonkerd, V., Klein, J. B., Pierce, W. M., Booze, R., Markesbery, W. R., Butterfield, D. A.: Proteomic identification of oxidatively modified proteins in Alzheimer's disease brain: part II. Dihydropyrimidinase-related protein 2, alpha-enolase and heat shock cognate. *J. Neurochem.* **2002**, 71:1524–1532.

69 Verbeek, M. M., De Jong, D., Kremer, H. P.: Brain-specific proteins in cerebrospinal fluid for the diagnosis of neurodegenerative diseases. *Ann. Clin. Biochem.* **2003**, 40:25–40.

70 Meier-Ruge, W., Iwangoff, P., Reichlmeier, K.: Neurochemical enzyme changes in Alzheimer's and Pick's disease. *Arch. Gerontol. Geriatr.* **1984**, 3:161–165.

71 Iwangoff, P., Armbruster, R., Enz, A., Meier-Ruge, W.: Glycolytic enzymes from human autoptic brain cortex: normal aged and demented cases. *Mech. Ageing Dev.* **1980**, 14:203–209.

72 Cumming, R. C., Schubert, D.: Amyloid-beta induces disulfide bonding and aggregation of GAPDH in Alzheimer's disease. *FASEB J.* **2005**, 19:2060–2062.

73 Rapoport, S. I., Horwitz, B., Grady, C. L., Haxby, J. V., DeCarli, C., Schapiro, M. B.: Abnormal brain glucose metabolism in Alzheimer's disease, as measured by position emission tomography. *Adv. Exp. Med. Biol.* **1991**, 291:231–248.

74 Vanhanen, M., Soininen, H.: Glucose intolerance, cognitive impairment and Alzheimer's disease. *Curr. Opin. Neurol.* **1998**, 11:673–677.

75 Bonnet, R., Pavlovic, S., Lehmann, J., Rommelspacher, H.: The strong inhibition of TPI by the natural β-carbolines may explain their neurotoxic actions. *Neuroscience* **2004**, 127:443–453.

76 Kuhn, W., Muller, T., Gerlach, M., Sofic, E., Fuchs, G., Heye, N., Prautsch, R., Przuntek, H.: Depression in Parkinson's disease: biogenic amines in CSF of "de novo" patients. *J. Neural. Transm.* **1996**, 103:1435–1440.

77 Hollán, S., Magócsi, M., Fodor, E., Horányi, M., Harsányi, V., Farkas, T.: Search for the pathogenesis of the differing phenotype in two compound heterozygote Hungarian brothers with the same genotypic triosephosphate isomerase deficiency. *Proc. Natl. Acad. Sci. USA* **1997**, 94:10362–10366.

78 Schneider, A. S.: Triosephosphate isomerase deficiency: historical perspectives and molecular aspects. *Bailliére's Clin. Haematol.* **2000**, 13:119–140.

79 Orosz, F., Oláh, J., Alvarez, M., Keserű, G. M., Szabó, B., Wágner, G., Kovári, Z., Horányi, M., Baróti, K., Martial, J. A., Hollán, S., Ovádi, J.: Distinct behavior of mutant triosephosphate isomerase in hemolysate and in isolated form: molecular basis of enzyme deficiency. *Blood* **2001**, 98:3106–3112.

80 Oláh, J., Orosz, F., Keserű, G. M., Kovári, Z., Kovács, J., Hollán, S., Ovádi, J.: Triosephosphate isomerase deficiency: a neurodegenerative misfolding disease. *Biochem. Soc. Trans.* **2002**, 30:30–38.

81 Oláh, J., Orosz, F., Puskás, L. G., Hackler, L., Horányi, M., Polgár, L., Hollán, S., Ovádi, J.: Triosephosphate isomerase deficiency: consequences of an inherited mutation at mRNA, protein and metabolic levels. *Biochem. J.* **2005**, 392:675–683.

82 Schuster, R., Holzhütter, H.-G.: Use of mathematical models for predicting the metabolic effect of large-scale enzyme activity alterations. Application to enzyme deficiencies of red blood cells. *Eur. J. Biochem.* **1995**, 229:403–418.

83 Heinrich, R., Rapoport, S. M., Rapoport, T. A.: Metabolic regulation and mathematical models. *Prog. Biophys. Mol. Biol.* **1977**, 32:1–82.

84 Martinov, M. V., Plotnikov, A. G., Vitvitsky, V. M., Ataullakhanov, F. I.: Deficiencies of glycolytic enzymes as a possible cause of hemolytic anemia. *Biochim. Biophys. Acta* **2000**, 1474:75–87.

85 Hollán, S., Fujii, H., Hirono, A., Hirono, K., Karro, H., Miwa, S., Harsányi, V., Gyódi, E., Inselt-Kovács, M.: Hereditary triosephosphate isomerase (TPI) deficiency: two severaly affected brothers one with and one without neurological symptoms. *Hum. Genet.* **1993**, 92:486–490.

86 Orosz, F., Wágner, G., Liliom, K., Kovács, J., Baróti, K., Horányi, M., Farkas, T., Hollán, S., Ovádi, J.: Enhanced association of mutant triosephosphate isomerase to red cell membranes and to brain microtubules. *Proc. Natl. Acad. Sci. USA* **2000**, 97:1026–1031.

87 Gibson, A. M., Edwardson, J. A., McDermott, J. R.: Post mortem levels of some brain peptidases in Alzheimer's disease: reduction in proline endopeptidase activity in cerebral cortex. *Neurosci. Res. Commun.* **1991**, 9:73–81.

88 Mantle, D., Falkous, G., Ishiura, S., Blanchard, P. J., Perry, E. K.: Comparison of proline endopeptidase activity in brain tissue from normal cases and cases with Alzheimer's disease, Lewy body dementia, Parkinson's disease and Huntington's disease. *Clin. Chim. Acta* **1996**, 249:129–139.

89 Ahmed, N., Battah, S., Karachalias, N., Babaei-Jadidi, R., Horányi, M., Baróti, K., Hollán, S., Thornalley, P. J.: Increased formation of methylglyoxal and protein glycation, oxidation and nitrosation in triosephosphate isomerase deficiency. *Biochim. Biophys. Acta* **2003**, 1639:121–132.

90 Shah, J. V., Cleveland, D. W.: Slow axonal transport: fast motors in the slow lane. *Curr. Opin. Cell Biol.* **2002**, 14:58–62.

91 Black, M. M., Lasek, R. J.: Slow components of axonal transport: two cytoskeletal networks. *J. Cell Biol.* **1980** 86:616–623.

92 Wojtas, K., Slepecky, N., von Kalm, L., Sullivan, D.: Flight muscle function in Drosophila requires colocalization of glycolytic enzymes. *Mol. Biol. Cell* **1997**, 8:1665–1675.

93 Sullivan, D. T., MacIntyre, R., Fuda, N., Fiori, J., Barrilla, J., Ramizel, L.: Analysis of glycolytic enzyme co-localization in Drosophila flight muscle. *J. Exp. Biol.* **2003**, 206:2031–2038.

94 Lehotzky, A., Pálfia, Z., Kovács, J., Molnár, A., Ovádi, J.: Ligand-modulated cross-bridging of microtubules by phosphofructokinase. *Biochem. Biophys. Res. Commun.* **1994**, 204:585–591.

95 Lehotzky, A., Telegdi, M., Liliom, K., Ovádi, J.: Interaction of phosphofructokinase with tubulin and microtubules. Quantitative evaluation of the mutual effects. *J. Biol. Chem.* **1993**, 268:10888–10894.

96 Vértessy, B. G., Kovács, J., Ovádi, J.: Specific characteristics of phosphofructokinase-microtubule interaction. *FEBS Lett.* **1996**, 379:191–195.

97 McNay, E. C., Fries, T. M., Gold, P. E.: Decreases in rat extracellular hippocampal glucose concentration associated with cognitive demand during a spatial task. *Proc. Natl. Acad. Sci. USA* **2000**, 97:2881–2885.

98 Waagepetersen, H. S., Sonnewald, U., Larsson, O. M., Schousboe, A.: Multiple compartments with different metabolic characteristics are involved in biosynthesis of intracellular and released glutamine and citrate in astrocytes. *Glia* **2001**, 35:246–252.

99 Sickmann, H. M., Schousboe, A., Fosgerau, K., Waagepetersen, S.: Compartmentation of lactate originating from glycogen and glucose in cultured astrocytes. *Neurochem. Res.* **2005**, 30:1295–1304.

100 Li, X., Yokono, K., Okada, Y.: Phosphofructokinase, a glycolytic regulatory enzyme has a crucial role for maintenance of synaptic activity in guinea pig hippocampal slices. *Neurosci. Lett.* **2000**, 294:81–84.

101 Ikemoto, A., Bole, D., Ueda, T.: Glycolysis and glutamate accumulation into synaptic vesicles. Role of glyceraldehyde phosphate dehydrogenase and 3-phosphoglycerate kinase. *J. Biol. Chem.* **2003**, 278:5929–5940.

102 Gorini, A., D'Angelo, A., Villa, R. F.: Energy metabolism of synaptosomal subpopulations from different neuronal systems of rat hippocampus: effect of L-acetylcarnitine administration in vivo. *Neurochem. Res.* **1999**, 24:617–624.

103 Walsh, M. J., Kuruc, N.: The postsynaptic density: constituent and associated proteins characterized by electrophoresis, immunoblotting, and peptide sequencing. *J. Neurochem.* **1992**, 59:667–678.

104 Jordan, B. A., Fernholz, B. D., Boussac, M., Xu, C., Grigorean, G., Ziff, E. B., Neubert, T. A.: Identification and verification of novel rodent postsynaptic density proteins. *Mol. Cell Proteomics* **2004**, 3:857–871.

105 Li, K. W., Hornshaw, M. P., Van der Schors, R. C., Watson, R., Tate, S., Casetta, B., Jimenez, C. R., Gouwenberg, Y., Gundelfinger, E. D., Smalla, K.-H., Smit, A. B.: Proteomics analysis of rat brain postsynaptic density. Implications of the diverse protein functional groups for the integration of synaptic physiology. *J. Biol. Chem.* **2004**, 279:987–1002.

106 Peng, J., Kim, M. J., Cheng, D., Duong, D. M., Gygi, S. P., Sheng, M.: Semiquantitative proteomic analysis of rat forebrain postsynaptic density fractions by mass spectrometry. *J. Biol. Chem.* **2004**, 279:21003–21011.

107 Yoshimura, Y., Yamauchi, Y., Shinkawa, T., Taoka, M., Donai, H., Takahashi, N., Isobe, T., Yamauchi, T.: Molecular constituents of the postsynaptic density fraction revealed by proteomic analysis using multidimensional liquid chromatography–tandem mass spectrometry. *J. Neurochem.* **2004**, 88:759–768.

108 Cheng, D., Hoogenraad, C. C., Rush, J., Ramm, E., Schlager, M. A., Duong, D. M., Xu, P., Rukshan, S., Hanfelt, J., Nakagawa, T., Sheng, M., Peng, J.: Relative and absolute quantification of postsynaptic density proteome isolated from rat forebrain and cerebellum. *Mol. Cell. Proteomics* **2006**, 5:1158–1170.

109 Collins, M. O., Yu , L., Coba, M. P., Husi, H., Campuzano, I., Blackstock, W. P., Choudhary, J. S., Grant, S. G.: Pro-

teomic analysis of in vivo phosphorylated synaptic proteins. *J. Biol. Chem.* **2005**, 280:5972–5982.

110 Collins, M. O., Husi, H., Yu, L., Brandon, J. M., Anderson, C. N. G., Blackstock, W. P., Choudhary, J. S., Grant, S. G. N.: Molecular characterization and comparison of the components and multiprotein complexes in the postsynaptic proteome. *J. Neurochem.* **2006**, 97[Suppl. 1]:16–23.

111 Siekevitz, P.: The postsynaptic density: a possible role in long-lasting effects in the central nervous system. *Proc. Natl. Acad. Sci. USA* **1985**, 82:3494–3498.

112 Rogalski-Wilk, A. A., Cohen, R. S.: Glyceraldehyde-3-phosphate dehydrogenase activity and F-actin associations in synaptosomes and postsynaptic densities of porcine cerebral cortex. *Cell Mol. Neurobiol.* **1997**, 17:51–70.

113 Wu, K., Aoki, C., Elste, A., Rogalski-Wilk, A., Siekevitz, P.: The synthesis of ATP by glycolytic enzymes in the postsynaptic density and the effect of endogenously generated nitric oxide. *Proc. Natl Acad. Sci. USA* **1997**, 94:13273–13278.

114 Schurr, A., Payne, R. S., Miller, J. J., Rigor, B. M.: Glia are the main source of lactate utilized by neurons for recovery of function posthypoxia. *Brain Res.* **1997**, 774:221–224.

115 Srere, P. A.: The metabolon. *Trends Biochem. Sci.* **1985**, 10:109–110.

116 Zanella, A., Izzo, C., Meola, G., Mariani, M., Colotti, M. T., Silani, V., Pellegata, G.: Metabolic impairment and membrane abnormality in red cells from Huntington's disease. *J. Neurol. Sci.* **1980**, 47:93–103.

9

Heart Simulation, Arrhythmia, and the Actions of Drugs

Denis Noble

Abstract

Biological modeling of cells, organs and systems has reached a very significant stage of development. Particularly at the cellular level, in the case of the heart there has been a long period of iteration between simulation and experiment. We have therefore achieved the levels of detail and accuracy that are required for the effective use of models in drug development, in toxicology and in health care. To be useful in this way, biological models must reach down to the level of proteins (receptors, transporters, enzymes, etc.) where drugs act, yet they must also reconstruct functionality right up to the levels of organs and systems, since problems like arrhythmia are cellular and organ-level phenomena.

9.1
The Problem

The side-effects of drugs on the heart are frequent and serious. According to some estimates around 40% of all new pharmaceutical compounds have such effects. They can cause serious arrhythmias that may be fatal. The cost to the industry is large. Withdrawal of a drug after approval represents a lost investment of around US $ 1 billion before one even begins to count lost sales and share values. A solution to this problem would therefore be beneficial to everyone. It would make the industry more successful, since many of the compounds withdrawn, or which never even make it to the market for this kind of reason, are effective drugs in other respects. The side-effects may be experienced by much less than 1% of the patient population. For the vast majority of patients the drugs are beneficial. Rescuing such drugs might therefore be worthwhile. And discovering combinations of actions that avoid the problem could improve the selection of lead compounds.

It would benefit health care for other reasons too. One reason that drug prices are high is that we, as patients, have to bear the cost of industry failures as well as industry successes. Unfortunately, the failures greatly outnumber the successes by around 50:1. One analyst in the industry told me: "all you have to do is to decrease

Biosimulation in Drug Development. Edited by Martin Bertau, Erik Mosekilde, and Hans V. Westerhoff
Copyright © 2008 WILEY-VCH Verlag GmbH & Co. KGaA, Weinheim
ISBN: 978-3-527-31699-1

our attrition rate from around 98 to 96% and you would make us twice as successful." That last remark underlines a message of this chapter: simulation does not have to be 100% successful to have a major impact on drug development. Even if just a fraction of the insights gained are valuable, that could be enough to make a significant difference.

It is also important to emphasize that these insights can be of many different kinds. The dream of being able to use computer models as though they were the equivalent of experimental animals, or even substitute virtual humans, is a long way off. Even the most sophisticated of computer models of the heart may represent the functionality of only 2% of the protein mechanisms involved; and many of those mechanisms need further refinement. We are not yet therefore in the position of the aircraft industry which can simulate a whole aircraft with reliable predictive capability on its ability to fly. But that analogy also provides the clue to a more positive message. Insights from computer models were valuable even before they replaced wind-tunnel experiments. Improving our quantitative understanding of complex interactions in biological systems, and refining the experimental approaches necessary to further refine the models, can have valuable practical spinoffs. In this chapter, I will show the extent to which simulation can help to address the cardiac arrhythmia problem.

Of course, people have looked for biological markers to give at least early warning of possible cardiac problems. But, at present we are stuck with very unreliable markers: QT and hERG. QT is the interval between the rapid QRS complex of the electrocardiogram, corresponding to the sharp depolarization wave as excitation spreads through the ventricle, and the T wave, which represents repolarization. To a first approximation, therefore, the QT interval is a measure of the duration of the ventricular action potentials. Since some of the drugs that cause arrhythmic side-effects prolong the action potential, measuring QT, perhaps together with other markers, such as action on the repolarizing current i_{Kr} or one of its proteins hERG, might identify the problem compounds. Measuring QT, right down to the last millisecond or two, has therefore become a refined technical art with many variations on how exactly to measure it. The T wave is hardly ever kind enough to give you a completely obvious point of measurement, so how to correct those measurements for unrelated variations, and how to automate all of this, has become necessary.

Unfortunately, drugs can cause arrhythmia without prolonging QT and some drugs that prolong QT do not cause arrhythmia. It is therefore a seriously flawed marker. A similar problem applies to hERG. Some drugs that target hERG (i_{Kr}) do not cause arrhythmia. That depends on what other actions they have.

Can we do better? That is the question I am going to try to answer.

9.2
Origin of the Problem

It is always best to understand a biological problem before trying to find ways to avoid it. The reason that action potential duration is important is that repolarization is a fragile process. It can fail, and when it does so, the action potential is followed by one or many oscillations. These are called early after-depolarizations (EADs). We will also encounter late after-depolarizations (DADs) later in this chapter.

Why is repolarization fragile, while depolarization is so robust? The answer lies in the quantities of electrical current involved. Depolarization is generated by an enormous increase in membrane conductance as the dense population of sodium channels is activated. Once the threshold for initiating the process is reached, it is exceedingly difficult to stop it running its full course. Depolarization is therefore extremely robust. By contrast, repolarization is brought about by very tiny currents resulting from a fine balance between almost equal depolarizing and repolarizing factors. Not only are the currents involved small, they are even smaller than you might expect because, once the fast depolarization phase is complete, the net membrane conductance actually falls well below its resting value, a fact that was first discovered by Weidmann [1]. This was a surprise since it was exactly the opposite result from that obtained during nerve excitation [2]. It was the reconstruction of Cole and Curtis' experimental result that formed one of the great successes of the Hodgkin–Huxley model of the nerve impulse [3].

The reason for this difference between heart and nerve is an evolutionary compromise of the form that I call "nature's pacts with the devil" [4]. These are what we, with hindsight, would call design faults but which, from an evolutionary point of view, are the inevitable price paid for many successful developments. They resemble Faust's pact with the devil in which Faust secured years of unlimited knowledge and power, but at the price of giving the devil his soul. The key to this kind of pact is that it is eventually fatal but for a long time it is of great benefit. This is just the kind of pact that nature stumbles upon when it finds a good combination of genes to transmit a function, at a price that may eventually be fatal. Evolution may take little notice of the fatality in individuals, particularly if it occurs well after the reproductive period of life. And evolution certainly did not anticipate the coming of the pharmaceutical industry.

The long-lasting cardiac action potential is a consequence of such an evolutionary compromise in the development of potassium currents in the heart. Early work on these channels [5] showed that they could be divided into two classes: channels that close on depolarization, i_{K1}, and channels that open during depolarization, including the various components of i_K and the transient outward current, i_{to}. At rest, i_{K1} is switched on and holds the resting potential at a very negative level, where the other K^+ currents are switched off. On depolarization i_{K1} rapidly switches off, while the other currents take time to activate and cause repolarization. This analysis of the potassium channels in cardiac muscle formed the basis of the first biophysically detailed model [6] and remains the basis of all subsequent models [7–9].

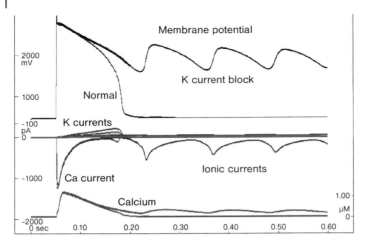

Fig. 9.1 Top: action potentials computed using a model of a guinea-pig ventricular cell. Middle: computed ionic currents. Bottom: computed calcium transients. 90% block of the fast component of i_K prevents repolariztion and generates multiple after-depolarizations [10].

The biological advantage of this potassium channel system is that it saves energy. By greatly reducing the potassium repolarizing current, very small sodium and calcium currents are sufficient to maintain depolarization during the long plateau of the action potential. The energy required to pump the ions back again during each beat is therefore minimized. Even with this economy, around 20% of the energy consumption of the heart is attributable to ion pumping. That figure would be much larger without the i_{K1} mechanism. That is the good side of nature's Faustian pact.

The bad side is that one of the proteins (hERG, that forms the main component of i_K) is highly promiscuous. The channel can be blocked by many pharmaceutical compounds. When that happens, repolarization fails and the action potential is followed by one or more waves of depolarization (Fig. 9.1). These can trigger cardiac arrhythmia that in some cases is fatal.

Many factors can interact with drugs to make this problem worse. These include genetic factors, such as mutations in sodium, potassium and calcium channel genes that predispose people to repolarization failure [11–13]. This is the main explanation for the fact that drugs with this side-effect have it in only a small fraction of the population. In principle, it should become possible to screen for such genetic predispositions to exclude such patients in clinical trials and to avoid treating them with drugs that interact in this way (Fig. 9.2).

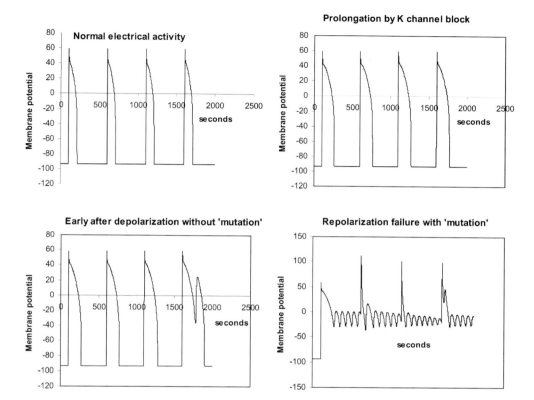

Fig. 9.2 Example of the use of simulation to understand the non-linear interaction between the effects of mutations affecting repolarization and drug actions. A guinea-pig cardiac ventricular cell model [14] was used to compute the threshold for initiation of repolarization failure for a known effect of one of the mutations predisposing towards arrhythmia. A voltage shift of 23 mV in the inactivation curve for the sodium channel (SCN5) was sufficient to initiate occasional repolarization failure in the form of single after-depolarizations (seen in the bottom left panel). The computations were then repeated (right) after simulating the action of a class III drug (i_{Kr} block by 90% – top right). The threshold "genetic mutation" shift was then reduced to 17 mV, resulting in much more severe multiple EADs (bottom right). This kind of computation explains several otherwise puzzling phenomena: 1. The interaction of both drug and mutation effects with the rest of the cellular components is highly non-linear. 2. Drug and mutation effects can interact in a way that would explain why a small fraction of the population may be particularly susceptible to drug-induced arrhythmia, as found in many clinical trials. 3. Below the threshold mutation effect, there is almost no effect on repolarization time, which would explain why, in the great majority of patients, no QT prolongation is seen.

9.3
Avoiding the Problem

Can drugs be designed to avoid this kind of problem? Clearly, there is no way in which we can correct nature's "mistake" of evolving repolarization channels that are blocked by so many drugs. Dreams of doing so by genetic manipulation are not just unimaginably improbable dreams, they would also be unethical. We can never be sure that a gene that we have identified with one particular function might not also be involved in many others we do not know about. And it is quite possible that the molecular properties that enable i_{Kr} to perform its role in the heart are also those that predispose it to drug sensitivity. The way forward therefore is to design better drugs. That this can be done is illustrated in Fig. 9.3. This shows the same computation as in Fig. 9.1, but with the addition of a computation in which 90% block of the potassium channel was combined with 20% block of the L-type calcium channels. The result is a smooth repolarization with no signs of EADs (see also Fig. 9.4). Clearly, a multi-receptor drug with this combination of properties would be expected to avoid arrhythmia. Such compounds exist. This particular computation was of a compound BRL-32872 that has exactly this profile of action [15]. Amiodarone, which is a multi-site drug, also includes this profile to which we can also add inhibition of sodium–calcium exchange [16]. This suggests that there may be many combinations of drug actions that could be effective. Drugs that include actions on persistent sodium current [17–20] are also in this category [21–23].

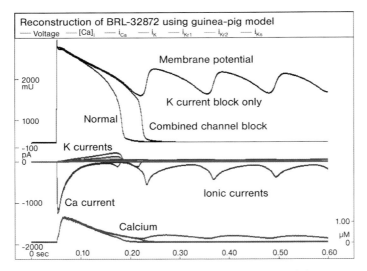

Fig. 9.3 The computation shown in Fig. 9.1 was extended by including 20% inhibition of the L type calcium current as well as 90% inhibition of i_{Kr}. The result is almost the same degree of action potential prolongation but with smooth repolarization [10]. This reproduces the action of BRL-32872 [15].

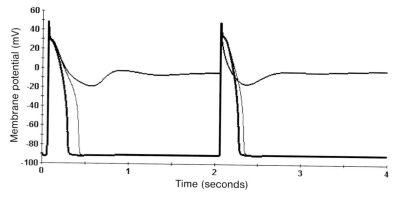

Fig. 9.4 Computations similar to those shown in Fig. 9.3, except that
in this case inhibition of persistent sodium current is added to that of
i_{Kr} to restore smooth repolarization. Control (thick line), 90% I_{Kr} block
(middle line) and 90% I_{Kr} + 50% I_{pNa} block (thin line) at 3.5 mM $[K]_o$
and 0.5 Hz. This result mimics the action of ranolazine [24].

9.4
Multiple Cellular Mechanisms of Arrhythmia

Early after-depolarizations (EADs) form just one of the several known cellular
mechanisms of arrhythmia. Delayed after-depolarizations (DADs) are a second
mechanism. These consist in spontaneous depolarizations arising after repolarization is complete and they are known to be caused by intracellular calcium oscillations in conditions of intracellular sodium overload. Such conditions are found in
a variety of pathological states, including heart failure and ischemic heart disease.
The initial causes vary but the common mechanism is reduced energy available
to pump sodium out of the cell via Na-K ATPase (the sodium pump). Intracellular sodium therefore rises above its normal range (around 5–10 mM) into a range
(12–20 mM) that can cause arrhythmias. The processes involved are now understood well enough to model them [25]. The rise in intracellular sodium acts via
sodium–calcium exchange to cause a rise in intracellular calcium. Above a certain
threshold this can stimulate release of calcium from the sarcoplasmic reticulum
via the same mechanism that underlies normal EC coupling, i.e. calcium-induced
calcium release. Finally, the oscillatory changes in intracellular calcium induce oscillatory inward current via sodium–calcium exchange which, if large enough, can
trigger extra action potentials (ectopic beats). Figure 9.5 shows an example of this
phenomenon in an atrial cell model.

Only some of the cell models succeed in reproducing this phenomenon, which
depends critically on the equations used to represent calcium-induced calcium release. There are still many gaps in our knowledge of this process [27], particularly
concerning the role of sub-sarcolemmal spaces in which the free calcium concentration may reach levels much higher than in the bulk cytosol. This is an area
where modeling needs to make much more progress by refining our understand-

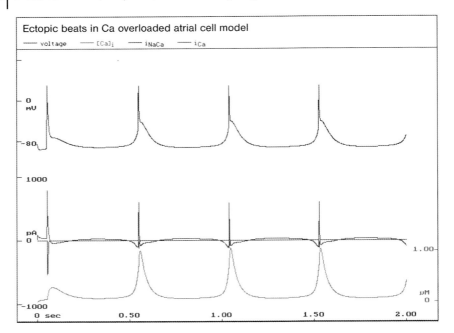

Fig. 9.5 The atrial cell model of Earm and Noble [26] was subjected to sodium loading to raise intracellular sodium to a level at which intracellular calcium oscillations occurred. The first action potential in this series was in response to an electrical stimulus. The subsequent action potentials were generated by calcium oscillations. The inward current produced by sodium–calcium exchange in response to each calcium transient was sufficient to reach threshold for action potential initiation [25].

ing of the EC coupling process, the role of "fuzzy" spaces and the mechanisms of calcium signaling within the cell.

This is a long cascade of events and therapeutic intervention can therefore be targeted at several different points. Some pharmaceutical approaches focus on the final stage, the generation of depolarizing electric current by sodium–calcium exchange. Inhibitors of sodium–calcium exchange have been developed. The object in this case is either to inhibit the electric current generated by the exchanger, or to reduce its contribution to calcium overload in conditions in which it operates in reverse mode. Neither approach has yet proven effective.

It might therefore be better to intervene at an earlier stage in the cascade and attempt to limit one of the earlier stages, i.e. sodium overload. This is the approach used in a new class of cardiac drugs that inhibit the persistent sodium current while having little or no effect on the peak sodium current. The first example of such a drug is Ranolazine [23]. As we have already seen in Fig. 9.4, inhibition of persistent sodium current is the basis of this compounds ability to avoid inducing EADs. Figure 9.6 shows that it would also be expected to reduce sodium loading. The top traces are experimental recordings from the work of Boyett et al. [28] showing the rise in intracellular sodium in a sheep Purkinje fiber on stimulating repetitively

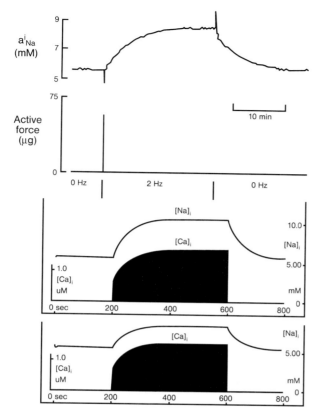

Fig. 9.6 Top: intracellular sodium activity and tension recorded in a cardiac Purkinje fiber during rest and during a period of repetitive stimulation at 2 Hz [28]. Middle: reconstruction using a model ventricular cell showing intracellular sodium and calcium concentrations [24]. The rise in intracellular sodium is similar to that seen experimentally. Bottom: same computation performed after 100% inhibition of persistent sodium current. The rise in sodium concentration is roughly halved (Noble and Noble, unpublished data).

after a long period of rest. Over a period of 400 s intracellular sodium activity increases from under 6 mM to nearly 9 mM. The middle traces show much the same degree of rise in internal sodium in the ventricular cell model. The lowest traces show a repeat of the computation with the persistent sodium current fully blocked. The rise in internal sodium is now roughly halved.

This mechanism is thought to be the basis of the therapeutic effect of this compound in cardiac ischemia, since one of the main causes of arrhythmia in this condition is attributable to sodium overload.

9.5
Linking Levels: Building the Virtual Heart

Although many of the mechanisms of cardiac arrhythmia can be studied and modeled at the cellular level, the question whether an arrhythmia is likely to be fatal requires analysis at multicellular levels, including that of the whole organ. A few ectopic beats, or a period of tachycardia, may be survived but if a re-entrant arrhythmia breaks down into fibrillation then it is usually fatal. The incorporation of the cellular models into models of cardiac tissue and of the whole organ is therefore essential. I have been privileged to collaborate with several of the key people involved in extending cardiac modeling to levels higher than the cell. The earliest work was with Raimond Winslow who used the "Connection Machine" at Minnesota, a huge parallel computer with 64 000 processors. We were able to construct models in which up to this number of cell models were connected together to form 2D or 3D blocks of atrial or ventricular tissue. This enabled us to study the factors determining whether ectopic beats generated during sodium overload would propagate

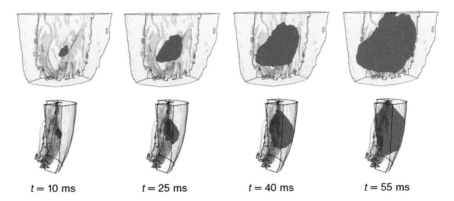

$t = 10$ ms $t = 25$ ms $t = 40$ ms $t = 55$ ms

Fig. 9.7 Snapshots of a re-entrant scroll wave in a $6\times6\times3$ cm^3 isotropic and homogeneous epicardial cuboid (left), and a wedge model of the human left ventricular free wall with similar dimensions, and fiber and sheet orientations giving orthotropy of propagation (right). Both examples use the ten Tusscher–Noble–Noble–Panfilov model [42] for excitation kinetics. In the wedge model these kinetics are spatially heterogeneous, with endocardial, mid-myocardial and epicardial tissue occupying approximately equal fractions of the transmural distance. In the cuboid, the diffusion coefficient was 0.154 mm^2 ms^{-1} in all directions. The diffusion coefficient in the wedge was set to 0.154 mm^2 ms^{-1} in the fiber axis direction, with a ratio of 36:9:1 in the fiber axis, sheet and sheet normal directions, respectively, to give a conduction velocity ratio of 6:3:1. Top panels show membrane potential color coded with the standard rainbow palette, from blue (resting) to red (excited). Bottom panels show the –40 mV isosurface in red. The wedge model highlights the effects of orthotropy and spatially heterogeneous excitation kinetics on propagation, that are not present in the cuboid model. The wedge geometry and architecture was extracted from a human DT-MRI dataset provided by P.A. Helm and R.L. Winslow at the Center for Cardiovascular Bioinformatics and Modeling and E. McVeigh at the National Institute of Health (from [43]).

across the tissue [29] and to study the possible interactions between sinus node and atrial cells [30].

Extension to the level of the whole organ came in collaboration with the University of Auckland, where Peter Hunter, Bruce Smaill, and their colleagues in bio-engineering and physiology constructed the first anatomically detailed models of a whole ventricle, including mechanics. These models include fiber orientations and sheet structure [31, 32], and are used to incorporate the cellular models in an attempt to reconstruct the electrical and mechanical behavior of the whole organ.

This work includes simulation of the activation wave front [7, 33, 34]. This is heavily influenced by cardiac ultra-structure, with preferential conduction along the fiber–sheet axes; and the result corresponds well with that obtained from multi-electrode recording from dog hearts *in situ*. Accurate reconstruction of the depolar-

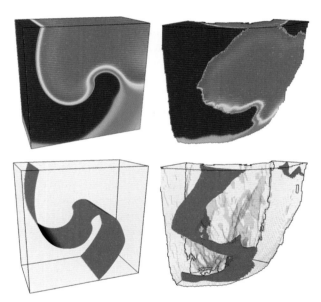

Fig. 9.8 Snapshots of orthotropic propagation of a single wave through a wedge model of the human end-diastolic (resting) left ventricular free wall, from an ectopic focus located on the endocardial surface. Spatially heterogeneous excitation kinetics are described using the ten Tusscher–Noble–Noble–Panfilov model [42], with endocardial, midmyocardial and epicardial tissue occupying approximately equal fractions of the transmural distance. The diffusion coefficient in the fiber axis direction was set to 0.154 mm^2 ms^{-1}, with a ratio of 36:9:1 in the fiber axis, sheet and sheet normal directions, respectively, to give a conduction velocity ratio of 6:3:1. Top panels show a view from the epicardial aspect, bottom panels show a transmural view with the endocardium on the left and the epicardium on the right. The spatial extent of the wedge geometry is indicated in light blue, excited tissue is in red. Times indicate duration since initial propagation from the ectopic focus. The architecture of the ventricular wall tissue results in complex wavefront geometries due to the rotational orthotropy inherent in the tissue. The wedge geometry and architecture was extracted from a human DT-MRI dataset provided by P.A. Helm and R.L. Winslow at the Center for Cardiovascular Bioinformatics and Modeling and E. McVeigh at the National Institute of Health (from [43]).

ization wavefront promises to provide reconstruction of the early phases of the ECG to complement work already done on the late phases [13] and, as the sinus node, atrium and conducting system are incorporated into the whole heart model [35], we can look forward to the first example of reconstruction of a complete physiological process from the level of protein function right up to routine clinical observation. The whole ventricular model has already been incorporated into a virtual torso [36], including the electrical conducting properties of the different tissues, to extend the external field computations to reconstruction of multiple-lead chest and limb recording. Incorporation of biophysically detailed cell models into whole organ models [8, 9, 32, 37] is still at an early stage of development, but it is essential to attempts to understand heart arrhythmias. So also is the extension of modeling to human cells [38, 39].

Work at the level of the whole ventricle is progressing rapidly as the necessary computing power becomes available (Fig. 9.7). This includes reconstructing some of the arrhythmic processes occurring during ischemia [40], the mechanisms of breakdown into fibrillation [41], modeling of the coronary circulation [34], and the mechanisms of defibrillation [37].

The multicellular and whole organ models are beginning also to be used in understanding the actions of drugs. A good example of this work within the BioSim network comes from Arun Holden's laboratory in Leeds. Figure 9.8 shows a reconstruction of the spread of ectopic excitation in a model of the left ventricular wall. This work has been used to define the liminal volume necessary for an ectopic focus to initiate a fully conducted wave of excitation.

These models have been used to study the mechanisms of arrhythmogenesis [43, 44] and to study the actions of drugs such as d-sotalol on propagation [43].

We can therefore look forward to testing the actions of drugs at multiple levels including that of the whole heart. One of the problems with computation at this level is that of computing resources. Even quite short simulations can require many hours of time on supercomputers. Work is progressing in attempting to solve these problems using more powerful computers and using networks of computers. We can therefore look forward to the day when it will be possible to have a complete suite of simulations in drug discovery and drug-tests running all the way from drug–receptor interactions to function in the whole organ.

References

1 Weidmann, S.: Effect of current flow on the membrane potential of cardiac muscle. *Journal of Physiology* **1951**, 115:227–236.

2 Cole, K. S. and Curtis, H. J.: Electric impedance of the squid giant axon during activity. *Journal of General Physiology* **1939**, 22:649–670.

3 Hodgkin, A. L. and Huxley, A. F.: A quantitative description of membrane current and its application to conduction and excitation in nerve. *Journal of Physiology* **1952**, 117:500–544.

4 Noble, D.: *The music of life*. OUP, Oxford, **2006**.

5 Hutter, O. F. and Noble, D.: Rectifying properties of heart muscle. *Nature* **1960**, 188:495.

6 Noble, D.: A modification of the Hodgkin–Huxley equations applicable to Purkinje

fibre action and pacemaker potentials. *Journal of Physiology* **1962**, 160:317–352.

7 Noble, D.: Modelling the heart: from genes to cells to the whole organ. *Science* **2002a**, 295:1678–1682.

8 Noble, D.: Modelling the heart: insights, failures and progress. *BioEssays* **2002b**, 24:1155–1163.

9 Noble, D.: The rise of computational biology. *Nature Reviews Molecular Cell Biology* **2002c**, 3:460–463.

10 Noble, D. and Colatsky, T. J.: A return to rational drug discovery: computer-based models of cells, organs and systems in drug target identification. *Emerging Therapeutic Targets* **2000**, 4:39–49.

11 Chen, Q., Kirsch, G. E., Zhang, D., Brugada, R., Brugada, J., Brugada, P., Potenza, D., Moya, A., Borggrefe, M., Breithardt, G., Ortiz-Lopez, R., Wang, Z., Antzelevitch, C., O'Brien, R. E., Schulze-Bahr, E., Keating, M. T., Towbin, J. A. and Wang, Q.: Genetic basis and molecular mechanism for idiopathic ventricular fibrillation. *Nature* **1998**, 392:293–296.

12 Clancy, C. E. and Rudy, Y.: Linking a genetic defect to its cellular phenotype in a cardiac arrhythmia. *Nature* **1999**, 400:566–569.

13 Antzelevitch, C., Nesterenko, V. V., Muzikant, A. L., Rice, J. J., Chien, G. and Colatsky, T.: Influence of transmural gradients on the electrophysiology and pharmacology of ventricular myocardium. Cellular basis for the Brugada and long-QT syndromes. *Philosophical Transactions of the Royal Society, Series A* **2001**, 359:1201–1216.

14 Noble, D., Varghese, A., Kohl, P. and Noble, P. J.: Improved guinea-pig ventricular cell model incorporating a diadic space, iKr and iKs, and length- and tension-dependent processes. *Canadian Journal of Cardiology* **1998**, 14:123–134.

15 Bril, A., Faivre, J. F., Forest, M. C., Cheval, B., Gout, B., Linee, P., Ruffolo, R. R. J. and Poyser, R. H.: Electrophysiological effect of BRL-32872, a novel antiarrhythmic agent with potassium and calcium channel blocking properties, in guinea pig cardiac isolated preparations. *Journal of Pharmacology and Experimental Therapeutics* **1995**, 273:1264–1272.

16 Watanabe, Y. and Kimura, J.: Inhibitory effect of amiodarone on Na+/Ca2+ exchange current in guinea-pig cardiac myocytes. *British Journal of Pharmacology* **2000**, 131:80–84.

17 Kiyosue, T. and Arita, M.: Late sodium current and its contribution to action potential configuration in guinea pig ventricular myocytes. *Circulation Research* **1989**, 64:389–397.

18 Maltsev, V. A., Sabbah, H. N., Higgins, R. S. D., Silverman, N., Lesch, M. and Undrovinas, A. I.: Novel, ultraslow inactivating sodium current in human ventricular myocytes. *Circulation* **1998**, 98:2545–2552.

19 Maltsev, V. A. and Undrovinas, A. I.: A multi-modal composition of the late Na+ current in human ventricular cardiomyocytes. *Cardiovascular Research* **2006**, 69:116–127.

20 Sakmann, B. F. A. S., Spindler, A. J., Bryant, S. M., Linz, K. W. and Noble, D.: Distribution of a persistent sodium current across the ventricular wall in guinea pigs. *Circulation Research* **2000**, 87:910–914.

21 Belardinelli, L., Shryock, J. and Fraser, H.: Inhibition of the late sodium current as a potential cardioprotective principle: effects of the selective late sodium current inhibitor, ranolazine. *Heart* **2005**, in press.

22 Bottino, D., Penland, R. C., Stamps, A., Traebert, M., Dumotier, B., Georgieva, A., Helmlinger, G. and Lett, G. S.: Preclinical cardiac safety assessment of pharmaceutical compounds using an integrated systems-based computer model of the heart. *Progress in Biophysics and Molecular Biology* **2005**, in press.

23 Undrovinas, A. I., Belardinelli, L., Nidas, A., Undrovinas, R. N. and Sabbah, H. N.: Ranolazine improves abnormal repolarization and contraction in left ventricular myocytes of dogs with heart failure by inhibiting late sodium current. *Journal of Cardiovascular Electrophysiology* **2006**, 17:S169–S177.

24 Noble, D. and Noble, P. J.: Late sodium current in the pathophysiology of cardiovascular disease: consequences of sodium-calcium overload. *Heart* **2006**, 92:iv1–iv5.

25 Noble, D. and Varghese, A.: Modeling of sodium-calcium overload arrhythmias and

their suppression. *Canadian Journal of Cardiology* **1998**, 14:97–100.

26 Earm, Y. E. and Noble, D.: A model of the single atrial cell: relation between calcium current and calcium release. *Proceedings of the Royal Society Series B* **1990**, 240:83–96.

27 Stern, M. D.: Theory of excitation-contraction coupling in cardiac muscle. *Biophysical Journal* **1992**, 63:497–517.

28 Boyett, M. R., Hart, G., Levi, A. J. and Roberts, A.: Effects of repetitive activity on developed force and intracellular sodium in isolated sheep and dog Purkinje fibres. *Journal of Physiology* **1987**, 388:295–322.

29 Winslow, R., Varghese, A., Noble, D., Adlakha, C. and Hoythya, A.: Generation and propagation of triggered activity induced by spatially localised Na-K pump inhibition in atrial network models. *Proceedings of the Royal Society B* **1993**, 254:55–61.

30 Noble, D., Brown, H. F. and Winslow, R.: Propagation of pacemaker activity: interaction between pacemaker cells and atrial tissue. In: Huizinga JD, ed., *Pacemaker activity and intercellular communication*. CRC Press, New York, **1995**, pp 73–92.

31 Hooks, D. A., Tomlinson, K. A., Marsden, S. G., LeGrice, I. J., Smaill, B. H., Pullan, A. J. and Hunter, P. J.: Cardiac microstructure: Implications for electrical propagation and defibrillation in the heart. *Circulation Research* **2002**, 91:331–338.

32 Crampin, E. J., Halstead, M., Hunter, P. J., Nielsen, P., Noble, D., Smith, N. and Tawhai, M.: Computational physiology and the physiome project. *Experimental Physiology* **2004**, 89:1–26.

33 Smith, N. P., Mulquiney, P. J., Nash, M. P., Bradley, C. P., Nickerson, D. P. and Hunter, P. J.: Mathematical modelling of the heart: cell to organ. *Chaos, Solitons and Fractals* **2001**, 13:1613–1621.

34 Smith, N. P., Pullan, A. J. and Hunter, P. J.: An anatomically based model of transient coronary blood flow in the heart. *SIAM Journal of Applied Mathematics* **2001**, 62:990–1018.

35 Garny, A., Kohl, P., Noble, D. and Hunter, P. J.: 1D and 2D models of the origin and propagation of cardiac excitation from the sino-atrial node into the right atrium. *Philosophical Transactions of the Royal Society of London B* **2000**.

36 Bradley, C. P., Pullan, A. J. and Hunter, P. J.: Geometric modeling of the human torso using cubic Hermite elements. *Annals of Biomedical Engineering* **1997**, 25:96–111.

37 Trayanova, N., Eason, J. and Aguel, F.: Computer simulations of cardiac defibrillation: a look inside the heart. *Computing and Visualization in Science* **2002**, in press.

38 Nygren, A., Fiset, C., Firek, L., Clark, J. W., Lindblad, D. S., Clark, R. B. and Giles, W. R.: A mathematical model of an adult human atrial cell: the role of K+ currents in repolarization. *Circulation Research* **1998**, 82:63–81.

39 Ten Tusscher, K. H. W. J., Noble, D., Noble, P. J. and Panfilov, A. V.: A model of the human ventricular myocyte. *American Journal of Physiology* **2003**, DOI 10.1152/ajpheart.00794.02003.

40 Rodriguez, B., Trayanova, N. and Noble, D.: Modeling cardiac ischemia. *Annals of the New York Academy of Sciences* **2007**, in press.

41 Panfilov, A. and Kerkhof, P.: Quantifying ventricular fibrillation: in silico research and clinical implications. *IEEE Trans. Biomed. Eng.* **2004**, 51:195–196.

42 Ten Tusscher, K. H. W. J., Noble, D., Noble, P. J. and Panfilov, A. V.: A model of the human ventricular myocyte. *American Journal of Physiology* **2004**, 286:H1573–H1589.

43 Benson, A. P.: *Computational electromechanics of the mammalian ventricles*. PhD thesis, University of Leeds, Leeds, **2006**.

44 Holden, A. V., Aslanidi, O. V., Benson, A. P., Clayton, R. H., Halley, G., Li, P. and Tong, W. C.: The virtual ventricular wall: a tool for exploring cardiac propagation and arrhythmogenesis. *Journal of Biological Physics* **2006**, in press.

Part III
Technologies for Simulating Drug Action and Effect

10

Optimizing Temporal Patterns of Anticancer Drug Delivery by Simulations of a Cell Cycle Automaton

Atilla Altinok, Francis Lévi, and Albert Goldbeter

Abstract

Determining the optimal temporal pattern of drug administration represents a central issue in chronopharmacology. Given that circadian rhythms profoundly affect the response to a variety of anticancer drugs, circadian chronotherapy is used clinically in cancer treatment. Assessing the relative cytotoxicity of various temporal patterns of administration of anticancer drugs requires a model for the cell cycle, since these drugs often target specific phases of this cycle. Here we use an automaton model to describe the transitions through the successive phases of the cell cycle. The model accounts for the progressive desynchronization of cells due to the variability in duration of the cell cycle phases, and for the entrainment of the cell cycle by the circadian clock. Focusing on the cytotoxic effect of 5-fluorouracil (5-FU), which kills cells exposed to this anticancer drug during the S phase, we compare the effect of continuous infusion of 5-FU with various circadian patterns of 5-FU administration that peak at either 4 a.m., 10 a.m., 4 p.m., or 10 p.m. The model indicates that the cytotoxic effect of 5-FU is minimum for a circadian delivery peaking at 4 a.m. – which is the profile used clinically for 5-FU – and is maximum for continuous infusion or a circadian pattern peaking at 4 p.m. These results are explained in terms of the relative temporal profiles of 5-FU and the fraction of cells in S phase.

10.1
Introduction

Multiple links exist between circadian rhythms and cancer. First, the rate of tumor growth in rodents increases as a result of: (a) mutations affecting the circadian clock [1] and (b) disruption of the neural pacemaker governing circadian rhythms [2]. Second, the cell cycle is directly controlled by the circadian clock [3–5]. This explains why progression through the cell cycle often displays a strong circadian dependence [6–9]. Third, the link between the circadian clock and cancer is further illustrated by the effect of circadian rhythms on a variety of anticancer medica-

Biosimulation in Drug Development. Edited by Martin Bertau, Erik Mosekilde, and Hans V. Westerhoff
Copyright © 2008 WILEY-VCH Verlag GmbH & Co. KGaA, Weinheim
ISBN: 978-3-527-31699-1

tions [10–12]. Each cancer medication is characterized, during the 24-h period, by a specific pattern of tolerance (chronotolerance) and efficacy (chronoefficacy) [12]. Moreover, the dosing time which results in the least toxicity of a drug for host cells usually achieves best antitumor efficacy [13]. The marked influence of circadian rhythms on chronotolerance and chronoefficacy has motivated the development of chronotherapeutic approaches, particularly in the field of cancer [10–15].

Assessing the effectiveness of various temporal schedules of drug delivery is central to cancer chronotherapeutics. Modeling tools can help to optimize time-patterned drug administration to increase effectiveness and reduce toxicity [16]. Probing the effect of circadian delivery of anticancer drugs by means of modeling and numerical simulations requires a model for the cell cycle. Different models for the cell cycle have been proposed. The complexity of these models increases as new molecular details are added [17–22]. Building on previous models for the embryonic and yeast cell cycles and for modules of the mammalian cycle [17–22], we are currently developing a model for the mammalian cell cycle in terms of a sequential activation of cyclin-dependent protein kinases, which behaves as a self-sustained biochemical oscillator in the presence of sufficient amounts of growth factors (C. Gérard and A. Goldbeter, in preparation). We have also developed a complementary, more pragmatic approach that shuns molecular details and relies on a simple phenomenological description of the cell cycle in terms of an automaton, which switches between sequential states corresponding to the successive phases of the cell cycle. In this model, the transition between some phases of the cell cycle, i.e. cell cycle progression or exit from the cycle, is affected by the presence of anticancer medications. The cell cycle automaton model is based on the perspective that the transitions between the various phases of the cell cycle entail a random component [23–25]; this model is directly inspired by our previous study of a follicular automaton model for the growth of human hair follicles [26, 27]. The model allows us to investigate how different temporal patterns of drug administration affect cell proliferation.

Anticancer medications generally exert their effect by interfering with the cell division cycle, often by blocking it at a specific phase. Thus, anticancer drugs exert most of their cytotoxicity on dividing cells through interactions with cell cycle or apoptosis-related targets [10–15]. Antimetabolites, such as 5-fluorouracil (5-FU), are primarily toxic to cells that are undergoing DNA synthesis, i.e. during the S-phase, while antimitotic agents, such as vinorelbine or docetaxel, are primarily toxic to cells that are undergoing mitosis, during the M phase. Conversely, alkylating agents such as cyclophosphamide or platinum complexes seldom display any cell cycle phase specificity. To illustrate the use of the cell cycle automaton model, we focus here on the chronotherapeutic scheduling of 5-FU, a reference drug for treating gastrointestinal, breast and various other cancers. The half-life of this medication is 10–20 min; thus, the exposure pattern is the only one considered here since it matches rather well the corresponding chronotherapeutic drug-delivery schedule [28].

A marked circadian dependence of the pharmacology of 5-FU has been demonstrated, both in experimental models and in cancer patients [29]. These data led to the development of intuitive chronomodulated delivery schedules aiming to minimize the toxic effects of 5-FU on healthy cells through its nighttime, rather than daytime, infusion. The most widely used chronomodulated schedule of 5-FU involves the sinusoidal modulation of its delivery rate from 10 p.m. to 10 a.m., with a peak at 4 a.m., in diurnally active cancer patients (see Fig. 10.3b below). This scheme improves patient 5-FU tolerability up to five-fold as compared with constant-rate infusion and makes possible a 40% increase in the tolerable dose and the near-doubling of antitumor activity in patients with metastatic colorectal cancer [30, 31]. The 5-FU chronomodulated schedule with peak delivery at 4 a.m. proves to be much less toxic than other circadian schedules, in which peak delivery differs from 4 a.m. by 6–12 h [32].

In this chapter we resort to the automaton model for the cell cycle to investigate the comparative cytotoxicity of different chronomodulated schedules of 5-FU administration. The analysis brings to light the importance of the circadian time of the peak in 5-FU as well as the effect of the variability in cell cycle phase durations in determining the response to this antiproliferative drug. The results explain why the least toxic schedule of 5-FU delivery for diurnally active cancer patients is a circadian modulated drug-administration pattern that peaks at 4 a.m., and why the most cytotoxic drug-administration schedule is either a circadian pattern that peaks at 4 p.m. or a continuous infusion [33]. Modeling the case of 5-FU illustrates an approach that can readily be extended to other types of anticancer drugs acting upon different stages of the cell cycle.

10.2
An Automaton Model for the Cell Cycle

10.2.1
Rules of the Cell Cycle Automaton

The automaton model for the cell cycle (Fig. 10.1a) is based on the following assumptions:

1. The cell cycle consists of four successive phases along which the cell progresses: G1, S (DNA replication), G2, and M (mitosis).

2. Upon completion of the M phase, the cell transforms into two cells, which immediately enter a new cycle in G1 (the possibility of temporary arrest in a G0 phase is considered elsewhere).

3. Each phase is characterized by a mean duration D and a variability V. As soon as the prescribed duration of a given phase is reached, the transition to the next phase of the cell cycle occurs. The time at which the transition takes place varies in a random manner according to a distribution of durations of cell cycle phases. In the case of a uniform probability distribution, the duration varies in the interval $[D(1 - V), D(1 + V)]$.

4. At each time step in each phase of the cycle the cell has a certain probability to be marked for exiting the cycle and dying at the nearest G1/S or G2/M transition. To allow for homeostasis, which corresponds to the maintenance of the total cell number within a range in which it can oscillate, we further assume that cell death counterbalances cell replication at mitosis. Given that two cells in G1 are produced at each division cycle, the probability P_0 of exiting the cycle must be of the order of 50% over one cycle to achieve homeostasis. When the probability of exiting the cycle is slightly smaller or larger than the value yielding homeostasis, the total number of cells increases or decreases in time, respectively, unless the probability of quitting the cycle is regulated by the total cell number.

We use these rules to simulate the dynamic behavior of the cell cycle automaton in a variety of conditions. Table 10.1 lists the values assigned in the various figures to the cell cycle length, presence or absence of cell cycle entrainment by the circadian clock, initial conditions, variability of cell phase duration, and probability of quitting the cell cycle.

(a)

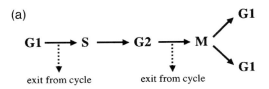

(b)

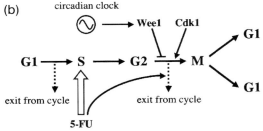

Fig. 10.1 (a) Scheme of the automaton model for the cell cycle. The automaton switches sequentially between the phases G1, S, G2, and M after which the automaton cell divides and two cells enter a new G1 phase. Switching from one phase to the next one occurs in a random manner as soon as the end of the preceding phase is reached, according to a transition probability related to a duration distribution centered for each phase around a mean value D and a variability V (see text). Exit from the cell cycle occurs with a given probability at the G1/S and G2/M transitions. (b) Coupling to the circadian clock occurs via the kinases Wee1 and cdc2 (Cdk1), which respectively inhibit and induce the G2/M transition. The scheme incorporates into the model the mode of action of the anticancer drug 5-FU. Cells exposed to 5-FU while in S phase have a higher probability of exiting the cell cycle at the next G2/M transition. The detailed operation of the automaton is schematized step by step in Fig. 10.7 below.

Table 10.1 Parameter values and initial conditions considered in the various figures based on numerical simulations of the cell cycle automaton model. All figures were established for a uniform distribution of durations of cell cycle phases around a mean value, with variability V. *Entrainment* The cell cycle is driven by the circadian clock through the circadian variation of Wee1 and Cdk1 (see text for further details). The cell cycle of 22 h duration consists of the following mean durations for the successive phases: G1 (9 h), S (11 h), G2 (1 h), and M (1 h).

Parameter			Conditions		
Cycle length	22 h		22 h		
Circadian entrainment	No entrainment		Entrainment		
Initial conditions	15 000 cells in G1		10 000 cells in steady state[a]		
Variability (V)	Probability of quitting the cycle (P_0; min^{-1})	Figures	Probability of quitting the cycle (P_0; min^{-1})	Figures	
0%	0.0005380	Fig. 10.2a	0.0004925	Figs. 10.2c, 10.6	
5%	0.0005380		0.0004930	Fig. 10.6	
10%	0.0005380		0.0005000	Fig. 10.6	
15%	0.0005380	Fig. 10.2b	0.0005125	Figs. 10.2d, 10.4 10.6	
20%	0.0005380		0.0005345	Fig. 10.6	

a)   Steady-state distribution of phases: 49.1% in G1, 44.2% in S, 3.9% in G2, 2.8% n M.

10.2.2
Distribution of Cell Cycle Phases

The variability in the duration of the cell cycle phases is responsible for progressive cell desynchronization. In the absence of variability, if the duration of each phase is the same for all cells, the population behaves as a single cell. Then, if all cells start at the same point of the cell cycle, e.g. at the beginning of G1, a sequence of square waves bringing the cells synchronously through G1, S, G2, M, and back into G1 occurs (see Fig. 10.2a) (A. Altinok and A. Goldbeter, in preparation). The drop in cell number at the end of the G1 and G2 phases reflects the assumption that exit from the cell cycle occurs at these transitions, to counterbalance the doubling in cell number at the end of M. These square waves continue unabated over time. However, as soon as some degree of variability of the cell cycle phase durations is introduced (Fig. 10.2b), the square waves transform into oscillations through the cell cycle phases, the amplitude of which diminishes as the variability increases. In the long term, these oscillations dampen as the system settles into a steady state distribution of cell cycle phases: the cells are fully desynchronized and have forgotten the initial conditions in which they all started to evolve from the same point of the cell cycle (A. Altinok and A. Goldbeter, in preparation).

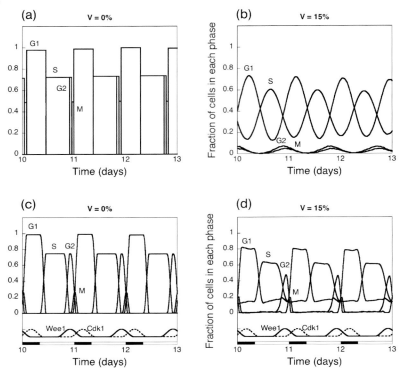

Fig. 10.2 Waves through cell cycle phases in absence (a, b) or presence (c, d) of entrainment by the circadian clock. The variability of durations for all cell cycle phases is equal to 0% (left column) or 15% (right column). The curves, generated by numerical simulations of the cell cycle automaton model, show the proportions of cells in G1, S, G2 or M phase as a function of time, for days 10–13. The time step used for simulations is equal to 1 min. The duration of the cell cycle before or in the absence of entrainment is 22 h. The successive phases of the cell cycle have the following mean durations: G1 (9 h), S (11 h), G2 (1 h), and M (1 h). As explained in the text and in Fig. 10.3a, entrainment by the circadian clock occurs in the model via a semi-sinusoidal rise in Wee1 (from 4 p.m. to 4 a.m.) and a similar, subsequent rise in Cdk1 (from 10 p.m. to 10 a.m.). The variations in Wee1 and Cdk1 from 0 acu to 100 acu (arbitrary concentration units) are represented schematically in panels (c) and (d) below the curves showing the fractions of cells in the various phases. The probability of premature G2/M transition in G2 depends on Cdk1

according to Eq. (2) where $k_c = 0.001$ acu^{-1} (in the simulations we consider that the probability goes to unity if k_c [Cdk1] $\geqslant 1$). The probability of the G2/M transition at the end of G2 depends on Wee1 according to Eq. (1) where $k_w = 0.015$ acu^{-1} (in the simulations we consider that the probability goes to zero when $(1 - k_w$ [Wee1]$) \leqslant 0$). The 24-h light/dark (L/D) cycle is shown as an alternation between an 8-h dark phase (black bar) and a 16-h light phase (white bar). Initial conditions are specified in Table 10.1. The probability of quitting the cycle (in units of 10^{-3} min^{-1}) is equal to 0.5380 for (a) and (c), 0.4925 for (b), and 0.5125 for (d); these values ensure homeostasis of the cell population, i.e. the number of cells in the population oscillates around and eventually reaches a stable steady state value. Panels (a) and (b) start initially with 15 000 cells in G1. Panel (c) and (d) start initially with 10 000 cells in steady state. The data in panels (a) and (b) are normalized by 15 000 cells, in panel (c) by 16 000 and in panel (d) by 14 000 (maximal cell number).

The top panels in Fig. 10.2 show the oscillations in the fraction of cells in the different cell cycle phases, as a function of time, in the absence of entrainment by the circadian clock. In the case considered, the duration of the cell cycle is 22 h, and the variability V is equal to 0% (Fig. 10.2a) or 15% (Fig. 10.2b). When variability is set to zero, no desynchronization occurs and the oscillations in the successive phases of the cell cycle are manifested as square waves that keep a constant amplitude in a given phase. Conversely, when variability increases up to 15% in the absence of entrainment (Fig. 10.2b), the amplitude of the oscillations decreases, reflecting enhanced desynchronization.

10.2.3
Coupling the Cell Cycle Automaton to the Circadian Clock

To determine the effect of circadian rhythms on anticancer drug administration, it is important to incorporate the link between the circadian clock and the cell cycle. Entrainment by the circadian clock can be included in the automaton model by considering that the protein Wee1 undergoes circadian variation, because the circadian clock proteins CLOCK and BMAL1 induce the expression of the *Wee1* gene (see Fig. 10.1b) [3–5]. Wee1 is a kinase that phosphorylates and thereby inactivates the protein kinase cdc2 (also known as the cyclin-dependent kinase Cdk1) that controls the transition G2/M and, consequently, the onset of mitosis.

In mice subjected to a 12:12 light–dark cycle (12 h of light followed by 12 h of darkness), the Wee1 protein level rises during the second part of the dark phase, i.e. at the end of the activity phase. Humans generally keep a pattern in which 16 h of diurnal activity are followed by 8 h of nocturnal sleep. Therefore, when modeling the link between the cell cycle and the circadian clock in humans, we consider a 16:8 light-dark cycle (16 h of light, from 8 a.m. to 12 p.m., followed by 8 h of darkness, from 12 p.m. to 8 a.m.; Fig. 10.3a) [7, 8, 12, 13]. To keep the pattern corresponding to the situation in mice (with a 12-h shift due to the change from nocturnal to diurnal activity) and in agreement with observations in human cells [8], the rise in Wee1 should occur at the end of the activity phase, i.e. with a peak at 10 p.m. The decline in Wee1 activity is followed by a rise in the activity of the kinase Cdk1, which enhances the probability of transition to the M phase. We thus consider that the rise in Wee1 is immediately followed by a similar rise in Cdk1 kinase (see Fig. 10.3a).

In the cell cycle model, we consider that the probability (P) of transition from G2 to M, at the end of G2, decreases as Wee1 rises, according to Eq. (1). Conversely, we assume that the probability of premature transition from G2 to M (i.e. before the end of G2, the duration of which was set when the automaton entered G2) increases with the activity of Cdk1 according to Eq. (2). The probability is first determined with respect to Cdk1; if the G2/M transition has not occurred, the cell progresses in G2. Only at the end of G2 is the probability of transition to M determined as a function of Wee1.

$$P(\text{transition G2} \rightarrow \text{M}) = 1 - k_w[\text{Wee1}] \tag{1}$$

$$P(\text{transition G2} \rightarrow \text{M}) = k_c[\text{Cdk1}] \tag{2}$$

(a)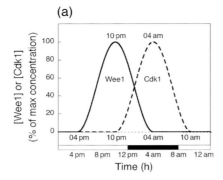

(b)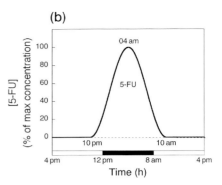

Fig. 10.3 (a) Semi-sinusoidal profile of Wee1 and Cdk1 used for entrainment of the cell cycle by the circadian clock (see text). (b) Semi-sinusoidal administration profile used clinically for 5-FU with peak time at 4 a.m. [30, 31]. Over the 24-h period, the 5-FU level is nil between 10 a.m. and 10 p.m., and rises in a sinusoidal manner between 10 p.m. and 10 a.m. according to Eq. (3), with

$A = 100$ and $d = 12$ h, with a peak at 4 a.m. In the model, the probability P of quitting the proliferative compartment at the next G2/M transition after exposure to the drug during the S phase is proportional to [5-FU], according to Eq. (4). At the maximum of [5-FU] reached at 4 a.m., the basal value of the exit probability is multiplied by a factor of 20.

In a previous study [33] we described the rise in Wee1 and Cdk1 by a step increase lasting 4 h. Here, instead of such a square-wave pattern, we will use a temporal pattern of semi-sinusoidal shape. Thus, we assume that Wee1 increases in a semi-sinusoidal manner between 4 p.m. and 4 a.m., with a peak at 10 p.m., while Cdk1 increases in the same manner between 10 p.m. and 10 a.m., with a peak at 4 a.m. (Fig. 10.3a).

Upon entrainment by the circadian clock, cells become more synchronized than in the absence of entrainment. In the case considered in Fig. 10.2c, d, the period changes from 22 h to 24 h, which corresponds to the period of the external LD cycle. When the variability is nil, we observe that the fraction of cells in S phase goes to zero at the trough of the oscillations (Fig. 10.2c). This does not occur when the variability is higher, e.g. 15% (Fig. 10.2d). The fraction of the S-phase cells then oscillates with reduced amplitude, reflecting again the effect of cell cycle desynchronization. However, in contrast to the progressive dampening of the oscillations in the absence of entrainment (Fig. 10.2b), when the cell cycle automaton is driven by the circadian clock oscillations appear to be sustained (Fig. 10.2d).

10.2.4
The Cell Cycle Automaton Model: Relation with Other Types of Cellular Automata

The automaton model for the cell cycle represents a cellular automaton. Because the latter term has been used in a partly different context, it is useful to distinguish the present model from those considered in previous studies. "Cellular automata" are often used to describe the spatiotemporal evolution of chemical or biological systems that are capable of switching sequentially between several discrete states [34–36]. One typical application of cellular automata pertains to excitable systems (e.g. neurons or muscle cells), which can evolve from a rest state to an excited state, then to a refractory state, before returning to the rest state. Spatially coupled automata can account for the propagation of waves in excitable media. Here, as in the model for the hair follicular cycles [26, 27], we consider spatially independent automata. Thus we assume that, within a population, the dynamics of a cycling cell will not be influenced by the state of neighboring cells. In contrast, in excitable systems, a cell in the rest state can be triggered to switch to the excited state when a neighboring cell is excited. The goal of the present study is to propose an automaton model for the cell cycle, to couple it to the circadian clock, and to use this model to determine the effect of anticancer drugs that kill cells at specific phases of the cell cycle.

10.3
Assessing the Efficacy of Circadian Delivery of the Anticancer Drug 5-FU

10.3.1
Mode of Action of 5-FU

Cells exposed in S phase to 5-FU arrest in this phase as a result of thymidilate synthase inhibition; then, they progress through the cell cycle or die through p53-dependent or independent apoptosis [11]. In the model we consider that cells exposed to 5-FU while in the S phase have an enhanced probability of quitting the proliferative compartment at the next G2/M transition (Fig. 10.1b). The probability of quitting the cycle is taken as proportional to the 5-FU concentration (see Section 10.3.2). We assume that the exit probability in the absence of 5-FU is multiplied by a factor of 20 when the level of 5-FU reaches 100% of its maximum value. Other hypotheses might be retained for the dose–response curve of the drug. Thus, larger or smaller slopes respectively correspond to stronger or weaker cytotoxic effects of 5-FU. A threshold dependence may also be introduced, in which case the linear relationship must be replaced by a sigmoidal curve which tends to a step function as the steepness of the threshold increases.

10.3.2
Circadian Versus Continuous Administration of 5-FU

In simulating the cell cycle automaton response to 5-FU, we impose a circadian profile of the anticancer medication similar to that used in clinical oncology [30, 31]: 5-FU is delivered in a semi-sinusoidal manner from 10 p.m. to 10 a.m., with a peak at 4 a.m. (Fig. 10.3b). During the remaining hours of the day and night, the drug concentration is set to zero. For comparison, we consider similar drug delivery patterns shifted in time, with peak delivery either at 10 a.m., 4 p.m., or 10 p.m.

The semi-sinusoidal delivery of the anticancer drug obeys the following equation which yields the concentration, [5-FU], as a function of time over the 24-h period:

$$[5\text{-FU}] = (A/2)[1 - \cos(2\pi(t - t_{\text{start}})/d)] \tag{3}$$

For delivery over a period d starting at 10 p.m. and ending at 10 a.m. and with a peak at 4 a.m., we take $t_{\text{start}} = 22$ h and $d = 12$ h, with $A = 100$. The probability (P) of exiting the cell cycle after exposure to a given level of 5-FU during the S phase is given by Eq. (4):

$$P = P_0(1 + k_f[5\text{-FU}]) \tag{4}$$

The value of P_0 is chosen so as to ensure tissue homeostasis, i.e. near constancy of the total cell number when cell proliferation is roughly compensated by cell death in the absence of anticancer drug treatment (such an assumption might not hold for tumoral tissues, in which case P_0 should be smaller than the value corresponding to homeostasis). The concentration of 5-FU, denoted [5-FU], varies from 0 to 100 (arbitrary units). Parameter k_f measures the cytotoxicity of 5-FU: the larger k_f, the larger the probability of quitting the cell cycle after exposure to the drug during the S phase. The values of P_0 used in numerical simulations are listed in Table 10.1. We choose the value $k_f = 0.19$ so that the probability of quitting the cell cycle is multiplied by 20 at the maximum 5-FU concentration.

It is useful to compare the circadian patterns of 5-FU delivery with the more conventional constant infusion drug-delivery pattern, in which the amount of 5-FU delivered over the 24-h period is the same as for the circadian delivery schedules. The quantity of 5-FU (Q_{5FU}) delivered over 24 h according to the semi-sinusoidal schedule defined by Eq. (3) is given by Eq. (5):

$$Q_{5FU} = \int_0^d \frac{A}{2}\left(1 - \cos\left(\frac{2\pi t}{d}\right)\right) dt = \frac{A}{2}(d + \sin 2\pi - \sin 0) = \frac{A}{2}d \tag{5}$$

For $A = 100$ [in arbitrary concentration units (acu)] and $d = 12$ h, this expression yields a mean 5-FU level of 25 acu, which is used for the case of constant infusion in Figs. 10.4b and 10.5e.

10.3.3
Circadian 5-FU Administration: Effect of Time of Peak Drug Delivery

The cytotoxic effect of the circadian administration of 5-FU depends on a variety of factors, which we consider in turn below. These factors include the mean duration D of the cell cycle phases, the variability V of cell cycle phase durations, entrainment by the circadian clock, and timing of the daily peak in 5-FU. For definiteness we consider the case where the cell cycle length, in the absence of entrainment, is equal to 22 h.

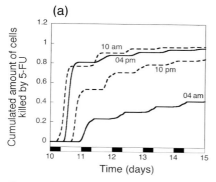
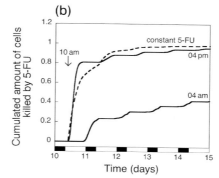

Fig. 10.4 (a) Cytotoxicity of chronomodulated 5-FU: effect of various circadian schedules of 5-FU delivery peaking at various times (4 a.m., 10 a.m., 4 p.m., 10 p.m.), when variability V is equal to 15%. (b) The circadian patterns peaking at 4 a.m. or 4 p.m. are compared with continuous delivery of 5-FU, which begins at 10 a.m. on day 10 (vertical arrow). The curves in (a) and (b) show the cumulated cell kill for days 10–15, in the presence of entrainment by the circadian clock. Prior to entrainment the cell cycle duration is 22 h. Parameter values and initial conditions are given in Table 10.1. The data are normalized by the mean cell number, 10 800.

To investigate the effect of the peak time of circadian delivery of 5-FU, we compare in Fig. 10.4a four circadian schedules with peak delivery at 4 a.m., 10 a.m., 4 p.m., and 10 p.m., for a cell cycle variability of 15%. The data on cumulated cell kill by 5-FU indicate a sharp difference between the circadian schedule with a peak at 4 a.m., which is the least toxic, and the other schedules. This difference is even more striking when cells are better synchronized, for smaller values of variability V (data not shown). The most toxic circadian schedules are those with a peak delivery at 4 p.m. or 10 a.m. We compare in Fig. 10.4b the least and most toxic circadian patterns of 5-FU delivery with the continuous infusion of 5-FU. Continuous delivery of 5-FU appears to be slightly more toxic than the circadian pattern with a peak at 4 p.m.

To clarify the reason why different circadian schedules of 5-FU delivery have distinct cytotoxic effects, we used the cell cycle automaton model to determine the time evolution of the fraction of cells in S phase in response to different patterns of circadian drug administration, for a cell cycle variability of 15%. The results, shown in Fig. 10.5, correspond to the case considered in Fig. 10.4, namely, entrainment of a 22-h cell cycle by the circadian clock. The data for Fig. 10.5a clearly indicate why the circadian schedule with a peak at 4 a.m. is the least toxic. The reason is that the fraction of cells in S phase is then precisely in antiphase with the circadian profile of 5-FU. Since 5-FU only affects cells in the S phase, the circadian delivery of the anticancer drug in this case kills but a negligible amount of cells.

When the peak delivery of 5-FU is at 4 p.m., the situation is opposite: now, the phase of 5-FU administration precisely coincides with the time period during which the majority of cells pass through the S phase (Fig. 10.5c). As a result, the first peak in S-phase cells is nearly annihilated following drug exposure. The remaining cells die after exposure to the second 5-FU pulse, which again coincides

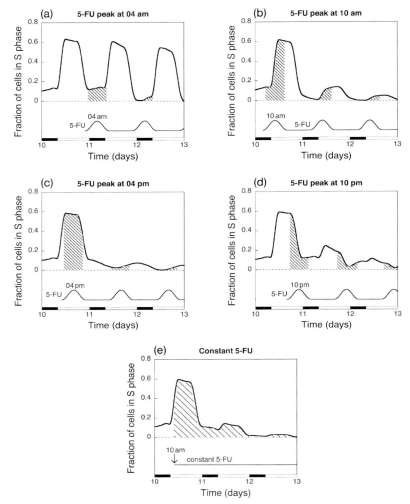

Fig. 10.5 Explanation of the cytotoxic effect of various circadian schedules of 5-FU delivery with peak at 4 a.m. (a), 10 a.m. (b), 4 p.m. (c), or 10 p.m. (d), and of continuous 5-FU delivery (e). Data are obtained for variability $V = 15\%$ and for a cell cycle duration of 22 h, in the presence of entrainment by the circadian clock. The hatched area shows the fraction of cells in S phase exposed to 5-FU and thus likely marked to exit the cell cycle at the next G2/M transition. The curves in Fig. 10.4 showing the cumulated number of cells killed indicate that the schedule with peak delivery at 4 a.m. is the one that causes minimal damage to the cells because the peak in 5-FU then coincides with the trough of the oscillations of S-phase cells. Continuous delivery of 5-FU is nearly as toxic as the most toxic circadian schedule of 5-FU delivery that peaks at 4 p.m. Because 5-FU is delivered at a constant, intermediate value in panel (e), the probability of exiting the cell cycle is enhanced but not as much as at the peak of the semi-sinusoidal delivery illustrated in the other panels. Hatching marks are thus more spaced in panel (e) to indicate this effect. Parameter values and initial conditions are given in Table 10.1. The data are normalized by the maximal cell number, 14 000.

with the next peak of S-phase cells. The latter peak is much smaller than the first one, because most cells exited the cycle after exposure to the first 5-FU pulse.

The cases of peak delivery at 10 a.m. (Fig. 10.5b) or 10 p.m. (Fig. 10.5d) are intermediate between the two preceding cases. Overlap between the peak of 5-FU and the peak of cells in S phase is only partial, but it is still greater in the case of the peak at 10 a.m., so that this pattern is the second most toxic, followed by the circadian delivery centered around 10 p.m. The comparison of the four panels Fig. 10.5a–d explains the results of Fig. 10.4a on the marked differences in cytotoxic effects of the four 5-FU circadian delivery schedules. The use of the cell cycle automaton helps clarify the dynamic bases that underlie the distinctive effects of the peak time in the circadian pattern of anticancer drug delivery.

The comparison between the curves in Figs. 10.4a and 10.5b and c raises the question of why the cytotoxic effect of 5-FU is nearly similar for the patterns in which 5-FU peaks at 4 p.m. and 10 a.m. (Fig. 10.4a), despite the fact that the peak in S phase cells presents a better overlap with the peak in 5-FU (the overlap is indicated by the dashed area under the fraction of S phase cells) when 5-FU peaks at 4 p.m. (Fig. 10.5c) compared with 10 a.m. (Fig. 10.5b). The reason is saturation in the cytotoxic effect of 5-FU. When the cytotoxic effect of 5-FU measured by parameter k_f is sufficiently large, most cells exposed to 5-FU during the first part of the S-phase peak in the case of Fig. 10.5c are already marked for exiting the cell cycle at the next G2/M transition, so that few additional cells are killed by 5-FU during the second part of the S-phase peak even though it corresponds to the 5-FU peak. Nearly the same amount of cells are marked for exiting the cycle after exposure to 5-FU during the first part of the S-phase peak in the case of Fig. 10.5b, because this fraction of the S-phase peak corresponds to the peak of 5-FU. To test this explanation we reduced the value of parameter k_f from 0.19 to 0.04, so that the maximum probability of quitting the cycle following exposure to 5-FU passes from $20 P_0$ to $5 P_0$ when 5-FU reaches its maximum value of 100 (see Eq. 4). As predicted, we observe a somewhat stronger differential effect between the patterns of 5-FU peaking at 4 p.m. and 10 a.m. when the value of k_f is reduced (data not shown).

When cells are more synchronized, e.g. in the presence of entrainment by the circadian clock for a variability of 5%, the results are similar, but the cytotoxic effect of the drug is decreased or enhanced depending on whether the peak of 5-FU occurs at 4 a.m. or 4 p.m., respectively (data not shown). Thus, for a peak delivery of 5-FU at 4 a.m., drug delivery is still in antiphase with the oscillation in S-phase cells, but because cells are more synchronized the fraction of S cells goes to zero at its trough. As a result, very few cells remain in S phase during the 5-FU pulse, so that only a minute cytotoxic effect is observed. For the pattern with peak drug delivery at 4 p.m., the situation is again close to the case of Fig. 10.5c: the peak of 5-FU precisely overlaps with the peak of cells in S phase, but because cells are more synchronized the amplitude of the peak in S cells is larger. The amount of cells killed after the first 5-FU pulse is thus larger than in the case when cells are less synchronized. Here again most cells killed by 5-FU exit the cycle after the first pulse of the drug.

The case of the continuous infusion of 5-FU is considered in Fig. 10.5e. Because the total amount of 5-FU administered over 24 h is the same as for the circadian semi-sinusoidal patterns, the level of 5-FU – and hence the cytotoxic effect of the drug – is sometimes below and sometimes above that reached with the circadian schedule. The numerical simulations of the automaton model indicate that the cytotoxicity is comparable to that observed for the most toxic circadian pattern, with peak delivery of 5-FU at 4 p.m.

A systematic investigation of the effect of the timing of the 5-FU peak on cell cytotoxicity indicates that the cytotoxic effect reaches its trough when the peak time of 5-FU is around 3 a.m., then rises progressively when the peak time increases up to 8 a.m. before reaching a plateau for peak times between 9 a.m. and 9 p.m., and finally drops when the peak time goes from 10 p.m. to 3 a.m. These results corroborate, with a higher degree of resolution, those illustrated in Figs. 10.4 and 10.5.

All the above results have been obtained for the case where the durations of the various cell cycle phases obey a probability distribution centered around the mean duration D, with a range of variation extending uniformly from $D - V$ to $D + V$. Similar results are obtained when assuming that the probability distribution obeys a lognormal distribution centered around the same mean value [33].

10.3.4
Effect of Variability of Cell Cycle Phase Durations

We have already alluded to the effect of synchronization governed by variability V. To further address this point, Fig. 10.6 shows, as a function of V, the cytotoxic effect of the 5-FU profile considered in Fig. 10.2b, with the peak at 4 a.m., in the presence of entrainment of the 22-h cycle by the circadian clock. The results indicate that the cumulated cell kill increases when V rises from 0% to 20%. For this circadian schedule of 5-FU, which is the least toxic to the cells (see above), we see that the better the synchronization, the smaller the number of cells killed. Here, in the presence of entrainment, a larger increase occurs between $V \leqslant 10\%$ and $V \geqslant 15\%$ in the number of cells killed by the drug. This jump is not observed in the absence of entrainment (data not shown). Entrainment by the circadian clock further enhances the synchronization of cells and protects them from the drug, as long as V remains relatively small, i.e. $V \leqslant 10\%$. Therefore, circadian entrainment magnifies the consequences of cell cycle variability, as it introduces a threshold in the effect of this parameter.

The effect of variability on drug cytotoxicity markedly depends on the temporal pattern of 5-FU delivery. When the peak in the circadian delivery of 5-FU occurs at 4 p.m., i.e. when the circadian schedule of 5-FU administration is most toxic to the cells, whether in the absence or presence of entrainment by the circadian clock, cytotoxicity increases as the degree of variability decreases. The effect is more marked in the conditions of entrainment: a threshold in cytotoxicity then exists between $V = 10\%$ and 15% (data not shown). Thus, in contrast to what is observed for the pattern of 5-FU peaking at 4 a.m. (Fig. 10.6), for the circadian 5-FU delivery sched-

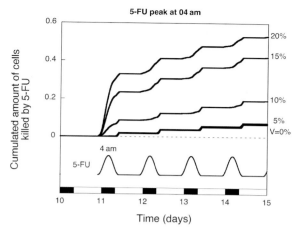

Fig. 10.6 Cytotoxicity of chronomodulated 5-FU: effect of variability of cell cycle phase durations. Shown is the cumulative cell kill when 5-FU is delivered in a circadian manner with peak at 4 a.m., in the presence of entrainment by the circadian clock, for different values of variability V indicated on the curves. The cell cycle duration is 22 h. Initial conditions and probabilities of exiting the cell cycle are specified in Table 10.1. Cell kill is normalized by the mean cell number, i.e. 13 000 for $V = 0\%$ and $V = 5\%$, 11 400 for $V = 10\%$, 10 800 for $V = 15\%$, and 10 700 for $V = 20\%$.

ule that peaks at 4 p.m, enhanced synchronization through decreased variability does not protect cells, but rather it increases their sensitivity to 5-FU cytotoxicity. This leads us to conclude that variability has opposite effects on cytotoxicity when the circadian delivery pattern of 5-FU peaks at 4 p.m. versus 4 a.m.

Numerical simulations therefore indicate that the least damage to the cells occurs when the peak of 5-FU circadian delivery is at 4 a.m., and when cells are well synchronized, i.e., when cell cycle variability V is lowest. In contrast, when the peak of 5-FU circadian delivery is at 4 p.m., cytotoxicity is enhanced when cells are well synchronized. The cytotoxic effect of the drug, therefore, can be enhanced or diminished by increased cell cycle synchronization, depending on the relative phases of the circadian schedule of drug delivery and the cell cycle entrained by the circadian clock. Continuous infusion of 5-FU is nearly as toxic as the most cytotoxic circadian pattern of anticancer drug delivery.

10.4
Discussion

The study of various anticancer drugs shows that many possess an optimal circadian delivery pattern, according to the phase of the cell cycle in which the cytotoxic effect is exerted. A case in point is provided by 5-FU. This widely used anticancer drug interferes with DNA synthesis and acts during DNA replication, the S phase of the cell cycle. Cells exposed to 5-FU during the S phase have an enhanced probability of dying from apoptosis at the next G2/M transition. In humans, the circadian pattern of 5-FU administration with peak delivery at 4 a.m. is the least cytotoxic; and the pattern with peak delivery at 4 p.m. is the most cytotoxic. In addition, for reasons which are still unclear, maximum 5-FU cytotoxicity to tumor cells occurs at the same time as best 5-FU tolerance, i.e. minimal damage to healthy tissues. In anticancer treatment, 5-FU is therefore administered according to a semi-sinusoidal pattern with peak delivery at 4 a.m. [30, 31].

It would be useful to base these empirical results on the effect exerted by the anticancer drug on the cell cycle in tumor and normal cells. This would not only help explain the dependence of cytotoxicity and tolerance on the temporal pattern of drug administration, but it would also provide firm foundations at the cellular level for the chronotherapeutical approach. To investigate the link between the cell cycle and the circadian clock and to assess the effect of circadian patterns of anticancer drug delivery, it is useful to resort to a modeling approach. Computational models allow a rapid exploration of a molecular or cellular mechanism over a wide range of conditions [16, 17, 33]. To assess the effect of various temporal patterns of anticancer drug administration we need a model for the cell cycle, allowing the study of its coupling to the circadian clock and of the effect of cytotoxic drugs. Rather than resorting to a detailed molecular model for the cell cycle in terms of cyclins and cyclin-dependent kinases and their control – models of this sort are available [17–22] and are currently being extended (C. Gérard and A. Goldbeter, in preparation) – we used here a phenomenological approach in which the progression between the successive phases of the cell cycle is described by a stochastic automaton.

The cell cycle automaton switches sequentially between the phases G1, S, G2, and M, with a probability P related to the duration of the various cell cycle phases. Each phase is characterized by its mean duration D and by its variability V. Upon mitosis (phase M), cells divide and enter a new cycle in G1. Exit from the cell cycle, reflecting cell death, occurs at the G1/S and G2/M transitions. Appropriate values of the exit probability allow for homeostasis of the total cell population. The anticancer drug 5-FU augments the exit probability for those cells that have been exposed to 5-FU during the S phase of DNA replication. An advantage of the stochastic automaton model is that it can readily be simulated to probe the cytotoxic effect of various circadian or continuous patterns of anticancer drug delivery. We showed that the cell cycle automaton model can be entrained by the circadian clock when incorporating a circadian block of the transition between the G2 and M phases, reflecting the circadian increase in the kinase Wee1. This increase takes place at the time set by the rise of the circadian clock protein BMAL1 in humans. Likewise, we incorporated the effect of the circadian increase in the kinase Cdk1, which immediately follows the peak in Wee1. The effect of Cdk1 corresponds in the model to an enhanced probability of transition between the G2 and M phases. The detailed operation of the cell cycle automaton model for a given cell i from G1 to mitosis is schematized step by step in Fig. 10.7.

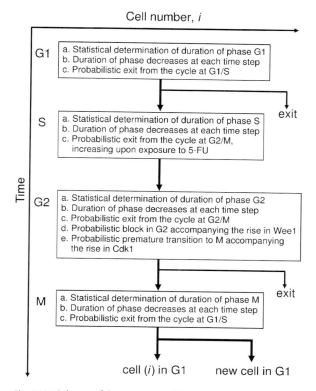

Cell number, i

Time

G1
a. Statistical determination of duration of phase G1
b. Duration of phase decreases at each time step
c. Probabilistic exit from the cycle at G1/S

exit

S
a. Statistical determination of duration of phase S
b. Duration of phase decreases at each time step
c. Probabilistic exit from the cycle at G2/M, increasing upon exposure to 5-FU

G2
a. Statistical determination of duration of phase G2
b. Duration of phase decreases at each time step
c. Probabilistic exit from the cycle at G2/M
d. Probabilistic block in G2 accompanying the rise in Wee1
e. Probabilistic premature transition to M accompanying the rise in Cdk1

exit

M
a. Statistical determination of duration of phase M
b. Duration of phase decreases at each time step
c. Probabilistic exit from the cycle at G1/S

cell (i) in G1   new cell in G1

Fig. 10.7 Scheme of the operation of the cell cycle automaton step by step, from G1 to M phase.

Coupling the cell cycle automaton to the circadian variation of Wee1 and Cdk1 permits entrainment of the cell cycle by the circadian clock, at a phase that is set by the timing of the peak of the circadian clock protein BMAL1. Entrainment strengthens cell synchronization. The peak of cells in S phase is of particular relevance for the action of 5-FU. The model predicts, upon entrainment by the circadian clock, that in humans the fraction of cells in S phase reaches a maximum during the light phase, around 4 p.m., while it reaches its minimum during night, around 4 a.m. (Fig. 10.2c, d).

We compared the effect of the continuous administration of 5-FU with various circadian patterns of 5-FU delivery peaking at 4 a.m, 10 a.m., 4 p.m., or 10 p.m. in the presence of entrainment by the circadian clock, by measuring the normalized, cumulative number of cells killed by 5-FU (Fig. 10.4). Several conclusions can be drawn from this comparison. First, the various circadian patterns of 5-FU delivery have markedly different cytotoxic effects on diurnally active cancer patients: the least toxic pattern is that which peaks at 4 a.m., while the most toxic one is that which peaks at 4 p.m. The other two patterns peaking at 10 a.m. or 10 p.m. exert intermediate cytotoxic effects. Conventional continuous infusion of 5-FU is nearly as toxic as the circadian pattern of 5-FU delivery peaking at 4 p.m.

The cell cycle automaton model permits us to clarify the reason why circadian delivery of 5-FU is least or most toxic when it peaks at 4 a.m. or 4 p.m., respectively. Indeed, the model allows us to determine the position of the peak in S-phase cells relative to that of the peak in 5-FU. As shown in Fig. 10.5, 5-FU is least cytotoxic when the fraction of S-phase cells oscillates in antiphase with 5-FU (when 5-FU peaks at 4 a.m.) and most toxic when both oscillate in phase (when 5-FU peaks at 4 p.m). Intermediate cytotoxicity is observed for other circadian patterns of 5-FU (when the drug peaks at 10 a.m. or 10 p.m.), for which the peak of 5-FU partially overlaps with the peak of S-phase cells. For the continuous infusion of 5-FU, the peak in S-phase cells necessarily occurs in the presence of a constant amount of 5-FU. Hence, the constant delivery pattern is nearly as toxic as the circadian pattern peaking at 4 p.m.

The goal of anticancer chronotherapies is to maximize the cytotoxic effect of medications on the tumor while protecting healthy tissues. The question arises as to how the above results might be used to predict the differential effect of an anticancer drug such as 5-FU on normal and tumor cell populations. This issue relates to the ways in which normal and tumor cells differ [13]. Such differences may pertain to the characteristics of the cell cycle, e.g. duration of the cell cycle phases and their variability, or entrainment of the cell cycle by the circadian clock. Such differences have been encountered in experimental tumor models [37, 38]. Thus both the molecular circadian clock and the 24-h pattern in cell cycle phase distribution depended upon the growth stage of Glasgow osteosarcoma in mice. In this rapidly growing tumor with a doubling time of ~ 2 days, both circadian and cell cycle clocks were present yet altered at an early stage and became ablated when the tumor grew bigger [39]. The clinical relevance of these findings is supported by heterogeneous and usually decreased expression of clock genes in human tumors [40–43].

The results of simulations indicate that, when the circadian delivery of 5-FU peaks at 4 a.m., differential effects of the drug on a population of healthy cells and on a population of tumor cells may be observed depending on whether the two cell populations are entrained or not by the circadian clock [33]. Another source of differential effect pertains to the degree of variability, given that, as previously noted, synchronization of the cells minimizes cytotoxic damage when the circadian 5-FU modulated delivery pattern peaks at 4 a.m. The results are markedly different when the circadian pattern of 5-FU delivery peaks at 4 p.m. [33]. Then the cytotoxic effect of the drug on the two populations is the inverse as that predicted for the circadian pattern peaking at 4 a.m. The effect of variability therefore depends on the circadian pattern of 5-FU delivery and on the possibility of entrainment of the cell cycle by the circadian clock.

The results presented here point to the interest of measuring, both in normal and tumor cell populations, parameters such as the duration of the cell cycle phases and their variability, as well as the presence or absence of entrainment by the circadian clock. As shown by the results obtained with the cell cycle automaton model, these data are crucial for using the model to predict the differential outcome of various anticancer drug delivery schedules on normal and tumor cell populations. In a subsequent step, we plan to incorporate pharmacokinetic–pharmacodynamic (PK-PD) aspects of 5-FU metabolism into the modeling approach. Thus, the enzymatic activities responsible for the catabolism of 5-FU and the generation of its cytotoxic forms display opposite circadian patterns in healthy tissues [29].

The results presented here show that the cell cycle automaton model displays a high sensitivity to the rate of spontaneous exit from the cell cycle. Progressive explosion or extinction of the cell population occurs for a value of the exit rate slightly above or below the value yielding homeostasis, i.e. stabilization of the cell count which oscillates in a constant range without displaying any oscillatory exponential increase or decrease. This result stresses the physiological importance of this parameter, which is likely controlled by the cell population as a function of total cell mass. Homeostasis may easily be guaranteed when such auto-regulation is implemented in the model by making the exit probability P_0 depend on the total cell number N, e.g. by replacing P_0 by the expression $P = P_0 + R[(N/N_s) - 1]$, where R is a parameter measuring the strength of regulation, and N_s denotes the number of cells in the population above or below which P_0 is respectively increased or decreased.

The present modeling approach to circadian cancer chronotherapy is based on an automaton model for the cell cycle. Continuous approaches to cell cycle progression have also been used to study the link between cell proliferation and circadian rhythms [44] and to determine, in conjunction with optimal control theory, the most efficient circadian schedules of anticancer drug administration [45]. Including more molecular details of the cell cycle in continuous models for cell populations represents a promising line for future research. Hybrid models incorporating molecular details into the automaton approach presented here will also likely be developed.

Besides circadian cancer chronotherapy, another line of research resorting to periodic schedules of anticancer drug delivery has been proposed and analyzed theoretically [46–49]. It is based on a resonance phenomenon between the period of drug administration and the cell cycle time of normal tissue. The goal of this approach is again to develop a strategy that limits, as much as possible, damage to normal sensitive tissue, while maximizing the destruction of tumor cells. While the assessment of circadian cancer chronotherapy has for long been the topic of multi-center clinical studies, the approach based on resonance in periodic chemotherapy is supported so far by a limited number of experimental studies in mice [50, 51], but has yet to be tested clinically. The main idea behind the latter approach is that the periodic scheduling of phase-specific cytotoxic agents can increase the selectivity of therapy when the treatment period is close to the mean cycle length of proliferation of normal susceptible cells, provided the cell cycle time of normal cells differs from that of malignant cells. Damage to the population of normal cells should thus remain limited when chemotherapy is administered with a period close to the normal cell cycle time. In contrast, each dose of chemotherapy should kill another fraction of the tumor cell population because the latter cells divide with a different periodicity.

The phenomenon of resonance in periodic chemotherapy has been analyzed further in more refined cell population models [52]. Potential difficulties inherent in this approach were examined by means of a theoretical model of acute myelogenous leukemia [53]. The authors concluded that chronotherapy based on the resonance effect is unlikely to be efficacious in the treatment of this particular disease. One reason is that the treatment itself may alter the kinetic parameters characterizing the tumor in such a way that the average intermitotic interval varies in the course of chemotherapy. The resonance-based efficiency of chronotherapy might wane if the difference of cell cycle length between normal and malignant cells declines as a result of drug administration.

To some extent the idea of resonance is also present in the case of circadian 5-FU delivery. Indeed, the circadian patterns of 5-FU which peak at 4 a.m. or 4 p.m. correspond to oscillations that are, respectively, in antiphase or in corresponding phase with the circadian variation of the fraction of cells in S phase. This effect can be seen even for cell cycle durations that differ from 24 h, because of the entrainment of the cell cycle by the circadian clock.

Here, as in a previous publication [33], we used the cell cycle automaton model to probe the cytotoxic effect of various patterns of circadian or continuous 5-FU delivery. The results provide a framework to account for experimental and clinical observations, and to help us predict optimal modes of drug delivery in cancer chronotherapy. By explaining the differential cytotoxicity of various circadian schedules of 5-FU delivery, the model clarifies the foundations of cancer chronotherapeutics. In view of its versatility and reduced number of parameters, the automaton model could readily be applied to probe the administration schedules of other types of anticancer medications active on other phases of the cell cycle.

Acknowledgments

This work was supported by the *Fonds de la Recherche Scientifique Médicale* (F.R.S.M., Belgium) through grant 3.4636.04 and by the European Union through the Network of Excellence BioSim, Contract No. LSHB-CT-2004-005137. This study was performed while one of the authors (A.G.) held a *Chaire Internationale de Recherche Blaise Pascal de l'Etat et de la Région Ile-de-France, gérée par la Fondation de l'Ecole Normale Supérieure* in the Institute of Genetics and Microbiology at the University of Paris-Sud 11 (Orsay, France).

References

1 Fu, L., Pelicano, H., Liu, J., Huang, P., Lee, C.C.: The circadian gene Period2 plays an important role in tumor suppression and DNA damage response in vivo. *Cell* **2002**, 111:41–50.

2 Filipski, E., King, V.M., Li, X.M., Granda, T.G., Mormont, M.C., Liu, X., Claustrat, B., Hastings, M.H., Levi, F.: Host circadian clock as a control point in tumor progression. *J. Natl Cancer Inst.* **2002**, 94:690–697.

3 Matsuo, T., Yamaguchi, S., Mitsui, S., Emi, A., Shimoda, F., Okamura, H.: Control mechanism of the circadian clock for timing of cell division in vivo. *Science* **2003**, 302:255–259.

4 Hirayama, J., Cardone, L., Doi, M., Sassone-Corsi, P.: Common pathways in circadian and cell cycle clocks: light-dependent activation of Fos/AP-1 in zebrafish controls CRY-1a and WEE-1. *Proc. Natl Acad. Sci. USA* **2005**, 102:10194–10199.

5 Reddy, A.B., Wong, G.K.Y., O'Neill, J., Maywood, E.S., Hastings, M.H.: Circadian clocks: neural and peripheral pacemakers that impact upon the cell division cycle. *Mutat. Res.* **2005**, 574:76–91.

6 Smaaland, R.: Circadian rhythm of cell division. *Prog. Cell Cycle Res.* **1996**, 2:241–266.

7 Bjarnason, G.A., Jordan, R.: Circadian variation of cell proliferation and cell cycle protein expression in man: clinical implications. *Prog. Cell Cycle Res.* **2000**, 4:193–206.

8 Bjarnason, G.A., Jordan, R.C.K., Wood, P.A., Li, Q., Lincoln, D.W., Sothern, R.B., Hrushesky, W.J.M., Ben-David, Y.: Circadian expression of clock genes in human oral mucosa and skin. *Am. J. Pathol.* **2001**, 158:1793–1801.

9 Granda, T.G., Liu, X.H., Smaaland, R., Cermakian, N., Filipski, E., Sassone-Corsi, P., Lévi, F.: Circadian regulation of cell cycle and apoptosis proteins in mouse bone marrow and tumor. *FASEB J.* **2005**, 19:304–306.

10 Focan, C.: Circadian rhythms and cancer chemotherapy. *Pharm. Ther.* **1995**, 67:1–52.

11 Lévi, F.: Chronopharmacology of anti-cancer agents. In: Redfern, P.H., Lemmer, B. (eds), *Physiology and Pharmacology of Biological Rhythms (Handbook of Experimental Pharmacology, Vol. 125)*, Springer-Verlag, Berlin, **1997**, pp 299–331.

12 Lévi, F.: Circadian chronotherapy for human cancers. *Lancet Oncol.* **2001**, 2:307–315.

13 Granda, T., Lévi, F.: Tumor based rhythms of anticancer efficacy in experimental models. *Chronobiol. Int.* **2002**, 19:21–41.

14 Lévi F.: From circadian rhythms to cancer chronotherapeutics. *Chronobiol. Int.* **2002**, 19:1–19.

15 Mormont, M.C., Lévi, F.: Cancer chronotherapy: principles, applications, and perspectives. *Cancer* **2003**, 97:155–169.

16 Goldbeter, A., Claude, D.: Time-patterned drug administration: Insights from a modeling approach. *Chronobiol. Int.* **2002**, 19:157–175.

17 Goldbeter, A.: *Biochemical Oscillations and Cellular Rhythms*, Cambridge University Press, Cambridge, **1996**.

18 Tyson, J.J., Novak, B.: Regulation of the eukaryotic cell cycle: molecular antagonism, hysteresis, and irreversible transitions. *J. Theor. Biol.* **2001**, 210:249–263.

19 Swat, M., Kel, A., Herzel, H.: Bifurcation analysis of the regulatory modules of the mammalian G1/S transition. *Bioinformatics* **2004**, 20:1506–1511.

20 Qu, Z., MacLellan, W.R., Weiss, J.N.: Dynamics of the cell cycle: checkpoints, sizers, and timers. *Biophys. J.* **2003**, 88:3600–3611.

21 Chassagnole, C., Jackson, R.C., Hussain, N., Bashir, L., Derow, C., Savin, J., Fell, D.A.: Using a mammalian cell cycle simulation to interpret differential kinase inhibition in anti-tumour pharmaceutical development. *BioSystems* **2006**, 83:91–97.

22 Csikasz-Nagy, A., Battogtokh, D., Chen, K.C., Novak, B., Tyson, J.J.: Analysis of a generic model of eukaryotic cell-cycle regulation. *Biophys J.* **2006**, 90:4361–4379.

23 Smith, J.A., Martin, L.: Do cells cycle? *Proc. Natl Acad. Sci. USA* **1973**, 70:1263–1267.

24 Brooks, R.F., Bennett, D.C., Smith, J.A.: Mammalian cell cycles need two random transitions. *Cell* **1980**, 19:493–504.

25 Cain, S.J., Chau, P.C.: Transition probability cell cycle model. I. Balanced growth. *J. Theor. Biol.* **1997**, 185:55–67.

26 Halloy, J., Bernard, B.A., Loussouarn, G., Goldbeter, A.: Modeling the dynamics of human hair cycles by a follicular automaton. *Proc. Natl Acad. Sci. USA* **2000**, 97:8328–8333.

27 Halloy, J., Bernard, B.A., Loussouarn, G., Goldbeter, A.: The follicular automaton model: effect of stochasticity and of synchronization of hair cycles. *J. Theor. Biol.* **2002**, 214:469–479.

28 Metzger, G., Massari, C., Etienne, M.C., Comisso, M., Brienza, S., Touitou, Y., Milano, G., Bastian, G., Misset, J.L., Levi, F.: Spontaneous or imposed circadian changes in plasma concentrations of 5-fluorouracil coadministered with folinic acid and oxaliplatin: relationship with mucosal toxicity in patients with cancer. *Clin. Pharmacol. Ther.* **1994**, 56:190–201.

29 Zhang, R., Lu, Z., Liu, T., Soong, S.J., Diasio, R.B.: Relationship between circadian-dependent toxicity of 5-fluorodeoxyuridine and circadian rhythms of pyrimidine enzymes: possible relevance to fluoropyrimidine chemotherapy. *Cancer Res.* **1993**, 53:2816–2822.

30 Lévi, F., Zidani, R., Vannetzel, J.M., Perpoint, B., Focan, C., Faggiuolo, R., Chollet, P., Garufi, C., Itzhaki, M., Dogliotti L.: Chronomodulated versus fixed-infusion-rate delivery of ambulatory chemotherapy with oxaliplatin, fluorouracil, and folinic acid (leucovorin) in patients with colorectal cancer metastases: a randomized multi-institutional trial. *J. Natl Cancer Inst.* **1994**, 86:1608–1617.

31 Lévi, F., Zidani, R., Misset, J.L.: Randomised multicentre trial of chronotherapy with oxaliplatin, fluorouracil, and folinic acid in metastatic colorectal cancer. *International Organization for Cancer Chronotherapy, Lancet* **1997**, 350:681–686.

32 Lévi, F., Focan, C., Karaboué, A., de la Valette, V., Focan-Henrard, D., Baron, B., Kreutz, M.F., Giacchetti, S.: Implications of circadian clocks for the rhythmic delivery of cancer therapeutics. *Adv. Drug Deliv. Rev.* **2007**, in press.

33 Altinok, A., Lévi, F., Goldbeter, A.: A cell cycle automaton model for probing circadian patterns of anticancer drug delivery. *Adv. Drug Deliv. Rev.* **2007**, doi: 10.1016/j.addr.2006.09.022.

34 Ermentrout, G. B., Edelstein-Keshet, L.: Cellular automata approaches to biological modeling. *J. Theor. Biol.* **1993**, 160:97–133.

35 Goldbeter, A.: Computational approaches to cellular rhythms. *Nature* **2002**, 420:238–245.

36 Blagosklonny, M. V., Pardee, A.B.: Exploiting cancer cell cycling for selective protection of normal cells (Review). *Cancer Res.* **2001**, 61:4301–4305.

37 Granda, T., Liu, X.H., Smaaland, R., Cermakian, N., Filipski, E., Sassone-Corsi, P., Lévi, F.: Circadian regulation of cell cycle and apoptosis proteins in mouse bone marrow and tumor. *FASEB J.* **2005**, 19:304–306.

38 You, S., Wood, P.A., Xiong, Y., Kobayashi, M., Du-Quiton, J., Hrushesky, W.J.: Daily coordination of cancer growth and circadian clock gene expression. *Breast Cancer Res. Treat.* **2005**, 91:47–60.

39 Iurisci, I., Filipski, E., Reinhardt, J., Bach, S., Gianella-Borradori, A., Iacobelli, S., Meijer, L., Lévi, F.: Improved tumor control through circadian clock induction by seliciclib, a cyclin-dependent kinase inhibitor. *Cancer Res.* **2007**, in press.

40 Yeh, K.T., Yang, M.Y., Liu, T.C., Chen, J.C., Chan, W.L., Lin, S.F., Chang, J.G.: Abnormal expression of period 1 (PER1) in endometrial carcinoma. *J. Pathol.* **2005**, 206:111–120.

41 Chen, S.T., Choo, K.B., Hou, M.F., Yeh, K.T., Kuo, S.J., Chang, J.G.: Deregulated expression of the PER1; PER2 and PER3 genes in breast cancers. *Carcinogenesis* **2005**, 26:1241–1246.

42 Gery, S., Gombart, A.F., Yi, W. S., Koeffler, C., Hofmann, W.K., Koeffler, H. P.: Transcription profiling of C/EBP targets identifies *Per2* as a gene implicated in myeloid leukaemia. *Blood* **2005**, 106:2827–2836.

43 Pogue-Geile, K.L., Lyons-Weiler, J., Whitcomb, D.C.: Molecular overlap of fly circadian rhythms and human pancreatic cancer. *Cancer Lett.* **2006**, 243:55–57.

44 Clairambault, J., Michel, P., Perthame, B.: Circadian rhythm and tumour growth *C.R. Acad. Sci. (Paris)* **2006**, 342:17–22.

45 Basdevant, C., Clairambault J., Lévi, F.: Optimisation of time-scheduled regimen for anti-cancer drug infusion. *Math. Mod. Numer. Anal.* **2005**, 39:1069–1086.

46 Dibrov, B.F., Zhabotinsky, A.M., Neyfakh, Y.A., Orlova, M.P., Churikova, L.I.: Mathematical model of cancer chemotherapy. Periodic schedules of phase-specific cytotoxic agent administration increasing the selectivity of therapy. *Math. Biosci.* **1985**, 73:1–31.

47 Dibrov, B.F.: Resonance effect in self-renewing tissues. *J. Theor. Biol.* **1998**, 192:15–33.

48 Agur, Z.: The effect of drug schedule on responsiveness to chemotherapy. *Ann. N.Y. Acad. Sci.* **1986**, 504:274–277.

49 Agur, Z., Arnon, R., Schechter, B.: Reduction of cytotoxicity to normal tissue by new regimens of cell-cycle phase-specific drugs. *Math. Biosci.* **1988**, 92:1–15.

50 Agur, Z., Arnon, R., Sandak, B., Schechter, B.: Zidovudine toxicity to murine bone marrow may be affected by the exact frequency of drug administration. *Exp. Hematol.* **1991**, 19:364–368.

51 Ubezio, P., Tagliabue, G., Schechter, B., Agur, Z.: Increasing 1-b-D-arabino-furanosylcytosine efficacy by scheduled dosing intervals based on direct measurements of bone marrow cell kinetics. *Cancer Res.* **1994**, 54:6446–6451.

52 Webb, G.F.: Resonance phenomena in cell population chemotherapy models. *Rocky Mountain J. Math.* **1990**, 20:1195–1216.

53 Andersen, L.K., Mackey, M.C.: Resonance in periodic chemotherapy: a case study of acute myelogenous leukemia. *J. Theor. Biol.* **2001**, 209:113–130.

11

Probability of Exocytosis in Pancreatic β-Cells: Dependence on Ca^{2+} Sensing Latency Times, Ca^{2+} Channel Kinetic Parameters, and Channel Clustering

Juris Galvanovskis, Patrik Rorsman, and Bo Söderberg

Abstract

The fusion of secretory vesicles and granules with the cell membrane prior to the release of their content into the extracellular space requires a transient increase of free Ca^{2+} concentration in the vicinity of the fusion site. Usually there is a short temporal delay in the onset of the actual fusion of membranes with reference to the rising free Ca^{2+} levels. This delay is described as a latency time of the Ca^{2+}-sensing system of the secretory machinery and has been observed in several cell types, including pancreatic β-cells. The presence of a delay time of a finite length inherent to the secretory machinery of the cell has an essential effect on the probability for a certain granule to fuse with the cell membrane and to release its contents into the extracellular space during the action potential. We investigate here, theoretically and by numerical simulations, the extent of this influence and its dependence on the parameters of Ca^{2+} channels, channel clustering, the Ca^{2+}-sensing system, and the length of depolarizing pulses. We use a linear probabilistic model for a random opening and closing of channels that yields an explicit expression for the Laplace transforms of the waiting time distributions for an event that at least one channel is open during the latency time. This allows one in principle to calculate the probability that a vesicle will fuse with the cell membrane during the action potential. We compare our theoretical results with numerical simulations.

11.1
Introduction

In many types of secreting cells such as neurons, pancreatic β-cells, and others, the secretion is released into extracellular space when Ca^{2+} entering through voltage-sensitive Ca^{2+} channels triggers the fusion of secretion-containing vesicles with the plasma membrane [1, 2]. The structural and functional organization of mechanisms involved in Ca^{2+} sensing, the fusion of vesicle and cell membranes, and the release of various substances into surrounding environment have been investigated recently in great detail, mainly in neurons [3–5] but also in other cells that

Biosimulation in Drug Development. Edited by Martin Bertau, Erik Mosekilde, and Hans V. Westerhoff
Copyright © 2008 WILEY-VCH Verlag GmbH & Co. KGaA, Weinheim
ISBN: 978-3-527-31699-1

use the same vesicular way of exporting their products [6, 7]. A set of proteins involved in these processes, for example the plasma membrane SNARE proteins syntaxin and SNAP-25 [8, 9] and the putative Ca^{2+}-sensing molecule synaptotagmin [10], have been identified. It is widely believed that these Ca^{2+} sensing proteins are localized in close vicinity to Ca^{2+} channels, since the triggering of release requires high local concentrations – 50 to 200 μM – of the ion. The importance of this is emphasized by the observation that exocytosis echoes Ca^{2+} channel activity and stops immediately upon re-polarization of cells [11]. Additionally, the sensing molecules must be subject to increased levels of Ca^{2+} during during a certain period of time – a latency time – for the fusion of a vesicle with the plasma membrane to occur. These latency times or the delay of the fusion vary from several milliseconds in neurons to several tens of milliseconds in other hormone-secreting cells [12, 13]. Obviously, during the latency period certain structural changes take place inside the docking complex that finally lead to the fusion of membranes.

The existence of latency times introduces into the secretory system an additional parameter that essentially alters the exocytosis probability, i.e. the probability that a vesicle will fuse with the plasmalemma and release its content during the action potential. If there is no delay in the sensing of increased Ca^{2+}, the exocytosis probability is determined by the mean first latency time of Ca^{2+} channels, since any opening of a channel situated close enough to the fusion site triggers the process. The latency of fusion activation asks that at least one channel must be open during a time period equal to the latency time. This inevitably leads to a decrease in the probability of fusion of the vesicular and cell membranes. The amount of change depends both on how large the latency time in the Ca^{2+} sensing system is and on the kinetics of the Ca^{2+} channels involved. In the case of simple two-state channels, this effect is fully defined by the mean open and shut times of the channel. Obviously, channel clustering in the vicinity of docked granules is a very effective way to increase the probability of exocytosis. In synapses, where the exocytosis speed is essential, more than 60 channels may open simultaneously per single granule. Such an arrangement is possible in places of high concentration of Ca^{2+} channels, as is the case in neuronal synapses. In other cell types, e.g. in pancreatic β-cells, the number of Ca^{2+} channels is small (450 channels per cell), latency times larger [12], and geometrical constraints on channel and granule placing may be much harder to satisfy because of the impossibility of using many channels to secure the release of hormone from a single granule. Therefore it is important to understand in detail how the probability of exocytosis depends on Ca^{2+} channel kinetic parameters such as open and shut times, respectively m_o and m_s, the latency time of the Ca^{2+} sensing system and the channel clustering around the docking sites of secretory granules or vesicles.

Here we investigate the distribution of waiting times for the event that at least one channel is open for a certain time. We also calculate the probability for a single exocytosis event in some special cases.

11.2
Theory

Throughout this paper it is assumed that all N ion channels localized in the vicinity of a secretory granule are identical and may be in two states: open or shut. Open times are assumed to be identically distributed non-negative random variables having a distribution function F_o, finite mean m_o and a density function f_o. Similarly, shut times are identically distributed non-negative random variables with distribution function F_s, finite mean m_s and a density function f_s. The Ca^{2+} sensing system of the secretory machinery has a latency time t_l. Having these initial conditions we can calculate the probability of a single exocytosis event assuming that the cell is subjected to the depolarizing pulse of duration t_p. As explained above, a single exocytosis event takes place if the vesicle is subjected to a high local Ca^{2+} concentration during at least the latency time t_l. The necessary increase in Ca^{2+} concentration is achieved only [11] when one or several Ca^{2+} channels in the vicinity of the granule stays open during t_l. Channels stay closed in a resting state. After a depolarizing pulse of duration $t_p > t_l$ has been applied, these channels start opening at random times. In this situation, the probability that an elementary exocytosis event occurs equals the probability of an event that the waiting time for at least one channel to be continuously open during t_l is less than t_p.

11.3
Mathematical Model

The waiting time distribution can be analyzed in a simple model, assuming each channel obeys a simple exponential decay law for its switching between the open and closed states. Consider a single channel and let $P_0(t)$ and $P_1(t)$ be the probability that the channel is open and closed, respectively, at a certain point t in time. Then P_0 and P_1 evolve according to:

$$\dot{P}_0 = -a P_0 + b P_1 \tag{1}$$

$$\dot{P}_1 = a P_0 - b P_1 \tag{2}$$

consistently with $P_0 + P_1 = 1$. Here a and b denote the switching rates, $a = 1/m_s$, $b = 1/m_o$.

For a collection of N identical (but independent) channels, define P_k as a probability that precisely k of the N channels are open at a certain time, defining a vector:

$$\mathbf{P} = (P_0, P_1, P_2, \dots, P_N)$$

Then, the evolution of $\mathbf{P}$ is governed by a linear equation:

$$\dot{\mathbf{P}} = M\mathbf{P} \tag{3}$$

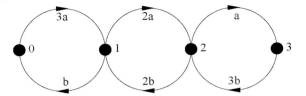

Fig. 11.1 Transition rates between the states of a system of three identical channels, in terms of the opening rate $a = 1/m_s$ and the closing rate $b = 1/m_o$. The small black circles represent states with 0, 1, 2, and 3 channels open, respectively.

in terms of a sparse matrix **M**:

$$M = \begin{pmatrix} -Na & b & 0 & \cdots & 0 & 0 \\ Na & -(N-1)a-b & 2b & \cdots & 0 & 0 \\ 0 & (N-1)a & -(N-2)a-2b & \cdots & 0 & 0 \\ \cdots & \cdots & \cdots & \cdots & \cdots & \cdots \\ 0 & 0 & 0 & \cdots -a-(N-1)b & Nb \\ 0 & 0 & 0 & \cdots & a & -Nb \end{pmatrix} \quad (4)$$

consistent with the conservation of probability, $\sum_k P_k = 1$.

The diagonal element $M_{kk} = -(N-k)a - kb$ determines the distribution $f_k(t)$ of dwell times in a state with k channels open:

$$f_k(t) = [(N-k)a + kb]e^{-[(N-k)a+kb]t} \quad (5)$$

while the off-diagonal elements yield the transition rates for $k \to k\pm 1$ (see Fig. 11.1 for an illustration).

Although not symmetric in itself, **M** is equivalent to a symmetric matrix (by means of a similarity transform), implying real eigenvalues. Indeed, it is easily diagonalized and has the eigenvalues $0, -(a+b), -2(a+b), \ldots, -N(a+b)$, with the eigenvalues associated with the zero eigenvalue corresponding to a binomial equilibrium distribution P_k^*:

$$P_k^* = \binom{N}{k} \frac{a^k b^{N-k}}{(a+b)^N} \quad (6)$$

for the number of k open channels.

11.4
Dwell Time Distributions

We next need to compute the distribution of dwell times in the aggregate of states with at least one channel open, i.e. $k \geqslant 1$. Therefore, define $g_k(t)$ as the distribution of dwell times in states with at least k units open. Obviously, $g_N = f_N$.

For lower k, g_k can be obtained recursively as follows. Denoting by d_k the rate for $k \to k-1$, and by u_k the rate for $k \to k+1$: they amount to:

$$d_k = kb \tag{7}$$

$$u_k = (N-k)a. \tag{8}$$

First note that each time interval with at least $k > 0$ channels open must start with a transition $k-1 \to k$. Then the next change in k is either, with probability $d_k/(d_k + u_k)$, back to $k-1$ and that is the end of that interval or, with probability $u_k/(d_k + u_k)$, a jump up to $k+1$, after which the system dwells with more than k channels open for a period of time distributed according to g_{k+1}, after which it will return to k and we are back at square one. Thus, after the initial transition from $k-1$ to k, the number of open channels may jump back and forth between k and more than k any number of times (possibly zero) and finally the interval ends with a transition from k back to $k-1$.

As a result, $g_k(t)$ must satisfy the integral equation:

$$g_k(t) = \frac{d_k}{u_k + d_k} f_k(t)$$

$$+ \frac{u_k}{u_k + d_k} \int_0^t dx \int_0^{t-x} dy \cdot f_k(x) g_{k+1}(y) g_k(t-x-y) \tag{9}$$

This is most easily solved by means of the Laplace transform, yielding:

$$\tilde{g}_k(s) = \frac{d_k}{d_k + u_k} \tilde{f}_k(s) + \frac{u_k}{u_k + d_k} \tilde{f}_k(s) \tilde{g}_{k+1}(s) \tilde{g}_k(s) \tag{10}$$

in terms of the Laplace-transformed distributions:

$$\tilde{g}_k = \int_0^\infty g_k(t) e^{-st} dt \tag{11}$$

etc. From this, $\tilde{g}_k(s)$ can be extracted as:

$$\tilde{g}_k(s) = \frac{d_k \tilde{f}_k(s)}{d_k + u_k(1 - \tilde{f}_k(s) \tilde{g}_{k+1}(s))} \tag{12}$$

Using the fact that f_k has a simple Laplace transform:

$$\tilde{f}_k(s) = \frac{u_k + d_k}{s + u_k + d_k} \tag{13}$$

that follows from Eq. (5), we obtain the recursive equation for $\tilde{g}_k$ at $0 < k < N$:

$$\tilde{g}_k(s) = \frac{d_k}{s + u_k + d_k - u_k \tilde{g}_{k+1}(s)} \tag{14}$$

which in $N - 1$ steps yields $\tilde{g}_1(s)$, starting from $\tilde{g}_N(s) = \frac{u_N + d_N}{s + u_N + d_N} = \frac{Nb}{s + Nb}$. Note that this must yield $\tilde{g}_1$ as a rational function of s, corresponding to $g_1(t)$ being a linear combination of exponentials:

$$\tilde{g}_1(s) = \sum_{i=1}^{N} \frac{C_i}{s + w_i} \tag{15}$$

$$g_1(t) = \sum_{i=1}^{N} C_i e^{-w_i t} \tag{16}$$

The coefficients C_i and the exponents w_i can be obtained by making a partial fraction expansion of $\tilde{g}_1(s)$.

Equation (16) can also be understood by altering the dynamics in Eq. (3) and replacing $\mathbf{M}$ by a modified matrix $\overline{\mathbf{M}}$ where the connection between $k = 0$ and $k = 1$ are set to zero, corresponding to cutting the two leftmost arcs in Fig. 11.1. As a result probability is no longer conserved and the probability, previously transferred between $k = 0$ and 1, is now lost. From an initial state with $k = 1$, the total probability decays according to $g_1(t)$. This process can be analyzed in terms of the eigenvectors of $\overline{\mathbf{M}}$, implying that g_1 must have the form of Eq. (16), with $w_i > 0$.

While $f_0(t)$ describes the length distribution of time intervals with all channels closed, $g_1(t)$ plays the same role for the interlaced intervals where at least one channel is on.

11.5
Waiting Time Distribution

The sought distribution $w(t)$ of waiting times, for at least one channel to be open during a period of time of length at least t_l, can now be derived from f_0 and g_1. As previously, this step is best done in terms of Laplace transforms. First, we must split g_1 in two parts, $g_<$ and $g_>$, according to:

$$g_<(t) = \begin{cases} g_1(t), & t < t_l \\ 0, & t > t_l \end{cases} ; \quad g_>(t) = \begin{cases} 0, & t < t_l \\ g_1(t), & t > t_l \end{cases} \tag{17}$$

Obviously, $g_< + g_> = g_1$. The corresponding Laplace transforms can be simply obtained once $\tilde{g}_1$ is written in the form of Eq. (15), yielding:

$$\tilde{g}_<(s) = \sum_{i=1}^{N} \frac{C_i}{s + w_i} \left(1 - e^{-t_l(s+w_i)} \right) \tag{18}$$

$$\tilde{g}_>(s) = \sum_{i=1}^{N} \frac{C_i}{s + w_i} \left(e^{-t_l(s+w_i)} \right) \tag{19}$$

There we can run the same trick again: the waiting time interval starts with $k = 0$ for a period of time distributed according to f_0, after which k stays non-zero for a period of time distributed according to g_1. If this is larger than t_l, we only count t_l and exit; otherwise, k becomes zero again and the whole process starts over. This implies the following integral equation for $w(t)$, in terms of the probability $p_< = \tilde{g}_>(0)$ that the dwell time with $k > 0$ is larger than t_l:

$$w(t) = p_> f_0(t - t_l)\Theta(t - t_l) + \int_0^t dx \int_0^{t-x} dy f_0(x) g_<(y) w(t - x - y) \quad (20)$$

For the Laplace transforms, this becomes:

$$\tilde{w}(s) = \tilde{g}(0)\tilde{f}_0(s)e^{-t_l s} + \tilde{f}_0(s)\tilde{g}_<(s)\tilde{w}(s) \quad (21)$$

and $\tilde{w}$ can be solved for, yielding:

$$\tilde{w}(s) = \frac{\tilde{g}_>(0)\tilde{f}_0(s)e^{-t_l s}}{1 - \tilde{f}_0(s)\tilde{g}_<(s)} \quad (22)$$

Note that in the limit $t_l \to 0$ of vanishing latency time, we obtain $\tilde{w}(s) \to \tilde{f}_0(s)$, as is to be expected. The only remaining problem then is to extract the waiting time distribution $w(t)$ from its Laplace transform $\tilde{w}(s)$, a task complicated by the appearance of exponential factors in $\tilde{g}_<$ in the denominator in Eq. (22).

11.6
Average Waiting Time

Here we limit ourselves to the simpler task of extracting the average waiting times. For a length distribution such as w, the average and higher moments are readily available by means of a Taylor expansion of $\tilde{w}$:

$$\tilde{w} = \sum_{n=0}^{\infty} \frac{(-s)^n}{n!} \langle t_w^n \rangle \quad (23)$$

yielding $\tilde{w}(0) = 1$, reflecting the normalization of $w(t)$, and:

$$\langle t_w \rangle = -\frac{d\tilde{w}}{ds}\bigg|_{s=0} \quad (24)$$

etc. Keeping in mind that $g_<$ and $g_>$ are not normalized to unity, but to $p_<$ and $p_>$, respectively, differentiating Eq. (21) with respect to s, and setting $s = 0$, yields:

$$\langle t_w \rangle = t_l + \frac{1}{p_>} \langle t_0 \rangle + \frac{p_<}{p^>} \langle t_< \rangle \quad (25)$$

where $\langle t_w \rangle = \frac{1}{Na}$ is the average interval length in the distribution f_0, while:

$$\langle t_< \rangle = \frac{1}{p_<} \sum_i \frac{C_i}{w_i^2} \left(1 - (1 + t_i w_i) e^{-t_i w_i} \right) \tag{26}$$

is the corresponding average in $g_<$. We also need:

$$p_> = \sum_i \frac{C_i}{w_i} e^{-t_i w_i} \tag{27}$$

and $p_< = 1 - p_>$. Thus we finally obtain the average waiting time as:

$$\langle t_w \rangle = t_l + \frac{\frac{1}{Na} + \sum_{i=1}^{N} \frac{C_i}{w_i^2} \left\{ 1 - (1 + t_i w_i) e^{-t_i w_i} \right\}}{\sum_{i=1}^{N} \frac{C_i}{w_i} e^{-t_i w_i}} \tag{28}$$

11.7
Cases $N = 1, 2$, and 3

For the case of a single unit or channel, we have $\tilde{g}_1(s) = \tilde{f}_1(s) = \frac{b}{s+b}$, yielding $C_1 = w_1 = b$, and we have for $N = 1$ the result:

$$\langle t_w \rangle = t_l + \frac{e^{bt_l}}{a} + \frac{e^{bt_l} - 1 - bt_l}{b} \tag{29}$$

For $N = 2$, we obtain

$$\tilde{g}_1(s) = \frac{b(s + 2b)}{s^2 + (a + 3b)s + 2b^2} \tag{30}$$

yielding

$$w_{1,2} = \frac{1}{2} \left(a + 3b \pm \sqrt{a^2 + 6ab + b^2} \right) \tag{31}$$

$$C_{1,2} = \frac{b \left(\pm(a - b) + \sqrt{a^2 + 6ab + b^2} \right)}{2\sqrt{a^2 + 6ab + b^2}} \tag{32}$$

Similarly, for $N = 3$, we obtain:

$$\tilde{g}_1 = \frac{b(s^2 + (a + 5b)s + 6b^2)}{s^3 + (3a + 6b)s^2 + (2a^2 + 7ab + 11b^2)s + 6b^3} \tag{33}$$

from which $w_{1,2,3}$ and $C_{1,2,3}$ can be extracted by solving a cubic equation.

11.8
Numerical Simulations

To investigate numerically the dependence of exocytosis probability on ion channel clustering and to test our theoretical results, we carried out numerical simulations on a system of identical and independent two-state channels. A set of dwell times for a single channel in shut and open states was generated with the help of a source equation:

$$\tau_k = -\tau \ln(rnd) \tag{34}$$

where τ is equal to m_o or m_s, $\tau_k(k = 1, 2, 3,\ldots, n)$ are the the consecutive dwell times in a given state, and rnd is a random number chosen uniformly inside the interval 0 to 1. At the beginning of simulation process all channels are assumed to be in the shut state. This corresponds to the experimental situation just before the application of a depolarizing pulse to a resting plasma membrane when all Ca^{2+} channels are shut. In the case of two-state channels, dwell times were calculated in two 10 s runs and the obtained single channel records were superimposed; for three channels the same calculations were repeated three times with the subsequent superimposition of single records, and so on. The resultant superposition process for three channels ($m_o = 1.9$ ms, $m_s = 5$ ms) is seen in Fig. 11.2.

For every individual simulation process, i.e. for each size of channel cluster, the waiting time for the event "at least one channel is open for t_l" was detected; this was repeated 400 times. Then the distribution of waiting times, $w(t)$ was estimated from binned data. An example of a histogram obtained is seen in Fig. 11.3 for a cluster that contains three channels, with $m_o = 1.9$ ms and $m_s = 5$ ms.

To compare simulation data with theoretical predictions, we calculated the average waiting time for the case of three channels, both from numerical data shown in

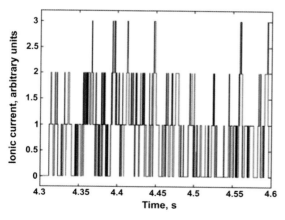

Fig. 11.2 A continuous segment (300 ms) of the simulated superposition process for the three-channel, two-state system with $m_o = 1.9$ ms and $m_s = 5$ ms.

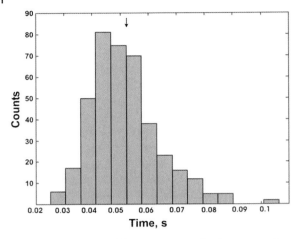

Fig. 11.3 A histogram of the waiting times for the superposition process shown in Fig. 11.2. The arrow marks the theoretical average, 0.0523 s.

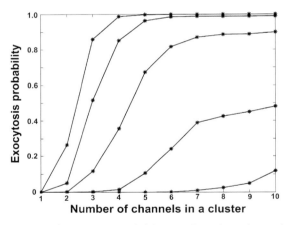

Fig. 11.4 The exocytosis probability as it depends on the number of channels in a cluster and the duration of the depolarizing pulse. The curves are for the following duration times of the depolarizing pulse (from bottom to top): 10, 20, 30, 40 and 50 ms. The delay time t_l is equal to 2 ms.

Fig. 11.3, yielding the value 0.0516, and theoretically, using Eq. (28), which yielded the value 0.0523, in good agreement with simulation data (see Fig. 11.3).

The simulation data were extended for other multiple channel systems, up to ten channels; these data are summarized in Figs. 11.4 to 11.6. The importance of channel clustering in case of large delay times is obvious from these figures, e.g. in the case of a 10 ms latency time, depolarizing pulses shorter than 20 ms practically do not induce exocytosis, even with clusters containing ten channels.

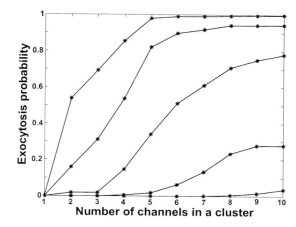

Fig. 11.5 The exocytosis probability as it depends on the number of channels in a cluster and the duration of the depolarizing pulse. The curves are for the following duration times of the depolarizing pulse (from bottom to top): 10, 20, 30, 40 and 50 ms. The delay time t_l is equal to 5 ms.

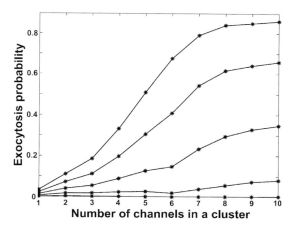

Fig. 11.6 The exocytosis probability as it depends on the number of channels in a cluster and the duration of the depolarizing pulse. The curves are for the following duration times of the depolarizing pulse (from bottom to top): 10, 20, 30, 40 and 50 ms. The delay time t_l is equal to 10 ms.

11.9
Discussion

We have shown here that a delay to the release of substances from secretory granules and vesicles, or a latency time that is inherent in the Ca^{2+}-sensing machinery at the exocytosis site, convincingly shown to be present in at least some cell types, affects essentially the probability of a unitary fusion event. The analysis of this phenomenon presented above provides insight into how the probability for a ready to merge granule to start forming the fusion pore depends on the size of the delay time, on the kinetic properties of Ca^{2+} channels, and on the character of the stimulus.

In a simple linear probabilistic model, we could derive expressions for the Laplace transform of the waiting time distribution, allowing us to readily calculate, e.g. the average waiting time, the most important parameter from an experimental point of view. Knowledge of this and other properties of the waiting time distribution allows one to estimate the percentage of granules that will be secreted by a depolarizing pulse.

A more complicated calculation of exocytosis probabilities requires more detailed knowledge about the distribution function of waiting times for the event that at least one channel is open during the latency time t_l. Work in this direction is in progress.

Besides giving the theoretical rates of exocytosis under different conditions, the results of this investigation provide also a rationale behind the structural organization of secreting machinery in various types of cells.

In synapses where the speed of exocytosis is usually most important, neutralization of the delay time in sensing the rising concentration of Ca^{2+} is achieved by arranging a large number of Ca^{2+} channels that open per individual secretory vesicle. Such an arrangement increases the probability of exocytosis practically to unity, even for short depolarizing pulses, i.e. comparable to the latency time t_l [14].

In other cell types where speed may be not the prime object, the morphological consequences of the delay time may take different forms. For example, in insulin-secreting β-cells of the pancreas, the main problem faced by the secreting machinery is the low number of Ca^{2+} channels available per individual insulin granule. In the case of diffuse and random distribution of these channels over the whole surface of β-cells, the functioning of the secreting machinery is not effective; the probability to induce the fusion pore by action potentials – 100 to 200 ms long – is low. As a consequence, large amounts of energy are needed to remove from the cytoplasm the excessive amounts of Ca^{2+} flooded into the cell in order to reach high local levels of the ion in the vicinity of releasable granules for a sufficiently long time. To counteract the effect of the latency time, which in these cells is about 10 ms, they organize Ca^{2+} channels in clusters of three [12].

The rate of exocytosis from any type of cell is a very important factor and is therefore subject to strict regulation. This rate can be changed through the modification of different parameters of the secreting machinery. Our investigation shows that the latency time in the Ca^{2+} sensing system and the clustering of channels may be two convenient targets for easily adjusting the rate of release for hormones or neurotransmitters from cells.

In conclusion it is to be mentioned that the kinetics of real ion channels is much more complicated than the two-state kinetics discussed in this chapter. Further analysis is necessary in order to assess the effect of more sophisticated kinetics on the rate of exocytosis.

11.10
Conclusions

The rate of exocytosis from any type of cell is a very important factor and is therefore subject to strict regulation. This rate can be changed through the modification of different parameters of the exocytotic machinery. Analysis of pancreatic β-cells shows that the latency time in the Ca^{2+} sensing system of the exocytotic machinery and the clustering of channels on the cell surface may be two convenient targets for adjusting the rate of the release of hormones or neurotransmitters from cells.

Acknowledgments

This work was in part supported by the Swedish Foundation for Strategic Research (B.S.) and by European Union programme Biosim (J.G., P.R.).

References

1 Katz, B.: *The release of neural transmitter substances.* Thomas, Springfield, Illinois, **1969**.

2 Rorsman, P.: The pancreatic B cell as a fuel sensor: an electrphysiologist's opinion. *Diabetologia* **1997**, 40:487–495.

3 An, S., Zenisek, D.: Regulation of exocytosis in neurons and neuroendocrine cells. *Curr. Opin. Neurobiol.* **2004**, 14:522–530.

4 Fernandez-Peruchena, C., Navas, S., Montes, M.A., et al.: Fusion pore regulation of transmitter release. *Brain Res. Rev.* **2005**, 49:406–415.

5 Bennett, M., Scheller, R.: A molecular description of synaptic vesicle membrane trafficking. *Annu. Rev. Biochem.* **1994**, 63:63–100.

6 Jackson, M.B., Chapman, E.R.: Fusion pores and fusion machines in Ca2+-triggered exocytosis. *Annu. Rev. Biophys. Biomol. Struct.* **2006**, 35:135–160.

7 Lang, J.: Molecular mechanisms and regulation of insulin secretion exocytosis as a paradigm of endocrine secretion. *Eur. J. Biochem.* **1999**, 259:3–17.

8 Kim, D., Catterall, W.: Ca2+-dependent and -independent interactions of the isoforms of the alpha 1A subunit of brain Ca2+ channels with pre-synaptic SNARE proteins. *Proc. Natl Acad. Sci. USA* **1997**, 94:14782–14786.

9 Sheng, Z., Westenbroek, R., Catterall, W.: Physical link and the functional coupling of pre-synaptic calcium channels and the synaptic vesicle docking/fusion machinery. *J. Bioenerg. Biomembr.* **1998**, 30:335–345.

10 Wiser, O., et al.: The voltage-sensitive L-type Ca2+ channels is functionally coupled to the exocytotic machinery. *Proc. Natl Acad. Sci. USA* **1999**, 96:248–253.

11 Ämmälä, C., et al.: Exocytosis elicited by action potentials and voltage-clamp calcium currents in individual mouse pancreatic B cells. *J. Physiol. (London)* **1993**, 474:665–688.

12 Barg, S., Ma, X., Eliasson, L., Galvanovskis, J., et al.: Fast exocytosis with few Ca2+ channels in insulin-secreting mouse pancreatic B cells. *Biophys. J.* **2001**, 81:3308–3323.

13 Yazejian, B., Sun, X.-P., Grinnell, A.: Tracking pre-synaptic Ca2+ dynamics during neurotransmitter release with Ca2+-activated K+ channels. *Nat. Neurosci.* **2000**, 19:566–571.

14 Wu, L.-G., et al.: Calcium channel types with distinct presynaptic localization couple differentially to transmitter release in single caly-type synapses. *J. Neurosci.* **1999**, 19:726–736.

12
Modeling Kidney Pressure and Flow Regulation

O. V. Sosnovtseva, E. Mosekilde, and N.-H. Holstein-Rathlou

Abstract

Self-sustained oscillations, synchronization, transitions to chaos, and other non-linear dynamic phenomena play an important role in the regulation and function of normal physiological systems as well as in the development of different states of disease. Pressure and flow regulation in the individual functional unit of the kidney (the nephron), for instance, tends to operate in an unstable regime involving several oscillatory modes. For normotensive rats, the regulation displays regular self-sustained oscillations, but for rats with high blood pressure the oscillations become chaotic. We explain the mechanisms responsible for this behavior and discuss the involved bifurcations. Experimental data show that neighboring nephrons adjust their pressure and flow regulation in accordance with one another. Accounting for both a hemodynamic coupling and a vascularly propagated coupling between nephrons that share a common interlobular artery, we analyze a model of the interaction of the pressure and flow regulations between adjacent nephrons. It is shown that this model, with physiologically realistic parameter values, can reproduce the different types of experimentally observed synchronization, including phase-shifted regimes and partial synchronization involving either slow tubuloglomerular feedback-mediated oscillations or fast myogenic dynamics.

12.1
Introduction

The kidneys play an important role in maintaining a proper environment for the cells in the body. By regulating the excretion of water, salts and metabolic end products, the kidneys control the plasma osmolality (i.e., the concentration of ions in the blood), the extracellular fluid volume, and the proportions of various blood solutes. The kidneys are also involved in the production of a set of hormones that make the blood vessels (arterioles) contract in the kidneys as well as in other parts of the body. These hormones can give rise to changes in the vascular structure, and

Biosimulation in Drug Development. Edited by Martin Bertau, Erik Mosekilde, and Hans V. Westerhoff
Copyright © 2008 WILEY-VCH Verlag GmbH & Co. KGaA, Weinheim
ISBN: 978-3-527-31699-1

disturbances in their production may cause hypertension, a prevalent disease in modern societies.

The arterial blood pressure varies significantly with time. There are two main periodicities in this variation, one at the frequency of the heart (about 1 Hz for humans) and the other associated with the 24 h daily activity rhythm. The various organs of the body dispose of mechanisms by which they can regulate the incoming blood flow according to their immediate needs. This control involves regulation of the diameter of the afferent arterioles in response to variations in the arterial blood pressure, the production of metabolites, and a variety of other signals. Over the bandwidth intermediate to the above two periodicities, the blood pressure typically displays a 1/f-like variation, caused by the more or less independent activities of the arterioles in the various organs, with the activation of large skeletal muscles making particularly strong contributions.

The kidneys are also perfused with blood and thus exposed to all the fluctuations present in the cardio-vascular system. By virtue of the special role that the kidneys serve, this perfusion significantly exceeds their own needs, and to protect the delicate, hormonally controlled absorption and secretion processes that take place in the distal tubule and the collecting duct (see Fig. 12.1), the kidneys dispose of a special mechanism that can ensure a relatively constant blood supply in spite of fluctuations in the arterial blood pressure. It has long been recognized that this control is associated with the so-called tubuloglomerular feedback (TGF), by which the individual nephron can regulate the incoming blood flow in dependence of the ionic composition of the fluid leaving the loop of Henle.

Figure 12.1 illustrates the main structure of a nephron [1]. The measurements to be reported in this chapter were performed on rats. A rat kidney contains approximately 30000 nephrons as compared to the one million nephrons in a human kidney. The process of urine formation starts with the filtration of plasma in the glomerulus, a system of 20–40 capillary loops. The presence of a relatively high hydrostatic pressure in this system allows water, salts and small molecules to pass out through the capillary wall and into the proximal tubule. Blood cells and proteins are retained, and the filtration process saturates when the protein osmotic pressure balances the hydrostatic pressure difference between the blood and the filtrate in the tubule. For superficial nephrons, the proximal tubule is visible in the surface of the kidney and easily accessible for pressure measurements by means of a thin glass pipette.

A small hydrostatic pressure difference drives the filtrate through the various sections of the tubular system: the descending and ascending limbs of the loop of Henle, the distal tubule, and the collecting duct where flows from several nephrons meet. During this transit, the volume and composition of the fluid are changed through selective reabsorption of water and salts, and by the time the fluid reaches the end of the ascending limb about 80% of glomerular filtrate and 90% of the filtrated salts have been returned to the blood. Under hormonal control from the adrenal gland, the brain, and the kidney itself, epithelial transport processes continue to adjust the composition of the filtrate as it passes through the distal tubule. This hormonal regulation has a limited dynamic range, and the solute load that the

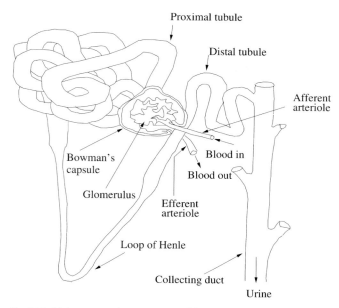

Proximal tubule

Distal tubule

Afferent
arteriole

Bowman's
capsule

Blood in

Blood out

Glomerulus

Efferent
arteriole

Loop of Henle

Collecting duct

Urine

Fig. 12.1 Main structural components of the nephron. Note
particularly how the terminal part of the loop of Henle passes within
cellular distances of the afferent arteriole. This forms the anatomical
basis for the tubuloglomerular feedback mechanism by which the
nephron regulates the incoming blood flow in response to variations
in the ionic composition of the fluid that leaves the loop of Henle.

epithelial cells receive via the tubular fluid must remain within relatively narrow
bounds.

The TGF regulation is made possible by the interesting anatomical feature that
the terminal part of the loop of Henle passes within cellular distances of the afferent arteriole associated with the same nephron. At the point of contact, specialized
cells, the macula densa cells, monitor the NaCl concentration in the tubular fluid
and produce a signal that activates the smooth muscle cells in the arteriolar wall.
The larger the glomerular filtration is, the faster the fluid will flow through the loop
of Henle, and the higher the NaCl concentration will be at the macular densa cells.
A high NaCl concentration causes the macular densa cells to activate the smooth
muscle cells in the arteriolar wall for contraction, thereby reducing the blood flow
to the glomerulus and lowering the rate of filtration.

The TGF mechanism produces a negative feedback control on the rate of
glomerular filtration. However, experiments performed on rats by Leyssac and
Baumback [2] and by Leyssac and Holstein-Rathlou [3] in the 1980s demonstrated
that the feedback regulation tends to be unstable and to generate large amplitude
self-sustained oscillations in the proximal intratubular pressure with a period of
30–40 s. With different amplitudes and phases, similar oscillations have subsequently been observed in the distal intratubular pressure and in the chloride concentration near the terminal part of the loop of Henle [4].

Destabilization of the TGF regulation is caused by the combination of a relatively long feedback delay, associated preliminary with the finite transit time for the fluid through the loop of Henle, and a high loop gain. The latter means that the response of the afferent arteriole is fairy strong, even to small variations in the glomerular filtration rate. As discussed in connection with the presentation of our single nephron model (see Section 12.3), the delay in the tubular system can be determined from the phase shift between the pressure oscillations in the proximal tubule and the oscillations in the chloride concentration near the macula densa.

For normotensive rats, the TGF-mediated oscillations typically have the appearance of a regular limit cycle (self-sustained oscillation) with a predominant spectral component corresponding to the period of the cycle. Spontaneously hypertensive rats (SHR), on the other hand display highly irregular oscillations with a more broadband spectral distribution. Hypertensive rats also show a steeper and stronger variation of the steady state feedback characteristic. Hence, it is natural to try to investigate the transition to irregular oscillations using the feedback delay and the loop gain as control parameters.

However, the steady state feedback characteristic only provides part of the picture. The afferent arteriole tends to respond to external stimuli in an oscillatory manner, and these vasomotoric (myogenic) oscillations may even become self-sustained if the activation of the smooth muscle cells is sufficiently strong [5, 6]. The vasomotoric oscillations typically have a period of 5–8 s, or about one-fifth of the period of the TGF oscillations. The presence of these two interacting oscillatory components allows for a variety of complex dynamic phenomena, including different forms of frequency locking between the modes. Moreover, it is likely that the vascular system is particularly sensitive in hypertensive rats and that the myogenic oscillations therefore play a stronger role in these rats. This could represent a, so far unexplored, explanation to the development of irregular oscillations in hypertensive rats.

Neighboring nephrons also communicate with one another. Experiments performed by Holstein-Rathlou show how nephrons that share a common interlobular artery tend to adjust their TGF mediated pressure oscillations so as to produce a state of in-phase synchronization [7]. Holstein-Rathlou also demonstrated how microperfusion of one nephron (with artificial tubular fluid) affects the amplitude of the pressure oscillation in a neighboring nephron. This provides a method to determine the strength of the nephron–nephron interaction.

Hemodynamic coupling represents a simple mechanism by which two neighboring nephrons can interact. This coupling depends on the fact that as the first nephron reduces the incoming blood flow, part of the blood flow will be displaced to the second nephron. One delay period later, when the increasing blood flow activates the feedback regulation of the second nephron, the blood flow to this nephron is reduced, and more blood will flow to the first nephron. We presume that this type of cross-talk can be significant for nephrons arranged in close pairs and sharing part of a common arteriole. The hemodynamic coupling tends to produce antiphase synchronization, i.e., a type of dynamics in which the blood flows to the two nephrons oscillate with opposite phases. A few experimental results witness the possibility of this type of behavior.

However, an alternative coupling, which we denote as vascular propagated coupling, is likely to be more important. This coupling is mediated through electrochemical signals, initiated by the TGF response of one nephron and propagated in a damped fashion along the smooth muscle cells of the arteriolar wall to the neighboring nephrons. Because of the relatively high speed at which such signals are transmitted (as compared with the length of the individual vessels and the period of the TGF oscillation), this coupling tends to produce in-phase synchronization. If the afferent arteriole of one nephron is stimulated by its TGF response to contract, the vascular propagated signal almost immediately reaches the neighboring nephrons and cause their TGF mechanisms to activate as well.

The main purpose of this chapter is to illustrate how one can integrate a variety of physiological mechanisms (glomerular filtration, delay in the loop of Henle, feedback from the macula densa cells, nonlinear oscillatory responses of the arteriole, and vascular propagated coupling between neighboring nephrons) into consistent models of the dynamics of individual nephrons and of systems of coupled nephrons. At the same time we try to explain some of the complicated nonlinear dynamic phenomena that the models display. Besides various types of bifurcations and transition to chaos, these phenomena include intra-nephron synchronization between fast (arteriolar) and slow (TGF-mediated) oscillations, and full and partial synchronization between neighboring nephrons. We believe that transitions between various states of synchronization can play an essential role in the overall regulation of the kidney.

12.2
Experimental Background

During the experiments to be discussed here, the rats were anesthetized, placed on a heated operating table to maintain the body temperature, and connected to a small animal respirator to ensure a proper oxygen supply to the blood. The frequency of the respirator was close to 1 Hz. This component is clearly visible in the frequency spectra of the observed tubular pressure variations. Also observable is the frequency of the freely beating heart, which typically gives a contribution in the 4–6 Hz regime. The frequencies involved in the nephron pressure and flow regulation are significantly lower and, presumably, not influenced much by the respiratory and cardiac forcing signals. When exposing the surface of the kidney, a small glass pipette could be inserted into the proximal tubule of a superficial nephron, allowing the intratubular pressure variations to be measured with a good temporal resolution.

Figure 12.2a shows an example of a recording of the proximal tubular pressure for a normotensive rat. This pressure is seen to exhibit relatively slow large scale oscillations with a period about 30 s and an amplitude close to 1.5 mmHg. On top of these fairly regular oscillations, various faster components can be observed.

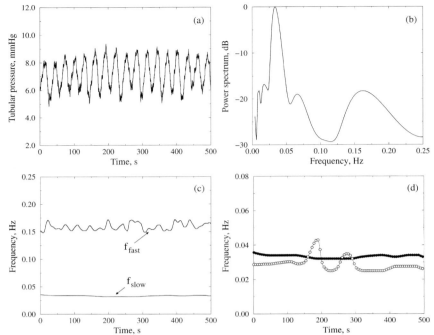

Fig. 12.2 (a) Experimental recording of the proximal tubular pressure in a single nephron of a normotensive rat. The power spectrum (b) clearly shows the TGF-mediated oscillations at $f_{slow}(t) \approx 0.034$ Hz and the myogenic oscillations at $f_{fast}(t) \approx 0.16$ Hz. The spectrum also displays harmonics and subharmonics of the TGF-oscillations. A plot of the instantaneous frequencies of the main components (c) reveals a modulation of the frequency of the myogenic oscillations by the TGF mode. Finally, a double-wavelet analysis (d) demonstrates the accordance between the instantaneous modulation frequency for the myogenic oscillations (open circles) and the TGF frequency (black circles).

Moreover, during part of the observation period (180 s < t < 480 s), the pressure variations show indications of period-2 dynamics, i.e., the minima of the pressure oscillations alternate between a low and a somewhat higher value. It is well known that period-2 dynamics is a typical precursor of deterministic chaos [8].

To analyze the spectral composition of the pressure variations in more detail we have made use of a wavelet approach [9]. This approach, which allows us to determine instantaneous values of the frequencies and amplitudes of the various oscillatory components, is particularly useful for biological time series that often are neither homogeneous nor stationary.

The wavelet transform $T_\varphi[x]$ of a signal $x(t)$ is defined by:

$$T_\varphi[x](a, b) = \frac{1}{\sqrt{a}} \int_{-\infty}^{\infty} x(t) \varphi^* \left(\frac{t - b}{a} \right) dt \tag{1}$$

where φ is referred to as the mother function, and a and b are parameters that define the time scale and temporal displacement of φ. In the present analysis we applied the Morlet function:

$$\varphi(\tau) = \exp(j 2\pi f_0 \tau) \exp\left[-\frac{\tau^2}{2} \right] \tag{2}$$

with $f_0 = 1$. j is the imaginary unit.

From the wavelet coefficients $T_\varphi[x](a, b)$ one can calculate the energy density $E_\varphi[x](a, t) = | T_\varphi[x](a, t) |^2$. The result of such a calculation is a three-dimensional surface $E_\varphi[x](a, t)$. Sections of this surface at fixed time moments $t = b$ define the local energy spectrum $E_\varphi[x](f, t)$ with $f = a^{-1}$. Finally, in order to obtain the mean spectral distribution of the time series $x(t)$ we may consider a so-called scalogram, i.e., the time-averaged energy spectrum. This is analogous to the classic Fourier spectrum.

Figure 12.2b shows such a power spectrum for the tubular pressure variations depicted in Fig. 12.2a. This spectrum demonstrates the existence of two main and clearly separated peaks: a slow oscillation with a frequency $f_{slow} \approx 0.034$ Hz that we identify with the TGF-mediated oscillations, and a significantly faster component at $f_{fast} \approx 0.16$ Hz representing the myogenic oscillations of the afferent arteriole. Both components play an essential role in the description of the physiological control system. The power spectrum also shows a number of minor peaks on either side of the TGF peak. Some of these peaks may be harmonics ($f \approx 0.07$ Hz) and subharmonics ($f \approx 0.017$ Hz) of the TGF peak, illustrating the nonlinear character of the limit cycle oscillations.

Figure 12.2c shows the temporal variation of the instantaneous frequencies for the two modes. It is interesting to observe how the frequency of the fast mode is modulated in a fairly regular manner. With about 17 modulation cycles for f_{fast} during the 500 s of observation time, we conclude that the frequency of the fast mode is modulated by the presence of the slow mode, indicating that the two modes interact with one another. If one compares the phase of the tubular pressure variations in Fig. 12.2a with the phase of the frequency modulation in Fig. 12.2c it appears that the maximum of f_{fast} occurs about 60° after the maximum of P_t. It is important to note, however, that the various steps of our wavelet analysis may have introduced a certain phase lag. We are presently trying to correct for such effects in order to obtain a better understanding of the instantaneous relation between the two variables.

To examine the modulation process in more detail we submitted the instantaneous frequency $f_{fast}(t)$ of the myogenic mode to a new wavelet analysis. Again the wavelet coefficients were calculated and the corresponding energy density obtained. This energy density contains information about all processes involved in the modulation of $f_{fast}(t)$. In a similar manner, we can consider variations in the instantaneous amplitude of the fast mode and thus examine the characteristics of the amplitude modulation. We refer to this approach as a double-wavelet analysis [10]. It allows us to characterize the nonstationary temporal dynamics of a modulated signal, i.e., to detect all components involved in the modulation, estimate their contributions, and analyze whether the modulation properties change during

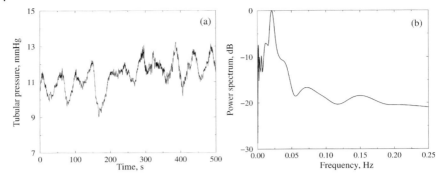

Fig. 12.3 Analysis of the tubular pressure data for a hypertensive rat. (a) is the original time series and (b) the corresponding power spectrum. Like the normotensive rat in Fig. 12.2 the hypertensive rat also displays spectral peaks corresponding to both the TGF-mediated oscillation and the myogenic oscillation of the afferent arteriole. We can also observe peaks at both the harmonic and the subharmonic frequencies of f_{slow}.

the observation time. Figure 12.2d represents the results of such an analysis for the time series shown in Fig. 12.2a. We can see that the modulation frequency for the fast oscillations (open circles) is close to the slow rhythm of the TGF oscillations (black circles).

Figure 12.3a shows a typical variation of the proximal tubular pressure for a hypertensive rat. This variation is clearly nonstationary and, like the tubular pressure variations in other hypertensive rats, it satisfies the criteria for deterministic chaos. In particular, the largest Lyapunov exponent is positive, and shuffling of the data series increases the entropy significantly. A positive Lyapunov exponent signals sensitivity to the initial conditions, i.e., two trajectories starting in neighboring points will diverge exponentially. The increasing entropy implies that the signal, though quite irregular, is not random. Figure 12.3b presents the corresponding power spectrum. Inspection of this figure shows that one can still identify both a slow TGF-mediated mode (at $f_{slow} \approx 0.020$ Hz) and fast myogenic mode ($f_{fast} \approx 0.015$ Hz). Moreover, the hypertensive rat also displays harmonics as well as subharmonics of the basic frequency. A main purpose of our modeling efforts is to understand the changes that take place in the nephron pressure and flow control when a rat develops hypertension.

12.3
Single-nephron Model

The purpose of this section is to illustrate how one can integrate different physiological processes into a consistent description of the feedback mechanisms that generate the complicated bimodal dynamics observed in the tubular pressure variations of individual nephrons. Let us start with the variation of the proximal tubular pressure in dependence of the glomerular filtration rate and the flow of fluid into the loop of Henle.

The tubule is a spatially extended structure, and it presents both elastic properties and resistance to the fluid flow. The dynamic pressure and flow variations in such a structure can be represented by a set of coupled partial differential equations [11]. An approximate description in terms of ordinary differential equations (a lumped model) consists of an alternating sequence of elastic and resistive elements, and the simplest possible description, which we will adopt here, applies only a single pair of such elements. Hence our model [12] considers the proximal tubule as an elastic structure with little or no flow resistance. The pressure P_t in the proximal tubule changes in response to differences between the in- and outgoing fluid flows:

$$\frac{dP_t}{dt} = \frac{1}{C_{tub}} \left[F_{filt} - F_{reab} - F_{Hen} \right] \tag{3}$$

where F_{filt} is the glomerular filtration rate and C_{tub} the elastic compliance of the tubule. The Henle flow:

$$F_{Hen} = \frac{P_t - P_d}{R_{Hen}} \tag{4}$$

is determined by the difference between the proximal (P_t) and the distal (P_d) tubular pressures divided by the flow resistance R_{Hen} in the loop of Henle. This description is clearly a simplification, since a significant reabsorption of water and salts occurs during passage of the loop of Henle. However, within the physiologically relevant flow range it provides an acceptable approximation to the experimentally determined pressure-flow relation for the loop of Henle [13].

As the filtrate flows into the descending limb of this loop, the NaCl concentration in the fluid surrounding the tubule increases by a factor of four, and osmotic processes cause water to be reabsorbed. At the same time, salts and metabolic products are secreted into the tubular fluid. In the ascending limb, in contrast, the tubular wall is nearly impermeable to water. Here, the epithelial cells contain molecular pumps that transport sodium and chloride from the tubular fluid into the space between the nephrons (the interstitium). These processes are accounted for in considerable detail in the spatially extended model developed by Holstein-Rathlou et al. [14]. In the present model, the reabsorption F_{reab} in the proximal tubule and the flow resistance R_{Hen} are treated as constants. Without affecting the composition much, the proximal tubule reabsorbs close to 60% of the ultrafiltrate produced by the glomerulus.

The glomerular filtration rate is expressed as [15]:

$$F_{filt} = (1 - H_a) \left(1 - \frac{C_a}{C_e} \right) \frac{P_a - P_g}{R_a} \tag{5}$$

where H_a is the hematocrit of the afferent arteriolar blood (i.e., the fraction that the blood cells constitute of the total blood volume at the entrance to the glomerular capillaries). C_a and C_e are the protein concentrations in the afferent and efferent plasma, respectively, and R_a is the flow resistance of the afferent arteriole. $(P_a - P_g)/R_a$ determines the incoming blood flow. Multiplied by $(1 - H_a)$ this gives the plasma flow V_a. Finally, the factor $(1 - C_a/C_e)$ relates the filtration rate to the change in protein concentration for the plasma remaining in the vessel. The rates at which protein enters and leaves the glomerular capillary system are V_aC_a and V_eC_e, respectively, with V_e representing the outgoing plasma flow. Conservation of protein in the blood flow gives $V_aC_a = V_eC_e$ leading finally to the expression in Eq. (5) for the rate of filtration $F_{filt} = V_a - V_e = V_a(1 - V_e/V_a) = V_a(1 - C_a/C_e)$.

The glomerular pressure P_g is determined by distributing the arterial to venous pressure drop between the afferent and the efferent arteriolar resistances, i.e., as the solution to the linear equation:

$$P_g = P_v + R_e \left(\frac{P_a - P_g}{R_a} - F_{filt} \right) \tag{6}$$

where the venous (P_v) and arterial (P_a) pressures and the efferent arteriolar resistance R_e are considered as constants. The efferent arteriolar resistance plays an important role for maintaining the relatively high glomerular pressure required for the filtration process to occur. In the direction of the blood flow, the efferent arteriole is followed by a second capillary system that embraces the tubule and returns the reabsorbed water and salts to the vascular system.

To determine the protein concentration C_e in the efferent blood we assume that filtration equilibrium is established before the blood leaves the glomerular capillaries, i.e., that the glomerular hydrostatic pressure minus the efferent colloid osmotic pressure P_{osm} equals the tubular pressure. The experimentally determined relation between the colloid osmotic pressure and the protein concentration C can be described as [16]:

$$P_{osm} = aC + bC^2. \tag{7}$$

Inserting the equilibrium condition $P_{osm} = P_g - P_t$, this equation leads to an expression of the form:

$$C_e = \frac{1}{2b} \left[\sqrt{a^2 - 4b(P_t - P_g)} - a \right] \tag{8}$$

In the computer model the simultaneous equations (5), (6) and (8) are combined into a single third-order equation for C_e, which is solved iterately for each integration step. For relevant values of the various parameters, the equation has a single positive solution.

The delay in the tubuloglomerular feedback is represented by means of three first-order coupled differential equations:

$$\frac{dx_1}{dt} = F_{Hen} - \frac{3}{T}x_1 \tag{9}$$

$$\frac{dx_2}{dt} = \frac{3}{T} (x_1 - x_2) \tag{10}$$

$$\frac{dx_3}{dt} = \frac{3}{T} (x_2 - x_3) \tag{11}$$

with T being the total delay time. This formulation implies that the delay is represented as a smoothed process, with x_1, x_2 and x_3 being intermediate variables in the delay chain. By using a delay of finite order we can implicitly account for dissipative phenomena in the form, for instance, of a damping of the pressure and flow oscillations from the proximal to the distal tubule.

The steady-state component of the glomerular feedback is described by a sigmoidal relation between the muscular activation ψ of the afferent arteriole and the delayed version x_3 of the Henle flow:

$$\psi = \psi_{max} - \frac{\psi_{max} - \psi_{min}}{1 + \exp\left[\alpha\left(3x_3 / T F_{Hen0} - S\right)\right]}. \tag{12}$$

Here, ψ_{max} and ψ_{min} denote, respectively, the maximum and the minimum values of the muscular activation. α determines the slope of the feedback curve, S is the displacement of the curve along the flow axis, and F_{Hen0} is a normalization value for the Henle flow. The relation between the glomerular filtration and the flow into the loop of Henle can be obtained from open-loop experiments in which a paraffin block is inserted into the proximal tubule and the rate of glomerular filtration (or, alternatively, the so-called tubular stop pressure at which the filtration ceases) is measured as a function of an externally forced rate of flow of artificial tubular fluid into the loop of Henle. Translation of the experimental results into a relation between muscular activation and Henle flow is performed by means of the model, i.e., the relation is adjusted such that it can reproduce the experimentally observed steady state relation. We have previously discussed the significance of the feedback gain α in controlling the dynamics of the system. α is one of the parameters that differ between hypertensive and normotensive rats, and α will also be one of the control parameters in our analysis of the simulation results.

Our next problem concerns the dynamic response of the arteriolar system to the signal from the mascula densa cells. This response is restricted to that part of the afferent arteriole that is closest to the glomerulus. Hence, the afferent arteriole is divided into two serially coupled sections of which the first (representing a fraction β of the total length) is assumed to have a constant flow (or hemodynamic) resistance, while the second (closer to the glomerulus) is capable of varying its diameter and hence the flow resistance in dependence of the tubuloglomerular feedback activation:

$$R_a = R_{a0}\left[\beta + (1 - \beta) r^{-4}\right] \tag{13}$$

Here, R_{a0} denotes a normal value of the arteriolar resistance and r is the radius of the active part of the vessel, normalized relatively to its resting value. In accordance with Poiseuille's law for laminar flows, the hemodynamic resistance of the active part is assumed to vary inversely proportional to r^4.

Experiments showed that arterioles tend to perform damped oscillatory contractions in response to external stimuli [17]. This behavior may be captured by a second-order differential equation of the form:

$$\frac{d^2r}{dt^2} + k\frac{dr}{dt} - \frac{P_{av} - P_{eq}}{\omega} = 0 \tag{14}$$

where k is a characteristic time constant describing the damping of the arteriolar dynamics. ω is a parameter that controls the natural frequency of the oscillations, and

$$P_{av} = \frac{1}{2}\left(P_a - (P_a - P_g)\beta\frac{R_{a0}}{R_a} + P_g\right) \tag{15}$$

is the average pressure in the active part of the arteriole. P_{eq} is the value of this pressure for which the arteriole is in equilibrium with its present radius at the existing muscular activation. P_{av} is obtained from a simple pressure distribution along the afferent arteriole.

The reaction of the arteriolar wall to changes in the blood pressure is considered to consist of a passive, elastic component in parallel with an active, muscular response. The elastic component is determined by the properties of the connective tissue, which consists mostly of collagen and elastin. The relation between strain ϵ and elastic stress σ_e for homogeneous soft tissue may be described as [18]:

$$\sigma_e = C_0\left(e^{\gamma\epsilon} - 1\right) \tag{16}$$

where C_0 and γ are constants characterizing the tissue. For very small values of ϵ ($\gamma\epsilon \ll 1$), we have an approximately linear strain-stress relation. However, for larger ϵ values, the stress rises exponentially with the strain, and the tissue becomes increasingly difficult to stretch.

At least for a first approach, the active component in the strain–stress relation may be treated in a simple manner. For some strain ϵ_{max} the active stress σ_a is maximum, and on both sides the stress decreases almost linearly with $|\epsilon - \epsilon_{max}|$. Moreover, the stress is proportional to the muscle tone ψ. By numerically integrating the passive and active contributions across the arteriolar wall, one can establish a relation among the equilibrium pressure P_{eq}, the normalized radius r, and the activation level ψ [19]. This relation is based solely on the physical characteristics of the vessel wall. However, computation of the relation for every time step of the simulation model is time-consuming. To speed up the process we have used the following analytic approximation [12]:

$$P_{eq} = 2.4 \times e^{10(r-1.4)} + 1.6(r-1) + \psi\left(\frac{4.7}{1 + e^{13(0.4-r)}} + 7.2(r + 0.9)\right) \tag{17}$$

where P_{eq} is expressed in kPa (1 kPa $= 10^3$ N/m$^2 \cong 7.5$ mmHg). The first two terms in Eq. (17) represent the pressure *vesus* radius relation for the nonactivated arteriole. This relation consists of an exponential and a linear term arising from the two terms in the expression Eq. (16) for σ_e. The terms proportional to ψ represent the active response. This is approximately given by a sigmoidal term superimposed onto a linear term. The activation from the TGF mechanism is assumed to be determined by Eq. (12). The expression in Eq. (17) closely reproduces the output of the more complex, experimentally based relation [12].

Figure 12.4 shows the relation between the equilibrium transmural pressure P_{eq} and the normalized arteriolar radius r for different values of the muscular activation ψ. Fully drawn curves represent the results of a numerical integration of the active and passive stress components across the arteriolar wall, and the dashed curves represent the analytic approximation Eq. (17).

The above discussion completes our description of the single-nephron model. In total we have six coupled ordinary differential equations, each representing an essential physiological relation. Because of the need to numerically evaluate C_e in each integration step, the model cannot be brought onto an explicit form. The applied parameters in the single-nephron model are specified in Table 12.1. They have all been adopted from the experimental literature, and their specific origin is discussed in Jensen et al. [13].

The parameters γ and ε relate to the model of nephron–nephron interaction to be discussed in Section 12.5. In order to represent the hemodynamic coupling it is necessary to also introduce a parameter C_{glo} that describes the elastic response of

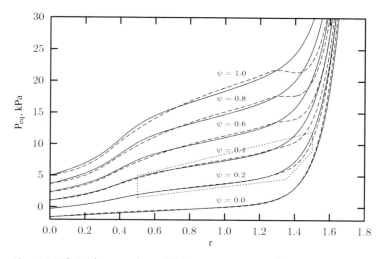

Fig. 12.4 Relation between the equilibrium pressure P_{eq} and the normalized radius r of the afferent arteriole for different values of the muscular activation level ψ. The solid curves represent our analytical approximation and the dashed curves represent the exact numerical solution. The area bounded by dotted lines corresponds approximately to the regime of operation for the model.

Table 12.1 Parameters used in the nephron model.

Arterial blood pressure	$P_a = 13.3$ kPa
Venous blood pressure	$P_v = 1.3$ kPa
Distal tubular pressure	$P_d = 1.3$ kPa
Flow resistance of afferent arteriole	$R_a = 2.3$ kPa/(nl/s)
Flow resistance of efferent arteriole	$R_e = 1.9$ kPa/(nl/s)
Flow resistance in loop of Henle	$R_{Hen} = 5.3$ kPa/(nl/s)
Elastic compliance of tubule	$C_{tub} = 3.0$ nl/kPa
Equilibrium flow in loop of Henle	$F_{Hen0} = 0.2$ nl/s
Proximal tubule reabsorption	$F_{reab} = 0.3$ nl/s
Arterial hematocrit	$H_a = 0.5$
Arterial plasma protein concentration	$C_a = 54$ g/l
Linear colloid osmotic coefficient	$a = 22 \cdot 10^{-3}$ kPa/(g/l)
Nonlinear colloid osmotic coefficient	$b = 0.39 \cdot 10^{-3}$ kPa/(g/l)^2
Lower activation limit	$\psi_{min} = 0.20$
Upper activation limit	$\psi_{max} = 0.44$
Equilibrium activation	$\psi_{eq} = 0.38$
Damping of arteriolar dynamics	$k = 0.04$ /s
Stiffness parameter	$\omega = 20$ kPa·s^2
Fraction of afferent arteriole	$\beta = 0.67$
Feedback delay (standard value)	$T = 16$ s
Feedback amplification (normal rats)	$\alpha = 12$
Vascular coupling parameter	$\gamma = 0.2$ (base case)
Hemodynamic coupling parameter	$\varepsilon = 0.2$ (base case)
Elastic compliance of glomerulus	$C_{glo} = 0.11$ nl/kPa
Displacement of feedback curve	$S = 1$

the capillary system in the glomerulus. Note, however, that as long as $C_{glo} \ll C_{tub}$, the precise value of C_{glo} is of little significance.

12.4
Simulation Results

The next step is now to investigate the model predictions for different parameter values to examine how well it can reproduce experimentally observed phenomena, and to suggest critical experiments that can be used to test the underlying hypotheses. In view of the large-amplitude oscillations observed in the tubular pressure for normotensive rats and the irregular oscillations observed for hypertensive rats, we emphasize the bifurcation structure of the model, i.e., the transitions through which the qualitative behavior of the model changes. In this connection we are particularly concerned with the synchronization phenomena that arise from interactions between the slow tubuloglomerular feedback-mediated oscillations and the faster myogenic oscillations produced by the arteriolar oscillator.

Figure 12.5 shows an example of a one-dimensional bifurcation diagram for the single-nephron model obtained by varying the slope α of the open-loop response

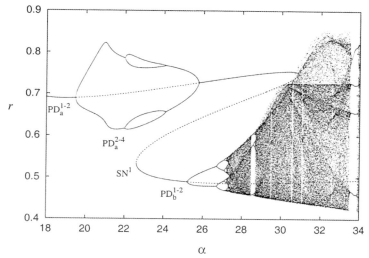

Fig. 12.5 One-dimensional bifurcation diagram for the single-nephron model obtained by varying the slope α of the open-loop response characteristics. r is the normalized arteriolar radius. The delay in the tubuloglomerular feedback is $T = 16$ s. Dotted curves represent unstable solutions determined by means of numerical techniques (continuation) that allow us to follow stable and unstable periodic orbits under variation of the system parameters. The study was performed by means of the in-house simulation package Simpack. Two saddle-node bifurcations of the period-1 cycle fold an incomplete period-doubling structure over a complete period-doubling transition to chaos.

characteristics [Eq. (12)] while keeping the other parameters constant. In particular, the delay in the feedback regulation is assumed to be $T = 16$ s, which agrees with physiological expectations [4]. The diagram was constructed by combining a so-called brute force bifurcation diagram with a bifurcation diagram obtained by means of continuation methods [20, 21]. Continuation methods allow us to follow stable as well as unstable periodic orbits under variation of a parameter and to find and identify the various bifurcations that the orbits undergo. Hence, in Fig. 12.5 fully drawn curves represent stable solutions and dotted curves represent unstable periodic solutions. (With two control parameters, the continuation technique can be used to follow curves of local bifurcations in parameter space as illustrated, for instance, in Fig. 12.9.)

For a given value of α, the brute force bifurcation diagram displays all the values of the relative arteriolar radius r that the model displays when the steady state trajectory intersects a specified hyperplane (the Poincaré section) in phase space. Due to the coexistence of several stable solutions, the brute force diagram must be obtained by scanning α in both directions.

For $T = 16$ s, the single nephron model undergoes a supercritical Hopf bifurcation at $\alpha \cong 11$ (outside the figure). In this bifurcation, the equilibrium point loses its stability, and stable periodic oscillations emerge as the steady-state solution. For $\alpha \cong 19.5$, at the point denoted PD_a^{1-2} in Fig. 12.5, this solution undergoes a period-doubling bifurcation, and in a certain interval of α-values the period-2 cycle is the only stable solution. In the bifurcation diagram, the presence of a period-2 cycle reveals itself by the existence of two different values of the arterial radius for the same value of α (e.g., $r \approx 0.66$ and $r \approx 0.76$ for $\alpha = 20$). The two different values of r represent the two intersection points of the limit cycle with the aforementioned hyperplane in phase space. As we continue to increase α, the period-2 solution undergoes a new period-doubling bifurcation at $\alpha \cong 22$ (i.e., at the point denoted PD_a^{2-4}). The presence of a stable period-4 cycle is revealed in Fig. 12.5 by the fact that r assumes four different values for the same value of α.

The term "bifurcation" denotes a sudden change in the steady state behavior of a system as a parameter is varied. In a Hopf bifurcation, a stable equilibrium point loses its stability as a pair of complex conjugated eigenvalues cross the imaginary axis and enter the positive half-plane. After the bifurcation, trajectories starting near the equilibrium point spiral away from it and, in nonlinear dissipative systems, they are typically attracted to a limit cycle (i.e., they approach a self-sustained oscillation of finite amplitude). In a supercritical Hopf bifurcation, the amplitude of the limit cycle grows continuously from nothing as the system is driven into the region of instability. (The alternative – referred to as a subcritical Hopf bifurcation – is that the loss of stability occurs as an unstable limit cycle, that exists before the bifurcation, contracts around the equilibrium point to finally merge with it.)

In the first period-doubling bifurcation, the simple period-1 limit cycle loses its stability and yields to a cycle of twice the period (a period-2 cycle). The period-2 cycle is still a regular solution. However, the system now completes two full oscillations with different amplitudes before the dynamics starts to retrace itself. In the temporal variation of a single variable (e.g., the tubular pressure), the period-2 oscillation reveals itself in the form of alternating high and low maxima. Our experimental results provide evidence for this type of dynamics in the nephron autoregulation (see, e.g., Fig. 12.2a). It is a major accomplishment of modern nonlinear dynamics to explain why such a period-doubling occurs and to show how an initial period-doubling may be succeeded by a cascade of similar bifurcations leading to a state of deterministic chaos [8].

With further increase of α, the stable period-4 orbit undergoes two consecutive backwards period-doublings, so that the original period-1 cycle again becomes stable around $\alpha = 26$. The stable period-1 cycle can hereafter be followed up to $\alpha \cong 31$ where it is destabilized in a saddle-node bifurcation. This is a type of bifurcation where the stable limit cycle (called a node in the Poincaré section of the system) meets an unstable limit cycle (a saddle in relation to the Poincaré section) of the same periodicity and annihilates with it. The stable limit cycle attracts trajectories from all sides whereas the saddle has a direction in which it repels trajectories. The saddle cycle can be followed backwards in the bifurcation diagram (dotted curve) to a point near $\alpha = 22.5$ where it undergoes a second saddle-node bifurcation, and a new stable period-1 orbit is born. This cycle has a considerably larger amplitude than the original period-1 cycle. As the parameter α is again increased, the new period-1 cycle undergoes a period-doubling cascade starting with the first period-doubling bifurcation at $\alpha \cong 25$ and accumulating with the development of deterministic chaos near $\alpha = 27$. At even higher values of α we notice the presence of a period-3 window near $\alpha = 28.5$ and the appearance of another stable period-4 cycle around $\alpha = 33.5$.

The above scenario is typical of nonlinear dynamical systems when the amplitude of the internally generated oscillations becomes sufficiently large. In the bifurcation diagram of Fig. 12.5 this occurs when the slope of the feedback characteristics exceeds a critical value. However, similar scenarios can be produced through variation of other parameters such as, for instance, the damping of the arteriolar oscillator.

For normotensive rats, the typical operation point around $\alpha = 10-12$ and $T \cong 16$ s falls near the Hopf bifurcation point. This agrees with the experimental finding that about 70% of the nephrons perform self-sustained oscillations while the remaining show stable equilibrium behavior [22]. We can also imagine how the system is shifted back and forth across the Hopf bifurcation by variations in the arterial pressure. This explains the characteristic temporal behavior of the nephrons with periods of self-sustained oscillations interrupted by periods of stable equilibrium dynamics.

Figure 12.6a shows the temporal variation of the proximal tubular pressure P_t as obtained from the single-nephron model for $\alpha = 12$ and $T = 16$ s. All other parameters attain their standard values as listed in Table 12.1. Under these conditions the system operates slightly beyond the Hopf bifurcation point, and the depicted pressure variations represent the steady-state limit cycle oscillations reached after the initial transient has died out. For physiologically realistic parameter values the model reproduces the observed self-sustained oscillations with characteristic periods of 30–40 s. The amplitudes in the pressure variation also correspond to experimentally observed values. Figure 12.6b shows the phase plot. Here, we have displayed the normalized arteriolar radius r against the proximal intratubular pressure. Again, the amplitude in the variations of r appears reasonable. The motion

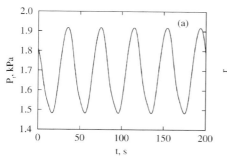 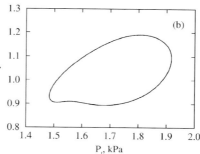

Fig. 12.6 (a) Temporal variation of the proximal tubular pressure P_t as obtained from the single-nephron model for $\alpha = 12$ and $T = 16$ s. (b) Corresponding phase plot. With the assumed parameters the model displays self-sustained oscillations in good agreement with the behavior observed for normotensive rats. The unstable equilibrium point falls in the middle of the limit cycle, and the motion along the cycle proceeds in the clockwise direction.

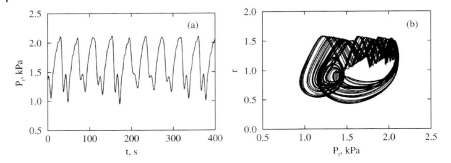

Fig. 12.7 (a) Pressure variations obtained from the single-nephron model for $\alpha = 32$ and $T = 16$ s. (b) Corresponding phase plot. With these parameters the model displays chaotic oscillations resembling the behavior observed for spontaneously hypertensive rats [13].

along the limit cycle proceeds in the clockwise direction. This implies that we first observe an increase the arterial radius r as the smooth muscle cells in the arteriolar wall relax and thereafter an increase in the tubular pressure P_t as the larger blood flow leads to a rising rate of filtration.

As previously noted, spontaneously hypertensive rats (SHR) have larger α-values than normal rats ($\alpha = 16.8 \pm 12.0$ vs $\alpha = 11.4 \pm 2.2$ for normotensive rats) [23]. On the other hand, it appears that the feedback delay is approximately the same for the two strains. Figure 12.7a shows an example of the chaotic pressure variations obtained for higher values of the TGF gain. Here, $\alpha = 32$ and $T = 16$ s. Under these conditions, the oscillations never repeat themselves, and calculations show that the largest Lyapunov exponent is positive [24]. As previously explained, this implies that the system displays sensitivity to the initial conditions. Two different initial conditions, however close they are, will after a while lead to clearly distinguishable trajectories and, even though the equations of motion are deterministic, long term prediction is impossible. The corresponding phase plot in Fig. 12.7b displays the characteristic picture of a chaotic attractor. One can interpret the behavior as resulting from a complicated interplay between the rapid modulations associated with the arteriolar dynamics and the slower TGF-mediated oscillations.

Figure 12.8 depicts simultaneous values of the arteriolar radius r and the equilibrium pressure P_{eq} in the afferent arteriole. By focusing on the arteriolar dynamics, this projection provides a somewhat different impression of the model behavior. In particular we notice how large oscillations across the curves of constant muscular activation ψ interfere with oscillations along these curves. Hence, the considered oscillation is actually a state of synchronization between the two oscillatory modes of the nephron autoregulation with the slow TGF-mediated oscillation performing precisely two cycles each time the myogenic oscillation completes eight cycles. This bimodal dynamics is clearly a major source of complexity in the nephron dynamics. We return to a discussion of the mode-to-mode interaction in Section 12.5.

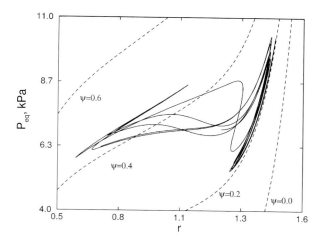

Fig. 12.8 Phase plot illustrating the oscillation in the variables associated with the afferent arteriole. The dynamics is interpreted as a 2:8 synchronization between the slow TGF-mediated mode and the fast myogenic mode. $T = 16$ s and $\alpha = 24.2$. This is the same solution that we referred to as a period-2 solution in connection with Fig. 12.5.

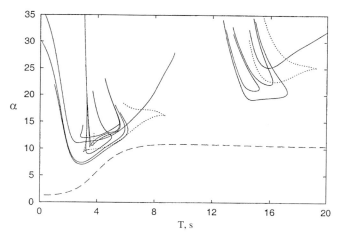

Fig. 12.9 Two-dimensional bifurcation diagram for the single-nephron model. The diagram illustrates the complicated bifurcation structure in the region of 1:1, 1:2, and 1:3 resonances between the arteriolar dynamics and the TGF-mediated oscillations. In the physiologically interesting regime around $T = 16$ s, another set of complicated period-doubling and saddle-node bifurcations occur. Here, we are operating close to the 1:4 and 1:5 resonances. The punctuated curve at low α-values is the Hopf bifurcation curve. Below this curve the TGF-mechanism displays a stable regulation, above the curve the regulation produces large amplitude self-sustained oscillations. Fully drawn curves represent period-doubling bifurcations, and punctuated curves are saddle-node bifurcations.

A more complete picture of the bifurcation structure of the single-nephron model is provided by the two-dimensional bifurcation diagram in Fig. 12.9. Here both the delay in the tubuloglomerular feedback and the slope α of the feedback characteristics are used as bifurcation parameters. Each time we pass one of the bifurcation curves in parameter plane, the steady-state solution of the nephron model undergoes a qualitative change. Starting around $\alpha = 1.3$ for $T = 0$ s, the lowest curve in the bifurcation diagram is the Hopf bifurcation curve. Below this curve, the model displays a stable equilibrium point, and above the curve the equilibrium point is unstable. The other curves represent either saddle-node bifurcations (dotted curves) or period-doubling bifurcations (fully drawn curves). To the left we observe a complicated structure of overlying period-doubling and saddle-node bifurcations. This structure is associated with 1:1, 1:2, and 1:3 resonances between the arteriolar dynamics and the TGF-mechanism. However, with feedback delays of the order of 4 s, the structure falls outside the physiologically relevant parameter region.

The region of physiological interest falls around $T = 16$ s. In this region we recover the bifurcation curves associated with the scenario described in connection with Fig. 12.5. As we scan vertically through the diagram for $T = 16$ s we first cross the Hopf bifurcation curve at $\alpha \cong 11$. In the interval from $\alpha \cong 19$ to $\alpha \cong 26$ we pass the period-doubling curves at which the period-2 and the period-4 solutions first emerge and subsequently disappear again. These solutions are not affected by the lower branch of the saddle-node bifurcation curve. However, the reestablished period-1 solution that exists after we have passed out through the period-doubling curves is destabilized at the upper branch of the saddle-node curve, and the unstable (saddle) solution can be followed down to the lower saddle-node bifurcation

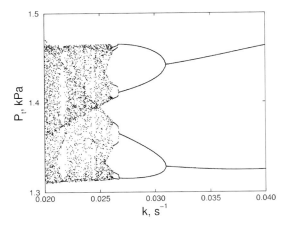

Fig. 12.10 Bifurcation diagram obtained from the model with the damping constant k of arteriolar oscillation as a parameter. If the dampling is reduced, chaotic phenomena can arise at relatively low values of the TGF gain. $T = 16$ s and $\alpha = 24$. For $k = 0.04$ s^{-1} the model displays a period-4 cycle. The figure only follows two of the four branches of this cycle.

curve where a high-amplitude stable period-1 solution is born. This solution then proceeds to chaos through the uppermost period-doubling curve.

The values of the TGF gain $\alpha \approx 30$ required to produce chaos in the nephron model with standard values of the other parameters exceeds reasonable physiological values by a factor of nearly two. Moreover, inspection of the temporal variation of the tubular pressure shows that the amplitude of the slow TGF-mediated oscillations often is larger for normotensive rats than for hypertensive rats. This indicates that the development of hypertension could be associated with parameters different from α. The most obvious choice of such a parameter is the damping constant k for the arteriolar oscillator. It is known the hypertensive state is accompanied by a sensitization of the arterioles in the kidney as well as in other organs of the body. Figure 12.10 shows a bifurcation diagram for $\alpha = 24$ and $T = 16.0$ with the damping constant k as a parameter. The standard value of the damping constant is $k = 0.04 \text{ s}^{-1}$ (see Table 12.1). With this damping the nephron model displays a period-4 cycle. However, as illustrated by the bifurcation diagram, reduction of the arteriolar damping lead to chaotic dynamics at values of α well below 30.

12.5
Intra-nephron Synchronization Phenomena

As demonstrated by the power spectra in Figs. 12.2a and 12.3b, regulation of the blood flow to the individual nephron involves several oscillatory modes. The two dominating time scales are associated with the period $T_{slow} \approx 30-40$ s of the slow TGF-mediated oscillations and the somewhat shorter time scale $T_{fast} \approx 5-10$ s defined by the myogenic oscillations of the afferent arteriolar diameter. The two modes interact because they both involve activation of smooth muscle cells in the arteriolar wall. Our model describes these mechanisms and the coupling between the two modes, and it also reproduces the observed multi-mode dynamics. We can, therefore, use the model to examine some of the phenomena that can be expected to arise from the interaction between the two modes.

For linear systems, the principle of superposition applies, and different oscillatory modes can evolve independently of one another. However, biological systems in general are not linear, and separation of different regulatory mechanisms may not be justified, even when they involve different time scales. One type of phenomenon that can arise from the interaction between two oscillatory modes is modulation of the amplitude and frequency of the faster mode in dependence of the phase of the slower mode. This type of phenomenon was demonstrated in Fig. 12.2c where the frequency of the myogenic mode f_{fast} changes in step with the amplitude of the TGF-mediated mode. Similar modulation phenomena can be expected to occur in many other biological systems such as, for instance, the interaction between the circadian and the ultradian rhythms of hormone secretion [25].

However, nonlinear interaction can also give rise to synchronization (mode-locking or entrainment) between the modes. This type of phenomenon arises as the interacting modes adjust their frequencies (and phases) relative to one another

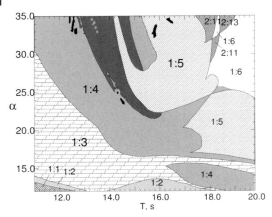

Fig. 12.11 Two-mode oscillatory behavior in the single nephron model. Black colored regions correspond to a chaotic solution. The figure shows different regions in which 1:4, 1:5 and 1:6 synchronization occurs in the interaction between the fast myogenic oscillations and the slower TGF-mediated oscillations. We can also observe regions with 2:11 and 2:13 synchronization. As discussed in connection with the bifurcation diagrams in Figs. 12.5 and 12.9, a given set of parameters may lead to two (or more) different solutions.

such that one mode completes precisely p cycles each time the other mode completes q cycles, with p and q being integers [26, 27]. It is well known, for instance, that the heart rate varies with the phase of the respiratory cycle and that this interaction can lead to a state of synchronization in which the heart beats three or four times during each respiratory period [28]. Synchronization implies that, instead of a so-called quasiperiodic motion where the two periodicities display an irrational ratio, the system adjusts its behavior such as to attain a regular period motion. The transition between synchronized and non-synchronized behavior represents a qualitative change in the regulation of the system and, as previously mentioned, we consider such transitions to play a significant role in the normal physiological control of the kidney.

To determine T_{slow} and T_{fast} in our numerical simulations we have calculated the mean return times of the trajectory to two appropriately chosen Poincaré sections

$$T_{fast} = < T_{ret}\Big|_{v_r=0} > \quad \text{and} \quad T_{slow} = < T_{ret}\Big|_{X_2=0} > \tag{18}$$

From these return times it is easy to obtain the intra-nephron rotation number (i.e., the rotation number associated with the two-mode behavior of the individual nephron)

$$n_{fs} = T_{fast}/T_{slow} \tag{19}$$

This measure can be used to characterize the various forms of frequency locking between the two modes. With varying feedback delay T and varying slope α of the open loop feedback curve, Fig. 12.11 shows how the two oscillatory modes can adjust their dynamics and attain states with different rational relations (p:q) between

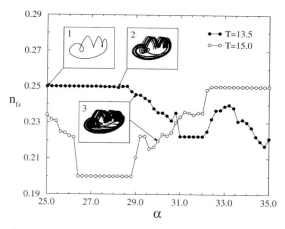

Fig. 12.12 Internal rotation number as a function of the parameter α calculated from the single-nephron model. Inserts present phase projections for typical regimes. Note how the intra-nephron synchronization is maintained through a complete period-doubling cascade to chaos.

the periods. Regions with 1:4, 1:5, and 1:6 resonances are seen to exist in the physiologically interesting range of the delay time $T \in [12 \text{ s}, 20 \text{ s}]$.

While the transitions between the different locking regimes always involve saddle-node bifurcations, bifurcations may also occur within the individual regime. A period-doubling transition, for instance, does not necessarily change n_{fs}, and the intra-nephron rotation number may remain constant through a complete period-doubling cascade and into the chaotic regime. This is illustrated in Fig. 12.12 where we have plotted n_{fs} as a function of the feedback gain α for different time delays $T_2 = 13.5$ s (filled circles) and $T = 15.0$ s (open circles). Phase projections (P_t, r) from the various regimes are shown as inserts. Inspection of the figure clearly shows that n_{fs} remains constant under the transition from regular 1:4 oscillations (black circles for $\alpha = 25$) to chaos (for $\alpha = 28$), see inserts 1 and 2. With further evolution of the chaotic attractor (insert 3), the 1:4 mode locking is destroyed. In the interval around $\alpha = 31.5$ we observe 2:9 mode locking. A similar transition is observed for $T = 15$ s (open circles). Periodic 1:5 oscillations ($\alpha = 27$) evolve into a chaotic attractor ($\alpha = 28.5$), but the rotation number maintains a constant value. For fully developed chaos, the 1:5 locking again breaks down.

To illustrate the occurrence of intra-nephron synchronization in our experimental results, Fig. 12.13 shows that ratio f_{slow}/f_{fast} as calculated from the time series in Fig. 12.2c. We still observe a modulation of the fast mode by the slow mode. However, the ratio of the two frequencies maintains a constant value of approximately 1:5 during the entire observation period, i.e., there is no drift of one frequency relative to the other. In full agreement with the predictions of our model, data for other normotensive rats show 1:4 or 1:6 synchronization. Transitions between different states of synchronization obviously represent a major source of complexity in the dynamics of the system. It is possible, for instance, that a nephron can display ei-

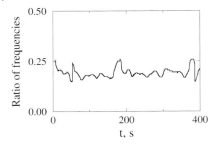

Fig. 12.13 Ratio of the internal time-scales for a normotensive rat. Note the 1:5 synchronization between the slow TGF-mediated oscillations and the myogenic oscillations. The 1:5 synchronization can be considered to arise from a 1:1 synchronization between the fifth harmonic of the TGF oscillations and the myogenic oscillations. This is normally a fairly weak interaction that will only take place in strongly nonlinear systems.

ther stable 1:4 or stable 1:5 synchronization for the same well-specified conditions. Small variations in the initial conditions may determine in which state the nephron ends up, and random external disturbances may switch the steady state from one mode of synchronization to the other.

We conclude, that besides being regular or chaotic, the self-sustained pressure variations in the individual nephron can be classified as being synchronous or asynchronous with respect to the ratio between the two time scales that characterize the fast (arteriolar) mode and the slow (TGF-mediated) mode, respectively. With this information we can now reinterpret the bifurcation diagram in Fig. 12.5. The branch that we observe at low values of α and along which the system undergoes an incomplete period-doubling cascade represents solutions that derive from the 1:4 synchronized state. The period-2 solution, for instance, is more appropriately referred to as 2:8 solution in which the myogenic oscillator completes precisely eight cycles each time the TGF-oscillator completes two cycles (with different amplitudes). The saddle-node bifurcation at $\alpha \approx 31$ represents the end of the region in which modes deriving from the 1:4 synchronized state occur, and the saddle-node bifurcation at $\alpha \approx 22.5$ represents the beginning of the region with dynamics related to the 1:5 synchronized state. In the interval between the two saddle-node bifurcations, solutions coexist that derive from the 1:4 state and from the 1:5 state. Here, the initial conditions will determine which type of solution the system displays. As we see in Section 12.6, this complexity in behavior may play an essential role in the synchronization between pairs of interacting nephrons.

12.6
Modeling of Coupled Nephrons

As illustrated in Fig. 12.14 the nephrons are arranged in a tree-like structure with their afferent arterioles branching off from a common interlobular artery [29]. This

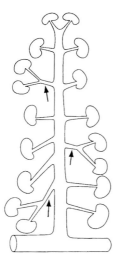

Fig. 12.14 Typical arrangement of a group of glomeruli with their afferent arterioles branching off from the same interlobular artery. Arrows point to couples of nephrons that share a piece of common arteriole. We suppose that hemodynamic coupling can be important for such nephrons. For other nephrons, the vascularly propagated coupling is likely to dominate.

arrangement allows neighboring nephrons to influence each others' blood supply, either through electrical or electrochemical signals that activate the vascular smooth muscle cells of the neighboring nephron or through a simple hemodynamic coupling. The two mechanisms depend differently on the precise structure of the arteriolar network. Hence, variations of this structure may determine which mechanism dominates in a given situation. The vascular propagated electrical signals travel as a wave along the smooth muscle cells of the vessel wall with exponentially decreasing amplitudes. This limits their immediate coupling range to distances of a few hundred microns. The hemodynamic coupling involves the displacement of blood flow from one nephron to a neighboring nephron as the first nephron responds to the signal from the macular densa cells by reducing its arteriolar diameter. Because of a relatively small pressure drop along the interlobular artery, it is likely that the hemodynamic coupling is particular significant for nephrons arranged in pairs (or triplets) and sharing a piece of common arteriole.

Let us start by considering the vascular coupling. The muscular activation ψ arises in the so-called juxtaglomerular apparatus and travels backwards along the afferent arteriole in a damped fashion. When it reaches the branching point with the arteriole from the neighboring nephron, it may propagate in the forward direction along that arteriole and start to contribute to its vascular response. In our model this type of cross-talk is represented by adding a contribution of the activation of one nephron to the activation of the other, i.e.:

$$\psi_{1,2tot} = \psi_{1,2} + \gamma \psi_{2,1} \tag{20}$$

where γ is the vascular coupling parameter, and ψ_1 and ψ_2 are the uncoupled activation levels of the two nephrons as determined by their respective Henle flows in accordance with Eq. (12).

As previously mentioned, the vascular signals propagate very fast as compared with the length of the vessels and the period of the TGF oscillations. As a first approach, the vascular coupling can therefore be considered as instantaneous. Experiments have shown [30] that the magnitude of the activation decreases exponentially as the signal travels along a vessel. Hence, only a fraction of the activation from one nephron can contribute to the activation of the neighboring nephron, and $\gamma = e^{-l/l_0} < 1$. Here, l is the propagation length for the coupling signal, and l_0 is the characteristic length scale of the exponential decay. As a base case value, we shall use $\gamma = 0.2$. As defined in Eq. (20), this parameter gives the fraction of the muscular activation of a given nephron that is propagated to a neighboring nephron.

To implement the hemodynamic coupling in our model, a piece of common afferent arteriole is included into the system, and the total length of the incoming blood vessel is hereafter divided into a fraction $\varepsilon < \beta$ that is common to the two interacting nephrons, a fraction $1 - \beta$ that is affected by the TGF signal, and a remaining fraction $\beta - \varepsilon$ for which the flow resistance is considered to remain constant. As compared with the equilibrium resistance of the separate arterioles, the piece of shared arteriole is assumed to have half the flow resistance per unit length.

Defining P_ε as the pressure at the branching point of the two arterioles, the equation of continuity for the blood flow reads:

$$\frac{P_a - P_\varepsilon}{\varepsilon R_{a0}/2} = \frac{P_\varepsilon - P_{g1}}{R_{a1}} + \frac{P_\varepsilon - P_{g2}}{R_{a2}} \tag{21}$$

with:

$$R_{a1} = (\beta - \varepsilon) R_{a0} + (1 - \beta) R_{a0} r_1^{-4} \tag{22}$$

and:

$$R_{a2} = (\beta - \varepsilon) R_{a0} + (1 - \beta) R_{a0} r_2^{-4} \tag{23}$$

Here, R_{a0} denotes the total flow resistance for each of the two nephrons in equilibrium. r_1 and r_2 are the normalized radii of the active parts of the afferent arterioles for nephron 1 and nephron 2, respectively, and P_{g1} and P_{g2} are the corresponding glomerular pressures. As a base value of the hemodynamic coupling parameter we shall use $\varepsilon = 0.2$. This parameter measures the fraction of the arteriolar length that is shared between the two nephrons.

Because of the implicit manner in which the glomerular pressure is related to the efferent colloid osmotic pressure and the filtration rate [Eqs. (5)–(8)], direct solution of the set of coupled algebraic equations for the two-nephron system becomes rather inefficient. Hence, for each nephron we have introduced the glomerular pressure P_g as a new state variable determined by

$$\frac{dP_{g,i}}{dt} = \frac{1}{C_{glo}} \left(\frac{P_\varepsilon - P_{g,i}}{R_{a,i}} - \frac{P_{g,i} - P_v}{R_e} - F_{filt,i} \right) \tag{24}$$

with $i = 1, 2$. This implies that we consider the capillary network as an elastic structure with a compliance C_{glo} and with a pressure variation determined by the imbalance between the incoming blood flow, the outgoing blood flow, and the glomerular filtration rate.

From a physiological point of view, this formulation is well justified. Compared with the compliance of the proximal tubule, C_{glo} is likely to be quite small, so that the model becomes numerically stiff. In the limit $C_{glo} \to 0$, the set of differential equations reduces to the formulation with algebraic equations. Finite values of C_{glo} will change the damping of the system, and therefore also the details of the bifurcation structure. In practice, however, the model will not be affected significantly as long as the time constant $C_{glo} R_{eff}$ is small compared with the periods of interest. Here, R_{eff} denotes the effective flow resistance faced by C_{glo}.

Introduction of a coupling tends to make similar nephrons synchronize their phases. Depending on the initial conditions and on the relative strength of the two coupling mechanisms this synchronization may be either in phase or in antiphase. The in-phase synchronization, which produces a symmetric motion for the coupled system, is favored if the vascular coupling is relatively strong. Anti-phase synchronization, in contrast, is more likely to occur in the presence of a strong hemodynamic coupling.

The ability to synchronize is obviously not restricted to the case where the two nephrons are identical. In the presence of a small parameter mismatch between the nephrons, a sufficiently strong coupling again forces the nephrons to synchronize their pressure variations so that the periods become the same. In the nonlinear system each nephron will adjust its pressure regulation relative to the other so as to attain a precise 1:1 relation between the periods. This explains the experimental observation that many pairs of adjacent nephrons are found to exhibit precisely the same period, even though they cannot be expected to have identical parameters [7]. As long as the mismatch is small, the coupling strength required to synchronize the nephrons tend to scale in proportion with the size of the mismatch.

Let us examine the situation for large values of α where the individual nephron exhibits chaotic dynamics. Figure 12.15a shows a phase plot for one of the nephrons in our two-nephron model for $\alpha = 32$, $T = 16$ s, $\varepsilon = 0.0$, and $\gamma = 0.2$. Here we have introduced a slight mismatch $\Delta T = 0.2$ s in the delay times between the two nephrons and, as illustrated by the tubular pressure variations of Fig. 12.15b, the nephrons follow different trajectories. However, the average period is precisely the same. This is a typical example of phase synchronization of two chaotic oscillators.

As discussed in Section 12.5, the two oscillatory modes associated with the pressure and flow regulation in the individual nephron may operate in synchrony or

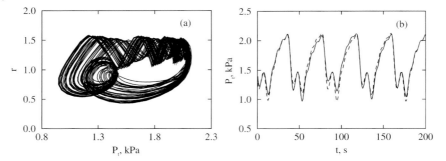

Fig. 12.15 (a) Phase plot for one of the nephrons and (b) temporal variation of the tubular pressures for both nephrons in a pair of coupled chaotically oscillating units. $\alpha = 32$, $T = 16$ s, and $\varepsilon = \gamma = 0.2$. The figure illustrates the phenomenon of chaotic phase synchronization. By virtue of their mutual coupling the two chaotic oscillators adjust their (average) periods to be identical. The amplitudes, however, vary incoherently and in a chaotic manner [27].

out of synchrony. When two nephrons interact, this allows for a variety of different phenomena: The nephrons can operate in complete synchrony (both modes entrain), in partial synchrony (only one mode is synchronized), or out of synchrony. Moreover, in the case of partial synchronization, either the fast or the slow mode may synchronize between the nephrons. Detailed analyses of our experimental results have shown that all of these possibilities occur in reality [31]. To illustrate how the same problem can be analyzed by means of our model we may introduce two rotation numbers:

$$n_f = T_{fast1}/T_{fast2}, \quad n_s = T_{slow1}/T_{slow2} \tag{25}$$

that measure the ratio of the periods in the two interacting nephrons of respectively the fast myogenic and the slow TGF-mediated mode. As before, the periods are determined as mean values over many oscillations.

Let us consider the case of $\alpha = 30$ corresponding to a weakly developed chaotic attractor in the individual nephron. The coupling strength $\gamma = 0.06$ and the delay time T_2 in the second nephron is considered as a parameter. Three different chaotic states can be identified in Fig. 12.16. For the asynchronous behavior both of the rotation numbers n_s and n_f differ from 1 and change continuously with T_2. In the synchronization region, the rotation numbers are precisely equal to 1. Here, two cases can be distinguished. To the left, the rotation numbers n_s and n_f are both equal to unity and both the slow and the fast oscillations are synchronized. To the right ($T_2 > 14.2$ s), while the slow mode of the chaotic oscillations remain locked, the fast mode drifts randomly. In this case the synchronization condition is fulfilled only for one of oscillatory modes, and we speak of partial synchronization. A detailed analysis of the experimental data series reveals precisely the same phenomena [31].

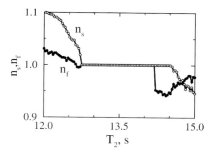

Fig. 12.16 Full and partial synchronization of the fast and slow motions between two interacting nephrons ($T_1 = 13.5$ s, $\alpha = 30.0$ and $\gamma = 0.06$). Full synchronization is realized when both the fast n_f and slow n_s rotation numbers equal to 1. To the right in the figure there is an interval where only the slow modes are synchronized. The delay T_2 in the loop of Henle for the second nephron is used as a parameter.

12.7
Experimental Evidence for Synchronization

Experiments on interacting nephrons were performed with normotensive as well as with spontaneously hypertensive rats at the Department of Medical Physiology, University of Copenhagen and at the Department of Physiology, Brown University [32]. When exposing the surface of a kidney, small glass pipettes, allowing simultaneous pressure measurements, could be inserted into the proximal tubuli of a pair of adjacent, superficial nephrons. After the experiment, a vascular casting technique was applied to determine whether the considered nephron pair were connected to the same interlobular artery. Only nephrons for which such a shared artery was found showed clear evidence of synchronization, supporting the hypothesis that the nephron–nephron interaction is mediated by the network of incoming blood vessels [29, 33].

Figure 12.17 shows an example of the tubular pressure variations that one can observe for adjacent nephrons for a normotensive rat. For one of the nephrons, the pressure variations are drawn in black, and for the other nephron in gray. Both curves show fairly regular variations in the tubular pressures with a period of approximately 31 s. The amplitude is about 1.5 mmHg and the mean pressure is close to 13 mmHg. Inspection of the figure clearly reveals that the oscillations are synchronized and remain nearly in phase for the entire observation period (corresponding to 25 periods of oscillation).

Figures 12.18a and b show examples of the tubular pressure variations in pairs of neighboring nephrons for hypertensive rats. These oscillations are significantly more irregular than the oscillations displayed in Fig. 12.17 and, as previously discussed, it is likely that they can be ascribed to a chaotic dynamics. In spite of this irregularity, however, one can visually observe a certain degree of synchronization between the interacting nephrons.

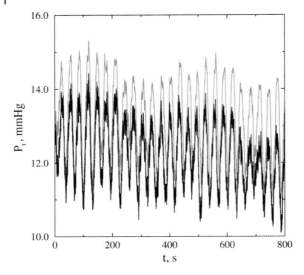

Fig. 12.17 Tubular pressure variations for a pair of coupled nephrons in a normotensive rat. The pressure variations remain nearly in phase for the entire observation time (or 25 periods of oscillation).

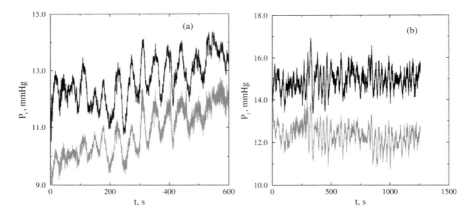

Fig. 12.18 Two examples (a and b) of the tubular pressure variations that one can observe in adjacent nephrons for hypertensive rats. In spite of the irregularity in the dynamics, one can see a certain degree of synchronization in the phases of the two oscillations. This synchronization is supported by formal investigations, e.g., by means of the wavelet technique.

In a recent study [34], we made use of wavelet and double-wavelet analysis to examine the relative occurrence of various states of synchronization in pairs of interacting nephrons. We showed that both full and partial synchronization occur for normotensive as well as for hypertensive rats, and that the partial synchronization can involve only the slow oscillations or only the fast oscillations. We also used the wavelet and double-wavelet techniques to extract additional information about the relation between the amplitude of the TGF-mediated oscillations and the frequency and amplitude modulation of the myogenic mode. Information of this type is essential to construct a more accurate model of the myogenic oscillator.

12.8
Conclusion and Perspectives

The models presented in this chapter were developed over a period of nearly 20 years. This development involved a continuous interaction of experimental work and data analysis with model validation, bifurcation analysis, and simulation. In several cases the models have pointed to the existence of phenomena that were subsequently verified experimentally. One example is the occurrence of anti-phase synchronization between the TGF-mediated oscillations in neighboring nephrons when the interaction takes place predominantly via hemodynamic coupling. Another example is the existence of states of partial synchronization in which only the fast modes synchronize between the nephrons. Most importantly, however, the models serve as an extremely valuable tool to integrate the complicated physiological mechanisms. We have demonstrated how one can combine existing physiological mechanisms of glomerular filtration, tubuloglomerular feedback, and arteriolar response into a consistent and quantitative model that can reproduce the observed oscillatory and chaotic phenomena with realistic parameter values. On several occasions model analyses have shown that initial hypotheses based on physiological intuition or generally accepted beliefs failed when subjected to a more quantitative analyses. A quantitative description of the pressure distribution along the interlobular artery has been instrumental, for instance, in allowing us to evaluate the relative importance of the vascula propagated *vis á vis* the hemodynamic coupling. The models also allow us to clearly reveal the importance of a variety of different nonlinear dynamic phenomena in the biological control systems. Particularly interesting is the observation of subharmonics to the slow TGF-mediated oscillations in several tubular pressure recordings. By contrast to harmonic components, subharmonics only arise through a bifurcation. This requires the system to operate far from equilibrium and supports our interpretation of the irregular dynamics in hypertensive rats as representing deterministic chaos.

In order to examine the synchronization phenomena that can arise in larger ensembles of nephrons, we recently developed a model of a vascular-coupled nephron tree [35], focusing on the effect of the hemodynamic coupling. As explained above, the idea is here that, as one nephron reduces its arteriolar diameter to lower the incoming blood flow, more blood is distributed to the other nephrons in accordance with the flow resistances in the network. An interesting aspect of this particular coupling is that the nephrons interact both via the blood flow that controls their tendency to oscillate and via the oscillations in this blood flow that control their tendency to synchronize. We refer to such a structure as a resource distribution chain, and we have shown that phenomena similar to those that we describe here

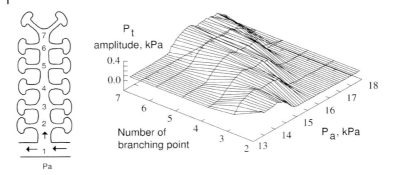

Fig. 12.19 Left: sketch of a vascular-coupled nephron tree including the interlobular artery, the afferent arterioles and the glomeruli. Right: oscillation amplitudes as function of the arterial pressure and the position of the branching point along the vascular tree.

arise in other resource distribution chains involving, for instance, electronic oscillators or predator–prey systems [35].

Our model of the vascular-coupled nephron tree consists of a set of afferent arterioles branching off from a single interlobular artery as shown schematically in Fig. 12.19. The nephron tree is described in terms of the lengths of the arteriolar and arterial branches together with their hemodynamic resistances. It is assumed that the glomerulus of each nephron is connected to the corresponding branching point via an arteriole of length L_i^g, and of hemodynamic resistance R_i^g, $i = 1 \ldots 12$. The arterial pressure P_a to be used in the model of individual nephron is now the blood pressure at the associated branching point P_j^b, $j, k = 2, \ldots 7$. Connection between the branching points is described in terms of the branch lengths L_{jk}^b and their hemodynamic resistances R_{jk}^b $j = 1 \ldots 7$. The same approach is used to describe the connection of branching point 1 to the terminal points with the constant pressure values P_{in} and P_{out}. This part of the vascular-nephron tree imitates the connection of the tree to higher-level arteries. The hemodynamic-coupled vascular tree is purely resistive. Transients associated with the distribution of the blood pressure among the branching points are negligible, and we can calculate the static pressure distribution for any state of connected nephrons using linear algebraic equations written for each branching point. An example of such an equation for the 6th branching point reads:

$$\frac{P_5^b - P_6^b}{R^b} - \frac{P_6^b - P_7^b}{R^b} + \frac{P_3^g - P_6^b}{R_3^g} + \frac{P_4^g - P_6^b}{R_4^g} = 0. \tag{26}$$

Here, P_3^g and P_4^g represent the blood pressure in the glomerulus of the 3th and 4th nephrons. P_5^b, P_6^b, and P_7^b represent the blood pressure in the 5th, 6th, and 7th branching points, respectively. R^b denotes the hemodynamic resistance between 6th and 7th branching points. This resistance is assumed to be the same for all branches. R_3^g and R_4^g are the hemodynamic resistances to the 3rd and 4th nephrons, respectively. Note that R_3^g and R_4^g are not constant because they include the resistances of the active parts of the afferent arterioles.

Equations of this type for the all branching points are obviously interdependent and, hence, produce a global hemodynamic coupling among nephrons in the vascular-coupled nephron tree. The strength of this coupling generally increases with R_{kj}^b, but decreases with increasing R_i^g.

As previously discussed, neighboring nephrons can also influence one another through vascular propagated electrical (electrochemical) signals. To account for this mechanism, the total activation potential for k-th nephron is assumed to be the sum of contributions from all other nephrons in the tree. Moreover, the electrical activation potentials are assumed to propagate along the vascular wall with an exponential decay. In this way, the vascular propagated interaction is delivered to each nephron as an additional part of its activation potential ψ:

$$\Delta\psi = \sum_{i=1, i \neq k}^{N} \psi_i \exp(-\gamma(L_{ji} + L_k^g)) \tag{27}$$

where j is the number of the branching point to which the considered nephron with number k is connected. The matrix L_{ji} contains the lengths from a given branching point j to all nephrons $i = 1 \ldots N$, and L_k^g is the length from the given nephron to the connected branching point.

The mathematical model of our vascular-coupled nephron tree thus consists of: (i) 12 sets of coupled ODEs describing individual nephrons, (ii) a set of linear algebraic equations that determines the blood pressure drop from one branching point to another, and (iii) algebraic relations for the vascular interaction.

Depending on the choice of control parameters, the amplitudes of the pressure oscillations in the nephron tree are found to be different at different positions in the tree. Due to model symmetry, two nephrons connected to the same node have the same oscillation amplitudes. Thus, we can refer to the number of the branching point to describe the amplitude properties. Branching points 2, 3, and 4 may correspond to deep nephrons and branching points 6 and 7 to superficial nephrons. Experimentally, only the pressure oscillations in nephrons near the surface of the kidney have been investigated. However, we suppose that deep (juxtamedullary) nephrons can exhibit oscillations in their pressures and flows as well.

With the parameters used in Fig. 12.19, a choice of the arterial pressure of $P_a = 13.3$ kPa allows all nephrons to be in the oscillatory regime. Here, the hemodynamic resistance is assumed to be $R^b = 0.002$ kPa·s/nl. The vascular coupling may then be varied by adjusting γ from 1.6 mm^{-1} (strong interaction) to 4.0 mm^{-1} (weak interaction). As defined above, R^b denotes the flow resistance between two successive branching points of the vascular tree (see Fig. 12.19), and the parameter γ measures the length constant associated with the exponential decay of the vascular propagated coupling along the arterioles.

In the year to come, we plan to perform a translational study in which we adjust the model of nephron dynamics developed for rats to predict the dynamics in other species (such as cats and dogs) and to finally predict the characteristics of nephron oscillations in man (if they exist). The idea is that if the model represents the physiological mechanisms well enough one should be able to replace the parameters used in the rat version with parameters pertaining to other species and obtain reasonable results.

In collaboration with Alexander Gorbach, NIH, we have initiated a study of the spatial patterns in the nephron synchronization. This study involves the use of infrared cameras or other types of equipment that can measure variations in the blood supply by small (0.01°C) fluctuations in the temperature at the surface of the kidney. It is also of interest to study how the large amplitude oscillations in pressure, fluid flow, and salt concentration at the entrance of the distal tubule influence the delicate hormonal processes in that part of the kidney, to establish a more quantitative description of some of the mechanisms involved in the development of hypertension, and to examine the effects of various drugs.

Acknowledgments

We wish to acknowledge long-lasting collaboration with Prof. Donald Marsh, Dept. of Physiology, Brown University, USA and Prof. Dmitry Postnov, Dept. of Physics, Saratov State University, Russia. Over the years they have contributed significantly to the development of an integrated model of nephron autoregulation and coupling through their deep insights in kidney function and complex systems theory. A significant number of students and other collaborators have also been involved in the development and analysis of the kidney model. We are particularly grateful to Alexey Pavlov who performed the wavelet analysis and to Mickael Barfred who performed some of the bifurcational analysis.

References

1 *The Kidney: Physiology and Pathophysiology*, 2nd ed., edited by D. W. Seldin and G. Giebisch. Raven, New York, 1992.

2 P.P. Leyssac and L. Baumbach, *An Oscillating Intratubular Pressure Response to Alterations in Henle Loop Flow in the Rat Kidney*, Acta Physiol. Scand. **117**, 415–419 (1983).

3 N.-H. Holstein-Rathlou and P.P. Leyssac, *TGF-mediated Oscillations in the Proximal Intratubular Pressure: Differences between Spontaneously Hypertensive Rats and Wistar-Kyoto Rats*, Acta Physiol. Scand. **126**, 333–339 (1986).

4 N.-H. Holstein-Rathlou and D.J. Marsh, *Oscillations of Tubular Pressure, Flow, and Distal Chloride Concentration in Rats*, Am.

J. Physiol. **256** (Renal Fluid Electrolyte Physiol. 25), F1007-F1014 (1989).

5 M. Lamboley, A. Schuster, J.-L. Bény, and J.-J. Meister, *Recruitment of Smooth Muscle Cells and Arterial Vasomotion*, Am. J. Physiol. Heart Circ. Physiol. **285**, H562-H569 (2003).

6 H. Gustafsson and H. Hilsson, *Rhythmic Contractions of Isolated Small Arteries from Rat: Role of Calcium*, Acta Physiol. Scand. **149**, 283–291 (1993).

7 N.-H. Holstein-Rathlou, *Synchronization of Proximal Intratubular Pressure Oscillations: Evidence for Interaction between Nephrons*, Pflügers Archiv **408**, 438–443 (1987).

8 S.H. Strogatz, *Nonlinear Dynamics and Chaos: With Applications to Physics, Biol-*

ogy, *Chemistry and Engineering* (Reading, Massachusetts: Addison-Wesley, 1994).

9 A. Grossmann and J. Morlet, S.I.A.M. J. Math. Anal. **15**, 723 (1984); I. Daubechies, *Ten Lectures on Wavelets* (S.I.A.M., Philadelphia, 1992).

10 O.V. Sosnovtseva, A.N. Pavlov, E. Mosekilde, N.-H. Holstein-Rathlou, and D. J. Marsh, *Double-Wavelet Approach to Study Frequency and Amplitude Modulation in Renal Autoregulation*, Phys. Rev. E 70, 031915-8 (2004).

11 N.-H. Holstein-Rathlou and D.J. Marsh, *A Dynamic Model of Renal Blood Flow Autoregulation*, Bull. Math. Biol. **56**, 411–430 (1994).

12 M. Barfred, E. Mosekilde, and N.-H. Holstein-Rathlou, *Bifurcation Analysis of Nephron Pressure and Flow Regulation*, Chaos **6**, 280–287 (1996).

13 K.S. Jensen, E. Mosekilde, and N.-H. Holstein-Rathlou, *Self-Sustained Oscillations and Chaotic Behaviour in Kidney Pressure Regulation*, Mondes en Develop. **54/55**, 91–109 (1986).

14 N.-H. Holstein-Rathlou and D.J. Marsh, *A Dynamic Model of the Tubuloglomerular Feedback Mechanism*, Am. J. Physiol. **258**, F1448–F1459 (1990).

15 N.-H. Holstein-Rathlou and P.P. Leyssac, *Oscillations in the Proximal Intratubular Pressure: A Mathematical Model*, Am. J. Physiol. **252**, F560–F572 (1987).

16 W.M. Deen, C.R. Robertson, and B.M. Brenner, *A Model of Glomerular Ultrafiltration in the Rat*, Am. J. Physiol. **223**, 1178–1183 (1984).

17 M. Rosenbaum and D. Race, *Frequency-Response Characteristics of Vascular Resistance Vessels*, Am. J. Physiol. **215**, 1397–1402 (1968).

18 Y.-C. B. Fung, *Biomechanics. Mechanical Properties of Living Tissues* (Springer, New York, 1981).

19 R. Feldberg, M. Colding-Jørgensen, and N.-H. Holstein-Rathlou, *Analysis of Interaction between TGF and the Myogenic Response in Renal Blood Flow Autoregulation*, Am. J. Physiol. **269**, F581–F593 (1995).

20 T.S. Parker and L.O. Chua, *Practical Numerical Algorithms for Chaotic Systems* (Springer-Verlag, Berlin, 1989).

21 M. Marek and I. Schreiber, *Chaotic Behavior in Deterministic Dissipative Systems* (Cambridge University Press, England, 1991).

22 N.-H. Holstein-Rathlou and D.J. Marsh, *Renal Blood Flow Regulation and Arterial Pressure Fluctuations: a Case Study in Nonlinear Dynamics*, Physiol. Rev. **74**, 637–681 (1994).

23 P.P. Leyssac and N.-H. Holstein-Rathlou, *Tubulo-Glomerular Feedback Response: Enhancement in Adult Spontaneously Hypertensive Rats and Effects of Anaesthetics*, Pflügers Archiv **413**, 267–272 (1989).

24 A. Wolf, *Quantifying Chaos with Lyapunov Exponents* in Chaos, edited by A.V. Holden (Manchester University Press, England, 1986).

25 E. Mosekilde, *Topics in Nonlinear Dynamics - Applications to Physics, Biology and Economic Systems* (World Scientific, 2002).

26 A. Pikovsky, M. Rosenblum, and J. Kurths, *Synchronization: A Universal Concept in Nonlinear Sciences* (The University Press, Cambridge, 2001).

27 E. Mosekilde, Yu. Maistrenko, and D. Postnov, *Chaotic Synchronization – Application to Living Systems* (World Scientific, Singapore, 2002).

28 C. Schäfer, M.G. Rosenblum, J. Kurths, and H.-H. Abel, *Heartbeat Synchronized with Ventilation*, Nature **392**, 239–240 (1998).

29 D. Casellas, M. Dupont, N. Bouriquet, L.C. Moore, A. Artuso and A. Mimran, *Anatomic Pairing of Afferent Arterioles and Renin Cell Distribution in Rat Kidneys*, Am. J. Physiol. **267**, F931–F936 (1994).

30 Y.-M. Chen, K.-P. Yip, D.J. Marsh and N.-H. Holstein-Rathlou, *Magnitude of TGF-Initiated Nephron-Nephron Interactions is Increased in SHR*, Am. J. Physiol. **269**, F198–F204 (1995).

31 O.V. Sosnovtseva, A.N. Pavlov, E. Mosekilde, and N.-H. Holstein-Rathlou, *Bimodal Oscillations in Nephron Autoregulation*, Phys. Rev. E., **66**, 1–7, 2002.

32 K.-P. Yip, N.-H. Holstein-Rathlou, and D.J. Marsh, *Dynamics of TGF-Initiated Nephron-Nephron Interactions in Normotensive Rats and SHR*, Am. J. Physiol. **262**, F980–F988 (1992).

33 Ö. Källskog and D.J. Marsh, *TGF-Initiated Vascular Interactions between Adjacent Nephrons in the Rat Kidney*, Am. J. Physiol. **259**, F60–F64 (1990).

34 O.V. Sosnovtseva, A.N. Pavlov, E. Mosek-ilde, K.P. Yip, N.-H. Holstein-Rathlou, and D.J. Marsh, *Coherence Among Mechanisms of Renal Autoregulation in Normotensive and Hypertensive Rats*, Am. J. Physiol., submitted.

35 D.E. Postnov, O.V. Sosnovtseva, and E. Mosekilde, *Oscillator Clustering in a Resource Distribution Chain*, CHAOS **15**, 013704–12 (2005).

13

Toward a Computational Model of Deep Brain Stimulation in Parkinson's Disease

A. Beuter and J. Modolo

Abstract

Parkinson's disease is a common neurodegenerative disorder whose aetiologies are still unknown today. Since the late 1980s chronic electrical neurostimulation also called deep brain stimulation or high frequency stimulation has offered long-term and drastic benefits for 5% of these patients. However, the mechanisms of action involved in this beneficial therapy remain to be discovered. Neurobiological studies have revealed that deep brain stimulation modulates the activity in a complex network involving the cortex, basal ganglia, and thalamus. Recently models inspired from mathematics and physics but incorporating results from neuroscience have started to appear. However, these models have not provided yet an integrated explanation of the phenomena occurring at large and multiple spatial scales while keeping in mind what is occurring at the cellular level. We present a preliminary computational model based on a large- and multi-scale approach, considering delays, relatively easy to manipulate, and able to generate synchronization. We also present a preliminary validation study based on a rudimentary artificial network. In the long term this approach should help understand why stimulating various targets in the motor loop leads to drastic improvements of PD motor symptoms.

13.1

Introduction

Parkinson's disease (PD) is a common neurodegenerative disorder characterized among other things by a progressive loss of dopaminergic neurons in the substantia nigra pars compacta (a nucleus located in the mesencephalon) followed by a depletion of dopamine in the neurons projecting to the striatum. The aetiologies of this disease are still unknown. It affects about 1.6% of the population over 65 years of age and the number of patients is estimated to reach more than six million in the world. Surprisingly more than 50% of these patients are found in ten occidental countries (study done between 1979 and 1997) and the direct link with pesticides has been established in California (1984–1994) [57]. This increase however appears

Biosimulation in Drug Development. Edited by Martin Bertau, Erik Mosekilde, and Hans V. Westerhoff
Copyright © 2008 WILEY-VCH Verlag GmbH & Co. KGaA, Weinheim
ISBN: 978-3-527-31699-1

to be slower in some countries such as Japan and France (maybe because of the food?). The estimated number of patients with PD is about 100 000 in France and 120 000 in the UK.

Cardinal signs of the disease include: (1) bradykinesia or slowness of movement or initiation of movement; (2) rigidity or resistance of limbs to passive movement; (3) tremor of hand and sometimes legs or head with a peak frequency in the 3–5 Hz range at rest or during action; (4) gait/balance instability manifested by difficulty walking with frequent falls; (5) dementia and depression in 20–40% of the patients. Motor signs increase in a nonlinear fashion as the disease progresses and they fluctuate over time within and between subjects. The proposed aim of treatments is to control the symptoms and no current therapeutic strategy is able yet to stop the progression of the disease. Therapeutic strategies include pharmacological treatments used to reinforce the dopaminergic tone (with precursors and agonists of dopamine), to augment cholinergic inhibitors, or to inhibit enzymes of dopamine catabolism. However, after a few years dopamine replacement therapy eventually produces undesirable side effects such as drug-induced dyskinesia and motor fluctuations dyskinesia. Thus, other treatments such as transplantation of embryonic dopaminergic cells have also been tested. During the 1970s and 80s lesions or surgical ablations including thalamotomy, pallidotomy, sub-thalamotomy were performed with relative success.

Since the late 1980s electrical stimulation also called deep brain stimulation (DBS) or high frequency stimulation has offered long-term benefits for patients with neurological disorders unresponsive to existing therapies. In this procedure electrodes (also called leads) with four active contacts are implanted uni- or bi-laterally in brain structures and electrical stimulation is delivered at high frequency via one or two programmable pulse generator(s) usually implanted under the skin near the collarbone (Fig. 13.1). The first nucleus stimulated electrically with DBS was the ventrointermediate nucleus of the thalamus (Vim) to relieve different types of tremors [4]). Later on the globus pallidus (GPi) was implanted and today the major and preferred target for PD is the subthalamic nucleus (STN) [5, 6]. However, patients must satisfy strict inclusion criteria including for example motor fluctuations not sufficiently controlled by medication, positive response to L-Dopa, satisfactory health status, no dementia or psychiatric complications, and less than 75 years of age. The lasting and usually drastic benefits of DBS are quantified on the basis of pre- and post-assessments of patients using standardized clinical rating scales such as the UPDRS (Unified Parkinson's Disease Rating Scale [16]).

While the idea of DBS was born in the 1930s [35] and again in the 1950s with the treatment of intractable pain [13] its rationale was only proposed in the 1980s by [3]. Initially, structures were lesioned but later, in the 1990s the same structures were electrically stimulated instead [5, 6]. Since the year 2000, DBS has become the neurosurgical procedure of choice and has replaced lesion or ablation procedures [67]. Brain stimulation therapies result in drastic improvements of symptoms in PD. DBS is officially approved by the Food and Drug Administration (FDA) for PD and PD-like symptoms in 1997 and for dystonia in 2003. However, the use of DBS has been extended to other neurologic and neuropsychiatric disorders such as ob-

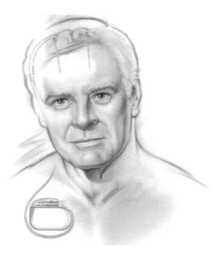

Fig. 13.1 See the two leads, the extension running under the skin and the neurostimulator. The clinican can program and adjust the settings externally using a hand held magnet (picture credit: Medtronic, Inc).

sessive compulsive disorders, epilepsy, depression, and headaches (for details see [13]). Other types of neuromodulation techniques used for a wide range of central nervous system disorders include Transcranial Magnetic Stimulation (TMS) and repetitive TMS, electroconvulsive therapy, and vagus nerve stimulation but are not discussed further in this chapter.

While the clinical effects of DBS are well recognized today, the mechanisms of action of DBS are still poorly understood and are currently the object of intense debate between clinicians, scientists, and recently from mathematicians and physicists. Recent advances in neuroscience and computational capabilities have opened the way for more realistic network models. Over the past few years several models have been proposed ([15, 22, 24, 26, 38, 47, 48, 58, 60, 62] to only name a few). However, these models do not seem able to provide an integrated explanation of the phenomena occurring at various brain spatial scales (see [53] for more details). The difficulty with the existing models is that to be really helpful they must take under consideration a multitude of physiological and computational constraints. First, they must include known neuroanatomy, network electrophysiology, pharmacology, pathophysiology, and behavioral findings [64]. Second, they must also incorporate several theoretical concepts and techniques from physics and mathematics (see the model presented below). Third, they must also be compatible with many assumptions and simplifications. Indeed the vast interconnectedness of the brain and the complexity of the signs and signals of the brain make the formulation of a complete model computationally challenging [11].

Despite these obvious difficulties we decided to develop a model of the effects of DBS using a systems biology approach. Our aim is to propose a large scale and multiscale computational model of DBS effects in PD based on physiology of individual neurons, population of neurons, and interacting populations of neurons.

The theoretical framework of this model is compatible with physiological delays and able to generate synchronization. Furthermore, this model is easy to manipulate. A preliminary validation of this model is presented with an artificial network inspired from biology. A long term goal is to use this model to propose a unified and coherent explanation of the local and more distant spectacular effects of DBS on motor signs.

13.2
Background

13.2.1
DBS Numbers, Stimulation Parameters and Effects

Since 1997 about 14000 patients have received DBS worldwide (Medtronic, unpublished data). There are close to 100 centers in Europe with 2–8 centers/country. Each center performs surgeries on 10–20 patients/year/centre giving about 1000 to 1500 patients operated each year in Europe. The cost of DBS surgery is estimated to be between 35 and 40 000 Euros [17]. Some side effects associated with DBS have been reported, including speech disturbances, weight gain, postural instabilities [23]. The DBS system consists of three components: the electrode, the extension, and the neurostimulator. The electrode is a thin, insulated wire which is inserted through a small opening in the skull and its tip is positioned within the targeted brain area. The extension is a wire connecting the electrode and the neurostimulator. It is passed under the skin of the head, neck, and shoulder. The neurostimulator contains the battery pack and controls the stimulation parameters.

Typical stimulation parameters selected include polarity (cathode), pulse amplitude ($\approx$ 3 V); pulse duration ($\approx$ 0.1 ms) and frequency ($\approx$ 150 Hz) [19]. Electrode placement is still not a completely resolved issue. For a discussion of the issues related to optimal electrode placement see [23, 37]. Today there are now numerous websites providing detailed explanations about DBS procedures. These websites are proposed by governmental agencies such as the NIH (USA) (http://www.ninds.nih.gov/disorders/deep_brain_stimulation), and also by companies commercializing stimulators such as Medtronic (http://www.medtronic.com/); by clinicians providing essential information to health professionals, students, parents, family members, caregivers and scientists (see for example: http: //www.neurology.wisc.edu/montgomery/) and by Parkinson's disease organizations such as http://www.apdaparkinson.org, http://www.parkinson.org, http://www.pdf.org, and http://www.parkinsonalliance.org. These web sites are detailed and provide all the explanations currently available to anyone concerned with DBS. As mentioned above DBS is associated with drastic and long-lasting benefits. Most patients continue to take their medication after the surgery. In most cases the amount of medication is significantly reduced and varies from patient to patient. This reduction in dose of medication leads in turn to a significant improvement in side effects induced by the medications.

The volume of tissue activated (VTA) by DBS is still the object of debate. Reference [12] developed a model to predict the effect of electrode location and stimulation parameters adjustments on the VTA and correlated the results of the model with clinical results recorded in a patient. They showed that therapeutic DBS of the STN is characterized by a VTA that spreads well outside the borders of the STN (toward the internal capsule and thalamus). By increasing the voltage and keeping other parameters constant they observed that while rigidity continued to decrease, bradykinesia improved and then worsened (inverted U-shape) and side effects such as paresthesia appeared suggesting that both sensorimotor signs and side effects react differently to stimulation voltage.

Regarding the delay of action and frequency of DBS, reference [8] examined tremor in patients with PD during successive transitions when the stimulator was off or on at efficient or inefficient frequencies (Fig. 13.2). Effects observed during DBS of the Vim, or GPi or STN included: the presence of a variable delay before tremor amplitude decreases (two top traces between E and O); a tendency for tremor amplitude to oscillate (top trace GPi, second "O" condition); and a tendency for tremor to escape (bursts) during efficient frequency stimulation (bottom trace, Vim second E condition).

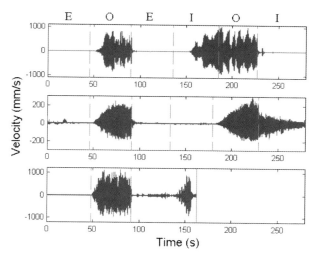

Fig. 13.2 X axis = time, Y axis = tremor intensity, approximately 45–50 s for each stimulation condition, E=effective (> 100 Hz), I=ineffective (< 100 Hz), O="off". Subject GPi (top): $E = 160$ Hz, $I = 60$ Hz; Subject STN (middle): $E = 185$ Hz, $I = 90$ Hz; Subject Vim (bottom): $E = 130$ Hz, $I = 65$ Hz.

13.2.2
DBS and Basal Ganglia Circuitry

Reference [2] proposed the existence of a network composed of segregated, largely closed and nonoverlapping re-entrant loops (motor, oculomotor, prefrontal, and limbic) involving the cortex, basal ganglia and thalamus. These loops are concerned with the contribution of action, motivation and cognition to behavior. Dopamine depletion in the early stages of PD affects preferentially the motor loop. The rationale for using DBS in PD is that movement disorders arise at least partially from dysfunctions occurring primarily within the motor loop [67]. Apparently Dopamine depletion causes increased activity and abnormal oscillations and synchronization in the motor loop and specifically in the GPi and STN.

In a highly simplified conceptual model of the motor circuitry of the basal ganglia, reference [50] illustrates the normal healthy state as compared to choreobalism (a hyperkinetic state) and PD (a hypokinetic state; Fig. 13.3). These models of the motor circuits while imperfect have guided the neurosurgeons in their search for the "right" target(s) to stimulate (for a historical review see [32]). Interestingly DBS along the motor loop largely reverses these abnormalities. This is the case for DBS in the posterolateral part of the GPi or in the anterodorsal part of the STN [67].

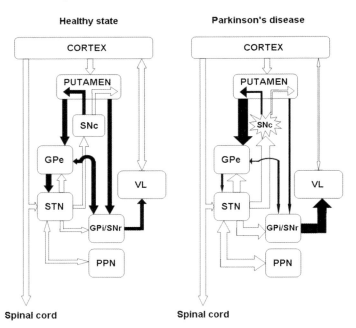

Fig. 13.3 Conceptual model summary of the main connections of the motor loop of the BG. VL=Ventrolateral Thalamus, PPN=PedonculoPontine Nucleus. In the classic model, Dopamine net effect is to reduce the output of the BG and to disinhibit the thalamus. In PD this model predicts an hyperactivity of the BG and an inhibition of the thalamus which in turn decreases the activity of the thalamo-cortical and brainstem centres (white arrows are for excitation and black arrows are for inhibition).

Other structures have recently been stimulated including the motor cortex [52], the pedunculopontine nucleus in the brain stem [54] and the zona incerta (a structure adjacent to the STN) [55]. This indicates that the stimulation of various structures along the motor loop can lead to improvement of parkinsonian signs but spreading to other adjacent structures cannot be excluded. In the present chapter we focus on DBS of the STN.

In the normal state, the putamen receives afferents from the motor and somatosensory cortical areas and communicates with the GPi/SNr through a direct inhibitory pathway and though a multisynaptic (GPe, STN) indirect pathway. In PD, dopamine deficiency leads to increased inhibitory activity from the putamen onto the GPe and disinhibition of the STN. STN hyperactivity by virtue of its glutamatergic action produces excessive excitation of the GPi/SNr neurons, which over-inhibit the thalamocortical and brain stem motor centers.

13.2.3
DBS: The Preferred Target Today

Overall, stimulating the STN is associated with fewer secondary effects (if we exclude cognitive secondary effects such as reduced lexical fluency and decline in executive functions) and more beneficial effects compared to other targets in PD. The STN is a small almond shaped nucleus located below the zona incerta, above the substantia nigra, medially to the red nucleus and laterally to the internal capsule. Its size and position are highly variable and differ systematically on magnetic resonance images and on atlases [56]. These authors found the size of the STN to be 3.7, 5.9 and 5 mm for coronal (x), sagittal (y), and axial (z) respectively. Other authors have reported $3 \times 8 \times 12$ mm [63]. The STN has $\pm 2 \times 10^5$ neurons on each side [40]. Of these, 92.5% have a glutamate mediated excitatory influence and 7.5% are GABAergic inhibitory interneurons which may play a role in abnormal synchronization. Under healthy conditions at rest the activity of the STN is almost random and not correlated whereas under diseased conditions its activity presents burst type discharges and synchronization. Output fibers passing from the GPi, STN projections neurons and afferent inputs represent possible therapeutic targets of DBS [12]. Afferent inputs may include the supplementary motor area, GPe, GPi, thalamus (centromedian/parafascicular), SNc. The subthalamic nucleus (STN) plays a pivotal role in controling the activity of both the GPe and GPi. Both nuclei receive monosynaptic excitatory and disynaptic GPe-mediated inhibitory inputs from the STN [33].

Figure 13.4 illustrates three aspects of the basal ganglia network. First, a reciprocal control exists between GPe and STN. Second the SNc acts not only on the striatum but also on the cortex, the STN and the GPi. Finally the location of the STN at the intersection between vertical and horizontal feedback loops is crucial. Reference [50] concludes that the BG can no longer be considered as a unidirectional linear system that transfers information based solely on a firing-rate code and must rather be seen as a highly organized network with operational characteristics that simulate a nonlinear dynamical system (Fig. 13.4).

13.2.4
DBS: Paradox and Mechanisms

The effects of DBS on the cortex-basal-ganglia-thalamus-cortex motor loop appear to be more complex than initially believed. The paradox of DBS is that electrical stimulation of brain tissue (which presumably induces brain activation), has a similar effect as that of a surgical lesion of that same structure (which effectively destroys brain tissue). These two realities are hard to reconcile. As indicated by [64] the ultimate elucidation of this paradox depends on the nature of the complex and interactive neural connections in the brain that communicate through electrical and chemical processes. There is an emerging view that DBS has both excitatory and inhibitory effects on how brain circuits communicate with one another depending on the distance from the electrode, the cell structures activated and the direction of the activation (ortho- versus anti-dromic). The effect appears to modulate the activity of a network as well as neural firing patterns. Long term effects on neurotransmitters and receptor systems cannot be excluded [64].

Recently [59] used a multiple-channel single-unit recording technique to investigate basal ganglia (BG) neural responses during behaviorally effective DBS of the STN in a rodent model of PD. Simultaneous recording of single unit activity in the striatum, globus pallidus, substantia nigra pars reticulate (SNr), and STN during effective DBS revealed a mixture of neural responses in the cortico-basal ganglia system. Predominant inhibitory responses appeared in the STN stimulation site.

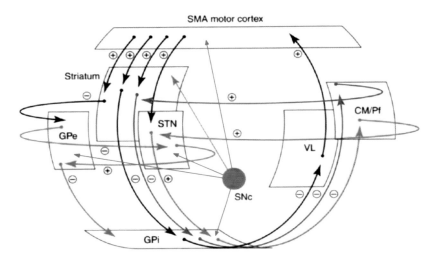

Fig. 13.4 Blue: CM/Pf-striatum-GPi-CM/Pf: a positive FB loop; and red:CM/Pf-STN-GPi-CM/Pf: a negative FB loop. Somatotopically organized parallel projections form closed loops (vertical, in black) between the motor cortex, the basal ganglia and the motor cortex. These loops signal movement preparation and execution. Concomitantly feedback (horizontal) loops stabilize basal ganglia activity. (From [50] with permission)

Nearly equal numbers of excitatory and inhibitory responses were found in the globus pallidus, and SNr, whereas more rebound excitatory responses were found in the striatum. Mean firing rate did not change significantly in some structures (striatum, globus pallidus, SNr) but decreased significantly on both sides of the STN confirming the clinical observation that unilateral DBS induces bilateral motor benefits. The results presented by [59] suggest that DBS induces the modulation and reorganization of the activity in the network involving the BG (see Fig. 13.4).

DBS appears to work by "freeing thalamocortical and brainstem motor systems from abnormal and disruptive basal ganglia influences" ([67], p. 202). At a local scale DBS appears to inhibit the soma of nerve cells and to activate the axons [42]. At an intermediate scale, DBS appears to alter the dynamics in the GPi-STN network [10]. At a distant scale, DBS appears to normalize the activity in the supplementary motor area and in the premotor cortex [67]. While many hypotheses have been formulated, the experimental results are difficult to reconcile especially when several physiological scales are considered. We now present a computational model based on a population density approach.

13.3
Population Density Based Model

13.3.1
Modeling Approach: Multiscaling

Modeling the effects of deep brain stimulation cannot be done by focusing on only one level of description. Applying a high-frequency current in a subcortical structure has an effect not only at the cellular level by modifying neuronal dynamics [19, 39] but also on the whole stimulated structure's behavior [18]. Furthermore, the interactions of the stimulated area with surrounding subcortical structures are modified and as a consequence of this change in the BG network's activity [59] a modification of motor behavior is observed at the periphery (for example, alleviating of tremor or rigidity observed in PD). At this point we can identify several functional levels going from the cellular to the behavioral level (Fig. 13.5).

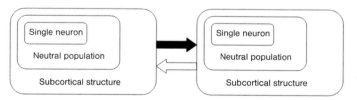

Motor behavior

Fig. 13.5 Example of the multi-scale modeling approach for a functional network of two subcortical substructures in interaction and implied in motor behavior (in this example, black arrow is for excitation and white arrow for inhibition).

Clearly, a model focusing only on one scale would provide an incomplete explanation. For example, cellular studies focusing on the behavior of neurons during the stimulation only quantify changes in cellular activity and so does not give much insight into the effect of DBS in the stimulated structure. They miss network effects. This multi-scale aspect appears as a necessity for a computational model designed to improve our understanding of the effects of DBS in PD: in order to be useful, a computational model must be integrative and not limited to a part of the complex circuitry concerned with DBS. The difficulty lies in the separation between the crucial elements and the non-essential ones in the area of neurophysiology, neuroanatomy, neural connectivity...

As indicated in the introduction we decided to build a computational model of the STN and its surrounding structures. Let us use as a neuron model the "leaky integrate and fire" (LIF) in which the membrane potential of the neuron is the only variable:

$$C_m \frac{dv}{dt} = -g_L(v - v_L) \tag{1}$$

where C_m is the membrane capacitance, v the membrane potential, g_L the leak conductance and v_L the membrane rest potential. The STN contains about 2×10^5 neurons on each side [40] which implies that 2×10^5 variables are needed at least as long as the connectivity matrix describing the connections between neurons in the network must be known too. Thus using a classic approach does not seem compatible with the multi-scale aspect that requires a computational model that would be of help in unveiling the mechanisms of DBS. Must we conclude that a computational model including the large-scale and multi-scale aspects is computationally prohibitive? Hopefully this is not the case. An alternative description of neural networks was initiated first by [69] who provided a continuous formulation of the classic discrete network activity formula. This approach has been extended by the authors in another paper [70] but even if their work constituted a solid basis for the continuous formulation of large-scale neural networks their formalism "missed" some short phenomena like "rapid transients" [49]. Another author [34] developed a continuous formulation of neural populations inspired from statistical mechanics. To illustrate this, let us consider a gas composed of identical molecules. How can this gas be described? The first idea is to describe each molecule by its position $x(t)$ and speed $v(t)$ which allows to precisely know the state of the gas at a given time. Of course, this is computationally prohibitive but a solution was proposed in 1867 by Maxwell who was not interested in the individual states of molecules in a gas but rather in the state distribution of molecules in the gas. This state distribution, or Maxwellian, permits to compute macroscopic variables of the gas knowing the properties of a single molecule.

Reference [34] followed by several groups ([49, 51] to name a few) inspired from this statistical approach initiated in physics by Maxwell to develop a similar theory applied to the description of a neural population. Basically a neural population is described by its population density $p(\vec{w}, t)$ where $\vec{w}$ is the state of a neuron (scalar in the case of the LIF model, a vector containing the state variables in the case of higher-dimension systems such as the Hodgkin–Huxley model). This states distribution is such that at time t, $p(\vec{w}, t)d\vec{w}$ neurons are in the state $[\vec{w}, \vec{w} + d\vec{w}]$. Even if the population density does not provide access to individual information about a given neuron's state, it allows to compute macroscopic and measurable variables such as the mean membrane potential (MMP) which is correlated with the average membrane potential of neurons in the population. The fact that this approach leads to compute measurable quantities will make experimental validations possible (see above).

The population density approach does not only permit large-scale and multi-scale modeling, but furthermore it is much more "computationally friendly" than a classical simulation: the states distribution is independent of the number of neurons (thus directly related to the population by a factor N, where N is the number of neurons in the population), so the computation time is exactly the same for ten or one billion neurons. Of course, the larger the number of neurons is, the better the agreement between the two approaches will be, as the population density approach is valid for a large amount of neurons, i.e., in the limit $N \rightarrow +\infty$. Computation time in direct simulation is an important issue and is proportional to the number of neurons simulated. Furthermore, let us note that a classic model using a large amount of parameters is subject to not being robust against parameter changes: as an example the model of [58] includes about 100 non-linear and coupled differential equations and 150 parameters, and a change of only 5% in two parameters dramatically affects the simulation's results (see [53]) as it completely changes the simulation results (instead of improving thalamic relay capacity, high-frequency stimulation of the STN worsens this capacity).

The fact that the population density approach only needs a few parameters is one of its advantages: only parameters of the single neuron (four in the Izhikevich model, see above) and connectivity (average number of afferents per neuron, distribution of synaptic delays) are required. Then, delays of axonal propagation can be included in population equations by introducing a delays distribution [49] which is a crucial feature as long as the balance of delays is a key parameter that can lead to synchronization in a network [14]. This last feature of the population density formalism is important as a number of cerebral pathologies are associated with an abnormal synchronization in or between several structures (e.g., epilepsy, PD, obsessive compulsive disorders). Finally, different types of synapses with their own kinetics can be taken into account [25] in order to include some more physiological and realistic features. The fact is that the difference between excitatory and inhibitory synapse time scales cannot reasonably be neglected with regards to their importance in neural network dynamics [65].

13.3.2
Model Equations

Now we derive population equations by considering a large population of identical neurons, which does not mean that they have necessarily the same morphology but only the same dynamics, which is still a strong hypothesis however. By "large" we mean that the number of neurons in the population is more than a few thousands, this assumption is needed by the fact that the statistical formalism is valid in the limit where the number of elements tends towards infinity. The conservation of the population density on the phase space implies that the population density verifies the following conservation equation:

$$\frac{\partial}{\partial t} p(v, t) = -\vec{\nabla} \vec{J}(v, t) \tag{2}$$

This equation describes the evolution over time of the number of neurons per state (a density) depending on a flux term. To illustrate this, let us imagine a fluid inside a basin: the higher the fluid is, the higher the molecular density is at this position. The evolution over time of this molecular density will depend of currents inside the fluid which represent the flux term. Our population density of neurons is very similar to this fluid; the difference is that the current inside the "fluid" is driven by individual neuronal dynamics and neuronal interactions.

Reference [51] proposes to split the flux term from the Fokker–Planck equation into two terms: one accounting for the dynamics of a single neuron independently of all other neurons in the population (the *streaming* term) and another including the interactions of a neuron with the other neurons in the network (the *interaction* term). Before going further, let us denote by $\vec{w}$ the state of a neuron, a vector in the general case. The proposed general expression for these two fluxes is [51]:

$$\vec{J}_s(\vec{w}, t) = \vec{F}(\vec{w}) p(\vec{w}, t) \tag{3}$$

$$\vec{J}_{imp}(\vec{w}, t) = \hat{e}_v \sigma(t) \int_{v-\epsilon}^{v} p(\tilde{v}, u, t) d\tilde{v} \tag{4}$$

In the streaming flux we can note the analogy of the term $d\vec{w}/dt$ with the speed of a fluid at a given point, $p(\vec{w}, t)$ being the number of neurons flowing through the state $\vec{w}$ at time t. $F(\vec{w})$ describes the dynamical behavior of a single neuron and can be viewed as a "speed field". Note that the interaction flux includes several new terms: the mean-field variable $\sigma(t)$ is the average individual spike reception rate, i.e. the number of action potentials received at time t in the population, and ϵ is the amplitude of the membrane potential "jump" that occurs when an action potential is received by a neuron (this jump is considered as instantaneous). This term must be modified when synaptic kinetics are considered as discussed below. We wrote general equations of the population density approach but at this point we have to go into details with regards to the individual neuron model. Most papers in the literature exposing a population density based model use the LIF model as a basis to describe the dynamical evolution of a single neuron. Here we chose to develop the equations with the Izhikevich model instead, which is much closer to a biological neuron's behavior using only two state variables. This model is composed of two differential and coupled equations with a non-linear term:

$$\frac{dv}{dt} = 0.04v^2 + 5v + 140 - u + I(t) \tag{5}$$

$$\frac{du}{dt} = a(bv - u) \tag{6}$$

and the model also includes the following reset mechanism for the two state variables:

$$v > 30\,mV \rightarrow v = c,\ u = u + d \tag{7}$$

A great advantage of this simple model is that it can reproduce all spiking patterns observed experimentally with the appropriate set of parameters (a, b, c, d) as exposed in Fig. 13.6.

These dynamical equations can be injected in the streaming flux term of the conservation equation as they describe the evolution of a single neuron only driven by its own dynamics. So the complete form of the conservation equation describing the dynamics of a neural population and using the Izhikevich model to describe the evolution of a single element is:

$$\frac{\partial}{\partial t}p(\vec{w}, t) = -\vec{\nabla}\left(\begin{bmatrix} 0.04v^2 + 5v + 140 - u + I(t) \\ a(bv - u) \end{bmatrix} p(\vec{w}, t)\right.$$

$$\left. + \hat{e}_v\sigma(t)\int_{v-\epsilon}^{v} p(\tilde{v}, u, t)d\tilde{v}\right) \tag{8}$$

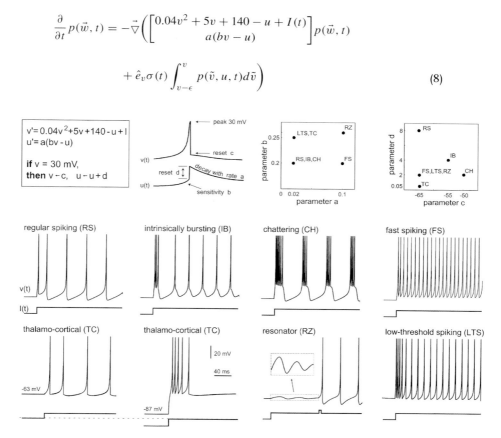

Fig. 13.6 Some examples of the richness of the Izhikevich model (electronic version of the figure and reproduction permissions are freely available at www.izhikevich.com).

Let us indicate that a supplementary flux must be added to take into account the neural flux which is re-injected at the discontinuity $v = c$ (reset potential) and $u = u + d$ after it flowed through the value $v = 30$ mV (maximum of the action potential) which adds the term $\vec{J}(s, u - d, t)$ on the right side of the conservation equation. This equation is complex and for the sake of progressive validation, we first consider an uncoupled neural population which means that the interaction flux is null and simplifies the conservation equation:

$$\frac{\partial}{\partial t} p(\vec{w}, t) = -\vec{\nabla} \left(\begin{bmatrix} 0.04v^2 + 5v + 140 - u + I(t) \\ a(bv - u) \end{bmatrix} p(\vec{w}, t) \right) \qquad (9)$$

Comparison of the integration of this last conservation equation with the direct simulation of an uncoupled neural population with a given distribution of states (v, u) at the initial time provides a validation for the simplified version of the model. Then adding the imposed flux accounting for connectivity allows us to simulate a large population of Izhikevich neurons with a given pattern of connectivity (number of afferents per neuron and delays kernel).

13.3.3
Synapses and the Population Density Approach

There is still a last key feature to introduce to capture the essential properties of a neural network: synaptic kinetics. The fact is that excitatory and inhibitory synapses have very different dynamics [25] and can have a huge impact on the network's behavior, so ignoring them would be neglecting an important biological aspect. Reference [25] developed a sophisticated method in which they first add two variables accounting for excitatory/inhibitory conductances, and then proceed to a dimensional reduction of the full equation. This allows them to obtain pretty good agreement between population equations and direct simulations results. We chose to introduce synaptic kinetics in a simpler way, though still capturing the essential of the phenomenon.

Let $\tilde{r}(t)$ be the average individual firing rate (i.e., the number of action potentials triggered per neuron at each time, in average) such as $\tilde{r}(t) = r(t)/N$, where $r(t)$ is the total firing rate of the population and N the number of neurons in the population. If G is the average number of afferents per neuron and if the spike conduction is instantaneous then the average individual reception rate expresses as $\sigma(t) = G \times \tilde{r}(t)$. Reference [49] introduced a kernel of conduction delays $\alpha(t)$ in the population to model the fact that conduction of action potentials along axons occur with a given delay which leads to:

$$\sigma(t) = G \int \alpha(\tau) \tilde{r}(t - \tau) d\tau \qquad (10)$$

where τ is the conduction delay. Here we present a formalism including synaptic kinetics without increasing the dimension of the Fokker–Planck equation using a mean-field expression for various synaptic conductances associated with each neurotransmitter. Let us consider that a synaptic input modifies the synaptic counductance associated to its neurotransmitter (NMDA, AMPA, GABA$_A$, GABA$_B$ [30]) and causes an applied current trough the neuron membrane (compact form): $I_{syn}(t) = \sum_i g_i(t) f_i(v)$. If each conductance follows first-order kinetics, then we can write the value of g at time t knowing that a spike occurred at time t' ($t > t'$):

$$g(t - t') = g_0 \left[\exp\left(-\frac{t - t'}{\tau} \right) \right] \tag{11}$$

In the past (time t' for example) a given neuron received spikes at a rate $\sigma(t')$, so the expression of the conductance at any time t depending of the action potentials received by this neuron is:

$$g(t) = g_0 \left[\int_{-\infty}^{t} \sigma(t') \exp\left(-\frac{t - t'}{\tau} \right) dt' \right] \tag{12}$$

where τ is the time constant of the synapse type. Injecting this expression in the compact form of the synaptic current previously given, we have:

$$I_{syn}(t) = \sum_i \left\{ g_{0i} f_i(v) \left[\int_{-\infty}^{t} \sigma(t') \exp\left(-\frac{t - t'}{\tau_i} \right) dt' \right] \right\} \tag{13}$$

where i indices the different conductances. Using this value of the applied synaptic current of a given neuron, we express the membrane potential variation $\epsilon(v, t) \approx I_{syn}(t) \times \Delta t$ caused by conductance changes during a time Δt:

$$\epsilon(v, t) \approx \Delta t \times \sum_i \left\{ g_{0i} f_i(v) \left[\int_{-\infty}^{t} \sigma(t') \exp\left(-\frac{t - t'}{\tau_i} \right) dt' \right] \right\} \tag{14}$$

Finally we can write the expression of the interaction flux using a mean-field approximation for the conductance and using the value of the average individual spike reception rate $\sigma(t)$ previously given, the complete expression of the imposed flux is:

$$\bar{J}_{imp}(\vec{w}, t) = \frac{G}{N} \left(\int_{\tau_-}^{\tau_+} r(t - \tau)\alpha(\tau)d\tau \right) \times \left(\int_{v-\epsilon(v,t)}^{v} p(v', u, t)dv' \right) \tag{15}$$

where τ_- and τ_+ are the extremum of the time conduction delays (τ_- different from 0). If we impose $r(t < t_0)$ where t_0 is the beginning of the simulation and then if we compute the total firing rate at each time as the flux through the maximum of the action potential [49], with $s = 30$ mV:

$$r(t) = \int_{u_-}^{u_+} J_v(s, u, t)du \tag{16}$$

where u_- and u_+ are the extrema values of the recovery variable u and J_v the total flux in the v direction, then the conservation equation is fully explicited for a population of Izhikevich neurons with first-order excitatory/inhibitory kinetics and an average pattern of connectivity.

13.3.4
Solving the Conservation Equation

The analogy between solving this two-dimensional conservation equation for a large collection of neurons and for a fluid led us to think that using a numerical method from fluid mechanics could be useful. Furthermore, the fact is that simple numerical methods like the first-order finite differences method gave poor results probably because of the important gradient in the v direction in the Izhikevich model. As a consequence we chose to use the finite volumes method commonly used in fluid mechanics or electromagnetism (and in general to solve transport equations).

Basically the finite volumes method consists first in discretizing the state space Ω in identical cells of volume $\Pi_i \Delta s_i$ where $\{s_i\}$ are the different state variables. On each cell of the phase space the Fokker–Planck equation is transformed using the Green–Ostrogradsky theorem: $\iiint_V div \vec{F} dv = \iint_S \vec{F} d\vec{s}$, which provides a balance equation implying incoming/outgoing flux on each cell of the discretized phase space. The method is thus conservative by construction as if some flux flows out from a cell then it flows into another one. Numerical expressions of the streaming flux on each cell at a given time are derived from the general expression provided by reference [51] applied in the particular case of the Izhikevich model, depending of the sign of the derivate of the state variables $\vec{w} = \{v, u\}$ with respect to time.

13.3.5
Results and Simulations

In order to validate the reduced model (uncoupled population of Izhikevich neurons) we chose to perform comparisons with a direct simulation model. In this last model the internal state of each neuron is computed at each time step (with a forth-order Runge–Kutta method) using the equations of the Izhikevich model so we have complete access to individual information as opposed to the population density formalism where only the states distribution can be computed. The simulation parameters were 50 000 Izhikevich neurons in tonic spiking mode ($a = 0.02$, $b = 0.2$, $c = -65$, $d = 6$ as provided in [29]), a gaussian form of $p(v, u, t)$ at $t = 0$ and a constant input current of $I = 60$ μA was applied to all neurons. The firing rate and the mean membrane potential (MMP) were computed at each time for the two methods during 15 ms.

Figure 13.7 presents the evolution over time of the firing rate and MMP in the case of a constant applied current for the two approaches.

And Fig. 13.8 presents the evolution over time of the firing rate and MMP for the two methods in the case of a high-frequency-current ($f = 320$ Hz).

These results indicate a good qualitative and quantitative agreement between the population based model and the direct simulation for an uncoupled neural pop-

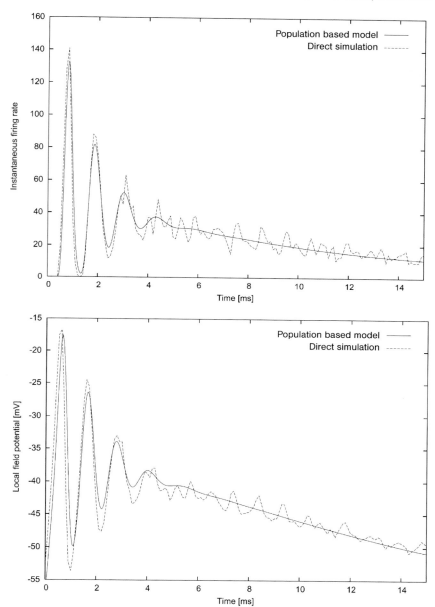

Fig. 13.7 Evolution over time of the mean firing rate and MMP for a constant current for the two approaches.

ulation. This validates the first step towards the resolution of the complete conservation equation including the interaction flux accounting for connectivity and synaptic kinetics in the network.

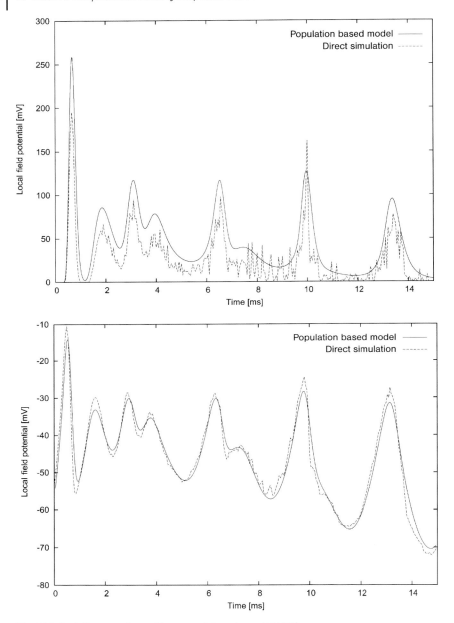

Fig. 13.8 Evolution over time of the mean firing rate and MMP for an alternative high frequency current for the two approaches.

13.4
Perspectives

We present here a mathematical model based on the population density approach describing a neural population with given pattern of discharge, connectivity and synaptic kinetics. Equations for independent neurons are derived, numerically solved, and validated by comparison with direct simulations and a similar process is in the works to validate the complete conservation equation including the imposed flux accounting for neural connectivity. Such a neural population will provide a basic element to build a network of neural populations mimicking the organization of the cortex-basal ganglia-thalamus-cortex loop and to study the dynamics in its various nuclei before and during stimulation. The possibility to stimulate in another structure of this motor loop for a less invasive surgery and more efficiency could be investigated too. This approach could be used to simulate other neurological disorders thanks to the modularity of the model.

This model should allow studying the validity of several mechanisms of DBS that have been proposed over the past few years, like the excitation or blockade of: (1) neurons in the stimulation area, (2) neurons projecting to the stimulated area via fibers of passage. Another hypothesis is the differential effects of stimulation between the soma and axon of neurons [41] and this could be tested by modifying the Izhikevich model. Frequency dependence, one of the most mysterious aspects of DBS, could be explored too as well as the possibility of modulation of synaptic kinetics by DBS.

Several extensions of this statistical formalism are possible and would allow taking into account more physiological and neuroanatomic features. For example, most of neural network models consider that every neuron receives exactly the same input current from the electrode, however in the case of DBS neurons do not receive the same current depending of their position with respect to the electrode [12]. This consideration is the first element in favor of adding a spatial state variable to the population density. Another example that goes in favor of this idea is the fact that peculiar features of small-world networks should be included. This class of complex networks was recently formalized [66] and some recent studies

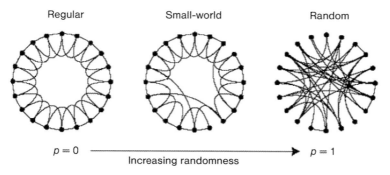

Regular   Small-world   Random

$p = 0$ ———— Increasing randomness ————▶ $p = 1$

Fig. 13.9 How to build a small-world network [66].

[1] showed that cerebral connectivity exhibits small-world features. In brief, small-world networks can be build by considering an homogeneous network where every connection has the same length, and then applying a long-range rewiring process by testing each connection with a probability p close to 0 (Fig. 13.9).

At the end of the rewiring process the network is almost homogeneous but contains some shortcuts and exhibits two main features which are (1) clustering (an element is almost connected with its neighbors) and (2) a low diameter (only a few jumps are needed to go from one node to another thanks to the shortcuts). These peculiarities have important consequences on the network's behavior: synchronization between elements is easier [66] and activity can propagate from one point to another one far away in the network quickly using the shortcuts. This last point suggests that the introduction of a spatial variable could take into account the faster transfer of neural activity within the network which influences synchrony.

13.5
Conclusion

Systems biology involves the study of mechanisms underlying complex biological processes as integrated networks of many interacting components. It requires the collection of large sets of experimental data, the proposal of mathematical models that might account for at least some significant aspects of these data sets; accurate computer solutions of the mathematical equations to obtain numerical predictions; and assessment of the quality of the model by comparing numerical simulations with the experimental data. Thus, systems biology goes beyond the application of computation to biological research. According to reference [43], the process that defines systems biology must be "hypothesis driven, quantitative, integrative, dynamic, and global (in the sense that it employs all relevant data in a unified and coherent theoretical structure". In this chapter we use mathematical modeling of a dynamical biological system (since it has been suggested that PD is a dynamical disease [7]) that integrates various functional levels. Our long term goal is to improve our understanding of the effects of DBS in PD. We hope that the present model will contribute not only to reach that goal but also to identify new therapeutic strategies.

References

1 Achard S., Salvador R., Whitcher B., Suckling J., Bullmore E. A resilient, low-frequency, small-world human brain functional network with highly connected association cortical hubs. *J Neurosci*, **2006**, *26*(1):63–72.

2 Alexander G.E., DeLong M.G. and Strick P.L. Parallel organization of functionnaly segregated circuits linking basal ganglia and cortex. *Annu Rev neurosci*, **1986**, *9*, 357–381.

3 Alexander G.E., Crutcher M.D., DeLong M.R. Basal ganglia-thalamocortical circuits: parallel substrates for motor, oculomotor, "prefrontal" and "limbic" functions. *Progr Brain Res*, **1990**, *85*, 119–146.

4 Benabid A.L., Pollak P., Louveau A., Henry S. and De Rougemont J. Combined

(thalamotomy and stimulation) stereotactic surgery of the VIM thalamic nucleus for bilateral Parkinson's disease. *Appl Neurophysiol*,1987, *50*, 344–346.

5 Benabid A.L., Pollak P., Gervason C., Hoffmann D., Gao D. M., Hommel M. et al. Long-term suppression of tremor by chronic stimulation of the ventral intermediate thalamic nucleus. *Lancet*, **1991**, *337*, 403–406.

6 Benabid A.L. Deep brain stimulation for Parkinson's disease. *Curr Opin Neurobiol*, **2003**, *13*(6):696–706.

7 Beuter A., Vasilakos K. Tremor: Is Parkinson's disease a dynamical disease? *Chaos*, **1995**, 5, 1, 35–42.

8 Beuter A., Titcombe M. S., Richer F., Gross C., Guehl D. Effect of deep brain stimulation on amplitude and frequency characteristics of rest tremor in Parkinson's disease. *Thalamus Relat Syst*, **2001**, 1:203–211.

9 Beuter A., Titcombe M. Modulation of tremor amplitude during deep brain stimulation at different frequencies. *Brain and Cognition*, **2003**, *53*, 210–212.

10 Brown P., Mazzone P., Oliviero A., Altibrandi M.G., Pilato F., Tonali P.A., Di Lazzaro V. Effects of stimulation of the subthalamic area on oscillatory pallidal activity in Parkinson's disease. *Exp Neurol*, **2004**, *188*, 480–490.

11 Bullock T.H. Signals and signs in the nervous system: the dynamic anatomy of electrical activity is probably information-rich. *Proc Natl Acad Sci*, **1997**, 94:1–6.

12 Butson C.R., Cooper S.E., McIntyre C.C. Deep brain stimulation of the subthalamic nucleus: Patient specific analysis of the volume of tissue activated. *10th Annual Conference of the International FES Society*, **July 2005**, Montreal, Canada.

13 Chang J.Y. Brain stimulation for neurological and psychiatric disorders, current staus and future directions. *J Pharmacol Exp Therapeut*, **2004**, *309*, 1, 1–7.

14 Dhamala M., Jirsa V.K., Ding M. Enhancement of neural synchrony by time delay. *Phys Rev Lett*, **2004** 92(7), 074–104.

15 Edwards R., Beuter A., Glass L. Parkinsonian tremor and simplification in network dynamics. *Bull Math Biol*,**1999**, *51*, 157–177.

16 Fahn S. and Elton R.L. Unified Parkinson's disease rating scale. In *Recent development in Parkinson's disease*, Fahn S. and Marsden C.D., Calne D.B., Goldstein (Eds). Macmillan Health Care, Florham Park, NJ, pp.153–164.

17 Fraix v., Houeto J.L., Lagrange C., Le Pen C. et al. Clinical and economic results of bilateral subthalamic nucleus stimulation in Parkinson's disease. *J Neurol Neurosurg Psychiatry*, **2006**, *77*, 443–449.

18 Gang L., Chao Y., Ling L., Lu S.C. Uncovering the mechanisms of Deep Brain Stimulation. *Journal of Physics: Conference Series*, **2005**, *13*, 336–344.

19 Garcia L., D'Alessandro G., Bioulac B., Hammond C. High-frequency stimulation in Parkinson's disease: more or less? *Trends in Neuroscience*, **2005**, *28*, 209–216.

20 Garenne A. and Chauvet G. A. A discrete approach for a model of temporal learning by the cerebellum: in silico classical conditioning of the eyeblink reflex. *J Integr Neurosci*, **2004**, *3*, 301–18.

21 Gerstner W. Population Dynamics of Spiking Neurons: Fast Transients, Asynchronous States, and Locking. *Neural Computat*, **1999**, 12:43–89.

22 Gillies A. J., Willshaw D. J. Models of the subthalamic nucleus: the importance of intranuclear connectivity. *Med Eng Phys*, **2004**, *26*(9):723–32.

23 Guehl D., Edwards R., Cuny E., Burbaud P., Rougier A., Modolo J., Beuter A. Statistical determination of optimal STN stimulation site in Parkinson's disease. *J Neurosurg*, (in press, **2007**).

24 Haeri M., Sarbaz Y., Gharibzadeh S. Modeling the Parkinson's tremor and its treatment. *J Theor Biol*, **2005**, *236*, 311–322.

25 Haskell E., Nykamp D.Q., Tranchina D. Population density methods for large-scale modelling of neuronal networks with realistic synaptic kinetics: cutting the dimension down to size. *Network: Comput Neural Syst*, **2001**; *12*, 141–174.

26 Hauptmann C., Popovych O., Tass P. A. Effectively desynchronizing deep brain stimulation based on a coordinated delayed feedback stimulation via several sites: a computational study. *Biol Cybern*, **2005**, *93*, 463–470.

27 Hines M. L. and Carnevale N.T. NEURON: a tool for neuroscientists. *Neuroscientist*, **2001**, *7*, 123–35.

28 Huertas M. A., Smith G. D. A multivariate population density model of the dLGN/PGN relay. *J Comput Neurosci*, **2006**, *21*(2):171–89

29 Izhikevich E. M. Simple model of spiking neurons. *IEEE Trans Neural Network*, **2003**, *14*, 1569–1572.

30 Izhikevich E. M., Gally J. A., and Edelman G. M. Spike-timing dynamics of neuronal groups. *Cereb Cortex*, **2004**, *14*, 933–44.

31 Izhikevich E. M. Which model to use for cortical spiking neurons? *IEEE Trans Neural Netw*, **2004**, *15*, 1063–70.

32 Kandel E.I. *Functional and stereotactic neurosurgery*, Plenum Medical Book Co, NY, London, **1989**.

33 Kita H., Tachibana Y., Nambu A., Chiken S. Balance of monosynaptic excitatory and disynaptic inhibitory responses of the globus pallidus induced after stimulation of the subthalamic nucleus in the monkey. *J Neurosci*, **2005**, *25*(38), 8611–8619.

34 Knight B.W., Manin D., Sirovich L. Dynamical models of interacting neuron populations. In *EC Gerf, ed. Symposium on Robotics and Cybernetics: Computational Engineering in Systems Applications*, **1996**, Cite Scientifique, Lille, France.

35 Kopell B.H., Rezai A.R., Chang J.W., Vitek J.L. Anatomy and physiology of the basal ganglia: implications for deep brain stimulation for Parkinson's disease. *Mov Disord*, **2006**, 21, 14, 238–246.

36 Kuhn A.A., Trottenberg T., Kivi A., Kupsch A., Schneider G.H., Brown P. The relation between local field potential and neural discharge in the subthalamic nucleus of patients with Parkinson's disease. *Exp Neurol*, **2005**, *194*, 212–220.

37 Kuncel A.M., Grill W.M. Selection of stimulus parameters for deep brain stimulation. *Clin Neurophysiol*, **2004**, *115*(11), 2431–2441.

38 Leblois A., Boraud T., Meissner W. et al. Competition between Feedback Loops Underlies Normal and Pathological Dynamics in the Basal Ganglia. *J Neurosci*, **2006**, *26*(13), 3567–3583.

39 Lee K.H., Chang S.Y., Roberts D.W., Kim U. Neurotransmitter release from high-frequency stimulation of the subthalamic nucleus. *J Neurosurg*, **2004**, *101*, 511–517.

40 Lévesque J.C., Parent A. GABAergic Interneurons in Human Subthalamic Nucleus. *Mov Dis*, **2005**, *20*(5), 574–584.

41 Grill W.M., McIntyre C.C. Extracellular excitation of central neurons: implications for the mechanisms of deep brain stimulation. *Thalamus Relat Syst*, **2001**, *1*, 269–277.

42 McIntyre C.C., Savasta M., Walter B.L., Vitek J.L. How does deep brain stimulation work? Present understanding and future questions. *J Clin Neurophysiol*, **2004**, 40–50.

43 Michelson S., Scherrer D., Bangs A. Target Identification and Validation Using Human Simulation Models. In *In Silico Technologies in Drug Target Identification and Validation*, Eds. Leon, D. and Markell, S. Taylor & Francis, **2006**.

44 Modolo J., Garenne A., Henry J., Beuter A. Probabilistic model of the subthalamic nucleus with small-world networks properties (*XXVIIIth Symposium of Computational Neuroscience*, Montréal (Canada), **2006**).

45 Modolo J., Garenne A., Henry J., Beuter A. Modélisation multi-échelles de la stimulation cérébrale dans la maladie de Parkinson (accepted, *NeuroComp, 1ère Conférence Française de Neurosciences Computationnelles*, Pont-à-Mousson (France), **2006**).

46 Modolo J., Garenne A., Henry J., Beuter A. A new population density approach using the Izhikevich model (in preparation for the *J Comput Neurosci*).

47 Montgomery E. B. Dynamically Coupled, High-Frequency Reentrant, Non-linear Oscillators Embedded in Scale-Free Basal Ganglia-Thalamic-Cortical Networks Mediating Function and Deep Brain Stimulation Effects. *Nonlinear Stud*, **2004**, *11*, 385–421.

48 Niktarash A. H. Transmission of the subthalamic nucleus oscillatory activity to the cortex; a computational approach. *J Comput Neurosci*, **2003**, *15*, 223–232.

49 Nykamp D. Q., Tranchina D. A population density approach that facilitates large-scale modeling of neural networks: analysis and an application to orientation tuning. *J Comput Neurosci*, **2000**, *8*(1), 19–50.

50 Obeso J.A., Rodríguez-Oroz M.C., Ro-
dríguez M., Arbizu J., Giménez-Amaya
J.M. The Basal Ganglia and Disorders of
Movement: Pathophysiological Mecha-
nisms. *News Physiol Sci*, **2002**, *17*, 51–55.

51 Omurtag A., Knight B. W., Sirovich L. On
the simulation of large populations of neurons. *J Comput Neurosci*, **2000**, *8*(1),51–63.

52 Pagni C.A., Altibrandi M.G., Bentivoglio
A., Caruso G., Cioni B., Fiorella C., Insola
A., Lavano A. et al. Extradural motor cortex stimulation (ECMS) for Parkinson's
disease. History and first results by the
study group of the Italian neurosurgical
society. *Acta Neurochir Suppl*, **2005**, *93*,
113–119.

53 Pascual A., Beuter A., Modolo J. Is a computational model useful to understand the
effect of deep brain stimulation in Parkinson's disease? *J Integr Neurosci* (submitted,
2006).

54 Plaha P., Gill S.S. Bilateral deep brain
stimulation of the pedoculopontine nucleus for Parkinson's disease. *Neuroreport*,
2005, *16*, 1883–1887.

55 Plaha P., Ben-Shlomo Y., Patel N.K., Gill
S.S. Stimulation of the caudal zona incerta
is superior to stimulation of the subthalamic nucleus in improving contralateral
parkinsonism. *Brain*; **2006**, *129*, 1732–
1747.

56 Richter E.O., Hoque T., Halliday W.,
Lozano A.M., Saint-Cyr J.A. Determining
the position and size of the subthalamic
nucleus based on magnetic resonance
imaging results in patients with advanced
Parkinson's disease. *J Neurosurg*, **2004**,
100, 541–546.

57 Ritz and Yu. Parkinson's Disease Mortality
and Pesticide Exposure in California 1984-
1994. *Int J of Epidemiol*, **2000**, *29*, 323–329.
UCLA.

58 Rubin J. E. and Terman D. High frequency
stimulation of the subthalamic nucleus
eliminates pathological thalamic rhythmicity in a computational model. *J Comput Neurosci*, **2004**, *16*, 211–235.

59 Shi L.H., Luo F., Woodward D.J., Chang
J.Y. Basal ganglia neural response during
behaviorally effective deep brain stimulation of the subthalamic nucleus in rats

performing a threadmill locomotion test.
Synapse, **2006**, *59*, 445–457.

60 Tass P. A. A model of desynchronizing
deep brain stimulation with a demand-
controlled coordinated reset of neural
subpopulations. *Biol Cybern*, **2003**, *89*,
81–88.

61 Terman D., Rubin J. E., Yew A. C. Wilson
C. J. Activity patterns in a model for the
subthalamopallidal network of the basal
ganglia. *J Neurosci*, **2002**, *22*, 2963–2976.

62 Titcombe M. S., Glass L., Guehl D., Beuter
A. Dynamics of Parkinsonian tremor during deep brain stimulation. *Chaos*, **2001**,
11, 766–773.

63 Voges J., Volkmann J., Allert N., Lehrke
R., Koulousakis A., Freund H.J. et al. Bilateral high-frequency stimulation in the
subthalamic nucleus for the treatment of
Parkinson's disease: correlation of therapeutic effect with anatomical electrode
position. *J Neurosurg*, **2002**, *96*, 269–279.

64 Voytek B. Emergent basal ganglia pathology within computational models. *J Neurosci*, **2006**, *26*, 28, 7317–7318.

65 Wang X.J. Synaptic basis of cortical persistent activity: the importance of NMDA
receptors to working memory. *J Neurosci*,
1999, *19*, 587–603.

66 Watts D.J., Strogatz S.H. Collective dynamics of "small-world" networks. *Nature*,
1998, *393*, 440–442.

67 Wichmann T., DeLong M.R. Deep brain
stimulation for neurologic and neuropsychiatric disorders. *Neuron*, **2006**, *52*, 197–
204.

68 Wielepp J.P., Burgunder J.M., Pohle T.,
Ritter E.P., Kinser J.A., Krauss J.K. Deac-
tivation of thalamocortical activity is responsible for suppression of parkinsonian
tremor by thalamic stimulation: a 99mTc-
ECD SPECT study. *Clin Neurol Neurosurg*,
2001, *103*, 228–231.

69 Wilson H. R., Cowan J. D. Excitatory and
inhibitory interactions in localized populations of model neurons. *Biophys J*, **1972**,
12, 1–24.

70 Wilson H. R., Cowan J. D. A mathematical theory of the functional dynamics of
cortical and thalamic nervous tissue. *Ky-
bernetik*, **1973**, *13*, 55–80.

14
Constructing a Virtual Proteasome
Alexey Zaikin, Fabio Luciani, and Jürgen Kurths

Abstract

The proteasome is a barrel-shaped multisubunit protease involved in the degradation of the ubiquitin-tagged proteins. Proteasomes play many important roles in the cellular metabolism, from regulation of the important proteins to generating antigenic peptides for the immune system. The prediction of the proteasome activity can help in the treatment of cancer, immune system disorders and neurodegenerative diseases, hence the constructing of a virtual proteasome is a necessary task. Here we review several model approaches which allow the analytical or numerical prediction of the proteasome functioning. The review sheds light on the-state-of-the-art of the proteasome modeling and discusses the perspectives of this research.

Proteasomes are multicatalytic enzyme complexes that are responsible for degradation of the majority of the intracellular proteins into smaller peptides. They are present in all eukaryotic cells, archaea, and certain bacteria [7, 20, 46]. Proteasomes are absolutely essential for the homeostasis because the removal of proteasome genes in eukaryotes is lethal [14]. Proteasomes have been found in the form of different, although similar molecular complexes that consist of the central part, the 20S proteasome, and regulating caps, the 19S [7] and PA28 particles [22, 38]. The most important 26S complex, which degrades ubiquitinated proteins, contains in addition to the 20S proteasome a 19S regulatory complex composed of multiple ATPases and components necessary for binding protein substrates [7]. The 20S proteasome is a barrel-shaped structure composed of four stacked rings of 28 subunits [7, 37]. The active cleavage sites are located within the central chamber of the 20S proteasome, into which protein substrates must enter through narrow openings of outer rings. The 20S proteasome degrades proteins by a highly processive mechanism [2], making many cleavages of the protein and digesting it into small products. This is important for the intracellular proteolytic system because the release of large protein fragments could interfere with the cell function and regulation [19]. Proteasomes can be found in its usual form as a constitutive proteasome, or as an immunoproteasome with modified cleavage centers [20, 21].

Biosimulation in Drug Development. Edited by Martin Bertau, Erik Mosekilde, and Hans V. Westerhoff
Copyright © 2008 WILEY-VCH Verlag GmbH & Co. KGaA, Weinheim
ISBN: 978-3-527-31699-1

Many roles in the cell's metabolism are played by proteasomes: they destroy abnormal and misfolded proteins tagged with ubiquitin and are an essential component of the ATP-ubiquitin-dependent pathway for protein degradation [6]. Proteasomes play an important role in the immune system by generating antigenic peptides of 8–12 residues to be presented by the MHC class I molecules and, hence, are the main supplier of peptides for its recognition by killer T-cells [12, 21, 26, 43]. Recently, proteasome inhibition was suggested as a promising new target for cancer treatment [1, 9, 34]. The proteasome function is also linked directly to the pathophysiology of malignancies, neurodegenerative disorders, type I diabetes, cachexia [11, 40] and to ageing [53].

Here we review several recently developed mathematical models which describe the influence of a length dependent cleavage, in- and efflux rates, and transport processes on the outcome of a digestion experiment. We start with a short introduction to available experimental results and proceed with a general description of aspects necessary for modeling the proteasome. Then we consider a microscopic model of the proteasome, review a possible translocation mechanism inside the proteasome [52], and show how one can compute length-dependent transport rates analytically or numerically. Next we review the transport model of the proteasome and discuss the influence of transport rates on the proteasome output [50, 51]. This approach is based on the idea that proteasome output significantly depends on length-dependent transport. After that we show how one can describe the kinetics of the protein degradation on the mesoscopic level without details of the transport mechanism [27, 28]. In all these approaches we neglect the sequence specificity of the incoming protein making the degradation dependent mainly on the protein length. Finally we discuss the future of the modeling approaches and show how one can include the sequence specificity in these models and how one can merge the models describing the protein translocation with models which described the in- and efflux.

14.1
Experiment and Modeling

Due to its significance in the cellular metabolism, simulation of the proteasome function is the central task in building a virtual immune system [29]. On the long road from the initial idea to the pharmacy and the start of a new drug, the simulation and prediction of the proteasome function seem to be possible now only in its early stages, i.e., for experiments *in vitro*. These experiments study the digestion of different substrates by the proteasome and the temporal dynamics of the fragment concentration over the course of time by mixing purified proteasomes and different substrates. The experimental results are analyzed, e.g., by mass spectroscopy methods, and provide us with information about the substrate cleavage pattern and the quantity of different fragments cut from the initial substrate. Even in these first stages of drug design, however, the simulation of the proteasome could signif-

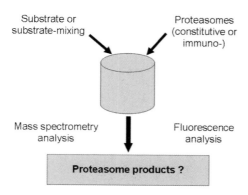

Fig. 14.1 A principal scheme of the experiment *in vitro*. The solution, where the substrate is mixed with proteasome, is analyzed for different moments of time. Mass spectroscopy or fluorescence analysis then shows how much substrate has been degraded and which fragments have been produced.

icantly decrease the experiment costs through the identification and prediction of proper parameter ranges.

A principle scheme of the experiment *in vitro* is shown in Fig. 14.1. The solution contains a mixture of constitutive or immuno-proteasomes with substrate or a mix of substrates. The solution is analyzed in different moments of time, hence giving the information how much substrate has been degraded and how much and which fragments have been produced. One of the most important experimental result is the cleavage pattern of the particular protein, which shows where protein has been cleaved. Additionally analysis of the fragment amount produced from this cleavage site gives an estimation of the cleavage strength at this protein position. Especially interesting is the production of a certain sequences, epitopes, which are then requested by the immune system.

The next important experimental result that describes the proteasome function is a length distribution of the fragments obtained from *in vitro* experiments. For long substrates it was found that this dependence typically is nonmonotonous and has a single peak around the length of 7–12 amino acids (aa) for practically all types of the proteasome [5, 19, 31, 32]. This length of peptides is the most requested length for normal functioning of the immune system. The mechanism behind such a length distribution is not completely clear. It was widely believed that the proteasome degrades proteins according to the "molecular ruler" to yield products of rather uniform size. It was proposed [49] that peptides of 7–9 residues were generated as a result of coordinated cleavages by neighboring active sites. However, evidence for the molecular ruler is quite limited because the maximum length distribution is smoothed and not pronounced as a sharp peak [19]. It is also interesting to note that, in some experiments, three peak length distributions have been found [23]. Several theoretical models for the kinetics of proteasome degradation have been published. Some of the models describe the degradation of short peptides with qualitatively different kinetics [41, 44, 45] or small number of cleavage positions

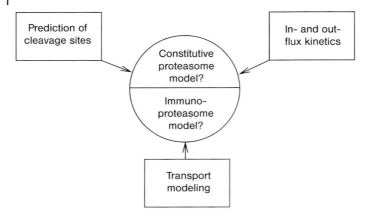

Fig. 14.2 Different factors which should be taken into account to model the proteasome function.

[16]. The theoretical model [16, 35] for the degradation of long substrates is applied to specific proteins with predefined cleavage sites and is fitted to experimental data describing the fragment quantity after proteasomal degradation.

Finally the amount of the fragments produced versus time evolution provides the information about the dynamics of fragment concentration. Interestingly, the concentration of a certain fragment can first increase and then decrease. This is connected with the fact that the fragment produced can compete with the initial substrate and reenter the proteasome, providing a decrease in the initial substrate degradation.

To model the proteasome mechanism, one should adequately describe three essential processes involved in the proteasome function: selection of cleavage sites, kinetics of generated fragments, and a peptide translocation inside the proteasome (see Fig. 14.2). At the moment there is no model which describe adequately all these factors. In this paper we describe these modeling approaches and discuss how one can merge them. It is important to note also that for different situations different models should be used, depending on which factor provides the largest influence on the phenomenon studied.

14.2
Finding the Cleavage Pattern

Several proteasomal cleavage prediction methods have been published. The first method, FragPredict, was developed by Holzhutter et al. [15] and is publicly available as a part of MAPPP service (www.mpiib-berlin.mpg.de/MAPPP/). It combines proteasomal cleavage prediction with MHC and TAP binding prediction. FragPredict consists of two algorithms. The first algorithm uses a statistical analysis of cleavage-enhancing and -inhibiting amino acid motifs to predict potential proteasomal cleavage sites [15]. The second algorithm, which uses the results of the first algorithm as an input, predicts which fragments are most likely to be generated. This model takes the time-dependent degradation into account based on a kinetic model of the 20S proteasome [16]. At the moment, FragPredict is the only method that can predict fragments, instead of only the possible cleavage sites. PA-ProC (www.paproc.de) is a prediction method for cleavages by human as well as by wild-type and mutant yeast proteasomes. The influences of different amino acids at different positions are determined by using a stochastic hill-climbing algorithm [24] based on experimentally *in vitro* verified cleavage and non-cleavage sites [33]. Recently, Tenzer et al. [47] published a method for predicting which peptides can be presented by the MHC class I pathway. In this work they characterise the cleavage specificity of the proteasome in terms of a stabilised matrix method (SMM) defining the specificity of the constitutive and immunoproteasome separately. This method is available at www.mhcpathway. net. All these methods make use of limited *in vitro* data for characterising the specificity of the proteasome. Moreover, both FragPredict and the matrix based methods by Tenzer et al. [47] are linear methods, and may not capture the nonlinear features of the specificity of the proteasome. Another prediction method called NetChop [18, 30] has two important extensions: first, the prediction system is trained on multilayered artificial neural networks. This allows the method to incorporate higher-order sequence correlations in the prediction scheme, making it potentially more powerful than both PAProC and the matrix-based methods, which use a linear method to predict proteasome cleavage. Second, the method is trained to predict proteasomal cleavage on *in vitro* digest data, similarly to the other previous methods, and using naturally processed MHC class I ligands.

In large-scale benchmark calculations, the predictive performance of the different methods are compared [47, 30]. According to these tests, at the moment, NetChop and SMM methods provide most reliable predictions of proteasomal cleavage.

14.3
Possible Translocation Mechanism

Considering the highly processive mechanism of the protein degradation by the proteasome, a question naturally arises: what is a mechanism behind such translocation rates? Let us discuss one of the possible translocation mechanisms. In [52] we assume that the proteasome has a fluctuationally driven transport mechanism and we show that such a mechanism generally results in a nonmonotonous translocation rate. Since the proteasome has a symmetric structure, three ingredients are required for fluctuationally driven translocation: the anisotropy of the proteasome–protein interaction potential, thermal noise in the interaction centers, and the energy input. Under the assumption that the protein potential is asymmetric and periodic, and that the energy input is modeled with a periodic force or colored noise, one can even obtain nonmonotonous translocation rates analytically [52]. Here we

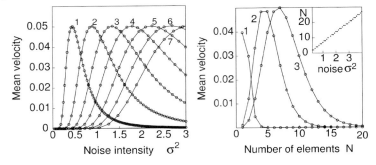

Fig. 14.3 Left: mean velocity of the protein as a function of noise intensity for different lengths of the protein (shown by numbers). The transport is possible only for a certain noise intensity. Right: mean velocity as a function of the peptide length for different noise intensities $\sigma^2 = 0.6$ (curve 1), 2 (curve 2), and 3 (curve 3). The inset plot shows the depth of the protein penetration for the velocity equal to 0.0005.

review these results and show that for different forms of the proteasome–protein interaction potential one gets translocation rates.

We assume that, after the protein has entered the proteasome, the protein–proteasome interaction is characterized by a spatially periodic asymmetric potential $U(x)$ with the period P equal to the distance between amino acids in the protein. In reality there is a basic periodicity, namely the periodicity of the protein (or peptide) backbone, that is superposed by a nonperiodic (in our sense irregular) part that is attributed to the amino acid-specific residues. Below we consider also the influence of the nonperiodic constituent. The spatial asymmetry results from breaking the symmetry by entering the proteasome from one end, as well as from the $C - N$ asymmetry of the protein (or peptide) backbone. Figure 14.3 (left) plots several examples of such asymmetric periodic potential. The detailed form of the asymmetric periodic interaction potential is of less importance for this qualitative study.

The proteasome acts upon the protein by a certain number of equidistant interaction centers. The dynamics of the protein inside the proteasome is, hence, governed by l interactions centers, where l is the number of protein elements (amino acids or multiples). There appear the following forces: potential force (protein–proteasome interaction) $-l\partial U(x)/\partial x$, fluctuations with collective $lF(t)$ and individual components $f_1(t) + \ldots + f_l(t)$, and protein friction forces $l\beta\dot{x}$ [4], where x is the coordinate of the protein with respect to the proteasome and β is the coefficient of friction. Due to small size of all protein particles, moving in the liquid cytosol, the motion occurs in the overdamped realm [3], hence we neglect inertia forces. Note that transport is possible only in the case of nonequilibrium fluctuations. In the simplified case, when fluctuations can be represented by the sum of a collective periodic force and are individual for every protein residue thermal noise, the model is analytically tractable, predicting the velocity dependence on the peptide size. Normalising all forces by friction and taking $\beta = 1$, the translocation of a protein in the proteasome is governed then by:

$$\frac{\partial x}{\partial t} = -\frac{\partial U(x)}{\partial x} + F(t) + \frac{1}{l}(f_1(t) + \dots + f_l(t)). \tag{1}$$

Analytical results are possible if we assume collective oscillations of the peptide elements, e.g., $F(t) = A\cos(\omega t)$, where A and ω stand for the amplitude and frequency of this oscillations. Additionally, each interaction center undergoes local thermal fluctuations, represented by mutually uncorrelated white noise of intensity σ^2: $f_i(t) = \xi_i(t)$, where $\langle \xi_i(t)\xi_j(t') \rangle = \sigma^2 \delta(t - t')\delta_{ij}$. In this case the stochastic term in Eq. (1) is white noise of intensity σ^2/l. The Fokker–Planck equation for the peptide coordinate probability distribution $w(x,t)$ associated with Eq. (1) is

$$\frac{\partial w}{\partial t} = -\frac{\partial}{\partial x}\left[\left(F(t) - \frac{\partial U}{\partial x}\right)w(x,t)\right] + \frac{\sigma^2}{2l}\frac{\partial^2 w(x,t)}{\partial x^2} \tag{2}$$

which may be solved in quasi-stationary adiabatic approximation $\partial w/\partial t = 0$ [17]. We obtain:

$$\frac{\sigma^2}{2l}\frac{\partial w(x,F)}{\partial x} - \left(F - \frac{\partial U}{\partial x}\right)w(x,F) = -G(F) \tag{3}$$

where $G(F)$ is the probability flux. For any periodic potential $U(x)$, the quasi-stationary solution of Eq. (3) is:

$$w(x,t) = \left[C(F) - \frac{2G(F)}{\sigma^2/l}\int_0^x \exp\left(\frac{U(x') - Fx'}{\sigma^2/2l}\right)dx'\right]$$
$$\times \exp\left(-\frac{U(x) - Fx}{\sigma^2/2l}\right) \tag{4}$$

where $C(F(t))$ and $G(F(t))$ are unknown functions of t. Using the periodicity condition $w(0,t) = w(P,t)$ and the normalisation of $w(x,t)$ we get $G(F)$. If the amplitude A meets the condition $LA \ll \sigma^2/l$, one can expand $G(F)$ and obtain

$$G(F) \approx G_{01}F + G_{02}F^2 \tag{5}$$

with the expansion coefficients $G_{01} = P/(I_{10}I_{20})$,

$$G_{02} = G_{01}\left(\frac{I_{11}}{I_{10}} - \frac{I_{21}}{I_{20}} - \frac{lP}{\sigma^2}\left(1 - \frac{2I_{30}}{I_{10}I_{20}}\right)\right), \quad I_{10} = \int_0^P e^{U'(x)}\,dx \tag{6}$$

$$I_{20} = \int_0^P e^{(-U'(x))}\,dx, \quad I_{11} = \frac{2l}{\sigma^2}\int_0^P xe^{U'(x)}\,dx, \quad U'(x) = \frac{2lU(x)}{\sigma^2} \tag{7}$$

$$I_{21} = \frac{2l}{\sigma^2}\int_0^P xe^{(-U'(x))}\,dx \quad I_{30} = \int_0^P\int_0^x e^{(U'(x')-U'(x))}\,dx'\,dx \tag{8}$$

Substituting Eq. (5) into $\overline{\langle \dot{x} \rangle} = \int_0^P \overline{G(x,t)}\,dx$, where $\overline{(\cdot)}$ denotes time averaging, we obtain the average protein transport velocity or the translocation, as a function of the noise intensity σ^2 and the peptide size l

$$R_t \approx \overline{\langle \dot{x} \rangle} \approx \frac{P^2 A^2}{2 I_{10} I_{20}} \left[\frac{I_{11}}{I_{10}} - \frac{I_{21}}{I_{20}} - \frac{lP}{\sigma^2} \left(1 - \frac{2 I_{30}}{I_{10} I_{20}} \right) \right]. \tag{9}$$

Using these formula, one can compute the velocity of the protein translocation as a function of the noise intensity, or, more importantly, as a function of the protein length (see Fig. 14.3). It is clearly seen that assuming the fluctuationally driven transport the lentgh can crucially change the velocity of the protein translocation. This function can be monotonous or nonmonotonous. The results are quite general, for example, if the noise intensity is large enough, the nonmonotonous transport rates can be obtained for a large variety of assymetric periodic potentials. These dependencies of the translocation rate on the peptide length, computed using Eq. (9), are shown in Fig. 14.4 (right) for different interaction potentials shown in the same figure. These results show that very different forms of the interaction potential, if this potential fullfils the conditions of the fluctuationally driven transport, result in the nonmonotonous translocation rate functions.

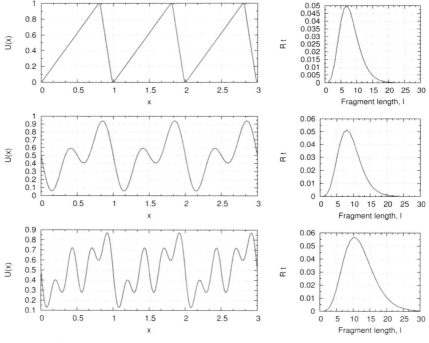

Fig. 14.4 Left: different examples of asymmetric periodic potential $U(x)$ with the period $P = 1$. Only three periods are shown. From top to bottom: $U(x)$ is a saw-tooth function with smoothed angles (for details see [25]; $U(x) = -(0.25 \sin(2\pi x/P) + 0.25 \sin(4\pi x/P) + 0.5; U(x) = -(0.166 \sin(2\pi x/P) + 0.166 \sin(4\pi x/P) + 0.166 \sin(8\pi x/P)) + 0.5$. Right: the corresponding dependencies of the velocity rate R_t on the peptide length l expressed in amino acids and $A = 1.15$.

14.4
Transport Model and Influence of Transport Rates on the Protein Degradation

14.4.1
The Transport Model

Next let us show how one can compute the proteasome output if the transport rates are given. In our model we assume that the proteasome has a single channel for the entry of the substrate with two cleavage centers present at the same distance from the ends, yielding in a symmetric structure as confirmed by experimental studies of its structure. In reality a proteasome has six cleavage sites spatially distributed around its central channel. However, due to the geometry of its locations, we believe that a translocated protein meets only two of them. Whether the strand is indeed transported or cleaved at a particular position is a stochastic process with certain probabilities (see Fig. 14.5).

The protein strand can be cleaved if it lies close to the cleavage center or it could be transported forward by one amino acid. We assume that the probability of transport depends only on the length of the strand inside the proteasome. The probability of transport is, therefore, given by a translocation rate function, $v(x + D)$ where $x + D$ is the length of the strand inside the proteasome, in terms of amino acids. The probability of cleavage is assumed to be a constant, denoted by γ. We also assume that the degradation of proteins by the proteasome is a highly *processive* mechanism [2], i.e., in other words, the protein is not released by the proteasome until it is completely processed. This leads to the possibility of the proteasome

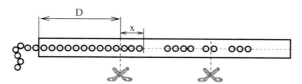

Fig. 14.5 Schematic diagram of the protein degradation by the proteasome. The protein strand (denoted by '0-0-0-0') enters the proteasome from the left to be cleaved by the cleavage centers (denoted by the scissors). The length of the strand after crossing the cleavage center is denoted by x.

Table 14.1 Parameter values used in the model.

Parameter	Description	Dimension	Default value
l	Protein length	Amino acids	200
L	Proteasome length	Amino acids	80
D	Distance between a cleavage center and the proteasome end	Amino acids	15
N	Number of degraded proteins	–	1000
γ	Cleavage rate	–	0.001

making several cuts in the same protein strand and in the formation of a greater number of smaller-length peptides. The model also does not allow cleavage products to overlap or outrun their predecessors. The standard parameters we have used for this particular model are given in Table 14.1.

14.4.2
Analytics – Distribution of Peptide Lengths

To derive analytically the proteasome length distribution, we apply the master equation approach for the distribution of the coordinate of the front end of the strand and find its stationary solution as well as the distribution of peptide length. We also assume that the products of cleavage leave the chamber immediately after cleavage. This assumption can be justified by the fact that the products of cleavage have rather small size and can really leave the proteasome very fast. We note that this assumption will be also motivated if the transport rate function is a monotonously decaying one, and the characteristic peptide length is small in comparison to the distance between the cleavage center and the proteasome end. In this case most of the peptides move faster than the incoming protein strand, and do not block it. Later on, using numerical simulations we compare whether this assumption can significantly change the results.

14.4.2.1 One Cleavage Center
We take the offset of the coordinate x (measured in amino acids) along the proteasome at the first cleavage center (see Fig. 14.5). During the time interval dt, the protein strand can move by one amino acid with the probability $v(x + D)dt$ and can be cut with the probability γdt.

Let us have an ensemble of identical proteasomes with proteins inside, and let $w(x)$ be the distribution of the proteins with the coordinate x. Then, at the x-th bond, the change in $dw(x)$ consists of the increase $v(x + D - 1)w(x - 1)dt$ due to the movement of protein ends from $x - 1$ to x, the decrease $-v(x + D)w(x)dt$ due to the movement of protein ends from x to $x + 1$, and the loss $-\gamma w(x)dt$ due to cleavage:

$$dw(x) = [v(x + D - 1)\, w(x - 1) - v(x + D)\, w(x)]\, dt - \gamma\, w(x)dt \qquad (10)$$

At $x = 0$ (the protein ends at the cleavage center) we have to set the boundary condition: no strands are to be cut, and no strands move from $x = -1$ because no proteins end there, but there is a gain $\sum_{x=1}^{\infty} \gamma\, w(x)dt$ from cleavage, i.e.:

$$dw(0) = \left[\gamma \sum_{x=1}^{\infty} w(x) - v(D)\, w(0)\right]dt \qquad (11)$$

Using the standard master equation technique, we consider the continuous limit, supposing that $w(x)$ and $v(x)$ vary slightly from one bond to another. Then,

$\sum_{x=1}^{\infty} w(x) \approx \int_0^{\infty} w(x)dx = 1$ (the last being the normalization condition), and the discrete derivative in Eq. (10) may be replaced with the continuous derivative:

$$\dot{w}(x, t) = -\frac{\partial}{\partial x}\left(v(x + D)\, w(x, t)\right) - \gamma\, w(x, t) \tag{12}$$

with the boundary condition Eq. (11), $\dot{w}(0) = \gamma - v(D)\, w(0)$, i.e. for stationary case:

$$w(0) = \frac{\gamma}{v(D)} \tag{13}$$

The stationary solution of Eq. (12) with the boundary condition Eq. (13) gives the asymptotic distribution of protein coordinates which coincides with the length distribution of peptides $\rho(x)$ (because probability of a strand to be cut is independent of its coordinate). The problem in Eqs. (12), (13) admits the stationary solution:

$$\rho(x) = w(x) = \frac{\gamma}{v(x + D)} \exp\left[-\gamma \int_0^x \frac{dx'}{v(x' + D)}\right] \tag{14}$$

thus providing us with the analytically found proteasome product length distribution.

14.4.2.2 Two Cleavage Centers

Let us now have a second cleavage center at $x_2 = L - 2D$ (see Fig. 14.5). If γdt is the probability of cleavage by a center during the time interval dt, then the probability of a peptide of length x to be cut by any center is $\gamma(x)dt$:

$$\gamma(x) = \begin{cases} \gamma, & 0 < x < x_2 \\ 2\gamma, & x \geq x_2 \end{cases}$$

In this case, the length distribution does not coincide with the distribution of protein coordinates. Let us start with the distribution of protein coordinates $w(x)$. Beyond the cleavage centers it obeys the equation which is similar to Eq. (12):

$$\dot{w}(x, t) = -\frac{\partial}{\partial x}\left(v(x + D)\, w(x, t)\right) - \gamma(x)\, w(x, t) \tag{15}$$

At the first cleavage center, again, $w(0) = \gamma / v(D)$; at the second one ($x = x_2$):

$$\dot{w}(x_2) = \gamma \int_{x_2}^{\infty} w(x)dx + v(x_2 + D - 1)\, w(x_2 - 1)$$
$$-v(x_2 + D)\, w(x_2) - \gamma w(x_2)$$

which provides:

$$w(x_2) = w(x_2 - 1) + \frac{u}{v(x_2 + D)} \tag{16}$$

where $u \equiv \gamma \int_{x_2}^{\infty} w(x)dx$.

So, for x in $(0, x_2)$, the protein coordinate distribution is:

$$w(x) = w_1(x) = \frac{\gamma}{v(x + D)} \exp\left[-\gamma \int_0^x \frac{dx'}{v(x' + D)} \right] \tag{17}$$

and for x in $[x_2, \infty)$, one can obtain $u = v(x_2 + D)w_1(x_2)$,

$$w_2(x) = \frac{2\gamma}{v(x + D)} \exp\left[-\int_0^x \frac{\gamma(x')\,dx'}{v(x' + D)} \right] \tag{18}$$

To note, the formulae in Eqs. (17) and (18) may be combined:

$$w(x) = \frac{\gamma(x)}{v(x + D)} \exp\left[-\int_0^x \frac{\gamma(x')\,dx'}{v(x' + D)} \right] \tag{19}$$

In its turn, the distribution of peptide lengths $\rho(x)$ is proportional to a superposition of $w(x)$ and shifted $w_2(x)$, i.e., $\rho(x) \propto w(x) + w_2(x + x_2)$, that gives us after normalization:

$$\rho(x) = \frac{w(x) + w_2(x + x_2)}{1 + e^{-\gamma \int_0^{x_2} \frac{dx'}{v(x' + D)}}} \tag{20}$$

14.4.2.3 Maximum in Peptide Length Distribution
As shown below, the presence of a second cleavage centre does not significantly change the results. Hence, to find the conditions for a maximum in the peptide length distribution, we use the analytical expressions for the case of one cleavage center. This peptide length distribution has extrema at the points where:

$$0 = \frac{dw(x)}{dx}$$
$$= \frac{\gamma}{v^2(x + D)} \exp\left[-\gamma \int_0^x \frac{dx'}{v(x' + D)} \right]\left(-\frac{dv(x + D)}{dx} - \gamma \right) \tag{21}$$

which gives the condition for extremum:

$$\frac{dv(x + D)}{dx} = -\gamma \tag{22}$$

This equation should be fulfilled at least in one point for $x + D > 0$. To note γ is here the rate of cleavage. Hence, there are no limitations for its value. Eq. (22) shows that the condition for obtaining a maximum when we have a single cleavage center is independent of the actual form of the transport rate function, but is rather dependent on its slope. This would suggest that it is possible to obtain a peak in the length distribution even when the transport rate function is monotonically decreasing, and indeed that is what we find from our numerical simulations.

14.5

Comparison with Numerical Results

We adapt our model for numerical simulation with the help of the Gillespie algorithm [10], which enables the system to jump to the next event via the calculation of the waiting time before any event will occur. Following the approach suggested by us [50], we stochastically model the system where several events can happen with different probabilities. Suppose that in some moment of time we have a set of N probable events with rates R_i, where the $i - th$ event has the rate R_i and $i = 1 \ldots N$. Then by generating two uniformly distributed in $(0, 1)$ random numbers RN_1 and RN_2, we estimate the time T after which the next event would occur as:

$$T = -\frac{\log(RN_1)}{\sum_{i=1}^{N} R_i} \tag{23}$$

The concrete event k that occurs after this time can then be found from:

$$\sum_{i=1}^{k} R_i < \sum_{i=1}^{N} R_i RN_2 \leqslant \sum_{i=1}^{k+1} R_i \tag{24}$$

The peptide or its part inside the proteasome can either be shifted by one amino acid or it can be cleaved if it is located near the cleavage center. Inside the proteasome, the translocation rates of the substrate or fragments depend only on their lengths and are described by the translocation rate function $v(x + D)$, see Fig. 14.5. The probability of cleavage is described by the function $R_c(p)$, where p is the position in the substrate sequence. We set the constant cleavage probability $R_c(p) = \gamma$ and discuss another situation later. When the protein is degraded, its fragments lengths are counted in the length distribution. The reliability of the results is ensured as we conduct the study over a large number of proteins N so that the trends observed in the length distribution have a basis in statistical results. To consider different situations, we study different relations between the geometry of the proteasome (the parameter D) and two qualitatively different transport rate functions: monotonous and nonmonotonous.

For the sake of generality, we choose the following qualitatively different transport rate function:

$$v_1(x) = e^{-0.2x}, \quad v_2(x) = 0.125 e^{-\alpha x} x^3 \tag{25}$$

where x is the peptide length and $\alpha = 0.54$ is a constant. It is important to note that such monotonous and nonmonotonous transport functions can be obtained also if we assume that the protein translocation is driven by nonequilibrium fluctuations, as we assumed [52].

14.5.1
Monotonously Decreasing Transport Rates

First we consider a case of the monotonously decreasing transport rate function v_1. This form of function (see Fig. 14.6a) may correspond to two different situations: when the condition in Eq. (22) can be fulfilled at some point and when it cannot. The first case occurs when the parameter D is smaller than the point where the derivative of the transport function is equal to the value $-\gamma$. As predicted by the theory in this case one should observe the maximum in the length distribution. This is indeed the case, as can be shown by a comparison of the numerical results with the theoretical curve for one cleavage center (see Fig. 14.6b) or for two cleavage centers (see Fig. 14.6c). It should be noted that addition of the second cleavage center does not change the results both as predicted from the theory or computed with the Gillespie algorithm. This holds for all situations considered in this paper, hence below we plot only the results for two cleavages centers. These plots clearly show a good matching between the analytical theory developed and the results of numerical simulations. One remembers that both the theory and numerical simulations have been made under the assumption that the cleavage products disappear

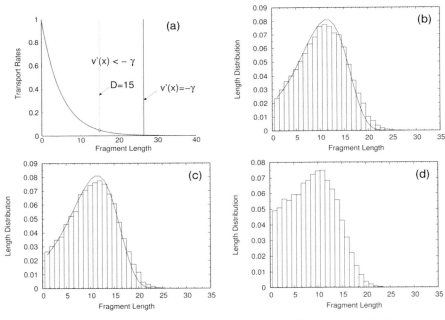

Fig. 14.6 (a) Monotonously decreasing transport rate function $v(x) = v_1(x)$. The vertical lines show the location of the point where the condition for the maximum in the length distribution (Eq. 22) holds and the location of the cleavage centre. The parameters are $D = 15$ and $\gamma = 0.001$.

As predicted by theory (solid line) the numerically computed peptide length distribution (denoted by boxes) has a maximum for one cleavage center (b), two cleavage centers with immediate disappearance of cleavage products (c), and in the case where the cleavage products do not disappear (d).

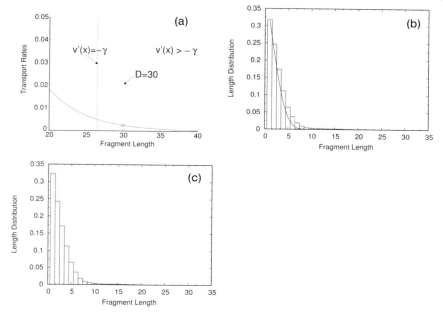

Fig. 14.7 (a) The case when D is large and the maximum condition holds nowhere. As predicted by the analytical theory (solid line) the length distribution does not have peaks under the assumption that the cleavage products disappear immediately (b) or without this assumption (c).

immediately after the cut. Also the numerical simulations also allow to check what happens if this assumption does not hold and the cleavage product still moves along the channel. As expected, for a monotonously decreasing transport function, the length distribution does not significantly change (see Fig. 14.6d). This happens because the shorter cleavage products have larger probability of transport and do not stack the incoming protein.

The situation qualitatively differs if after the cleavage centre the condition of the maximum is never fulfilled. This happens if D is relatively large (see Fig. 14.7a). In this case the length distribution would not have a maximum which is in good correspondence to the result predicted by the theory. Figure 14.7b also illustrates that in this case the analytical results have a good match with the numerics. The length distributions also practically do not change if the cleavage products do not disappear after cleavage but are transported with the same rules as the incoming protein (see Fig. 14.7c).

14.5.2
Nonmonotonous Transport Rates

To cover several possible forms of the transport function, next we analyse the transport function $v_2(x)$ with one peak (see Fig. 14.8e). The derivative of this function

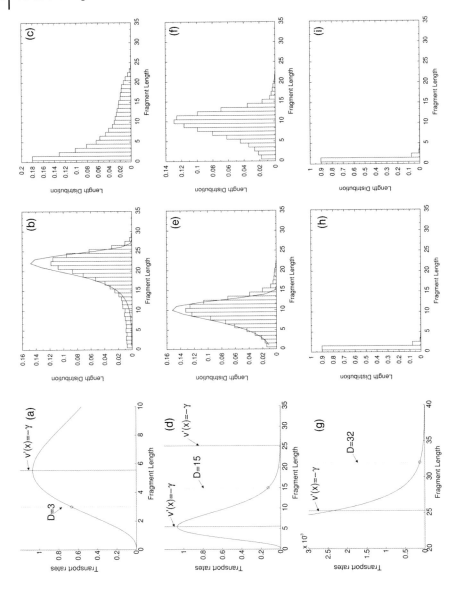

Fig. 14.8 The case of nonmonotonous transport rate function $v_2(x)$, shown by a solid line in (a,e,h). From top to bottom: different location of the cleavage centre $D = 3, 15, 32$ (shown the vertical line) with respect to two points where the condition of maximum holds (shown vertical lines); (b,f,i) the corresponding length distributions for the case of two cleavage centres. Numerics (boxes) is always in good agreement with theoretical predictions (solid line); (c,g,j) the corresponding length distribution if computed without the assumption of immediate disappearance. Only in one case (c) of rather unrealistic too small D this length distribution changes a lot from distributions (b,f,i).

can be equal to $-\gamma$ in two points dividing all possible coordinates into three regions where the cleavage centre D can be located. Let us study these three possibilities.

First, let us consider the case when the location of the cleavage center allows the maximum condition to be fulfilled in two points (see Fig. 14.8a). In this case the theory predicts the non-monotonous length distribution, as confirmed also by the numerics both for one or two cleavage centers (see Fig. 14.8b). However, if we let the cleavage products not disappear, then the length distribution is monotonously decreasing, as found by numerical simulations (see Fig. 14.8c). This is the only case when the use of this assumption changes the results significantly and the theory does not predict the form of length distribution without this assumption. One should note that this case of $D = 3$ seems to be rather unrealistic, because it means that the cleavage centre is too close to the entrance what is not the case in the reality [7, 37].

Next, we analyze the case when the cleavage centre $D = 15$ is between two points where the derivative is equal to $-\gamma$, and hence the maximum condition can be fulfilled in one point (see Fig. 14.8e). As predicted by theory and confirmed by numerics, the length distribution has a maximum in this case (see Fig. 14.8f). It is noteworthy that the theory in this case works sufficiently also in the case when the cleavage products do not disappear immediately (see Fig. 14.8g). If we believe in the nonmonotonous transport rate function hypothesis [50], this case seems to be the most adequate for protein degradation by the proteasome. In the last case, the cleavage centre is so deep in the proteasome, $D = 32$, that the maximum condition can be fulfilled nowhere. As expected, the length distribution has no maxima in all cases computed for this relation between the transport function and the geometry of the proteasome.

14.6
Kinetic Model of the Proteasome

14.6.1
The model

Here we review the mathematical description of the proteasome degradation in the case when the translocation plays not the main role and can be neglected in comparison with the kinetics of the in- and efflux rates. The model describes the rates at which the concentrations of fragments of length k change over time. The concentrations change by proteasomal cleavage, making two short fragments out of a long one, and by the influx and efflux of fragments through the gates. The dynamics does not depend on the amino acid sequence and orientation of the fragment, but only depends on the length of the fragment. Let n_k and N_k be the concentration of fragments of length k inside and outside the proteolytic chambers. Then:

$$\frac{dN_k}{dt} = -a(k)\left[1 - v\sum_{j=1}^{L} jn_j\right]N_k + e(k)n_k, \tag{26}$$

$$\frac{dn_k}{dt} = a(k)\left[1 - v\sum_{j=1}^{L} jn_j\right]N_k - e(k)n_k$$

$$-c\sum_{i=1}^{k-1} F_{k,i}n_k + c\sum_{j=k+1}^{L} (F_{j,k} + F_{j,j-k})n_j \tag{27}$$

for $k = 1, 2, \ldots, L$. The substrate N_L is an outside fragment of length $k = L$. The first term of Eq. (26) describes the influx of fragments into the proteasome. For the influx function $a(k)$ we consider the case where there is no re-entry of fragments other than the substrate, i.e., we set $a(k) = \hat{a}$ for $k = L$, and $a(k) = 0$ otherwise. The influx of substrate into the proteolytic chambers is a rate limiting factor in protein degradation. Experimental works have suggested that the influx is limited by the maximum amount of amino acids that can be accommodated in the proteasome (see [27] and refs therein). In our model the influx rate therefore decreases when the total amount of amino acids inside, $\sum_{k=1}^{L} kn_k$, increases. The maximum filling of the proteasome is normalised to one by a scaling parameter v determining the maximum number of amino acids that can be accommodated within the CP (Table 14.2).

We assume that the influx does not strongly depend on the amino acid composition of the substrate. Based on the intuition that each peptide binds with a probability p to the gate subunits, hence impairing the passage through the narrow pore, it is proposed that the efflux rate is a negative exponent of the length $exp(-\gamma n)$ where $\gamma = \frac{1}{1-p}$ and n is the fragment's length [16]. In contrast, the analysis of *in vitro* digestion of 25- and 27-aa substrates performed with the 20S proteasome suggests a length-dependent reprocessing rate which decreases with increasing substrate length. These data suggest an increasing Hill function with a high exponent to describe the reduced cleavage rate for short substrates [36]. Hence, we describe the

Table 14.2 Parameters values of the kinetic model.

Parameter	Description	Dimension	Default value
L	Length of the substrate	amino acids	100
$N_L(0)$	Initial substrate concentration	mol	100
$\hat{a}$	Rate of influx	time^{-1}	0.1
$\hat{e}$	Rate of efflux	time^{-1}	1
c	Cleavage rate	time^{-1}	1
θ	Critical fragment length	amino acids	25
μ	Preferred cleavage position	amino acids	9
σ	Std of cleavage position	amino acids	3
v	Scaling factor	–	1/200

efflux rate with a phenomenological Hill function with high exponent and a critical length $\theta = 25$ aa $e(k) = \hat{e}/(1 + (k/\theta)^{10})$. The efflux rate switches at a fragment length of $k \simeq \theta$ from the maximal efflux rate $\hat{e} = 1$ for short fragments to an efflux close to zero for long fragments (see Fig. 14.9a).

The first two terms of Eq. (27) are the same influx and efflux terms as discussed above. The last terms describe the cleavage machinery located in the core of the proteasome. Fragments of length k are cut at a maximum rate c and with probability $0 < F_{k,i} < 1$ into two fragments of length i and $k - i$. Two terms account for the loss and for the gain of each fragment of length k. The negative term corresponds to a loss for fragments of length k which are cut into shorter fragments, and the positive term is a gain because fragments of length $j > k$ can be cleaved into a fragment of length k. For parameters see Table 14.2.

The main assumption for the cleavage mechanism is that the proteasome cleaves proteins starting around their N-termini or C-termini. It is suggested that there is a

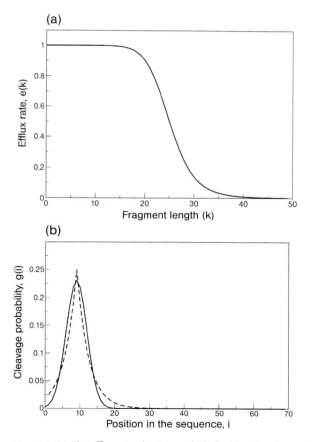

Fig. 14.9 (a) The efflux rate. (b) Binomial (dashed lines) and gaussian (solid lines) distributions for the cleavage probability.

preferred length of 7–9 aa for an optimal docking of the substrate with the binding grooves during the cleavage process inside the CP [13]. We therefore assume that the proteasome starts at a distance $m \simeq 9$ from one end of the protein/peptide, and scans the substrate chain in both directions until a cleavage site is found. One can model [27] the cleavage probability with a phenomenological binomial or a Gaussian distribution, as here.

The model has three rate parameters: the cleavage rate c, the maximum influx rate $\hat{a}$, and the maximum efflux rate $\hat{e}$. A normal time scale of proteasome experiments is minutes. However, experimental results on proteasome degradation are typically compared for a certain level of substrate degradation, rather than at a specific point in time. Since time is not an important issue, one can always rescale the time such that $c = 1$ per time unit. Increasing the cleavage rate will therefore be the same as decreasing the flux through the gates (i.e., as decreasing $\hat{a}$ and $\hat{e}$). For details of the simulation algorithm see [27].

14.6.2
Kinetics

Experimental data suggest that the *in vitro* degradation rate of substrates by the proteasome obeys Michaelis–Menten kinetics (see [27] and refs therein). For long substrates the maximum degradation rate and the Michaelis–Menten constant are known to decrease with the length of the substrate. Our model also exhibits Michaelis–Menten kinetics (see Fig. 14.10). For various initial substrate concentrations, 14.10 depicts the depletion of the substrate ($L = 100$) in the solution (Fig. 14.10a), and the corresponding filling of the proteasome (Fig. 14.10b). There is a rapid initial phase during which the proteasome fills up by influx of the substrate. At the very early stage of degradation, due to the filling of the proteasome, the substrate loss is not linear. When the initial substrate concentration is low this initial phase accounts for a significant depletion of the substrate concentration N_L (see Fig. 14.10a). Otherwise, the substrate concentration remains high and the filling of the proteasome approaches a quasi-steady state corresponding to a maximum degradation rate.

To study the Michaelis–Menten kinetics, we fix the substrate concentration by fixing $N(t) = N(0)$ and then let the model approach the corresponding steady state. At the steady state we measure the degradation rate as the number of substrate molecules in solution which are lost per unit time, and we depict that as a function of the substrate concentration and the length of the substrate, L (see Fig. 14.10c). This reveals a family of Michaelis–Menten curves for the various lengths of the substrate. The longer the substrate, the smaller the maximum degradation rate, V_{max}, and the smaller the Michaelis–Menten constant, K_m. The degradation rate at low substrate concentrations is fairly independent on the length of the substrate (see Fig. 14.10c).

Figure 14.10d shows the maximum degradation rate V_{max}, calculated numerically from the full model in Eqs. (26), (27), as a non-linear function of the substrate length. It increases for small substrates with a maximum at ca. 10 aa and de-

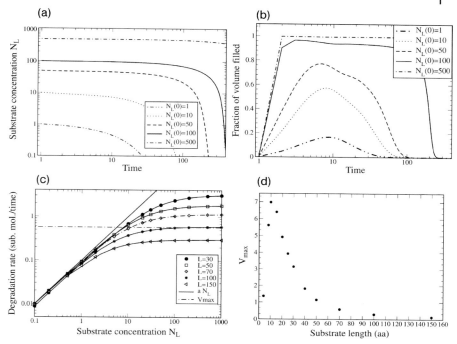

Fig. 14.10 (a) Substrate consumption for a different initial concentration, $N_L(0)$. (b) Filling of the proteasome core particle (in amino acids). (c) Michaelis–Menten log-log plot; each symbol-curve denotes a substrate of a different length obtained from the model presented in Eqs. (26), (27) in conditions where the proteasome is in a bath of substrate in order to prevent substrate limiting effects. The solid line is given by the initial slope $\hat{a}N_L$ and the dashed-dot line is the V_{max}, both predicted with the simplified model (see [27]) for the standard parameter values (see Table 14.2) and an average efflux $\bar{e} = 0.2$. (d) V_{max} as a function of the substrate length. It increases for substrate shorter than 10 aa and decreases for longer substrates. The initial substrate concentration is $N_L(0) = 6000$.

creases with the inverse of the length for longer substrates. Very small fragments are weakly degraded because of the low cleavage rate for fragments shorter than 9 aa. This explains why the degradation rate decreases for very short fragments. For longer fragments, V_{max} was found to decrease with the inverse of the length. This is due to the fact that the CP volume is finite and thus the influx decreases with the increase of the substrate length. Therefore, the degradation rate decreases accordingly. This compares well with experimental results (see [27] and refs therein). Interestingly, the ratio of V_{max}/K_m increases with the increase of the substrate length [8] and saturates for long substrates reaching 50% of its maximum for substrate of 23 aa, which indicates that both constants decrease with the length of the substrate in the same ratio.

14.6.3
Length Distribution of the Fragments

In vitro experiments generate cleavage products that range from 2 aa to 35 aa with an average length of 7–8 aa (see [27] and refs therein). Using size exclusion chromatography and on-line fluorescence detection, Köhlet et al. [23] showed that the products generated by the wild-type (WT) proteasome have a length distribution with three broad peaks corresponding to lengths of 2–3, 8–10, and 20–30 aa, respectively.

Figure 14.11 shows how the fragment length distribution depends on the size of the gate, i.e., on the influx and efflux rates $\hat{a}$ and $\hat{e}$, as calculated with the model in Eqs. (26), (27). For an intermediate efflux rate, we obtain three-peaked distributions similar to those observed in experiments [23] for a wide range of influx rates. Note that the first peak has its maximum at 1 aa, but we call this decreasing slope "peak" for simplicity. In our model, the three-peaked distributions are the result of the cleavage machinery, which tends to cut fragments of 8–10 aa, and the efflux of products, which favors short fragments.

In the Fig. 14.11a the efflux is slow compared with the cleavage ($c/\hat{e} = 10$). As a consequence, most substrate molecules are fragmented extensively before they are exported, and one observes short fragments in the solution. Increasing the efflux rate 10-fold (Fig. 14.11b) gives a similar time scale to the efflux and to the cleavage, and allows for a three-peak distribution. Another 10-fold increase of the efflux rate (see Fig. 14.11c) makes cleavage the limiting factor. The ratio of long to short fragments increases. Because the residence time of fragments in the CP is short, there is less fragmentation, and the first peak at 1–3 aa decreases.

When the efflux rate and the cleavage rate have a similar time scale we observe three peaks in the distribution of fragments (see Fig. 14.11b). Similar to what is observed experimentally [23], the third peak is much smaller than the other two, and the second peak is larger than the first peak. In our model, the first peak corresponding to the small fragments reflects an efficient cleavage mechanism where fragments are repeatedly cleaved before they are released from the CP. These "rest"

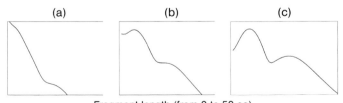

Fragment length (from 0 to 50 aa)

Fig. 14.11 Length distributions of the fragments outside the proteasome. From left to right the efflux rate $\hat{e}$ increases $\hat{e} = (0.1, 1, 10)$. Each distribution has been taken at the time 170 (a), 86 (b), 76 (c) when 20 % of substrate degraded. The vertical on each plot is the log frequency (from 0 to 0.25). The influx rate $\hat{a} = 0.01$. Note that the distributions are insensitive to the variation in the influx rate $\hat{a}$ [27].

products do not collapse to single amino acids because the cleavage of very short fragments is improbable in our model (see Fig. 14.9b). The second peak corresponding to fragments with a length of 8–10 aa, is the result of the preference to cut at $\mu = 9$ aa. Fragments are found in a broad peak around 9 aa, because of the variation in the cleavage (i.e., standard deviation of the Gaussian function). The third peak around 25 aa found in the WT distribution is due to the efflux function. It results from the high probability of a fraction of intermediate 25–35 aa fragments to exit the proteasome. As they would have been a source for fragments of length 15–25 aa, the production of fragments of this length drops. This intuitive explanation elucidates the presence of the third peak.

14.7
Discussion

14.7.1
Development of Modeling

This work provides insights on the mathematical modeling of the kinetic and translocation properties of proteasome degradation. A realistic choice for the models parameters is hard to obtain. Both model are phenomenological and designed to qualitatively capture the main features of kinetics and transport. It is important to note that there is no contradiction between these two models, the kinetic model operates on the mesoscopic scale whereas the transport model operates on the microscopic one. Hence, the transport model can be included inside the parameters of the kinetic model. At the present stage the models discussed cannot be used for quantitative predictions and serves as an explanation of the proteasome product size distribution. We identify two possible directions which can be taken in order to develop these models to be able to perform quantitative predictions. A first possibility is to implement a specific protein sequence, which certainly influences a cleavage pattern and the proteasome–protein interaction potential, and description of in- and outfluxes, taking into account the possibility of gate opening and closing [23].

It is important to note that both the transport and kinetic models of the proteasome allows the inclusion of the sequence specific cleavage strengths. In the transport model the cleavage specificity can be included if we make the cleavage probability dependent on the position in the peptide chain. Let us simulate the degradation of casein as in [23] with constant cleavage rates and with sequence specific cleavage rates computed with Netchop algorithms. Casein has a length of $L_p = 188$ aa and a cleavage pattern as in Fig. 14.12, right. All other parameters are the same as in the case of the three-peak length distribution, see Fig. 14.12, left. We rescaled the cleavage strength to have the same mean value of 0.001. To our surprise the length distributions of the not sequence-specific case and sequence specific case are practically identical (see Fig. 14.12, left). Hence to model the length distribution as a result of different translocation rates, it is not so important to

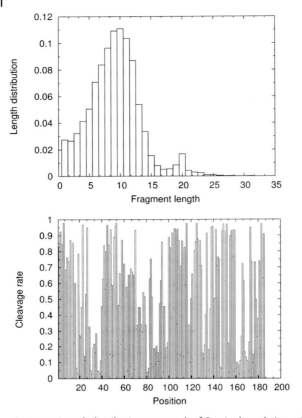

Fig. 14.12 Length distribution as a result of Casein degradation, with non sequence specific cleavage rates (left). Casein cleavage strength pattern computed with Netchop (right).

consider the cleavage pattern of the substrate. However, for shorter substrates the cleavage pattern can significantly change the length distribution.

Cleavage specificity can be also included in the kinetic model. For this, instead of the cleavage matrix which depends only on the distance from the protein end, one should include the cleavage coefficients which depend on the protein sequence. This would be possible if one consider, separately the concentration equations for all fragments generated from the initial substrate. Then taking the cleavage strength and the substrate concentration as an input for the mathematical model, one can predict the dynamics of the fragment concentration. The results which illustrate this approach will be published by us elsewhere.

Finally for effects where both the translocation and kinetic properties are important one can merge both models. This can be done by including the transport corrections in the kinetic model described here. In this case different transport rate functions result in different influx coefficients and, hence, different kinetic properties of the proteasome. This modeling approach is especially important for the study of the proteasome role in the neurodegenerative diseases, as we discuss in the next subsection.

14.7.2
Kinetics Models and Neurodegenerative Associated Proteasome Degradation

A broad array of human neurodegenerative diseases share strikingly similar histopathological features, e.g., the presence of insoluble protein deposits, such as the neurofibrillary tangles and neuritic plaques of Alzheimer's disease, the Lewy bodies of Parkinson's disease, and the intranuclear inclusions of Huntington's disease [42].

As the 26S proteasome is the cellular proteolytic machinery involved in the clearance of uniquitinated proteins, this led to the suggestion that a chronic imbalance between their generation and processing may be the primary cause for the formation of protein deposits. Recent experimental data revealed an impaired 26S proteasome activity in neurodegnerative diseases [39, 48]. Using our kinetics model, several possible scenarios to explain proteasome dysfunction in neurodegenerative diseases can be envisaged. For instance the transport of elongated molecules, such as the substrates found in neurodegenerate brains, may require a longer time. Therefore, the delayed transport can be responsible for keeping the proteasome sequestrated for a long time and lowering the degradation rate.

Kinetics models could also predict and provide quantitative results of proteasome inhibition through a reduced catalytic activity of the 20S proteasome chamber. A reduced cleavage rate c can result in a filling up of the proteasome chambers, therefore impairing the degradation process. Moreover, a non-trivial scenario arises in the kinetics model presented here when a regime of reduced cleavage activity is considered. In this condition the degradation rate decreases with the increase in the influx rate of substrate within the proteasome (data not shown). This unexpected result is a direct consequence of the filling of the proteasome and the internal dynamics of undegraded substrates and intermediate long fragments. In fact, when both the influx and the cleavage rate are very low, the proteasome is almost empty and the substrates entering the chamber are immediately degraded and fragments are ejected. Increasing the influx rate the proteasome fills up much faster. Consequently, because of the low cleavage rate, many long intermediate fragments are generated. The proteasome fills up at a high rate, reducing the number of substrate molecules entering the proteasome chamber per unit time. A quasi-steady-state regime is reached, where proteasomes are filled up with long intermediate fragments. This unexpected result might be experimentally explored: experiments with different substrate concentrations in a low cleavage regime could prove this theoretical expectation.

Modeling the ATP-dependent activity of the 19S base and the mechanical transport of substrate molecules within the 20S core are suggestive and intriguing phenomena that can help the understanding of proteasome blockage and the impaired activity observed in neurodegenerative associated protein degradation. Unfortunately to date, no such model has been proposed, the main reason being the very scant information available about these mechanisms. One model candidate for analyzing this problem is the ratchet-kinetics model, intented to describe the function of the transport through the 19S and the kinetics of cleavage within the 20S core particle. attach to the towards the proteasome core particle. This complete model can unravel the dynamics of transport of a given substrate through the 19S subunits and the successive kinetics within the core particle, where substrate-specific cleavage machinery is implemented.

Therapeutic application of these future results can be envisaged. For instance, the deficiency of ATP as major cause of deficient transport for elongated and/or modified molecules can be tested quantitatively. Providing an ATP source in this impaired system, such as Alzheimer's disease or Parkinson's disease, to the proteasome system can result in a reduction of substrate accumulation. Predictions with mathematical models can be tested by manipulating the activity of the proteasome *in vivo*. This is a challenging task but will undoubtedly provide more insight into the pathogenesis of neurodegenerative diseases and new therapeutic options.

Acknowledgments

A.Z. acknowledges financial support from the VW-Stiftung. A.Z. and J.K. acknowledges financial support from the European Union through the Network of Excellence BioSim, Contract No. LSHB-CT-2004-005137. F.L. acknowledges financial support from the Australian Research Council through the Discovery Grant scheme (DP0556732, DP0664970).

References

1 J. Adams, V. J. Palombella, and P. J. Elliott. Proteasome inhibition: a new strategy in cancer treatment. *Investigational New Drugs*, 18:109–121, 2000.

2 T. N. Akopian, A. F. Kisselev, and A. L. Goldberg. Processive degradation of proteins and other catalytic properties of the proteasome from *thermoplasma acidophilum*. *The Journal of Biological Chemistry*, 272:1791–1798, 1997.

3 M. Bier. Processive motor protein as an overdamped brownian stepper. *Phys. Rev. Lett.*, 91:148104, 2003.

4 C. J. Brokaw. protein-protein ratchets: stochastic simulation and application to processive enzymes. *Biophys. J.*, 81:1333–1344, 2001.

5 P. Cascio, C. Hilton, A. F. Kisselev, K. L. Rock, and A. L. Goldberg. 26S proteasomes and immunoproteasomes produce mainly n-extended versions of an anti-

genic peptide. *EMBO J.*, 20:2357–2366, 2001.

6 A. Ciechanover. The ubiquitin-proteasome proteolytic pathway. *Cell*, 79:13–21, 1994.

7 O. Coux, K. Tanaka, and A. L. Goldberg. Structure and functions of the 20S and 26S proteasomes. *Annu. Rev. Biochem.*, 65:801–847, 1996.

8 I. Dolenc, E. Seemuller, and W. Baumeister. Decelerated degradation of short peptides by the 20S proteasome. *FEBS. Lett.*, 434:357–361, 1998.

9 Q. P. Dou, D.M. Smith, K.G. Daniel, and A. Kazi. Interruption of tumor cell cycle progression through proteasome inhibition: implications for cancer therapy. *Progress in Cell Cycle Research*, 5:441–446, 2003.

10 D. T. Gillespie. A general method for numerically simulating the stochastic time evolution of coupled chemical reactions. *J. Comput. Phys.*, 22:403–434, 1976.

11 M. Glickman and A. Ciechanover. The ubiquitin-proteasome proteolytic pathway: destruction for the sake of construction. *Physiol. Rev.*, 82:373–428, 2002.

12 A. L. Goldberg, P. Cascio, T. Saric, and K. L. Rock. The importance of the proteasome and subsequent proteolytic steps in the generation of antigenic peptides. *Mol. Immunology*, 39:147–164, 2002.

13 M. Groll and R. Huber. Substrate access and processing by the 20S proteasome core particle. *Int. J. Biochem. Cell. Biol.*, 35:606–616, 2003.

14 W. Hilt and D. H. Wolf. Proteasomes of the yeast s. cerevisiae: genes, structure and functions. *Mol. Biol. Rep.*, 21:3–10, 1995.

15 H. G. Holzhuetter, C. Frommel, and P.M. Kloetzel. A theoretical approach towards the identification of cleavage-determining amino acid motifs of the 20s proteasome. *J. Mol. Biol.*, 286:1251–1265, 1999.

16 H. G. Holzhütter and P. M. Kloetzel. A kinetic model of vertebrate 20S proteasome accounting for the generation of major proteolytic fragments from oligomeric peptide substrates. *Biophysical Journal*, 79:1196–1205, 2000.

17 P. Jung and P. Hänggi. Resonantly driven brownian motion: Basic concepts and exact result. *Phys. Rev. A*, 41:2977, 1990.

18 C. Kesmir, A.K. Nussbaum, H. Schild, V. Detours, and S. Brunak. Prediction of proteasome cleavage motifs by neural networks. *Protein Engineering*, 15:287–296, 2002.

19 A. F. Kisselev, T. N. Akopian, and A. L. Goldberg. Range of sizes of peptide products generated during degradation of different proteins by archaeal proteasomes. *J. Biol. Chem.*, 273:1982–1989, 1998.

20 P. M. Kloetzel. Antigen processing by the proteasome. *Nat. Rev. Mol. Cell. Biol.*, 2:179–187, 2001.

21 P. M. Kloetzel. Generation of major histocompatibility complex class I antigens: functional interplay between proteasomes and TPPII. *Nature Immunology*, 5:661–669, 2004.

22 P. M. Kloetzel. The proteasome and MHC class I antigen processing. *Biochem. Biophys. Acta.*, 1695:225–233, 2004.

23 A. Köhler, P. Cascio, D. S. Leggett, K. M. Woo, A. L. Goldberg, and D. Finley. The axial channel of the proteasome core particle is gated by the rpt2 ATPase and controls both substrate entry and product release. *Molecular Cell*, (7):1143–1152, 2001.

24 C. Kuttler, A. K. Nussbaum, T. P. Dick, H. G. Rammensee, H. Schild, and K. P. Hadeler. An algorithm for the prediction of proteasomal cleavages. *J. Mol. Biol.*, 298:417–429, 2000.

25 P. S. Landa, A.A. Zaikin, and L. Schimansky-Geier. Effect of the potential shape and of a brownian particle mass on noise-induced transport. *Chaos, Solitons & Fractals*, 12:1459–1471, 2001.

26 B. Lankat-Buttgereit and R. Tampe. The transporter associated with antigen processing: function and implications in human diseases. *Physiol Rev*, 82:187–204, 2002.

27 F. Luciani, C. Kesmir, M. Mishto, M. Or-Guil, and R. J. de Boer. A mathematical model of protein degradation by the proteasome. *Biophysical Journal*, 88:2422–2432, 2005.

28 F. Luciani and A. Zaikin. Mathematical models of the proteasome product size distribution. 2006. (submitted).

29 O. Lund, S. Brunak, and et. al. Creating a virtual immune system. *Journal of Biological Physics*, 2006. (in press).

30 M. Nielsen, C. Lundegaard, O. Lund, and C. Kesmir. The role of the proteasome in generating cytotoxic t-cell epitopes: insights obtained from improved predictions of proteasomal cleavage. *Immunogenetics*, 57:33–41, 2005.

31 A. Nussbaum. *From the test tube to the World Wide Web: The cleavage specificity of the proteasome.* PhD thesis, Eberhard-Karls-Universitaet Tuebingen, 2001.

32 A. K. Nussbaum, T. P. Dick, W. Kielholz, M. Schirle, S. Stevanovic, K. Dietz, W. Heinemeyer, M. Groll, D. H. Wolf, R. Huber, H. G. Rammensee, and H. Schild. Cleavage motifs of the yeast 20S proteasome β subunits deduced form digests of enolase 1. *Proc. Natl. Acad. Sci. USA*, 95:12504–12509, 1998.

33 A.K. Nussbaum, C. Kuttler, K.P. Hadeler, H.G. Rammensee, and H. Schild. PA-ProC: a prediction algorithm for protea-

somal cleavages available on the WWW. *Immunogenetics*, 53:87–94, 2001.

34 R. Z. Orlowski. The role of the ubiquitin-proteasome pathway in apoptosis. *Cell Death Differ.*, 6:303–313, 1999.

35 B. Peters, K. Janek, U. Kuckelkorn, and H. G. Holzhütter. Assessment of proteasomal cleavage probabilities from kinetic analysis of time-dependent product formation. *J. Mol. Biol.*, 318:847–862, 2002.

36 B. Peters, K. Janek, U. Kuckelkorn, and H. G. Holzhutter. Assessment of proteasomal cleavage probabilities from kinetic analysis of time-dependent product formation. *J. Mol. Biol.*, 318:847–862, 2002.

37 J. M. Peters, Z. Cejka, J. R. Harris, J. A. Kleinschmidt, and W. Baumeister. Structural features of the 26S proteasome complex. *J. Mol. Biol.*, 234:932–937, 1993.

38 M. Rechsteiner, C. Realini, and V. Ustrell. The proteasome activator 11S REG (PA28) and class I antigen presentation. *Biochem. J.*, 345:1–15, 2000.

39 C.A. Ross and M.A. Poirier. What is the role of protein aggregation in neurodegeneration? *Nat. Rev. Mol. Cell. Biol.*, 6:891–898, 2005.

40 K. Sakamoto. Ubiquitin-dependent proteolysis: its role in human diseases and the design of therapeutic strategies. *Mol. Genet. Metab.*, 77:44–56, 2002.

41 G. Schmidtke, S. Emch, M. Groettrup, and H. G. Holzhuetter. Evidence for the existence of a non-catalytic modifier site of peptide hydrolysis by the 20S proteasome. *J. Biol. Chem.*, 275:22056–22063, 2000.

42 A. L. Schwartz and A. Ciechanover. The ubiquitin-proteasome pathway and pathogenesis of human diseases. *Annu. Rev. Med.*, 50:57–74, 1999.

43 N. Shastri and S. Schwab. Producing nature's gene-chips: the generation of peptides for display by MHC class I molecules. *Annu. Rev. Immunol.*, 20:463–493, 2002.

44 R. L. Stein, F. Melandri, and L. Dick. Kinetic characterization of the chymotrypic activity of the 20S proteasome. *Biochemistry*, 35:3899–3908, 1996.

45 R. Stohwasser, U. Salzmann, J. Giesebrecht, P. M. Kloetzel, and H. G. Holzhuetter. Kinetic evidence for facilitation of peptide channelling by the proteasome activator pa28. *Eur. J. Biochem.*, 267:6221–6230, 2000.

46 T. Tamura, I. Nagy, A. Lupas, F. Lottspeich, Z. Cejka, G. Schoofs, K. Tanaka, R. De Mot, and W. Baumeister. The first characterization of a eubacterial proteasome: the 20S complex of rhodococcus. *Curr. Biol.*, 5:766–774, 1995.

47 S. Tenzer, B. Peters, S. Bulik, O. Schoor, C. Lemmel, M. M. Schatz, P. M. Kloetzel, H. G. Rammensee, H. Schild, and H. G. Holzhuetter. Modeling the MHC class I pathway by combining predictions of proteasomal clavage, TAP transport and MHC class I binding. *CMLS, Cell. Mol. Life Sci.*, 62:1025–1037, 2005.

48 L. G. Verhoef, K. Lindsten, M.G. Masucci, and N.P. Dantuma. Aggregate formation inhibits proteasomal degradation of polyglutamine proteins. *Hum. Mol. Genet.*, 11:2689–2700, 2002.

49 T. Wenzel, C. Eckerskorn, F. Lottspeich, and W. Baumeister. Existence of a molecular ruler in proteasomes suggested by analysis of degradation products. *FEBS Letters*, 349:205–209, 1994.

50 A. Zaikin and J. Kurths. Optimal length transportation hypothesis to model proteasome product size distribution. *Journal of Biological Physics*, 2006. (in press, DOI: 10.1007/s10867-006-9014-z).

51 A. Zaikin, A. Mitra, D. Goldobin, and J. Kurths. Influence of transport rates on the protein degradation by the proteasome. *Biophysical Reviews and Letters*, 1:375–386, 2006.

52 A. Zaikin and T. Pöschel. Peptide-size-dependent active transport in the proteasome. *Europhysics Letters*, 69:725–731, 2005.

53 B. Y. Zeng, A. D. Medhurst, M. Jackson, S. Rose, and P. Jenner. Proteasomal activity in brain differs between species and brain regions and changes with age. *Mech Ageing Dev*, 126:760–766, 2005.

Part IV
Applications of Biosimulation

15

Silicon Cell Models: Construction, Analysis, and Reduction

Frank J. Bruggeman, Hanna M. Härdin, Jan H. van Schuppen,
and Hans V. Westerhoff

Abstract

Biosimulation has a dominant role to play in systems biology. In this chapter, we briefly outline two approaches to systems biology and the role that mathematical models has to play in them. Our focus is on kinetic models, and silicon cell models in particular. Silicon cell models are kinetic models that are firmly based on experiment. They allow for a test of our knowledge and identify gaps and the discovery of unanticipated behavior of molecular mechanisms. These models are very complicated to analyze because of the high level of molecular–mechanistic detail included in them. To facilitate their analysis and understanding of their behavior, model reduction is an important tool for the analysis of silicon cell models. We present balanced truncation as one method to perform model reduction and apply it to a silicon cell model of glycolysis in *Saccharomyces cerevisiae*.

15.1
Introduction

The molecular biosciences, including molecular biology, genetics, biophysics, and biochemistry, have discovered and characterized many of the molecules and processes that constitute living organisms. The many genome sequences and their perhaps unexpected similarities have greatly accelerated the functional annotation of genes and the prediction of generic intracellular networks [1]. Entirely new phenomena continue to be found [2], and new approaches to measuring cellular phenomena continue to be developed leading to even more discoveries [3]. Much needs to be discovered, but perhaps not just much more of the same. With new gene sequences being "discovered" every minute, too much appears to be discovered, thereby confounding understanding with data [4]. Then what are the most important aspects of living organisms that need to be discovered? And which are the ones that may be discovered with the methods we now have at hand?

Biosimulation in Drug Development. Edited by Martin Bertau, Erik Mosekilde, and Hans V. Westerhoff
Copyright © 2008 WILEY-VCH Verlag GmbH & Co. KGaA, Weinheim
ISBN: 978-3-527-31699-1

The more recently developed "Omics" approaches aim at obtaining a complete characterization of the state of a cell under a given condition. They lead to a shift from molecule-centered bioscience to a more network-oriented approach. The emerging large datasets have been subjected to sophisticated multivariate data analyses (chemometrics). Correlations in the networking and behavior of macromolecules have been observed. Because Omics is about to look at all that is present in living organisms, the urge is to understand everything that is not understood, but it is this understanding that is slow in coming. The paradox is that there may be more in the collection of components of the living organism than what really matters for its existence and function. Yet, what really matters is *not* present in the components as they appear in Omics studies, i.e. in the parts list. Much of biological function is not in the parts list of *Life* but emerges from the nonlinear interactions of the parts and is not listed therefore. This emergence is not just the self-organization proposed in the 1950s and 1960s [5, 6] but self-organization coached by prespecification and perpetration such as in the propagation of biological structure and activity at cell division.

In order to get to grips with the emergence, maintenance, and perpetration of biological function in systems of hundreds of simultaneous processes, one needs to integrate biological knowledge about the interactive properties of components into mathematical models and then compute the emergence of system function [7]. This change of focus in biological research has nurtured a young discipline in the sciences, defined as *Systems Biology*.

Two approaches can be distinguished in systems biology: top-down and bottom-up [7, 8]. They relate to the different histories of the biological sciences [9]. They have in common that they aim to understand how the properties of systems (or networks) as a whole derive from the properties and the interactions between their molecular constituents. Such networks may include signaling, genetic, and metabolic networks. The approaches differ in their perspective, in the size of the networks they consider, and in their precision. And, the two methods are used to different purposes.

Top-down systems biology emerged from the Omics revolution. Omics generate large datasets, either deriving from fluxomics, metabolomics, proteomics, or transcriptomics, upon perturbation of a living cell. These perturbations may involve additions of toxins or nutrients, changes in temperature, additions of signaling ligands, and mutations. Using a variety of tools, such datasets are analyzed with the aims of extracting information about the role particular molecules play within the cell and of further characterizing its networks, e.g. by predicting new interactions. Then experiments test the predicted roles and interaction strengths. Most of the experimental techniques are not yet sufficiently quantitative to test the predictions rigorously, but this aspect is now beginning to attract attention. Whereas the early experiments concentrated more on snapshot data – at a single time point – nowadays experiments are carried out with multiple samples over time, which is more costly but also much more insightful. Top-down systems biology is an approach that may be well suited to hunt for unknown interactions and molecular intermediates, i.e. to finish the characterization of the molecular network inside cells that was initiated by the molecular biosciences.

Bottom-up systems biology does not rely that heavily on Omics. It predates top-down systems biology; and it developed out of the endeavors associated with the construction of the first mathematical models of metabolism in the 1960s [10, 11], the development of enzyme kinetics [12–15], metabolic control analysis [16, 17], biochemical systems theory [18], nonequilibrium thermodynamics [6, 19, 20], and the pioneering work on emergent aspects of networks by researchers such as Jacob, Monod, and Koshland [21–23].

Some bottom-up systems biology involves the construction of a detailed kinetic model of the 10–30 processes constituting a well characterized subnetwork in a cell. Here processes are considered to include diffusion, facilitated or active transport, and enzyme-catalyzed reactions. Ideally an experiment-driven modeling study should follow where the model is improved by testing its predictions and by adding newly discovered interactions or more accurately determined parameter values. Eventually, model behavior should resemble experimentally determined molecular behavior closely enough. Then the model has turned into a "silicon cell" [72]. A "silicon cell" or "silicon pathway" is a detailed kinetic model of cellular processes constructed on the basis of the experimental data on kinetic characteristics and parameter values determined by *in vitro* or *in vivo* experiments with individual reactions, i.e. the kinetic parameters are not fitted to data obtained by experiments on the predicted integral behavior of the silicon pathway. One important characteristic of silicon cell models is that they are detailed. All their parameters should resemble parameters of molecular processes that can be determined in the laboratory with molecular methods (enzyme kinetics or binding studies). This means that silicon cell models are successful in prediction of system behavior that emerges out of the experimentally known molecular interactions. They do lead to "understanding" in the sense of predictions of behavior that withstand experimental tests, but they do not immediately yield human *understanding* of systemic phenomena in terms of molecular processes and properties. Perfect silicon cells are no closer to such *understanding* than the real cell they represent, except that the silicon cells enable *in silico* experimentation that should greatly facilitate achieving such true understanding.

With "true" *understanding* we here refer to the aspect of identifying the molecular interactions that are most important for the emergence of the functional behavior. True understanding may refer to the ability to explain complex behavior in such simple terms that actual modeling is not necessary to follow the essence of the explanation. For some cases such a type of understanding is achievable. For others it might not be and the silicon cell might be the only level at which understanding is complete, or a robot scientist able to "understand" networks of higher complexity than the human mind is able to understand [24].

Model reduction aims at simplifying without losing the essence of the dynamic behavior of a model. Reduction of silicon cells should thereby facilitate the understanding of real cells. Strategies for model reduction, pinpointing molecular organizational properties that are essential for network behavior, are essential to make silicon cell models *understandable*.

This chapter addresses how silicon cell models can be used in biosimulation for systems biology. We first describe the process of model building, as well as its purpose and how it fits in systems biology. Then we compare the use of silicon cell models with the use of the less-detailed "core" models. We briefly discuss various simulation methods used to model phenomena involving diffusion and/or stochasticity as well as methods for model analysis. Finally we discuss balanced truncation as a method for model reduction. This method is illustrated by applying it to a silicon cell model of yeast glycolysis.

15.2
Kinetic Models in Cell Biology: Purpose and Practice

The construction of a kinetic model describing a cellular phenomenon manifested by some intracellular network composed of interacting molecules involves the integration of kinetic data bearing on the molecular constituents. This integration continues until each process (interaction) – each binding and catalytic event – is described by a rate or equilibrium equation. A rate equation describes the rate of a process as a function of the concentration of the reactants and effectors, if present, and in terms of kinetic parameters characterizing the time scales of the processes as well as binding constants (or their look-alikes) measuring the concentrations at which reactants and effectors influence enzyme activity [15]. In biochemical networks virtually every process in the network is catalyzed by an enzyme. Exceptions are nonfacilitated transmembrane transport (which is relevant for a very low number of normal molecules and for some xenobiotics) and diffusion within cellular compartments. When all processes have been characterized by a rate equation, then for each type of molecule that is consumed and/or produced in the system, a mass balance can be defined on the basis of the network structure. A mass balance describes the rate of change of the concentration of a molecule within the network as the difference between the rates, denoted by v_i, of its production and consumption reactions. Often rates are nonlinear functions of the concentration of the molecules, i.e. ratios or powers appear in the rate equation such as

$$v_1 = V_{\max} S/(S + K_m)$$

or

$$v_2 = kA^2$$

The integration of these mass balances, often with the help of the computer because the nonlinearity of the equation does not allow for an analytical solution, allows for a calculation of the changes in the concentrations of the molecules over time, given all concentration at time zero, values of all the kinetic parameters, and a description of the interaction of the network (the "system") with its environment.

The purpose of the construction of a kinetic model can be manifold:

- a test of the "completeness" of molecular knowledge – whether the molecular knowledge indeed leads to the measured system behavior [25–30];
- optimization of the functioning of a network for applied purposes in bioengineering or medicine [28, 31–33];
- to find principles that describe the system – "unification" or classification of molecular mechanisms [34, 35];
- to test the function of a design property or physicochemical aspect of the network [36–39];
- to help analyze experimental data.

Many of these purposes apply not only to the construction of kinetic models but also to the construction of models describing solely the interaction structure or stoichiometry of networks [40–46].

Kinetic models exhibit particularities that lead to rather invariant properties not found for dynamical systems in general. Before we discuss these, we remind the reader of the general description of the dynamics of a kinetic model (see also [47, 48]). The mass balances describing the rate of change in the concentrations of the variable molecular species in the network are linear combinations of the rates of the processes in the network, assuming the network can be modeled as a well stirred environment in the absence of noise. In matrix format this leads to:

$$\frac{d}{dt}\mathbf{x}(t,\mathbf{p}) = \mathbf{N} \cdot \mathbf{v}(\mathbf{x},\mathbf{p}) \tag{1}$$

Vectors, such as $\mathbf{x}$, are denoted by bold lower case font. Matrices, such as $\mathbf{N}$, are denoted by bold upper case fonts. The vector $\mathbf{x}$ contains the concentration of all the variable species; it represents the state vector of the network. Time is denoted by t. All the parameters are compounded in vector $\mathbf{p}$; it consists of kinetic parameters and the concentrations of constant molecular species which are considered buffered by processes in the environment. The matrix $\mathbf{N}$ is the stoichiometric matrix, which contains the stoichiometric coefficients of all the molecular species for the reactions that are produced and consumed. The rate vector $\mathbf{v}$ contains all the rate equations of the processes in the network. The kinetic model is considered to be in steady state if all mass balances equal zero. A process is in thermodynamic equilibrium if its rate equals zero. Therefore if all rates in the network equal zero then the entire network is in thermodynamic equilibrium. Then the state is no longer dependent on kinetic parameters but solely on equilibrium constants. Equilibrium constants are thermodynamic quantities determined by the standard Gibbs free energies of the reactants in the network and do not depend on the kinetic parameters of the catalysts, enzymes, in the network [49].

Equation (1) is a special case of a dynamical system, which is described by: $d\mathbf{s}/dt = \mathbf{f}(\mathbf{s}, \mathbf{p})$.

Equation (1) gives rise to properties of kinetic systems through the multiplication of $\mathbf{N}$ and $\mathbf{v}$. The (right) nullspace of $\mathbf{N}$ and its transpose play a dominant role in determining the independent fluxes in steady state and the independent intermediates respectively [50]. Combined with convex analysis, a sophisticated form of linear programming, this leads to the decomposition of the steady-state flux vector into a positive linear combinations of flux modes, either elementary flux modes or extreme pathways [51–53]. The fact that the rates of change are linear combination of reactions allows for the application of linear programming techniques to determine optimal flux distribution in steady state [54–61], provided that the performance to be optimized is a linear function of the flux modes or metabolite concentrations. In addition, this structure leads naturally to the emergence of summation laws for control coefficients in biochemical reaction networks as described by metabolic control analysis, both for steady and transient states [17, 34, 62]. In addition, it decomposes regulatory effects into stoichiometric and effector contribution. All regulatory effects are described by the entries in the jacobian matrix:

$$\mathbf{N} \partial \mathbf{v} / \partial \mathbf{x}$$

that correspond to the nonzero entries in $\mathbf{N}$ whereas the remaining interactions are then effector interactions.

15.3
Silicon Cell Models

Depending on the purpose of the model and the status of the molecular knowledge of the network, either simplified kinetic core models or kinetic detailed models can be constructed. Core models are most useful for showing principles of regulation or dynamics [5, 37, 41, 63–66] or to study a network in more phenomenological terms when it is poorly characterized [67–70]. In what follows we only consider detailed kinetic models.

A detailed kinetic model is meant to describe the structure of the network as well as the relevant kinetic properties of all its components as close to reality as possible, given current experimental data. This only leads to a complete kinetic model if the kinetic parameters of all the processes in the network are given a value. Two extreme strategies for parameterization may be distinguished. The first method relies solely on parameter estimation methods. Parameter estimation involves the search of the value of a parameter, using a dedicated algorithm, that makes the output of the model resemble experimental data as closely as possible [71–73]. This method has problems: (1) very often multiple parameter sets can be found that lead to model–experiment resemblance, i.e. the model can give good results even though it is wrong, (2) not all parameters can be identified, only dimensionless groups can be, and (3) each time when a new process or subnetwork is added to the network, all kinetic parameters of the rate equations need to be fitted again (if the boundary conditions are not considered properly while fitting [74]). Despite these pitfalls parameter estimation remains a valuable method for model construction as, very often, a model is considered good or bad on the basis of its predictive value.

The second method relies on the experimental determination of the kinetic parameters using techniques from biophysics or enzymology. Also in this case problems exist: (1) the kinetic parameters are often determined under conditions different from the conditions in the cytoplasm; (2) an enormous number of experiments need to be done, even for a network of moderate size, to determine all kinetic parameters experimentally. When the second method is used to parameterize a kinetic model then the resulting model is considered a silicon cell model. A number of silicon cell models exist [25–27, 29, 75–77].

Very often a mixture of these two approaches is used to determine the values of the parameters. Good examples are Dano et al. [78] and Chassagnole et al. [79]. In these studies many parameters were taken from the literature and, in a parameter estimation approach, were allowed to vary within experimental error to fit the unknown parameters. When considering dynamics, the boundary conditions of the network have to be supplied as explicit functions of time, and therefore they have to be measured in order to give good values for parameter with parameter estimation [74, 79]. Many detailed and core models can be interrogated online at JWS online (www.jjj.bio.vu.nl) [80].

After the kinetic model for the network is defined, a simulation method needs to be chosen, given the systemic phenomenon of interest. The phenomenon might be spatial. Then it has to be decided whether in addition stochasticity plays a role or not. In the former case the kinetic model should be described with a reaction–diffusion master equation [81], whereas in the latter case partial differential equations should suffice. If the phenomenon does not involve a spatial organization, the dynamics can be simulated either using ordinary differential equations [47] or master equations [82–84]. In the latter case but not in the former, stochasticity is considered of importance. A first-order estimate of the magnitude of stochastic fluctuations can be obtained using the linear noise approximation, given only the ordinary differential equation description of the kinetic model [83–85, 87].

Many methods have been developed for model analysis: for instance, bifurcation and stability analysis [88, 89], parameter sensitivity analysis [90], metabolic control analysis [16, 17, 91] and biochemical systems analysis [18]. One highly important method for model analysis and especially for large models, such as many silicon cell models, is model reduction. Model reduction has a long history in the analysis of biochemical reaction networks and in the analysis of nonlinear dynamics (slow and fast manifolds) [92–104]. In all cases, the aim of model reduction is to derive a simplified model from a larger ancestral model that satisfies a number of criteria. In the following sections we describe a relatively new form of model reduction for biochemical reaction networks, such as metabolic, signaling, or genetic networks.

15.4
Model Reduction by Balanced Truncation

Most methods used thus far for model reduction of biochemical systems are based on some *a priori* knowledge about the system, such as knowledge about the dispersion of time scales. Because of the size and complexity of silicon cell models, this kind of *a priori* information is rarely available. This motivated us to consider alternative reduction methods, perhaps deriving from other areas of science.

In mathematical system theory, the subject of model reduction has been studied for about 30 years. The focus is on model reduction of linear systems, in particular methods based on singular value decomposition. One of the best known of these methods is balanced truncation. It is used extensively for various engineering purposes, such as electronic chip design and the reduction of models of aerospace structures. This method does not require the type of *a priori* information about the system mentioned above. Only recently has it been tried out on biochemical systems [105, 106].

The aim of the project reported here is to develop system reduction methods for large biochemical systems, including silicon cell models. Here we present our first approach using balanced truncation. The plan is to develop reduction methods custom-made for biochemical systems. To use balanced truncation is a natural first step towards the development of finer methods, since it is a basic method in system theory, and many other methods are variants of this.

We now introduce some concepts that will be needed for the later description of the method of balanced truncation. In Section 15.2 kinetic systems were introduced of the form:

$$dx(t)/dt = f(x(t))$$

The vector **x** contains the *n* time-dependent concentrations, which are called *state variables* from now on. A *state* is the value of **x** at a certain time *t*, **x**(*t*), and *n* is called the *order* of the system. *Inputs* **u** and *outputs* **y** feature as follows:

$$\frac{dx(t)}{dt} = f(x(t), u(t))$$

$$y(t) = g(x(t), u(t))$$

The input **u** is a vector that contains time-dependent variables that may be controlled by the engineer, or alternatively, some variables that are naturally to be regarded as input to the system. For example, to the citric acid cycle, [acetyl-CoA] may serve as input variable. The output **y** is a vector containing either variables that we can measure, or variables that are suitably regarded as output. In the case of glycolysis, [pyruvate] or [NADH] could be chosen, since these are Gibbs energy-rich compounds that are produced in glycolysis and used in subsequent pathways for the production of ATP. For the coming description of balanced truncation, the concept *input–output behavior* play a central roll. The term refers to the output's total dependence on the input, and can be considered as the function **y**(**u**). Note that in the system the output can depend directly on **u**:

$$y = g(x, u)$$

but also indirectly via **x**, since the dynamics of **x** depends on **u**:

$$d\mathbf{x}/d\mathbf{t} = \mathbf{f}(\mathbf{x}, \mathbf{u})$$

In order to use balanced truncation the system must be a linear system. Since most biological systems are highly nonlinear, they have to be linearized before balanced truncation can be implemented. *Linearization* is the procedure of constructing a linear system of which the state variables are approximations of the deviations of the state variables in the original system from some selected state, for example the steady state. The approximation will be good when the state variables are close to the steady state, but less good when far away from this state.

Balancing should ensure that the variables that will be truncated away are of small importance for the input–output behavior of the system. Balancing is nothing else than a state space transformation of the system, i.e. new state variables are constructed as linear combinations of the old state variables. The transformation matrix is determined according to an algorithm which yields that the new state variables are ordered in decreasing order of importance for the input–output behavior. The algorithm involves making two matrices equal, the controllability and observability gramians, hence the name balancing. The input–output behavior of the balanced system is the same as that of the linearized system and the order of the systems are the same. Thus, the balancing is neither an approximation nor a reduction of the linearized system, but only a preparation for truncation.

Truncation is the exclusion of some of the last state variables of the balanced system, which are the state variables of least importance for the input–output behavior. This results in a system of lower order than the original system, with an input–output behavior similar to that of the linearized, but not exactly the same. In the coming section we will describe the mathematical procedure of balanced truncation, but for more details we refer to Chapter 7 in the book by Athanasios [107].

15.5
Balanced Truncation in Practice

Here the procedures of linearization, balancing, and truncation are described such that the reader is enabled to use it. To start with, one has a nonlinear biochemical system, for example a silicon cell model, in the form of differential equations.

Step 1. *Determine a constant state* $\mathbf{x}_s$, *and a constant input* $\mathbf{u}_s$, *around which the system will be linearized*

For example a steady state can be chosen. As mentioned, a steady state is a state for which the mass balances are zero. For a system with input, the steady state $\mathbf{x}_s$ depends on the chosen steady input $\mathbf{u}_s$, and satisfies:

$$\mathbf{0} = \mathbf{f}(\mathbf{x}_s, \mathbf{u}_s)$$

Step 2. *Linearization*

Determine the matrices A, B, C, D as follows:

$$\mathbf{A} = \left.\frac{\partial \mathbf{f}}{\partial \mathbf{x}}\right|_{x_s,u_s}, \mathbf{B} = \left.\frac{\partial \mathbf{f}}{\partial \mathbf{u}}\right|_{x_s,u_s}, \mathbf{C} = \left.\frac{\partial \mathbf{g}}{\partial \mathbf{x}}\right|_{x_s,u_s}, \mathbf{D} = \left.\frac{\partial \mathbf{g}}{\partial \mathbf{u}}\right|_{x_s,u_s}$$

Considering that $\mathbf{f}$, $\mathbf{g}$, $\mathbf{x}$, and $\mathbf{u}$ are vectors, the differentiation leads to formation of matrices. The matrix $\mathbf{A}$ is well known in stability analysis as "the jacobian matrix"; it quantifies the effects of all state variables on their rates of change. A matrix similar to $\mathbf{B}$ turns up in metabolic control analysis, as $\mathbf{N}\partial \mathbf{v}/\partial \mathbf{p}$ [48, 108], where it denotes the immediate effects of parameter perturbations on the rates of change of all variables. If the function $\mathbf{y}$ is scalar and denotes a rate, then $\mathbf{C}$ becomes a row vector $\mathbf{c}$ harboring unscaled elasticity coefficients and $\mathbf{D}$ becomes a row vector $\mathbf{d}$ containing so-called π-elasticities – sensitivities of the rates with respect to the parameters [109]. The linearized system is:

$$\frac{d\mathbf{x}_l(t)}{dt} = \mathbf{A}\mathbf{x}_l(t) + \mathbf{B}\mathbf{u}_l(t)$$

$$\mathbf{y}_l(t) = \mathbf{C}\mathbf{x}_l(t) + \mathbf{D}\mathbf{u}_l(t)$$

The new state variables $\mathbf{x}_l(t)$ are approximations of $\mathbf{x}(t) - \mathbf{x}_s$ for the input $\mathbf{u}_l(t) = \mathbf{u}(t) - \mathbf{u}_s$, and the output $\mathbf{y}_l(t)$ is an approximation of $\mathbf{y}(t) - \mathbf{y}_s$, where $\mathbf{y}_s = \mathbf{g}(\mathbf{x}_s, \mathbf{u}_s)$. Note that the linearized system has the same order as the original system.

Step 3. *Balancing*

Determine the so-called controllability and observability gramians, the matrices $\mathbf{P}$ and $\mathbf{Q}$. These are solutions to the equations:

$$AP + PA^T = -BB^T, \quad QA + A^TQ = -C^TC$$

Perform the two singular value decompositions and compute T, A_b, B_b, C_b, and D_b:

$$P = U_1 D_1^2 U_1^T, D_1 U_1^T Q U_1 D_1 = U_2 D_2^4 U_2^T, T = D_2 U_2^T D_1^{-1} U_1^T$$

$$A_b = TAT^{-1}, B_b = TB, C_b = CT^{-1}, D_b = D$$

The balanced system is then:

$$\frac{dx_b(t)}{dt} = A_b x_b(t) + B_b u_b(t)$$

$$y_b(t) = C_b x_b(t) + D_b u_b(t)$$

This is a transformation of the linearized system, and can be derived by substituting $x_b = Tx_l$ in the linearized system. The gramians $\mathbf{P_b}$ and $\mathbf{Q_b}$ of the balanced system satisfy:

$$A_b P_b + P_b A_b^T = -B_b B_b^T, Q_b A_b + A_b^T Q_b = -C_b^T C_b,$$

$$P_b = Q_b = \text{Diag}(\lambda_1, \lambda_2, \ldots, \lambda_n)$$

Thus $\mathbf{P_b}$ and $\mathbf{Q_b}$ are now equal and diagonal.

Step 4. *Truncation*

First, the number of state variables has to be chosen, i.e. the order $n_r, 0 < n_r < n$, of the reduced system. If n_r is chosen too low, i.e. if we are reducing too much, too much of the dynamics might be lost. However, it may be possible to reduce the number of state variables considerably without losing much of the dynamics. One wishes to choose n_r as small as possible, but still capture as much as possible of the dynamics. What is suitable in terms of choice of n_r may be investigated by calculating the so-called *Hankel singular values* (HSVs). The HSVs are the diagonal elements $\lambda_1, \lambda_2, \ldots, \lambda_n$ of the gramians of the balanced system. These are of decreasing order and are reflecting the level of importance for the input–output behavior.

There are n HSVs, each corresponding to a state variable in the balanced system. The states that correspond to low HSVs, also have little influence on the input–output behavior of the system, and can hence be truncated away without much loss. It is not the absolute values that are of interest, but the difference between the values. The principle is illustrated in the the next section.

Once the order n_r of the reduced system has been chosen, one should define:

$$T_r = \begin{pmatrix} I_{n_r} & 0 \end{pmatrix}$$

where I_{n_r} is the $n_r \times n_r$ identity matrix. Then one should compute:

$$A_r = T_r A_b T_r^T, B_r = T_r B_b, C_r = C_b T_r^T, D_r = D$$

The reduced system is then:

$$\frac{dx_r(t)}{dt} = A_r x_r(t) + B_r u_r(t)$$

$$y_r(t) = C_r x_r(t) + D_r u_r(t)$$

This system has n_r state variables. Since the reduced system is an approximation of the linearized system, and the linearized system variables are approximations of the deviation from steady state of the original system, the states of the original system are approximated by:

$$x_s + T^{-1} T_r^T x_r(t)$$

The output is approximated by:

$$y_s + C_r x_r(t) + D_r u_r(t)$$

This is the final result of the reduction.

15.6
Balanced Truncation in Action: Reduction of a Silicon Cell Model of Glycolysis in Yeast

Here a model of glycolysis in yeast [26] is reduced using the model reduction method outlined above. Glycolysis is the well known process that uses glucose to produce Gibbs energy-rich compounds such as pyruvate, and is depicted in Fig. 15.1.

The mathematical model consists of 13 differential equations of the form:

$$\frac{dx(t)}{dt} = f(x(t))$$

where $\mathbf{x}(t)$ is a vector with 13 state variables, as listed in Table 15.1. The differential equations are derived from enzyme kinetics and network topology, as explained in Section 15.2 [26].

To enable linearization and balanced truncation, inputs and outputs must be defined. As input $\mathbf{u}$, we choose extracellular glucose concentration, since glucose is feeding the process and can be varied in an experiment. As output $\mathbf{y}$, the rate of production of pyruvate was chosen, since pyruvate is a Gibbs energy-rich end-product of the glycolysis that is used in further steps in Gibbs energy metabolism. It is also a concentration that is easy to measure. The system with inputs and outputs is then:

$$\frac{dx(t)}{dt} = f(x(t), u(t))$$

$$y(t) = g(x(t))$$

Note that the output $\mathbf{y}$ does not directly depend on the input $\mathbf{u}$. This is because there is no direct dependence between external glucose concentration and pyruvate flow, only indirectly through the state variable $\mathbf{x}$. The system with inputs and outputs above is called the *original system* in order to later distinguish it from the linearized, balanced, and truncated system.

The original system was reduced by linearization, balancing, and truncation. The steady state was determined by simulation in Mathematica. The steady input con-

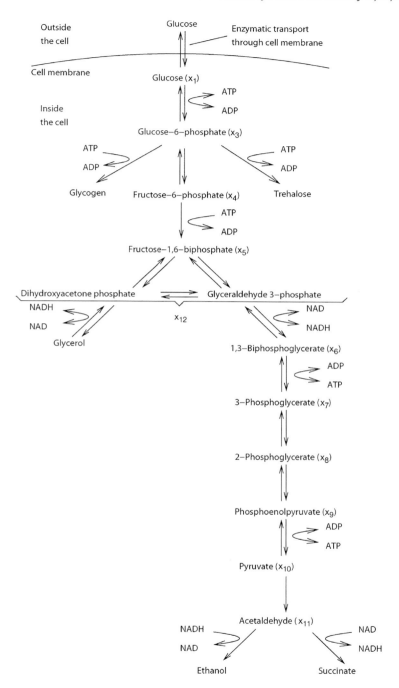

Fig. 15.1 Glycolysis as modeled by Teusink et al. [26].

Table 15.1 The state variables and their meaning are listed in the first two columns. Names of molecules written in square brackets stand for concentrations of the molecules. All concentrations refer to intracellular concentrations. The values of the steady-state concentrations are listed to the right.

State variable	Meaning of the state variable	Steady state values [mM]
x_1	[Glucose]	0.0987441
x_2	2*[ATP]+[ADP]	6.3094
x_3	[Glucose-6-phosphate]	1.03346
x_4	[Fructose-6-phosphate]	0.11285
x_5	[Fructose-1-6-phosphate]	0.60195
x_6	[1,3-Biphosphoglycerate]	0.000329693
x_7	[3-Phosphoglycerate]	0.35652
x_8	[2-Phosphoglycerate]	0.044849
x_9	[Phosphoenolpyruvate]	0.0736333
x_{10}	[Pyruvate]	8.52346
x_{11}	[Acetaldehyde]	0.170117
x_{12}	[Dihydroacetone phosphate] + [glyceraldehyde 3-phosphate]	0.77756
x_{13}	[NADH]	0.0444398

centration was 50 mM, and the steady state is displayed in Table 15.1. The order of the reduced system was determined by investigating the Hankel singular values. The first four singular values were 0.0459, 0.0046, 0.0006, and 0.0002, while the remaining nine values were so small with respect to the first that they are regarded as zero. Thus, a system of order two, three, or four should very well describe the dynamics of the reduced system. If we choose to reduce the model to order three, the reduced model becomes:

$$\frac{dx_r(t)}{dt} = \begin{pmatrix} -2.0259 & -4.6900 & -1.0217 \\ 4.69004 & -21.6664 & -9.2004 \\ 1.0218 & -9.2002 & -9.7734 \end{pmatrix} x_r(t) + \begin{pmatrix} -0.4313 \\ 0.4487 \\ 0.1074 \end{pmatrix} u_r(t)$$

$$y_r(t) = \begin{pmatrix} -0.4313 & -0.4487 & -0.1074 \end{pmatrix} x_r(t)$$

where $x_r = (x_{r,1} \; x_{r,2} \; x_{r,3})^T$ is the vector of the new three state variables. The state variables of the reduced system are approximating the deviation from steady state of linear combinations of state variables. The three state variables that were kept after the truncation are representing the following linear combinations of the original variables:

$$x_{r,1} : -9.63x_1 - 0.0637x_2 - 11.3x_3 - 11.6x_4 - 12.0x_5 - 6.10x_6 + 2.51x_7 \\ -9.28x_8 - 9.24x_9 - 9.27x_{10} - 9.37x_{11} - 9.36x_{12} - 2.40x_{13}$$

$$x_{r,2} : 10.0x_1 + 3.54x_2 + 14.6x_3 + 12.3x_4 + 13.6x_5 + 6.03x_6 - 1.21x_7 \\ +3.49x_8 - 0.154x_9 - 0.650x_{10} - 2.18x_{11} - 6.85x_{12} + 1.19x_{13}$$

$$x_{r,3} : 2.40x_1 - 7.46x_2 - 4.55x_3 - 1.92x_4 - 7.48x_5 - 3.73x_6 + 0.501x_7$$
$$-5.20x_8 + 2.23x_9 + 1.33x_{10} - 1.51x_{11} + 3.81x_{12} - 0.409x_{13}$$

This can be read out from the transformation matrix of the balancing, since $x_b = Tx_l$, and the coefficients above are the entries in the three first rows of the matrix $\mathbf{T}$. The coefficients of the linear combinations are all in a similar range, which indicates that most of the state variables seem to have quite similar importance for the input–output behavior. However, we see one interesting feature: the state variable of the reduced system that is of most importance for the input–output behavior, $x_{r,1}$, depends less on one of the state variables, i.e. x_2 (ATP concentration), than on the others. It should be noted that, given the choices of input and output, the input–output relation to consider is the dependence of glycolytic flux on external glucose concentrations. The low coefficient in front of x_2 in the first linear combination then suggests the interesting fact that the flow's dependence on external glucose depends less on [ATP] than the other state variables such as $x_1 = $ [intracellular glucose]. Observe that this conclusion can only be drawn for concentrations near the steady state, since we linearized the system.

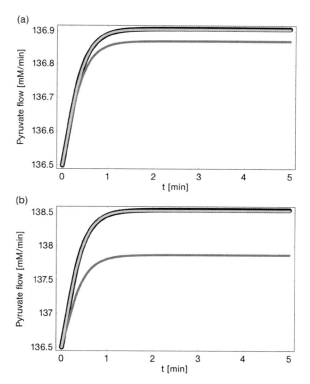

Fig. 15.2 Simulation results. Pyruvate flow in the original (mid-gray line), linearized (black line), and truncated (pale gray line) systems, as responses to (a) 10%, and (b) 50% step changes in external glucose concentration at time 0.

To compare the original system with the reduced system, the pyruvate flow in the original, the linearized, and the reduced model was simulated. Starting at steady state, with the constant input $\mathbf{u(t)} = 50$ mM, 10% and 50% step increases in input glucose concentration were applied, resulting in increases in the flow of pyruvate by only 0.3% and 1.5% respectively (Fig. 15.2). Notwithstanding the small increase in flux, the balanced truncation introduced a 10% overestimation of the increase in pyruvate flux.

However, the approximation error resulted from the linearization and not from the reduction of the linearized model. It was expected that linearization would produce a substantial error as the system operated in a highly nonlinear domain with respect to the perturbing parameter (external glucose concentration). The flow in glycolysis hardly depended on the glucose concentration outside the cell, which seems to be a robustness property. This is not surprising since glucose outside was taken as 50 mM and therefore should have saturated the glucose transporter, and hence changes in it had only a weak influence on the flux.

15.7
Conclusions

The method of linearization, balancing, and truncation was applied to a silicon cell model of glycolysis in yeast, and a significant reduction of complexity was possible. In addition to receiving a highly reduced system, we could read out which molecule concentrations have more influence on the flow of glycolysis than others. It turned out that, according to the model we used, ATP seems to have much less influence on the flow's dependence of glucose than any of the metabolites in glycolysis. We found that the errors of the approximation were much more due to the linearization than to the truncation.

This work is a first approach to system reduction for biochemical systems. The research perspective is to apply system reduction to a large biochemical system, possibly using decomposition of the system into small subsystems or modules. Moreover, system reduction methods have to be developed which apply to nonlinear systems with state variables that remain positive when starting with positive initial concentrations, and produce systems with the same properties, which admit a biological interpretation.

Acknowledgment

F.J.B. is supported by the European Union through the Network of Excellence BioSim, Contract No. LSHB-CT-2004-005137.

References

1 Francke, C.; Siezen, R. J.; Teusink, B.: Reconstructing the metabolic network of a bacterium from its genome. *Trends Microbiol* **2005**, 13:550–558.

2 Gottesman, S.: The small RNA regulators of Escherichia coli: roles and mechanisms*. *Annu Rev Microbiol* **2004**, 58:303–328.

3 Kell, D. B.: Metabolomics and systems biology: making sense of the soup. *Curr Opin Microbiol* **2004**, 7:296–307.

4 Lazebnik, Y.: Can a biologist fix a radio? – Or, what I learned while studying apoptosis. *Cancer Cell* **2002**, 2:179–182.

5 Turing, A.: The mathematical basis of morphogenesis. *Philos Trans R Soc Lond Ser B* **1952**, 237:37–47.

6 Nicolis, G., Prigogine, I.: *Self-organization in nonequilibrium systems: from dissipative structures to order through fluctuations.* John Wiley & Sons, New York, **1977**.

7 Alberghina, L.; Westerhoff, H. V.: Systems biology: definitions and perspectives (Topics in current genetics), Springer-Verlag, Heidelberg, **2005**.

8 Bruggeman, F. J.; Westerhoff, H. V.: The nature of systems biology. *Trends Microbiol* **2006**, accepted for publication.

9 Westerhoff, H. V.; Palsson, B. O.: The evolution of molecular biology into systems biology. *Nat Biotechnol* **2004**, 22:1249–1252.

10 Chance, B.; Garfinkel, D.; Higgins, J.; Hess, B.: Metabolic control mechanisms. 5. A solution for the equations representing interaction between glycolysis and respiration in ascites tumor cells. *J Biol Chem* **1960**, 235:2426–2439.

11 Garfinkel, D.; Hess, B.: Metabolic Control mechanisms. vii. A detailed computer model of the glycolytic pathway in ascites cells. *J Biol Chem* **1964**, 239:971–983.

12 Cleland, W. W.: The kinetics of enzyme-catalyzed reactions with two or more substrates or products. I. Nomenclature and rate equations. *Biochim Biophys Acta* **1963**, 67:104–137.

13 Cleland, W. W.: The kinetics of enzyme-catalyzed reactions with two or more substrates or products. II. Inhibition: nomenclature and theory. *Biochim Biophys Acta* **1963**, 67:173–187.

14 Cleland, W. W.: The kinetics of enzyme-catalyzed reactions with two or more substrates or products. III. Prediction of initial velocity and inhibition patterns by inspection. *Biochim Biophys Acta* **1963**, 67:188–196.

15 Segel, I. H.: *Enzyme kinetics: behavior and analysis of rapid equilibrium and steady-state enzyme systems.* John Wiley & Sons, New York, **1993**.

16 Heinrich, R.; Rapoport, T. A.: A linear steady-state treatment of enzymatic chains. General properties, control and effector strength. *Eur J Biochem* **1974**, 42:89–95.

17 Kacser, H.; Burns, J. A.: The control of flux. *Symp Soc Exp Biol* **1973**, 27:65–104.

18 Savageau, M. A.: *Biochemical systems analysis: a study of function and design in molecular biology.* Addison–Wesley, New York, **1976**.

19 Prigogine, I.; Lefever, R.; Goldbeter, A.; Herschkowitz-Kaufman, M.: Symmetry breaking instabilities in biological systems. *Nature* **1969**, 223:913–916.

20 Westerhoff, H. V.; Van Dam, K.: *Thermodynamics and control of biological free-energy transduction.* Elsevier Science, Amsterdam, **1987**.

21 Monod, J.: From enzymatic adaptation to allosteric transitions. *Science* **1966**, 154:475–483.

22 Monod, J.: *Chance and necessity: an essay on the natural philosophy of modern biology.* Collins, London, **1972**.

23 Koshland, D. E. Jr.: A model regulatory system: bacterial chemotaxis. *Physiol Rev* **1979**, 59:811–862.

24 King, R. D.; Whelan, K. E.; Jones, F. M.; Reiser, P. G.; Bryant, C. H.; et al.: Functional genomic hypothesis generation and experimentation by a robot scientist. *Nature* **2004**, 427:247–252.

25 Bakker, B. M.; Michels, P. A. M.; Opperdoes, F. R.; Westerhoff, H. V.: Glycolysis in bloodstream form Trypanosoma brucei can be understood in terms of the kinetics of the glycolytic enzymes. *J Biol Chem* **1997**, 272:3207–3215.

26 Teusink, B.; Passarge, J.; Reijenga, C. A.; Esgalhado, E.; van der Weijden, C. C.; et al.: Can yeast glycolysis be understood in

terms of in vitro kinetics of the constituent enzymes? Testing biochemistry. *Eur J Biochem* **2000**, 267:5313–5329.

27 Rohwer, J. M.; Meadow, N. D.; Roseman, S.; Westerhoff, H. V.; Postma, P. W.: Understanding glucose transport by the bacterial phosphoenolpyruvate:glucose phosphotransferase system on the basis of kinetic measurements in vitro. *J Biol Chem* **2000**, 275:34909–34921.

28 Hoefnagel, M. H.; Starrenburg, M. J.; Martens, D. E.; Hugenholtz, J.; Kleerebezem, M.; et al.: Metabolic engineering of lactic acid bacteria, the combined approach: kinetic modelling, metabolic control and experimental analysis. *Microbiology* **2002**, 148:1003–1013.

29 Bruggeman, F. J.; Boogerd, F. C.; Westerhoff, H. V.: The multifarious short-term regulation of ammonium assimilation of Escherichia coli: dissection using an in silico replica. *FEBS J* **2005**, 272:1965–1985.

30 Kholodenko, B. N.; Demin, O. V.; Moehren, G.; Hoek, J. B.: Quantification of short term signaling by the epidermal growth factor receptor. *J Biol Chem* **1999**, 274:30169–30181.

31 Hornberg, J. J.; Binder, B.; Bruggeman, F. J.; Schoeberl, B.; Heinrich, R.; et al.: Control of MAPK signalling: from complexity to what really matters. *Oncogene* **2005**, 24:5533–5542.

32 Visser, D.; Schmid, J. W.; Mauch, K.; Reuss, M.; Heijnen, J. J.: Optimal redesign of primary metabolism in Escherichia coli using linlog kinetics. *Metab Eng* **2004**, 6:378–390.

33 Schmid, J. W.; Mauch, K.; Reuss, M.; Gilles, E. D.; Kremling, A.: Metabolic design based on a coupled gene expression-metabolic network model of tryptophan production in Escherichia coli. *Metab Eng* **2004**, 6:364–377.

34 Hornberg, J. J.; Bruggeman, F. J.; Binder, B.; Geest, C. R.; de Vaate, A. J.; et al.: Principles behind the multifarious control of signal transduction. ERK phosphorylation and kinase/phosphatase control. *FEBS J* **2005**, 272:244–258.

35 Bluethgen, N., Bruggeman, F.J., Sauro, H.M., Kholodenko, B.N., Herzel, H.: Sequestration limits zero-order ultrasensitivity in signal transduction. **2007**, in preparation.

36 Bruggeman, F. J.; Libbenga, K. R.; Van Duijn, B.: The diffusive transport of gibberellins and abscisic acid through the aleurone layer of germinating barley grain: a mathematical model. *Planta* **2001**, 214:89–96.

37 Teusink, B.; Walsh, M. C.; van Dam, K.; Westerhoff, H. V.: The danger of metabolic pathways with turbo design. *Trends Biochem Sci* **1998**, 23:162–169.

38 Bakker, B. M.; Mensonides, F. I.; Teusink, B.; van Hoek, P.; Michels, P. A.; et al.: Compartmentation protects trypanosomes from the dangerous design of glycolysis. *Proc Natl Acad Sci U S A* **2000**, 97:2087–2092.

39 Rohwer, J. M.; Postma, P. W.; Kholodenko, B. N.; Westerhoff, H. V.: Implications of macromolecular crowding for signal transduction and metabolite channeling. *Proc Natl Acad Sci U S A* **1998**, 95:10547–10552.

40 Shen-Orr, S. S.; Milo, R.; Mangan, S.; Alon, U.: Network motifs in the transcriptional regulation network of Escherichia coli. *Nat Genet* **2002**, 31:64–68.

41 Mangan, S.; Alon, U.: Structure and function of the feed-forward loop network motif. *Proc Natl Acad Sci USA* **2003**, 100:11980–11985.

42 Watts, D. J.; Strogatz, S. H.: Collective dynamics of 'small-world' networks. *Nature* **1998**, 393:440–442.

43 Jeong, H.; Tombor, B.; Albert, R.; Oltvai, Z. N.; Barabasi, A. L.: The large-scale organization of metabolic networks. *Nature* **2000**, 407:651–654.

44 Schuster, S.; Hilgetag, C.; Woods, J. H.; Fell, D. A.: Reaction routes in biochemical reaction systems: algebraic properties, validated calculation procedure and example from nucleotide metabolism. *J Math Biol* **2002**, 45:153–181.

45 Fell, D. A.; Wagner, A.: The small world of metabolism. *Nat Biotechnol* **2000**, 18:1121–1122.

46 Price, N. D.; Reed, J. L.; Palsson, B. O.: Genome-scale models of microbial cells: evaluating the consequences of constraints. *Nat Rev Microbiol* **2004**, 2:886–897.

47 Heinrich, R., Schuster, S.: *The regulation of cellular systems, 1st edn.* Chapman & Hall, New York, **1996**.

48 Reder, C.: Metabolic control theory: a structural approach. *J Theor Biol* **1988**, 135:175–201.

49 Cleland, W. W.: An analysis of Haldane relationships. *Methods Enzymol* **1982**, 87:366–369.

50 Sauro, H. M.; Ingalls, B.: Conservation analysis in biochemical networks: computational issues for software writers. *Biophys Chem* **2004**, 109:1–15.

51 Papin, J. A.; Stelling, J.; Price, N. D.; Klamt, S.; Schuster, S.; et al.: Comparison of network-based pathway analysis methods. *Trends Biotechnol* **2004**, 22:400–405.

52 Schuster, S.; Fell, D. A.; Dandekar, T.: A general definition of metabolic pathways useful for systematic organization and analysis of complex metabolic networks. *Nat Biotechnol* **2000**, 18:326–332.

53 Schilling, C. H.; Letscher, D.; Palsson, B. O.: Theory for the systemic definition of metabolic pathways and their use in interpreting metabolic function from a pathway-oriented perspective. *J Theor Biol* **2000**, 203:229–248.

54 Varma, A.; Palsson, B. O.: Metabolic capabilities of Escherichia coli. 1. Synthesis of biosynthetic precursors and cofactors. *J Theor Biol* **1993**, 165:477–502.

55 Varma, A.; Palsson, B. O.: Metabolic capabilities of Escherichia coli. 2. Optimal-growth patterns. *J Theor Biol* **1993**, 165:503–522.

56 Varma, A.; Palsson, B. O.: Stoichiometric flux balance models quantitatively predict growth and metabolic by-product secretion in wild-type Escherichia coli W3110. *Appl Environ Microbiol* **1994**, 60:3724–3731.

57 Varma, A.; Palsson, B. O.: Metabolic flux balancing – basic concepts, scientific and practical use. *Bio-Technology* **1994**, 12:994–998.

58 Savinell, J. M.; Palsson, B. O.: Network analysis of intermediary metabolism using linear optimization. II. Interpretation of hybridoma cell metabolism. *J Theor Biol* **1992**, 154:455–473.

59 Savinell, J. M.; Palsson, B. O.: Network analysis of intermediary metabolism using linear optimization. I. Development of mathematical formalism. *J Theor Biol* **1992**, 154:421–454.

60 Fell, D. A.; Small, J. R.: Fat synthesis in adipose tissue. An examination of stoichiometric constraints. *Biochem J* **1986**, 238:781–786.

61 Bonarius, H. P. J.; Schmid, G.; Tramper, J.: Flux analysis of underdetermined metabolic networks: The quest for the missing constraints. *Trends Biotechnol* **1997**, 15:308–314.

62 Acerenza, L.; Sauro, H. M.; Kacser, H.: Control analysis of time-dependent metabolic systems. *J Theor Biol* **1989**, 137:423–444.

63 Mangan, S.; Zaslaver, A.; Alon, U.: The coherent feedforward loop serves as a sign-sensitive delay element in transcription networks. *J Mol Biol* **2003**, 334:197–204.

64 Kalir, S.; Mangan, S.; Alon, U.: A coherent feed-forward loop with a SUM input function prolongs flagella expression in Escherichia coli. *Mol Syst Biol* **2005**, 1:2005–2006.

65 Mangan, S.; Itzkovitz, S.; Zaslaver, A.; Alon, U.: The incoherent feed-forward loop accelerates the response-time of the gal system of Escherichia coli. *J Mol Biol* **2006**, 356:1073–1081.

66 Bluthgen, N.; Bruggeman, F. J.; Legewie, S.; Herzel, H.; Westerhoff, H. V.; et al.: Effects of sequestration on signal transduction cascades. *FEBS J* **2006**, 273:895–906.

67 Selkov, E.: Stabilization of energy charge, generation of oscillations and multiple steady states in energy metabolism as a result of purely stoichiometric regulation. *Eur J Biochem* **1975**, 59:151–157.

68 Goldbeter, A.; Lefever, R.: Dissipative structures for an allosteric model – application to glycolytic oscillations. *Biophys J* **1972**, 12:1302–1308.

69 Huang, C. Y.; Ferrell, J. E. Jr.: Ultrasensitivity in the mitogen-activated protein kinase cascade. *Proc Natl Acad Sci USA* **1996**, 93:10078–10083.

70 Wolf, J.; Heinrich, R. Effect of cellular interaction on glycolytic oscillations in yeast: a theoretical investigation. *Biochem J* **2000**, 345:321–334.

71 Moles, C. G.; Mendes, P.; Banga, J. R.: Parameter estimation in biochemical pathways: a comparison of global optimization methods. *Genome Res* **2003**, 13:2467–2474.

72 Kremling, A.; Fischer, S.; Gadkar, K.; Doyle, F. J.; Sauter, T.; et al.: A benchmark for methods in reverse engineering and model discrimination: problem formulation and solutions. *Genome Res* **2004**, 14:1773–1785.

73 Gadkar, K. G.; Gunawan, R.; Doyle, F. J. 3rd: Iterative approach to model identification of biological networks. *BMC Bioinformatics* **2005**, 6:155.

74 Snoep, J. L.; Bruggeman, F.; Olivier, B. G.; Westerhoff, H. V.: Towards building the silicon cell: a modular approach. *BioSystems* **2006**, 83:207–216.

75 Hoefnagel, M. H. N.; Starrenburg, M. J. C.; Martens, D. E.; Hugenholtz, J.; Kleerebezem, M.; et al.: Metabolic engineering of lactic acid bacteria, the combined approach: kinetic modelling, metabolic control and experimental analysis. *Microbiology* **2002**, 148:1003–1013.

76 Chassagnole, C.; Rais, B.; Quentin, E.; Fell, D. A.; Mazat, J. P.: An integrated study of threonine-pathway enzyme kinetics in Escherichia coli. *Biochem J* **2001**, 356:415–423.

77 Poolman, M. G.; Fell, D. A.; Thomas, S.: Modelling photosynthesis and its control. *J Exp Bot* **2000**, 51[Spec No]:319–328.

78 Hynne, R.; Dano, S.; Sorensen, P. G.: Full-scale model of glycolysis in Saccharomyces cerevisiae. *Biophys Chem* **2001**, 94:121–163.

79 Chassagnole, C.; Noisommit-Rizzi, N.; Schmid, J. W.; Mauch, K.; Reuss, M.: Dynamic modeling of the central carbon metabolism of Escherichia coli. *Biotechnol Bioeng* **2002**, 79:53–73.

80 Olivier, B. G.; Snoep, J. L.: Web-based kinetic modelling using JWS Online. *Bioinformatics* **2004**, 20:2143–2144.

81 Baras, F.; Mansour, M. M.: Reaction-diffusion master equation: a comparison with microscopic simulations. *Phys Rev E* **1996**, 54:6139–6148.

82 Gillespie, D. T.: A general method for numerically simulating the stochastic time evolution of coupled chemical reactions. *J Comput Phys* **1976**, 22:403–434.

83 Van Kampen, N. G.: *Stochastic processes in chemistry and physics.* North–Holland: Amsterdam, **1992**.

84 Keizer, J.: *Statistical thermodynamics of nonequilibrium processes.* Springer-Verlag, Berlin, **1987**.

85 Elf, J.; Ehrenberg, M.: Fast evaluation of fluctuations in biochemical networks with the linear noise approximation. *Genome Res* **2003**, 13:2475–2484.

86 Scott, M.; Ingalls, B.; Kaern, M.: Estimations of intrinsic and extrinsic noise in models of nonlinear genetic networks. *Chaos* **2006**, 16:026107.

87 Paulsson, J.: Summing up the noise in gene networks. *Nature* **2004**, 427:415–418.

88 Guckenheimer, J., Holms, P.: *Nonlinear oscillations, dynamical systems, and bifurcations of vector fields.* Springer-Verlag, New York, **1983**.

89 Stucki, J. W.: Stability analysis of biochemical systems – practical guide. *Prog Biophys Mol Biol* **1978**, 33:99–187.

90 Larter, R.; Clarke, B. L.: Chemical-reaction network sensitivity analysis. *J Chem Phys* **1985**, 83:108–116.

91 Westerhoff, H. V.; Chen, Y. D.: How do enzyme activities control metabolite concentrations? An additional theorem in the theory of metabolic control. *Eur J Biochem* **1984**, 142:425–430.

92 Klonowski, W.: Simplifying principles for chemical and enzyme reaction kinetics. *Biophys Chem* **1983**, 1983:73–87.

93 Schuster, R.; Schuster, S.: Relationships between modal-analysis and rapid-equilibrium approximation in the modeling of biochemical networks. *Syst Anal Model Simul* **1991**, 8:623–633.

94 Kholodenko, B. N.; Schuster, S.; Garcia, J.; Westerhoff, H. V.; Cascante, M.: Control analysis of metabolic systems involving quasi-equilibrium reactions. *Biochim Biophys Acta* **1998**, 1379:337–352.

95 Liao, J. C.; Lightfoot, E. N. Jr.: Extending the quasi-steady state concept to analysis of metabolic networks. *J Theor Biol* **1987**, 126:253–273.

96 Segel, L. A.; Slemrod, M.: The quasi-steady-state assumption – a case-study in perturbation. *SIAM Rev* **1989**, 31:446–477.

97 Stiefenhofer, M.: Quasi-steady-state approximation for chemical reaction networks. *J Math Biol* **1998**, 36:593–609.

98 Schneider, K. R.; Wilhelm, T.: Model reduction by extended quasi-steady-state approximation. *J Math Biol* **2000**, 40:443–450.

99 Rao, C. V.; Arkin, A. P.: Stochastic chemical kinetics and the quasi-steady-state assumption: application to the Gillespie algorithm. *J Chem Phys* **2003**, 118:4999–5010.

100 Heinrich, R.; Rapoport, S. M.; Rapoport, T. A.: Metabolic regulation and mathematical models. *Prog Biophys Mol Biol* **1977**, 32:1–82.

101 Delgado, J.; Liao, J. C.: Control of metabolic pathways by time-scale separation. *Biosystems* **1995**, 36:55–70.

102 Fraser, S. J.: The steady-state and equilibrium approximations. *Geometrical Picture* **1988**, 195:99-PHYS.

103 Roussel, M. R.; Fraser, S. J.: Geometry of the steady-state approximation. *Perturbation and Accelerated Convergence Methods* **1990**, 93:1072–1081.

104 Roussel, M. R., Fraser, S.J.: Invariant manifold methods for metabolic model reduction. *Chaos* **2001**, 11:196–206.

105 Liebermeister, W.; Baur, U.; Klipp, E.: Biochemical network models simplified by balanced truncation. *FEBS J* **2005**, 272:4034–4043.

106 Hardin, H.; Van Schuppen, J. H.: *System reduction of nonlinear positive systems by linearization and truncation. (Lecture notes in control and information sciences).* Springer, Berlin/Heidelberg, **2006**, pp 431–438.

107 Athanasios, C. A.: *Approximation of large-scale dynamical systems. (Advances in design and control, number DC06).* SIAM, Philadelphia, **2005**.

108 Bruggeman, F. J.; Westerhoff, H. V.; Hoek, J. B.; Kholodenko, B. N.: Modular response analysis of cellular regulatory networks. *J Theor Biol* **2002**, 218:507–520.

109 Heinrich, R.; Schuster, S.: *The regulation of cellular systems.* Chapman and Hall, New York, **1996**.

16

Building Virtual Human Populations:
Assessing the Propagation of Genetic Variability in Drug Metabolism to Pharmacokinetics and Pharmacodynamics

Gemma L. Dickinson and Amin Rostami-Hodjegan

Abstract

New challenges in drug development, including the rising cost, are increasing the need to apply *in silico* approaches with potential to: (a) improve the design of studies and (b) avoid unnecessary clinical studies. Although modeling and simulation (M&S) is gradually gaining popularity within the management of drug development, the majority of applications are limited to the so-called phase I–III stages (e.g. modeling disease progression, extrapolating pharmacological observations from Phase II to Phase III). Thus, one of the most attractive features of M&S is often ignored, namely the possibility of extrapolating observations from *in vitro* systems to the *in vivo* behavior of new compounds (IVIVE). Many of the previous chapters in this book indicate the systems biology approach to modeling the physiology and pathophysiology of the human body with an implication for a better understanding and treatment of different diseases. A similar approach can be used to create a framework that enables drug developers to integrate information on human body functions in relation to the handling of drugs within the body ("what the body does to drugs"). The knowledge requirement for IVIVE and testing compounds in "virtual populations of patients" is vast. However, many pieces of required information are independent of the specific drugs and can be applied to all drug candidates. Some implications of the IVIVE approach are shown in this chapter including the assessment of the propagation of genetic differences in drug-metabolizing enzymes in different populations, although some other aspects are briefly discussed too.

16.1
Introduction

This chapter is concerned with the simulation and prediction of the fate of the drugs in the human body as part of assessing pharmacokinetics (PK) and pharmacodynamics (PD) in virtual populations of individuals. It is indicated that how the clearance of various drugs as well as the fraction of drug escaping first pass hepatic

Biosimulation in Drug Development. Edited by Martin Bertau, Erik Mosekilde, and Hans V. Westerhoff
Copyright © 2008 WILEY-VCH Verlag GmbH & Co. KGaA, Weinheim
ISBN: 978-3-527-31699-1

and gut metabolism can be estimated for each individual by utilizing *in vitro* data on drug metabolism. The simulations can incorporate variability from a number of demographic and physiological sources. In addition, the approach may consider incorporation of any known genetic differences in drug metabolism. Examples are presented where the clearance values obtained via the above simulations are used in conjunction with established PK/PD models to allow the investigation of the impact of genetic polymorphisms in drug metabolism on the PK and PD of model drugs. These investigations are used as the basis for clinical trial simulations (CTS) to assess the importance of study size and other aspects of study design in identifying genetic differences in drug metabolism.

16.2
ADME and Pharmacokinetics in Drug Development

The traditional view of drug development represents the process as 4 discrete stages designated as: (a) discovery, (b) preclinical research and development, (c) clinical research and development and (d) 'post-marketing' pharmacovigilance [1].

During drug discovery, a drug that might act on a certain biological pathway or bind to certain receptors is identified with the hope that it may be useful in preventing or treating a particular disorder or group of disorders. Once a potentially useful compound is recognized, the drug proceeds to preclinical testing. During this stage, *in vitro* and *in vivo* (in animal) tests are carried out to elucidate some of the PK, PD and toxicological properties of the drug. The clinical research and development stage is the longest (around 10 years) and it is further divided into three phases of studies, commonly known as phases I, II and III. Phase I studies comprise the first clinical investigations to be carried out in humans and they generally involve small groups of healthy volunteers. The main aim of the phase I tests are to characterize the PK properties of the drug and to assess its safety for further testing although, whenever possible, obtaining information on the pharmacologic effects via monitoring biomarkers is becoming popular. Phase II studies are carried out in patients who suffer from the disease or disorder that the drug is intended to treat. In this stage the safety, tolerability, effectiveness, and appropriate dosage of the drug are studied in greater depth. Upon obtaining successful results from phase II, the clinical development proceeds to phase III with appropriate dosing schemes and with the aim of gathering information that secures drug approval by regulatory bodies (i.e. demonstration of the drug's efficacy and safety in large numbers of people). The scheme described above reflects the traditional structure of drug development; however it has been recognized recently that in order to optimize the speed and cost of drug development it is advantageous to have separate stages running simultaneously rather than sequentially. Modeling and simulation of the processes that define the plasma concentration–time course of a drug, namely, absorption, distribution, metabolism, and elimination (ADME) is an indispensable tool in integrating the available information and accelerating decision making.

The exposure of an individual to a certain drug can be measured by the *area under the concentration time curve* (AUC). The AUC after an oral dose is dependent on the proportion of the dose that is absorbed and is available in systemic circulation after passage from the gut and through the liver (the bioavailability of the drug – F), the clearance (CL) and the dose of the drug (D) [2]:

$$AUC = \frac{F \cdot D}{CL} \tag{1}$$

Total CL is defined as the volume of blood which is completely cleared of drug per unit time and encompasses clearance by the liver, the kidneys and biliary excretion. In the absence of variation in pharmacological response (e.g. genetic differences in receptors, underlying effect of the disease itself, pharmacological effects of concomitant drugs) maximum exposure, as measured by maximum concentration of a drug in blood circulation (C_{max}), and total exposure, as measured by AUC, are often closely linked to the pharmacological effects as well as concentration-related toxicity or adverse effects. Thus, both early determination of the PK characteristics of a drug in the process of drug development, and, defining the influence of individual covariates are desirable. A brief description of ADME processes that determine the overall PK is provided below. This will enable separation of elements which are related to human biology and physiology (i.e. system parameters) from those related to the drug that, in combination, lead to observed ADME behavior.

16.2.1
Absorption

Bioavailability (F) is a term often used to describe the absorption of a drug. It is defined as the proportion of an oral dose of a drug that reaches the systemic circulation. It is dependent on a number of factors which are described by the following equation:

$$F = f_a \times F_G \times F_H \tag{2}$$

Where f_a is the fraction of the dose of the drug which enters the gut wall (drug may be lost by decomposition in the gut lumen [2] or it may not be able to get across the gut wall), F_G is the fraction of drug which escapes metabolism in the gut wall and enters the portal vein and F_H is the fraction of the drug that enters the liver and escapes metabolism, entering systemic circulation.

16.2.2
Distribution

Distribution refers to the reversible transfer of drug from one location to another within the body [2]. Distribution of drugs to and from blood and other tissues occurs at various rates and to various extents. Several factors are responsible for the distribution pattern of a drug within the body over time. Some of these are related to the nature of the drug (such as the ability of the drug to cross membranes, plasma protein binding, partitioning into the red blood cells, tissues, and fat) and others relate to characteristics of the individual (such as the perfusion rate of different tissues by blood, concentration of plasma proteins, hematocrit, body composition). Since the dosage regimen of a given drug depends on the amount of drug reaching the site of action and the rate at which drug is removed, the above factors are important determinants of dosage regimens.

16.2.3
Drug Metabolism

For the majority of new (and mostly lipophilic) drug compounds, metabolism is a major route of elimination from the body [3]. Therefore, several important questions need to be answered, in full or in part, during the development of a new drug:

1. What are the major (and minor) routes of metabolism?
2. Are there any active and/or toxic metabolites produced?
3. Are any drug-drug interactions possible and if so, what are they?

Some factors influencing these questions are discussed below.

During preclinical development, the quantitative prediction of *in vivo* metabolic clearance from *in vitro* data is arguably one of the most important objectives of *in vitro* metabolism studies. The reason for this being that the successful prediction of *in vivo* drug clearance from *in vitro* data allows a large amount of important information to be gathered early on during development. Assessing the possible impact of genetic polymorphisms in drug metabolism is a vital issue that will be discussed in details later. The information on metabolic routes can also be used to exclude the likelihood of certain drug–drug interactions (DDIs).

16.2.4
Excretion

All drugs are ultimately removed from the body, either in the form of their metabolites or in their unchanged form, by excretory organs (mainly the kidneys) and excreted in the urine (although in some instances, excretion may be via the biliary route). Compromised renal function may effect the PK of the drugs if urinary excretion is a substantial part of the overall elimination.

16.3
Sources of Interindividual Variability in ADME

Interindividual differences in drug response are a well recognized issue in many areas of drug therapy. In the 1950s it was discovered that African-American sol-

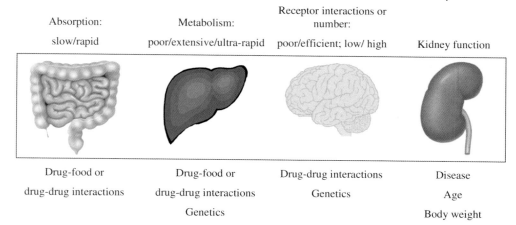

| Absorption:
slow/rapid | Metabolism:
poor/extensive/ultra-rapid | Receptor interactions or
number:
poor/efficient; low/ high | Kidney function |

| Drug-food or
drug-drug interactions | Drug-food or
drug-drug interactions
Genetics | Drug-drug interactions
Genetics | Disease
Age
Body weight |

Fig. 16.1 Sources of variability in PK and PD and how these variabilities are manifest [6].

diers who developed sever anemia after taking the anti-malarial drug, primaquine were deficient in the enzyme glucose-6-phosphate dehydrogenase [4]. Almost simultaneously, it was noticed that patients who experienced prolonged effects of the anesthetic drug succinylcholine possessed an atypical from of a cholinesterase enzyme [5]. Since then, an enormous amount of information has become available on the physiological, demographic, ethnic and genetic basis of individual variation in drug response. However, so far, examples of the administration of drugs based on the characteristics of the patient have been limited to consideration of renal function (via assessment of serum creatinine) and the effect of body size (i.e. weight or body surface area); and it has applied mainly to the hospital setting (e.g. chemotherapy in oncology). The philosophy of 'one size fits all' was followed by many pharmaceutical companies until recently. However, this is now shifted to a new policy that involves identifying subgroups that will benefit from the new candidate drug most, and adjusting the dose according to patient characteristics. This approach also helps with *a priori* identification of some 'outlier' individuals who are most at 'risk' from either adverse events or ineffective therapy. Interindividual variability arises from a number of sources (Fig. 16.1). These could include environmental factors (e.g. food, pollutants, time of day and season, location, etc.), genetics, disease, age, concomitant medication and compliance [2]. With the exception of compliance, each of these factors affects variability by either altering the PK or the PD of a particular drug (Fig. 16.1).

A small number of examples now exist for which genetic tests are available which precede treatment with a handful of drugs and inform the clinician on how best to proceed with treatment for a particular individual. Pharmacogenetic information is contained in about 10% of labels for drugs approved by the FDA and a significant increase of labels containing such information has been observed over the past decade (http://www.fda.gov/cder/genomics/genomic_biomarkers_table.htm). For example, in November 2004, the labeling of Camptosar (irinotecan; a drug that is used to treat colon/rectal cancer) was updated to include information on the main enzyme responsible for its clearance, UGT1A. Nearly 70% of patients require a dose reduction in order to prevent neutropenia. Since exposure to the drug is dependent on its metabolism by UGT1A1, patients with certain enzyme mutations are expected to require a lower dose of the drug.

Despite the above examples, obstacles with regard to personalizing medicine are still relevant. For example, it is important to consider a strategy for individualizing a dosage regimen in phase III if smaller phase I or phase II studies have shown that individuals deficient in a certain enzyme will have an increased exposure to a particular drug candidate, and they therefore may require a lower dose. Regulators will be interested in information on necessary dose reduction that is proven to be relevant in large clinical studies however clinical trials can help with the design of the most effective studies.

16.3.1
Pharmacokinetic Variability

Pharmacokinetics (PK) can be a major source of variability in the dose response relationship. It manifests itself in interindividual differences in the plasma concentration–time profile of a drug. Factors which lead to variability in the ADME parameters are therefore of importance in understanding overall variability in PK.

16.3.1.1 Variability in Absorption

As described previously, absorption is affected by three parameters: f_a, F_G, and F_H. Each of these parameters is sensitive to differences in physiological characteristics between individuals. An important factor affecting f_a is variation in gastrointestinal motility (Fig. 16.2), in particular for drugs with low permeability. F_G is sensitive to the genetics of an individual in relation to drug metabolizing enzymes, diet, and active secretion of the drug back into the gut by the multidrug efflux pump, P-glycoprotein (P-gp), which may be susceptible to interindividual fluctuations in both abundance and activity. Variability in the first pass metabolism of the drug by

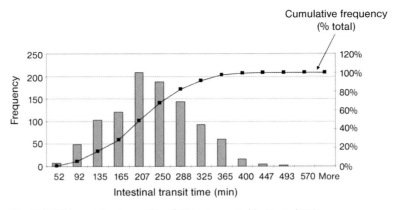

Fig. 16.2 Wide variation in motility of GI tract reported by Yu et al. [40].

the liver (F_H) is a result of genetic differences in drug metabolism and differences in the physiology of the individual affecting liver blood flow and the size of the liver.

16.3.1.2 Variability in Distribution

The fraction of drug which is unbound in blood (fu_B) is often variable among individuals. For very highly bound drugs this variability can be large [8]. However, interindividual variability in the binding of a drug tends to be less variable than such variability for other PK parameters [7]. The fu_B of a drug in an individual is a function of the affinity of the drug to plasma proteins and red blood cells as well as the individual levels of circulating plasma proteins and hematocrit. The levels of protein in the plasma vary due to many different factors such as age, liver cirrhosis, pregnancy, trauma, and stress. Similarly, age, sex and environmental factors affect the hematocrit.

Variability in the distribution of drugs is also related to differences in the size and composition of various organs. For example, differences in body fat content between individuals in particular between males and females may lead to differences in the volume of distribution. A highly lipid soluble drug may distribute into the adipose tissue of a female with high body fat and become less available to the eliminating organs. Consequently, the half life of the drug may increase in females compared to males.

16.3.1.3 Variability in Metabolism

The hepatic clearance of drugs which are extracted inefficiently from the blood is sensitive to changes in the activity of drug-metabolizing enzymes in the liver. Also, variation in activity of drug-metabolizing enzymes can affect oral clearance of all drugs regarding the efficiency of extraction (i.e. by affecting the hepatic first pass clearance or the systemic clearance). Induction and inhibition of enzymes by environmental substances or toxins contributes to interindividual differences in drug metabolism as much as the genetic makeup of the individual. Genetic variants of genes coding for drug-metabolizing enzymes result in enzymes with higher, lower or no activity, or they may lead to a complete absence of the enzyme [8]. Such genetic variation is termed a 'polymorphism' when the monogenic trait occurs at a single gene locus with a frequency of more than (arbitrarily) 1% [9].

The potential consequences of genetic polymorphisms of the type mentioned above for PK processes in individuals who lack activity in an enzyme include: (1) either unwanted responses (toxic effects) or sub-therapeutic responses associated with the average, 'normal' doses, (2) a lack of pro-drug activation, or (3) a dependence on alternative routes of elimination. This may be a problem if these routes are also compromised (e.g. by renal impairment, drug–drug or drug–food interactions) [10]. Depending upon which of the above examples explains how the polymorphism influences the drug's systemic exposure, affected individuals may require either very low or very high doses for their therapy to be effective.

Polymorphisms of drug-metabolizing enzymes are of central importance during drug development. However, the seriousness of the consequences of such polymorphisms for the PK of a drug depends on a number of factors [10]: (1) whether

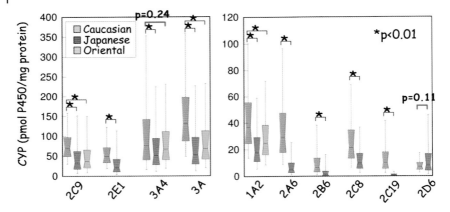

Fig. 16.3 Large interindividual variability in abundance of different CYP450. (Adapted from data given by Inoue et al. [41]).

the polymorphic enzyme metabolizes the parent drug or the metabolite(s) or both, (2) the contribution of the polymorphic enzyme to overall elimination of the drug, (3) the potency of any active metabolites and (4) the viability of the drug's other, competing pathways of elimination.

Drugs that are metabolized by polymorphic enzymes are often actively screened out of the drug development process. However, such action may be unnecessary when the clinical relevance of the polymorphism in relation to the drug is assessed more carefully. It is also important to realize that there is wide variation in the abundance of enzymes even within the same genotype, as shown in Fig. 16.3.

16.3.1.4 Variability in Excretion
The renal drug clearance, as the major excretion route, is influenced to different extents by urine flow, urine pH and plasma protein binding. The extent of sensitivity to these parameters depends on the nature of compounds and mechanisms involved in its renal elimination (e.g. glumerular filtration, active secretion, passive re-absorption). Thus, interindividual variation in these parameters together with renal function determines the overall renal clearance, The urine flow is sensitive to individual fluid intake and administration of diuretic drugs. Interindividual differences in urine pH are mainly related to differences in diet and physical activity.

Biliary excretion plays smaller role in elimination of most drugs. However, pathophysiological conditions that cause cholestasis [11] or the genetics of transporters (e.g. OATP [12]) may influence PK of the compounds if biliary excretion is a significant component of their PK.

16.3.2
Pharmacodynamic Variability

The consequences of polymorphisms of drug response depend on a further number of contributing factors [10]: (1) at which point on the concentration-effect relationship typical concentrations arising from 'normal' doses fall, (2) the therapeutic index and utility of the drug, and (3) whether PK variability is outweighed by PD variability in receptor sensitivity or density or in the turnover of a particular endogenous receptor ligand.

Drug response is likely to be the result of a complex function of the influence of many genes interacting with environmental and behavioral factors. Whether PK-PD variability translates into clinically relevant differences in drug response depends on further issues including compliance, the availability of alternate drugs and doctor/patient perception of side-effects [10].

Ideally, the overall benefit to risk ratio would be carefully considered before making a go/no-go decision to prevent the premature termination of drugs which were potentially safer and perhaps more effective than the alternative. Sources of pharmacodynamic variability include genetics (e.g. of transporters or receptors) and demographics (e.g. developmental differences in the abundance of receptors or hormonal influence on the regulation of receptors).

There is substantial evidence that genetic variation in drug receptors is associated with variable drug response [34, 35]. Examples of genes for which genetic polymorphisms have been reported include those encoding β-adrenergic receptors [13], μ-opioid receptors [14, 15], and G proteins [16]. All of these polymorphisms have the effect of altering response to drug therapy independent of dose and exposure levels and increasing interindividual variability in response.

16.3.3
Other Sources of Variability in Drug Response

There are a number of sources of variability in human drug response which do not fit into either of the above two categories. These include the disease state of the individual, their compliance with the prescribed drug regimen, and any exercise which the patient may or may not undertake. These variables will not be discussed in detail here since they are not considered as sources of variability in the PKPD models used within this chapter. The *in vivo* studies which the PKPD models are chosen to represent are usually carried out either under controlled conditions in healthy individuals where such variables do not apply, or in hospitalized patients where compliance should not be an issue.

16.4
Modeling and Simulation of ADME in Virtual Human Population

The purpose of building a "virtual human population" is to capture all the complex interplay between different sources of variability mentioned above in the form of a network of system information that interacts with drug characteristics measured using *in vitro* studies and indicates potential co-variations determining the drug therapy outcome (for an example, see Fig. 16.4).

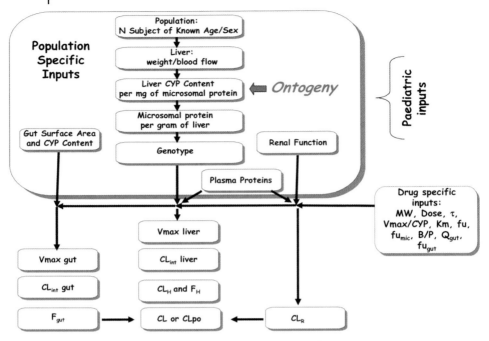

Fig. 16.4 A schematic representation of a computer network of models used by Johnson, et al. [34] to capture the elements of pharmacokinetics that determine the age-related changes of clearance in pediatric populations.

The advent of computer techniques for use in the pharmaceutical industry has radically changed drug development. However, it is the view held by many that we are far from utilizing such techniques to their full potential and 'in silico' processes still have a long way to go [17]. Some argue that at some point in the future we will be able to model each and every one of the body's systems and we will administer drugs 'in silico' to see if they work and allow us to judge what are likely to be the problems and what is the ideal dose to give before we even contemplate administering the drug to an animal or a human being. The focus of modeling and simulation in drug development within this chapter is the integration of preclinical data to PK and then into "Clinical Trial Simulations" (CTS). The main aim of CTS is to maximize the information content obtained during the drug development process to ensure the greatest chance of 'success' in a clinical trial [18]. The concept was first utilized by statisticians working within the field of drug development [18]. Indeed, the term CTS is still often used in its narrower sense to describe the use of pharmacostatistical techniques within drug development.

16.4.1
The Need for More Efficient Clinical Trials

Declining productivity has been a concern of the pharmaceutical industry for some time. However, expenditure on drug development appears to be simultaneously increasing at an almost exponential rate. There seems to be a general consensus in the industry that there is much room for improvement. Change is needed to improve the efficiency of drug development [19] as the process could be made faster, and more efficient, hence reducing costs and increasing the number of new drug launches. For example, recently, a report commented on the large number of inadequately powered studies that are carried out to determine the effect of a particular drug-metabolizing enzyme polymorphism on the PK of drugs [20]. The authors indicated that published studies which claim to either prove or disprove the functional relevance of polymorphisms on the PK of drugs may be misleading the scientific field as a result of underpowered studies and suggested that CTS is a valuable tool for ensuring the use of adequately sized samples for clinical studies.

16.4.2
Current Clinical Trial Simulation in Drug Development

CTS are built on mathematical and statistical models which essentially capture the complex systems involved in drug PK and PD. CTS is a term encompassing the use of premises and assumptions to generate a large number of replications of virtual clinical trials which represent the actual clinical trials [21].

Much of the basis for modeling and simulation during early development is provided by the use of computer methods for predicting human PK characteristics from preclinical data. PK/PD models developed from information gathered from preclinical studies also come into play at this stage. Information collected using these methods can be used to optimize dosage regimens and to design subsequent phase I, phase II and phase III studies. As discussed earlier, it is possible to synthesize numerous PK parameters by combining the *in vitro* data with all known individual attributes relevant to the absorption, distribution, metabolism, and elimination of the drug. This information can be used to predict the time course of drug in plasma in humans.

In the very early stages of drug development the information known about a compound can be little and somewhat 'patchy'. It is important that the available information is integrated so that coherent and comprehensive conclusions may be drawn from it. The information can be incorporated into physiologically based PK (PBPK) or PK/PD models to allow investigators to ask 'what if' questions and evaluate the impact of certain study design features of further studies and what impact they will have on study outcomes [21]. Once PBPK and/or PK/PD models have been developed for a drug it is possible to develop these into more comprehensive models that can form the basis of much more extensive CTS. If physiological models are made available in the early stages of drug development, then patient differences in the parameters which determine variability can be identified by proactive planning to record such covariates as part of the study design. It is, therefore, possible to define the PK of a drug *a priori* in patients who are to go through phase II and III studies. This approach is rarely utilized in practice due to the wide spread use of alternative compartmental fitting of PK profiles.

16.4.3
Incorporation of *In Vitro* Preclinical Data Into CTS

It is clear that the majority of CTS currently carried out during drug development rely heavily on data already available from *in vivo* studies. However, if modeling and simulation techniques are to benefit the drug development process to their full advantage, then CTS must be carried out much earlier. Conversely, it is difficult to envisage how this is possible earlier on in the drug development process when the majority of CTS rely on information gathered in clinical studies with the rest being heavily reliant on data gathered from preclinical animal studies. Clearly, there is a need for utilizing PK information as early as possible during drug development.

It is theoretically possible to simulate all the parameters which determine the ADME properties of a drug in humans. Indeed, this has already been put into practice successfully for many of these parameters. However, at the current time, such simulations are still in the early stages of development and they are rarely put to their full use by incorporation into CTS. Moreover, the focus of such extrapolations has been to obtain values in the average man rather than investigating the possible range of parameters expected in a large population of virtual individuals.

16.4.3.1 Prediction of Absorption
Numerous *in vitro* systems (e.g. animal tissue, Caco-2 cell lines) are utilized in assessing the intestinal absorption (f_a) potential of new drug candidates. However, a major disadvantage of all these techniques is that they do not incorporate the effect of interindividual variability in physiological factors such as gastrointestinal transit time and emptying rate, or gastric pH [22]. The prediction of F_G can be achieved by assuming that the affinity of a drug for drug-metabolizing enzymes in the gut wall is equivalent to their metabolism in the liver. Metabolism can be extrapolated from *in vitro* data. The estimation of F_H also involves the use of *in vitro* systems to estimate liver metabolism.

16.4.3.2 Prediction of Metabolism
In vitro systems can provide an inexpensive, high-throughput method for the early prediction of drug clearance and drug–drug interactions during development. Various systems are available, including human liver microsomes, recombinantly expressed CYP enzymes, purified and reconstituted CYP enzymes, and isolated hepatocytes. Each of the different systems is associated with its own advantages and disadvantages; and these are discussed below. The extrapolation of *in vitro* information to the *in vivo* situation is discussed later.

Human Liver Microsomes Hepatic microsomes are widely available and can be easily stored at −80 °C. However, a major disadvantage associated with the use of human liver microsomes in these studies is the artefactual variability in CYP protein expression and function caused by: (1) the time delay between tissue harvesting and freezing, (2) differences in the cause of death and medications taken by the donor, and (3) differences in environment, diet and lifestyle between the donors. Variability in the levels of CYP expression and function also arise from genetic differences due to polymorphisms of the CYP enzymes. Unless genotyping of the microsomes is undertaken, investigators cannot be sure of the reasons underlying variability between samples. A further disadvantage of the use of human liver microsomes is a result of their lipid and protein content. This can lead to protein binding and cause a decrease in the free concentration of the drug in the medium [23].

Recombinantly Expressed Enzymes Recent advances in molecular biology have led to the stable expression of CYP450 enzymes in a wide range of expression systems including yeast, bacteria, insect and mammalian cells (rCYPs). The wide availability of such systems offers investigators a viable alternative to the use of human tissue preparations as a source of hepatic enzymes. An important advantage of rCYPs is the reproducibility of information gained from them. Since the introduction of these enzymes, it has become apparent that although they are highly similar to their human counterparts in terms of amino acid sequence, they may differ widely in terms of intrinsic catalytic activity compared to human preparations. However, this problem can be rectified through the use of intersystem extrapolation factors (ISEFs) [24]. An important advantage of producing enzymes in this way is the availability of different genotypes of the same CYP isoform.

Hepatocytes Human hepatocytes are available commercially and are a popular and well established tissue for drug-metabolism studies. As they are whole, living cells containing the full complement of drug-metabolizing enzymes and transport systems, they potentially allow for any concentration gradients mediated by transporters that may affect exposure of substrate/inhibitor to enzymes. However, hepatocytes systems are expensive and some transporters rapidly lose their functional activity after isolation of hepatocytes. Furthermore, maintenance of hepatocyte cultures can be problematic [23, 25]. An important disadvantage of the use of human hepatocytes is that the results gained from experiments which utilize them are dependent on the source of the hepatocytes. That is, because the hepatocytes originate from a human, the activity of any enzymes and transporters are dependent on the genotype and environment of the source.

To make use of the data gathered from *in vitro* systems, there must be a system for 'scaling up' from what is observed *in vitro* to what is expected to happen in humans. This process is called *in vitro–in vivo* extrapolation (IVIVE) and involves the use of calculations which utilize scaling factors to extrapolate from *in vitro* to *in vivo*. Methods for IVIVE were first described more than 25 years ago by Rane et al., who used *in vitro* intrinsic clearance data collected from *in vitro* experiments on isolated perfused rat liver to predict the hepatic extraction ratio of seven drugs in humans [26]. More recently, with the increased availability of human liver samples and the advent of new technology for developing recombinant drug-metabolizing enzymes, methods for the prediction of drug clearance and metabolic drug–drug interactions (mDDIs) have been refined and widely implemented [27, 28].

16.4.3.3 Prediction of Efficacy/Toxicity

The extrapolation of *in vitro* data on drug response to human drug efficacy and toxicity is a relatively new area of drug research. However, a number of successful studies have demonstrated its potential usefulness for drug development [29–31]. For example, Cox et al. [31] demonstrated the extrapolation of *in vitro* μ-receptor binding information to *in vivo* opioid EEG response. In 2003, Visser and colleagues utilized data on the *in vitro* receptor binding affinity of benzodiazepines to $GABA_A$ receptors to successfully describe the *in vivo* properties of these drugs [29].

Methods such as these could ultimately allow the efficacy and toxicity of a drug molecule to be predicted in man before the drug leaves the laboratory. Mechanism-based PK/PD models also allow us to gain a better understanding of the PD basis of interindividual variability in human drug response [29]. A further advantage to be gained from incorporating *in vitro* PD information into CTS is the potential for simulating the influence of certain receptor polymorphisms such as those in the β_2-adrenoceptors [32] or the HERG [33] channel protein on drug response. This would be of great use in assessing the individuals most 'at risk' from new drug candidates. Currently, information on receptor binding is not routinely available in the early stages of drug development where it has the potential to be most useful. However, if the current paradigm was altered, the list of possible benefits to be gained from the utilization of such techniques is long.

16.5
The Use of Virtual Human Populations for Simulating ADME

Three different cases are used to illustrate the various implications of building virtual human population for better drug development practices. These include applications to show the population variability instead of point estimates, incorporating age related changes in drug elimination, and investigating the propagation of genetic variability in drug metabolism to pharmacological effects.

16.5.1
Assessing the Interindividual Variability of *In Vivo* Drug Clearance from *In Vitro* Data

Methods for predicting *in vivo* drug clearance from *in vitro* data are not new [26]. However, a recent increase in the availability of human liver samples has led to an amplification of *in vitro* to *in vivo* extrapolation (IVIVE) efforts to predict clearance (CL). Although IVIVE attempts in all recent reports have been carried out using average parameter values (leading to prediction of average clearances), Howgate et

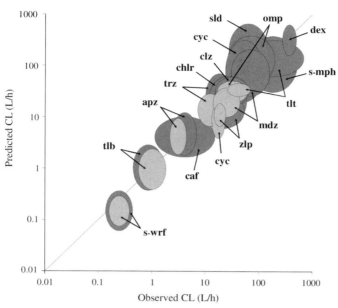

Fig. 16.5 Predicted versus observed clearances of 15 drugs. The ellipses delineate the 90% confidence intervals of both predicted and observed values. Dark gray indicates data for oral drug clearance, light gray for clearance after intravenous administration. The line is the line of identity. Drug abbreviations are as follows: Apz = alprazolam, Caf = caffeine, Chlr = chlorzoxazone, Clz = clozapine, Cyc = cyclosporine; Dex = dextromethorphan, S-mph = S-mephenytoin, Mdz = midazolam, Omp = omeprazole, Sld = sildenafil, Tlb = tolbutamide, Tlt = tolterodine, Trz = triazolam, S-Wrf = S-warfarin, Zlp = zolpidem.

al. [36] developed a mechanistic model that can simulate and predict CL and mDDI in virtual populations using a Monte Carlo approach. The algorithms combine information on genetic, physiological, and demographic variability with preclinical *in vitro* data to allow extrapolation to *in vivo* pharmacokinetics. Predicted values of median CL fall within 2-fold of observed values for 93% of the drugs that are administered orally (p.o.) and 100% of the drugs that are administered intravenously (i.v., Fig. 16.5). However, the most significant observation, perhaps, is related to the predicted *x*-fold variability which falls within 2-fold of the observed variability for 80% (p.o.) and 67% (i.v.) of the drugs.

16.5.2
Prediction of Clearance and its Variability in Neonates, Infants, and Children

Prediction of the exposure of neonates, infants and children to xenobiotics is likely to be more successful using physiologically based pharmacokinetic models than simplistic allometric scaling, particularly in younger children [37]. Building a virtual human population that captures the required comprehensive information on the ontogeny of anatomical, physiological and biochemical variables is a substantial task. Johnson et al. [34] recently integrated demographic, genetic, physiological

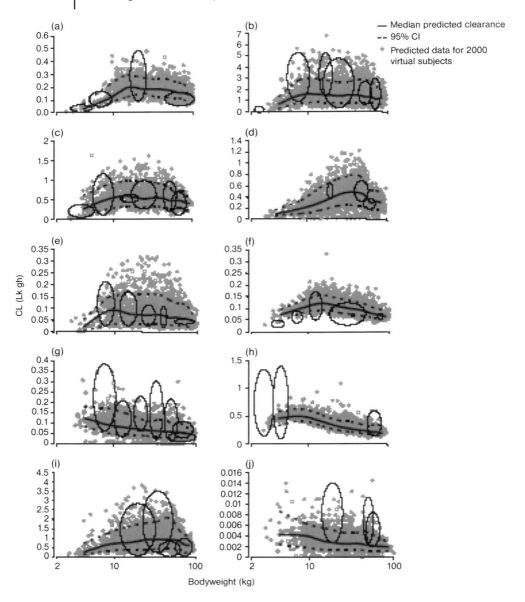

Fig. 16.6 Changes in bodyweight-normalized clearance (CL) with bodyweight for nine metabolized drugs: (a) midazolam, intravenous, (b) midazolam (oral), (c) caffeine (oral), (d) diclofenac (IV), (e) omeprazole (oral), (f) cisapride (oral), (g) carbamazepine (oral), (h) theophylline (oral), (i) phenytoin (oral), (j) S-warfarin (oral). The ellipses encompass data from actual *in vivo* studies. The ellipses at the highest weights in each graph indicate data from adult studies.

and pathological information in pediatric populations with *in vitro* data on human drug-metabolism and transport to predict population distributions of drug clearance (CL) and its variability in pediatric populations. Additional information to that used by Howgate et al. [36] included information on developmental physiology and the ontogeny of specific cytochromes P450.

In neonates, 70% (7/10) of predicted median CL values were within 2-fold of the observed values. Corresponding results for infants, children and adolescents were 100% (9/9), 89 (17/19), and 94% (17/18), respectively. Predicted variability (95% CI) was within 2-fold of the observed values in 70 (7/10), 67 (6/9), 63 (12/19), and 55% (10/18) of cases, respectively. The accuracy of the physiologically based model was superior to that of simple allometry, especially in under 2-year-olds.

The *in silico* prediction of pharmacokinetic behavior in pediatric patients indicated the complexity of age related changes such that the clearance of drugs metabolized by the same enzymes took different patterns (Fig. 16.6), possibly due to other elements of PK such as differences in protein binding, red blood cell distribution, etc.

16.5.2.1 Incorporating Information on Population Variability into Mechanistic DM-PK-PD Modeling to Assess the Power of Pharmacogenetic Studies

Dickinson et al. [38, 39] recently used (S)-warfarin and dextromethorphan as examples to demonstrate the value of incorporating mechanistic IVIVE models into population PK-PD models. Based on *in vitro* drug metabolism data and information on the frequency and activity of the different allelic forms of relevant CYPs, the statistical power of *in vivo* studies needed to discern the effect of genotypes on PK and PD was estimated. This approach represents a paradigm for assessing the impact of genetic polymorphisms on the PK and PD of new drugs prior to costly

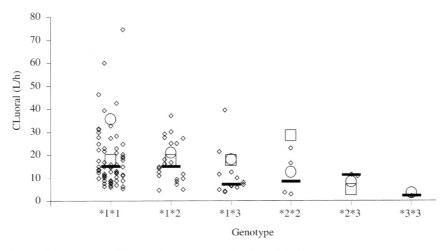

Fig. 16.7 Comparison of observed (○ [42], □ [43]) and simulated (◇; $n = 100$) median values of the unbound oral clearance of (S)-warfarin [38]. Solid lines indicate the medians of the simulated data.

Table 16.1 The outcomes of reported studies of the PK and/or PD of warfarin with respect to CYP2C9 genotype. ('Combo' signifies comparison between wild type versus the combination of all other genotypes.) Cases where prediction matched observation are underlined. Crosses (×) indicate failure of the study to show a statistically significant difference between the corresponding genotype and wild type Ticks (✓) indicate success of the study in showing a statistically significant difference between the corresponding genotype and wild type.

Reference	Type of study	Outcome measure	n	Significant difference from *1/*1					
				*1/*2	*2/*2	*1/*3	*2/*3	*3/*3	Combo
[43]	PK	Unbound oral clearance	47	× 25	× 0	× 48	× 20	× 0	
[42]	PK	Unbound oral clearance	93	× 20	× 0	× 82	✓ 44	✓ 10	
[44]	PK	Plasma (S)-warfarin concentration	121	✓ 20	× 0	× 92	× 48	× 12	
[45]	PK	Plasma clearance	156	× 18	× 0	✓ 94	× 52	× 16	
[43]	PD	Weight normalized maintenance dose	47	× 15	× 0	× 48	× 15	× 0	
[46]	PD	Maintenance dose	53	× 16	× 0	✓ 50	× 16	× 0	
[47]	PD	Maintenance dose/bleeding rate	73	× 12	× 0	× 64	× 26	× 4	✓ 42
[42]	PD	Maintenance dose	93	× 10	× 0	✓ 76	✓ 45	✓ 10	
[44]	PD	Maintenance dose	121	× 10	× 2	✓ 82	× 45	× 12	
[48]	PD	Maintenance dose	126	× 11	× 2	× 82	× 45	× 13	
[45]	PD	Maintenance dose	156	✓ 12	× 3	✓ 86	× 45	× 16	
[49]	PD	Maintenance dose	153	✓ 12	× 3	✓ 86	× 45	× 17	
[50]	PD	Maintenance dose	159	× 12	× 3	✓ 88	× 45	× 18	
[51]	PD	Mean maintenance dose	175	× 12	× 3	✓ 88	× 45	× 20	✓ 60
[52]	PD	Mean maintenance dose	180	✓ 12	× 3	✓ 88	× 45	× 20	
[53]	PD	Maintenance dose/bleeding rate	185	× 12	× 3	× 90	× 45	× 21	✓ 62
[54]	PD	INR > 3	219	× 14	✓ 4	× 92	× 45	× 26	✓ 68
[55]	PD	Maintenance dose	297	✓ 20	✓ 8	✓ 94	× 50	× 32	✓ 78
[56]	PD	Maintenance dose	350	× 24	× 12	× 94	× 54	× 35	✓ 80

population studies, which is in line with the rationale for clinical trial simulation (CTS) [25, 26].

The results of a study on (*S*)-warfarin [38], using virtual human populations, indicated that only 20% of cases (19 of 95 comparisons within 11 PD and 4 PK-PD studies) could show statistically significant differences between wild-type and other individual genotypes (Table 16.1). This was very similar to the percentage observed in reported experimental studies (21%, Chi-squared test, $P = 0.80$).

The consistency between the observed and predicted clearances in different genotypes was exceptionally high despite the fact that only *in vitro* data was used to predict clearances (Fig. 16.7).

16.6
Conclusions

Collecting adequate prior information on the abundance of an enzyme (e.g. cytochrome P450) involved in the metabolism of the different compounds, the relative activity and population frequency of different genotypic forms, physiological and anatomical factors determining clearance (e.g. blood flows, plasma protein levels, liver size) and their covariates is no insignificant task. However, the applications of creating virtual human populations relevant to ADME show that it may be possible to predict *in vivo* drug clearance as a function of genotype with reasonable accuracy. Although *in vivo* information on relevant PD models are currently combined with the PK data, models based on *in vitro* receptor binding data and, indeed, any information on genetic variability in the target receptors could be informative in a model building exercise. Thus, in the near future, it would be possible to utilize prior *in vitro* data within fully mechanistic PK-PD models built in virtual population of patients to carry out virtual clinical trials and evaluate statistical power of certain designs. Such *in silico* trials are not intended to replace clinical studies. Nonetheless, they provide a valuable tool that helps drug developers avoid assumptions about the distributions and variation of primary PK parameters and reduce the number of inconclusive studies by early preclinical data (IVIVE) and CTS.

Acknowledgment

The authors would like to thank Mr. Ben Meakin for his help in the preparation of this chapter.

References

1 Lipsky, M.S. and Sharp, L.K.: From idea to market: the drug approval process. *J Am Board Fam Pract* **2001**, 14:362–367.

2 Rowland, M. and Tozer, T.N.: *Clinical Pharmacokinetics: Concepts and Applications*, Lippincott Williams & Wilkins, **1995**.

3 Williams, J.A. et al.: Drug–drug interactions for UDP-glucuronosyltransferase substrates: a pharmacokinetic explanation for typically observed low exposure (AUCi/AUC) ratios. *Drug Metab Dispos* **2004**, 32:1201–1208.

4 Beutler, E.: The hemolytic effect of primaquine and related compounds: a review. *Blood* **1959**, 14:103–139.

5 Kalow, W. and Staron, N.: On distribution and inheritance of atypical forms of human serum cholinesterase, as indicated by dibucaine numbers. *Can J Biochem Physiol* **1957**, 35:1305–1320.

6 Ingelman-Sundberg, M.: Pharmacogenetics: an opportunity for a safer and more efficient pharmacotherapy. *J Intern Med* **2001**, 250:186–200.

7 Yacobi, A. et al.: Frequency distribution of free warfarin and free phenytoin fraction values in serum of healthy human adults. *Clin Pharmacol Ther* **1977**, 21:283–286.

8 Lin, J.H. and Lu, A.Y.: Role of pharmacokinetics and metabolism in drug discovery and development. *Pharmacol Rev* **1997**, 49:403–449.

9 Meyer, U.A.: Genotype or Phenotype: the Definition of a Pharmacogenetic Polymorphism. *Pharmacogenetics* **1991**, 1:66–67.

10 Dickins, M. and Tucker, G.T.: Drug disposition: to phenotype or genotype. *Int J Pharm Med* **2001**, 15:70–73.

11 Shargel, L. and Yu, A.B.C.: *Applied Biopharmaceutics and Pharmacokinetics*, Appleton and Lange, **1999**.

12 Yamashiro, W. et al.: Involvement of transporters in the hepatic uptake and biliary excretion of valsartan, a selective antagonist of the angiotensin II AT1-receptor, in humans. *Drug Metab Dispos* **2006**, 34:1247–1254.

13 Liggett, S.B.: The pharmacogenetics of beta2-adrenergic receptors: relevance to asthma. *J Allergy Clin Immunol* **2000**, 105:S487–S492.

14 Hollt, V.: A polymorphism (A118G) in the mu-opioid receptor gene affects the response to morphine-6-glucuronide in humans. *Pharmacogenetics* **2002**, 12:1–2.

15 Lotsch, J. et al.: The polymorphism A118G of the human mu-opioid receptor gene decreases the pupil constrictory effect of morphine-6-glucuronide but not that of morphine. *Pharmacogenetics* **2002**, 12:3–9.

16 Johnson, J.A. and Lima, J.J.: Drug receptor/effector polymorphisms and pharmacogenetics: current status and challenges. *Pharmacogenetics* **2003**, 13:525–534.

17 Smith, D.A.: Hello *Drug Discovery*, I am from *insilico*, take me to your president. *Drug Discov Today* **2002**, 7:1080–1081.

18 Bonate, P.L.: Clinical trial simulation in drug development. *Pharm Res* **2000**, 17:252–256.

19 Rooney, K.F. et al.: Modelling and simulation in clinical drug development. *Drug Discov Today* **2001**, 6:802–806.

20 Williams, J.A. et al.: So many studies, too few subjects: establishing functional relevance of genetic polymorphisms on pharmacokinetics. *J Clin Pharmacol* **2006**, 46:258–264.

21 Blesch, K.S. et al.: Clinical pharmacokinetic/pharmacodynamic and physiologically based pharmacokinetic modeling in new drug development: the capecitabine experience. *Invest New Drugs* **2003**, 21:195–223.

22 Balimane, P.V. et al.: Current methodologies used for evaluation of intestinal permeability and absorption. *J Pharmacol Toxicol Methods* **2000**, 44:301–312.

23 Venkatakrishnan, K. et al.: Drug metabolism and drug interactions: application and clinical value of in vitro models. *Curr Drug Metab* **2003**, 4:423–459.

24 Proctor, N.J. et al.: Predicting drug clearance from recombinantly expressed CYPs: intersystem extrapolation factors. *Xenobiotica* **2004**, 34:151–178.

25 Tucker, G.T. et al.: Optimizing drug development: strategies to assess drug metabolism/transporter interaction potential – toward a consensus. *Pharm Res* **2001**, 18:1071–1080.

26 Rane, A. et al.: Prediction of hepatic extraction ratio from in vitro measurement of intrinsic clearance. *J Pharmacol Exp Ther* **1977**, 200:420–424.

27 Iwatsubo, T. et al.: Prediction of in vivo drug disposition from in vitro data based on physiological pharmacokinetics. *Biopharm Drug Dispos* **1996**, 17:273–310.

28 Rostami-Hodjegan, A. and Tucker, G.: 'In silico' simulations to assess the 'in vivo' consequences of 'in vitro' metabolic drug–drug interactions. *Drug Discovery Today: Technologies* **2004**, 1:441–448.

29 Visser, S.A. et al.: Mechanism-based pharmacokinetic/pharmacodynamic modeling of the electroencephalogram effects of GABAA receptor modulators: in vitro–in vivo correlations. *J Pharmacol Exp Ther* **2003**, 304:88–101.

30 Cleton, A. et al.: Mechanism-based modeling of functional adaptation upon chronic treatment with midazolam. *Pharm Res* **2000**, 17:321–327.

31 Cox, E.H. et al.: Pharmacokinetic–pharmacodynamic modeling of the electroencephalogram effect of synthetic opioids in the rat: correlation with the interaction at the mu-opioid receptor. *J Pharmacol Exp Ther* **1998**, 284:1095–1103.

32 Liu, J. et al.: Gly389Arg polymorphism of beta1-adrenergic receptor is associated with the cardiovascular response to metoprolol. *Clin Pharmacol Ther* **2003**, 74:372–379.

33 Bezzina, C.R. et al.: A common polymorphism in KCNH2 (HERG) hastens cardiac repolarization. *Cardiovasc Res* **2003**, 59:27–36.

34 Johnson, J.A.: Drug target pharmacogenomics: an overview. *Am J Pharmacogenomics* **2001**, 1:271–281.

35 Evans, W.E., McLeod, H.L.: Pharmacogenomics – drug disposition, drug targets, and side effects. *N Engl J Med* **2003**, 348:538–549.

36 Howgate, E.M. et al. : Prediction of *in vivo* drug clearance from *in vitro* data. I: impact of inter-individual variability. *Xenobiotica* **2006**, 36:473–497.

37 Johnson, T.N. et al. : Prediction of the clearance of eleven drugs and associated variability in neonates, infants and children. *Clin Pharmacokinet* **2006**, 45:931–956.

38 Dickison, G.L. et al. : The use of mechanistic DM-PK-PD modelling to assess the power of pharmacogenetic studies – CYP2C9 and Warfarin as an example. *Brit J Clin Pharmacol* **2007**, in press.

39 Dickinson, G.L. et al. : Clinical trial simulation of the effect of CYP2D6 polymorphism on the antitussive response to dextromethorphan. *J Pharm Sci* **2007**, 47:175–186.

40 Yu, H.C. et al.: Estimation of standard liver volume for liver transplantation in the Korean population. *Liver Transpl* **2004**, 10:779–783.

41 Inoue, S. et al.: Prediction of in vivo drug clearance from in vitro data. II: potential inter-ethnic differences. *Xenobiotica* **2006**, 36:499–513.

42 Scordo, M.G. et al.: Influence of CYP2C9 and CYP2C19 genetic polymorphisms on warfarin maintenance dose and metabolic clearance. *Clin Pharmacol Ther* **2002**, 72:702–710.

43 Takahashi, H. et al.: Population differences in S-warfarin metabolism between CYP2C9 genotype-matched Caucasian and Japanese patients. *Clin Pharmacol Ther* **2003**, 73:253–263.

44 Kamali, F. et al.: Contribution of age, body size, and CYP2C9 genotype to anticoagulant response to warfarin. *Clin Pharmacol Ther* **2004**, 75:204–212.

45 Loebstein, R. et al.: Interindividual variability in sensitivity to warfarin- nature or nurture? *Clin Pharmacol Ther* **2001**, 70:159–164.

46 Khan, T. et al.: Dietary vitamin K influences intra-individual variability in anticoagulant response to warfarin. *Br J Haematol* **2004**, 124:348–354.

47 Joffe, H.V. et al.: Warfarin dosing and cytochrome P450 2C9 polymorphisms. *Thromb Haemost* **2004**, 91:1123–1128.

48 Siguret, V. et al.: Cytochrome P450 2C9 polymorphisms (CYP2C9) and warfarin maintenance dose in elderly patients. *Rev Med Interne* **2004**, 25:271–274.

49 Tabrizi, A.R. et al.: The frequency and effects of cytochrome P450 (CYP) 2C9 polymorphisms in patients receiving warfarin. *J Am Coll Surg* **2002**, 194:267–273.

50 King, B.P. et al.: Upstream and coding region CYP2C9 polymorphisms: correlation with warfarin dose and metabolism. *Pharmacogenetics* **2004**, 14:813–822.

51 Peyvandi, F. et al: CYP2C9 genotypes and dose requirements during the induction phase of oral anticoagulant therapy. *Clin Pharmacol Ther* **2004**, 75:198–203.

52 Margaglione, M. et al.: Genetic modulation of oral anticoagulation with warfarin. *Thromb Haemost* **2000**, 84:775–778.

53 Higashi, M.K. et al.: Association between CYP2C9 genetic variants and anticoagulation-related outcomes during warfarin therapy. *Jama* **2002**, 287:1690–1698.

54 Lindh, J.D. et al.: Several-fold increase in risk of overanticoagulation by CYP2C9 mutations. *Clin Pharmacol Ther* **2005**, 78:540–550.

55 Sconce, E.A. et al.: The impact of CYP2C9 and VKORC1 genetic polymorphism and patient characteristics upon warfarin dose requirements: proposal for a new dosing regimen. *Blood* **2005**, 106:2329–2333.

56 Aquilante, C.L. et al.: Influence of coagulation factor, vitamin K epoxide reductase complex subunit 1, and cytochrome P450 2C9 gene polymorphisms on warfarin dose requirements. *Clin Pharmacol Ther* **2006**, 79:291–302.

17

Biosimulation in Clinical Drug Development

Thorsten Lehr, Alexander Staab, and Hans Günter Schäfer

Abstract

Modeling and simulation are important aspects of the current clinical drug development program in large pharmaceutical companies. Particularly four types of models are of special interest: pharmacokinetic, pharmacodynamic, disease progression and patient models. The level of detail of these models can differentiate between empirical and mechanistic. However, the level of detail and the model type applied in clinical drug development is highly variable and depends on the stage of drug development, the *a priori* knowledge available and the question that needs to be answered. The majority of models applied today in drug development are at the level of empirical to semi-mechanistic details. The parameters used in these models are quite often estimated based on data obtained in clinical studies by the nonlinear mixed effect modeling approach. Nevertheless, trends towards more mechanistic and physiological models are observable and might show their value in the future. Overall, there is still an urgent need for more efficient drug development processes and the application of modeling and simulation might be a cornerstone to achieve this.

17.1
Introduction

The application of modeling and simulation (M&S) in clinical drug development is nothing completely new. In the area of pharmacokinetics these techniques were already in frequent use during the 1970s. However, with the advance of computer technologies and the introduction of the population approach using the nonlinear mixed effect modeling technique (NLME) M&S has developed into a scientific cornerstone of modern (clinical) drug development. Since the first application of this technique to pharmacokinetic (PK) and pharmacodynamic (PD) data analysis by Sheiner et al. [1] this approach has received considerable attention. In the meantime the United States Food and Drug Administration (FDA) recommends the use of M&S [2] for a better understanding of the dose–concentration (exposure)–

Biosimulation in Drug Development. Edited by Martin Bertau, Erik Mosekilde, and Hans V. Westerhoff
Copyright © 2008 WILEY-VCH Verlag GmbH & Co. KGaA, Weinheim
ISBN: 978-3-527-31699-1

response relationship of the drug [2, 3]. Also, the *Common Technical Document*, which is a standard format for submissions to the authorities responsible for the regulation of pharmaceutical products in Europe, USA, and Japan, contains sections dedicated to the results of population PK and population (PK/)PD analyses. A recently published review showed the remarkable impact of M&S on the drug development process [4].

Despite this established role and proven benefit of M&S in the drug development process, there is an even stronger need within the pharmaceutical industry to embark on these modern *in silico* technologies due to the current inefficiency of the drug development process (i.e. reduced number of innovative products approved) and the overall increase in development costs [5]. Regulatory agencies, particularly the FDA, support M&S. In a recent paper [6] entitled "*Innovation/Stagnation*", the FDA proposed a modified drug development process. One key element of this proposal is the application of computer-based predictive models along the entire drug development value chain.

The overall idea is to see the M&S activities during drug development as a continuum starting in early preclinical development and carried through until approval. This means that the various M&S technologies (i.e. systems biology, preclinical and clinical empirical and mechanism-based (population) PK/PD models) need to be integrated to provide maximal possible benefit to the development teams. The current challenge is this integration, as the models used in the various stages are quite different: early models are much more mechanism-based and try to resemble the biological reality and combine data from various sources, but ignoring intersubject variability, whereas later models based on clinical trial data are more empirical and able to include and estimate various sources of variability. As a response to the challenge of "model-based drug development", several major pharmaceutical companies have implemented modeling and simulation teams supporting the development teams in applying M&S to evaluate nonclinical and clinical data and to provide additional insight, which can help to optimize and streamline subsequent development. The aim of this chapter is to give an overview of the currently used methods of M&S in the context of clinical drug development, the application of M&S, and an outlook on future trends.

17.2
Models in Clinical Development

A model is a theoretical construct that represents biological processes, with a set of parameters and a set of logical and quantitative relationships between them. The models used in clinical drug development can be categorized with respect to the model types used and the degree of mechanistic level included, resulting in a huge number of possible models. In the following section the model types as well as the degree of mechanistic levels (empirical vs fully mechanistic vs semi-mechanistic) are briefly described.

17.2.1
Model Types

In clinical drug development the major interest is to develop a so-called dose–exposure–response model, which describes the relationship between the dose and the achieved exposure in blood or other body fluids of interest and the observed pharmacodynamic response (e.g. biomarker, efficacy, safety measurements). The observations of interest (i.e. any measurement) are obtained in healthy subjects and/or patients, who participate in the clinical trials. Observations are generally taken before, during and after treatment with the investigational drug. The length of a clinical trial can vary between a few days, several months and up to several years. In order to adequately model the dose–exposure–response relationship in the target population over a longer time interval the following four different kinds of models need to be considered:

1. A pharmacokinetic (PK) model to describe the relationship between administered dose, route of administration, dosing schedule, and concentrations of the parent drug as well as active metabolites (if necessary) in various body fluids (mainly plasma and urine). Time dependent changes (e.g. decreased clearance due to given treatment) and influence of intrinsic (e.g. influence of creatinine clearance on clearance) and extrinsic factors should also be reflected in the model.

2. A pharmacodynamic (PD) model describing the relationship between the observed concentration/exposure measure (e.g. the area under the plasma concentration–time profile: AUC) and the observed drug effects on biomarkers, efficacy or safety measurements (or endpoints). Time dependent changes (e.g. development of tolerance) and influence of intrinsic and extrinsic factors should also be reflected in the model.

3. A disease progression model describing the change in underlying disease during the course of the trial without treatment.

4. A patient model describing relevant intrinsic (e.g. race, gender, age) and extrinsic factors (e.g. food, smoking, life style, compliance, drop-out) and their correlation in the patient population. Furthermore, behavioral aspects of the patient population like compliance and probability of discontinuing a trial (i.e. drop-out rates) can be reflected in the patient model.

All four models influence each other considerably, as shown in Fig. 17.1. It depends on the question asked which model or which combination of models is needed to answer a specific question. The models can be further differentiated with regard to the complexity with which they reflect the biological system (from empirical to fully mechanistic models), which are described in detail in the following sections.

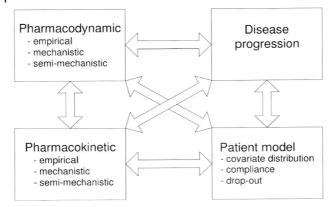

Fig. 17.1 Schematic representation of model types in clinical drug development.

17.2.1.1 Complexity of Models

There are multiple questions raised during the clinical drug development process where the availability of a model would be helpful. The two key objectives for the modeling analysis are: (1) to quantitatively describe the dose–exposure–response relationship and the patient characteristics, which influence this relationship and (2) to simulate so-called "what if" scenarios (= virtual trials) to help in optimizing the design of subsequent studies and thereby increasing the probability of technical success for the new drug candidate. It is generally desirable to make the models resemble as much of the underlying biology as possible, i.e. making the models "mechanism-based", instead of using "empirical" mathematical functions to describe the relationship (see Section 17.5.2). Mechanism-based models (= mechanistic models) in a pure understanding would be models as described in the field of systems biology where each known step of the biological system is described by mathematical equations. In the field of clinical M&S the terminology mechanism base/mechanistic models is used for models that describe the functioning of the biological system and the interaction of the drug with the system in a simplified way as compared to system biology (i.e. lumping steps together that can not be differentiated by measurements). Semi-mechanistic models show a further reduction in complexity as they often concentrate only on a key element of the mechanistic model and lump other elements together. A further reduction in complexity results in the empirical models that often reduce the biological system and the interaction of the drug to a few descriptive elements. However these models are most often able to describe the observed data very well. The transit between these kinds of models is fluent and a clear distinction almost impossible. Table 17.1 contrasts the main features of empirical and mechanistic models.

In clinical trials only a limited amount of data can be obtained. Thus, if parameters of models describing these data should be estimated often empirical models are completely sufficient. These models allow a thorough investigation of the influence of extrinsic and intrinsic factors on the observed pharmacokinetic or pharmacodynamic data.

Table 17.1 Empirical versus mechanism-based models.

	Empirical model	Mechanism-based model
Dependency	Drug-dependent	Biological system parameters = drug independent Drug impact on biological system = drug dependent
Aim	Quantitative description of the dose–exposure–response relationship	Mechanism-based description of the processes underlying the dose–response relationship
Number of model parameters	Few (as many as identifiable by estimation)	Many
Value/origin of model parameters	Estimated based on data available	Estimated using combined datasets or taken from different sources (e.g. databases, literature, own experiments)
Variability and standard error of model parameters	Estimated	Estimated or approximated or unknown
Model background	Purely data-driven	Biology/physiology-driven and data-driven
Model evaluation	Required	Required
Complexity	Low	High
Acceptance	Wide → increasing	Moderate → increasing

In order to better understand the mechanism of the drug mechanistic models are required. If parameters should be estimated it is often necessary that either different data sets are combined to provide enough information for parameter estimation or a subset of parameters is not estimated and values obtained from other sources (e.g. literature, public databases, former clinical trials) are used for these parameters. In an extreme case all parameters are taken from different sources and it is tested whether the model can describe the data by simulation. If this is not the case one or more hypotheses employed when building the model need to be adjusted.

In general, models used to justify dose adjustments in special situations (e.g. renally impaired) and supporting label claims, should be based on clinical trial data obtained during the development of the new drug candidate. Here the use of other less valid data sources is not acceptable. In contrast, if a model is used to support internal decision making in a pharmaceutical company, missing model components/parameters might be taken from other data sources. Based on the fact that clinical trial data frequently do not support highly mechanism-based models, most of the models derived from and applied in the clinical development phase are empirical or semi-mechanism-based.

17.3
Clinical Drug Development

The drug development program is divided into preclinical development, clinical development, and post-approval surveillance. Several preclinical and clinical trials are conducted during this program. In the following the phases of the clinical drug development process are described with a special focus on the data available for modeling and simulation and some objectives are exemplarily listed for the respective phases.

17.3.1
Phase I

Phase I trials are the first-stage of testing a new investigational drug in human subjects ("first in man studies"). At the end of this phase, normally a small group of about 80–100 healthy volunteers are dosed with the investigational drug. The primary purpose is to assess the safety, tolerability, pharmacokinetics, and pharmacodynamics after single and multiple administration of ascending doses of the investigational drug. These trials are always conducted in a specialized clinic, where the subjects are under continuous close supervision by full-time medical staff until they are discharged from the clinic. The range of doses administered is usually wider than in later stages of development. Phase I trials most often include healthy volunteers, however there are circumstances when patients are used, such as with oncology (cancer) and HIV investigational drugs.

In general, multiple (up to 30–40) blood samples can be obtained per subject to measure drug and metabolite concentrations as well as biomarkers in these phase I clinical trials. Furthermore, pharmacodynamic measurements can be included to get a first impression on the drug effect in humans, however, limited by the fact that healthy volunteers were studied and not patients. As strict inclusion and exclusion criteria are used, the demographic characteristics of the healthy volunteers do not provide sufficient spread to investigate the effect of intrinsic factors. Therefore, phase I trials provide very "rich" data to develop pharmacokinetic and pharmacodynamic models on biomarker, but cannot be used to develop models for efficacy, safety, influence of patient factors on PK and/or PD and disease progression.

In addition to the first in man studies there are subsequent studies in healthy volunteers, which are performed in parallel to the phase II/III clinical development. These studies address issues like drug–drug interactions, special sub-populations (e.g. patients with liver and renal impairment), or bioavailability/bioequivalence issues. These studies are also conducted under the same well controlled conditions and therefore contribute "rich" data.

Objectives of M&S in phase I are in particular [7]:

- understanding of the pharmacokinetic (and if possible pharmacodynamic) characteristics in healthy subjects or initial patient population;
- support design of further studies by simulating dose–exposure(–response) for the phase Ib and phase II studies.

17.3.2
Phase II

Once the initial safety/tolerability of the investigational drug has been demonstrated in the initial Phase I trials, the compound is progressed into Phase II trials. These studies are performed in larger, quite often multi-center trials in the target patient population with the objective of providing so-called "proof of concept" information. The number of patients to be enrolled into these phase II trials is very much dependent on the therapeutic area and can range from 50–200 patients up to several hundreds.

Sometimes, Phase II clinical development is sub-divided into Phase IIA and Phase IIB. During Phase IIA several dosing regimens are tested in a smaller patient population, whereas in Phase IIB only a few dosing regimens are tested in a larger patient population. In these trials, the number of observations which can be obtained per patient is reduced compared to the initial Phase I studies, as patients in most cases come only for a limited time, so-called "visits", to the clinic; and obtaining blood samples over several hours is difficult. The number of observations for pharmacokinetics and pharmacodynamics (biomarker, efficacy) per patient is about 3–12 obtained over several clinical visits. The duration of these trials is in the range from days to approximately 12 weeks and the variability in patient characteristics is larger, which allow one to build initial models for the effect of intrinsic and extrinsic factors on pharmacokinetics and pharmacodynamics. For most indications, data on the clinical endpoint for approval cannot be obtained in this development phase. Also information on disease progression is limited due to the relatively short duration of the trials. Therefore, models based on Phase II clinical trial data are limited to pharmacokinetics, pharmacodynamics (biomarker, efficacy), and initial models to describe the influence of intrinsic and extrinsic factors. However, these are the first models describing the data of the investigational drug in the target population.

Objectives of M&S in phase II are in particular [7]:

- understand the dose–exposure–response relationship in the target population by developing a PK/PD model using phase II data;
- explorative evaluation of the need for dose adjustment based on intrinsic and/or extrinsic patient factors;
- support design and dose selection of further studies especially phase III by appropriate simulations using also literature data (e.g. for the relationship of biomarker to clinical endpoint, for drop-out rates etc.).

17.3.3
Phase III

Phase III studies are large in general, double-blind, placebo- or active drug-controlled, randomized trials. The number of patients depends on the therapeutic area and the clinical endpoint in question, but in general the number ranges from several hundred up to several thousand. These trials aim at being the definitive assessment of the efficacy and safety of the new therapy, especially in comparison with currently available alternatives. Phase III trials are very expensive, time-consuming, and organizationally difficult trials to design and run, particularly for therapies in chronic diseases.

The number of observations per patient with respect to pharmacokinetics and pharmacodynamics is very limited. If at all, blood samples for pharmacokinetics can be obtained only at a few visits and a maximum of 2–4 samples per patient per visit are achievable. Quite often, biomarker data are not measured at all, which is a real disadvantage, as these data would provide the basis for a biomarker to be qualified for surrogate endpoint status and would be usable for development of other drugs in the respective therapeutic area. The treatment duration of the phase III trials is also longer compared to phase II clinical trials and ranges in general from several months up to 1–2 years. Therefore, Phase III clinical trials provide sound data on clinical efficacy endpoints, patient factors, and disease progression, but limited data on pharmacokinetics and pharmacodynamics. After successful Phase III development, the pharmaceutical company submits the *Common Technical Document* to the regulatory authorities [e.g. European Medicines Agency (EMEA) or FDA] for marketing approval of the new drug.

Objectives of M&S in phase III are in particular [7]:

- establish/confirm the dose–exposure–response relationship in the target population;
- advice on need for dose adjustment (label) based on extrinsic and/or intrinsic patient factors.

17.3.4
Phase IV

Post-marketing studies, also called phase IV studies, often have several objectives:

1. These studies are often performed in special patient populations not previously studied (for example, pediatric or geriatric populations). Here, modeling and simulation is applied to guide dose selection in these populations.
2. The studies are often designed to monitor a drug's long-term effectiveness and safety and impact on patients' quality of life.
3. Many studies are designed to determine the cost-effectiveness of a drug therapy relative to other traditional and new therapies. With respect to model-based drug development these studies are in general of lower importance.

17.4
Modeling Technique: Population Approach

The so-called population approach is a standard technique, which is currently used to model data from clinical trials. It can be defined as a tool to study the variability in drug concentrations and/or pharmacological effects between individuals [8].

Analysis of variability in PK or PD is nothing new. Indeed, it has been the purpose of most PK and PD studies to estimate the variability in PK and PD parameters (e.g. clearance, volume of distribution) and explain the variability by physiological, genetic or environmental factors. Those studies are usually performed in a small number of individuals under well controlled conditions. Calculation of the population parameter under these conditions can be performed using the standard two-stage method (STS) [9]. At the first stage, the estimates from each individual will be obtained by analyzing each individual separately. Subsequently, the actual population parameters are derived by calculating descriptive statistics, e.g. the mean and the standard deviation of the individual parameter estimates. Using STS for the population approach shows several drawbacks. First, it can only be used if the number of observations in each individual is sufficiently large to allow estimation of individual parameters. Second, the number of samples has to be balanced across the individuals to avoid bias in the calculation of the mean parameters.

Using the nonlinear mixed effects (NLME) technique, an improved method for the calculation of population parameters became available to overcome the limitations of the STS approach. This method estimates structural model parameters (e.g. pharmacokinetic model parameters like clearance) and variability (inter-, intraindividual as well as residual) simultaneously using the data of all individuals [10]. The population approach using NLME technique allows combining data from different development phases. It can be applied in data rich as well as in data sparse and unbalanced situations. This is a very important advantage considering the data structure generated from clinical trials and the fact that the dose–exposure–response relationship needs also to be quantified in so-called "special populations", where multiple measurements are quite often not justifiable and technically not possible like in, e.g. children, elderly, liver or kidney impaired patients or severely ill patients. Furthermore, data obtained following different administration routes and schedules can be used by the NLME approach. Thus, this flexibility makes the population approach so valuable for this type of data and enables the investigation of a large number of individuals.

In addition to the STS and the NLME approach, other population approaches are available, including the Bayesian and the nonparametric modeling methods. These techniques are less frequently applied in drug development. Thus, the following section will refer to the NLME approach.

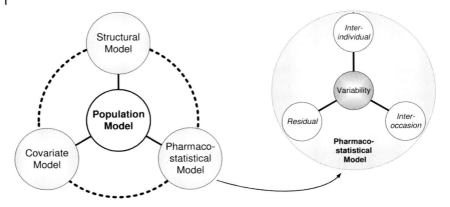

Fig. 17.2 Population model.

17.4.1
Model Structure

Using the NLME, the population model contains three components: the structural model, the statistical model and – if necessary – the (integrated) patient covariate model (Fig. 17.2).

17.4.1.1 Structural Model

The structural model describes the typical time profile of the observations (e.g. plasma concentrations, pharmacodynamic measurements) as a function of the corresponding model parameters (e.g. typical clearance, E_{max}), the dosing schedule and the administered dose. The structural model can be described by the following function:

$$Y_{ij} = f(\phi_i, x_{ij})$$

where $f()$ is the function describing the structural model (PK or PD) that relates the independent variables, x_{ij} (e.g. time and dose at time point j of subject i), to the measured response (Y_{ij}) given the ith individual vector of the model parameters ϕ_i.

17.4.1.2 Statistical Model

The variability in the observations made during clinical trials can be attributed to differences between patients (intersubject variability), differences occurring within a patient from day to day (interoccasion variability) and residual variability, which can be attributed to assay variability or other unknown sources. The pharmacostatistical model characterizes these sources of variability and estimates the magnitude of the different sources of variability.

The *interindividual variability* (IIV) covers the inexplicable differences of model parameters between individuals. This means that the individual parameter value varies from the typical population parameter value to a random extent (e.g.

$CL_{individual}$ vs $CL_{population}$). This difference can be described by different kinds of models resulting in different distributions of the individual parameters. An exponential model is exemplarily listed in the following:

$$P_{ki} = \theta_k \cdot e^{\eta_{ki}}$$

where P_{ki} denotes the value of parameter k of individual i ($=$ individual parameter), θ_k is the typical value of the population parameter k and η_{ki} is the difference between the natural logarithms of P_{ki} and θ_k ($\eta_{ki} = \ln P_{ki} - \ln \theta_k$). It is commonly assumed that η_ks are independently and symmetrically distributed with zero mean and the variance ω_k^2. This means that an exponential model results in a log-normal distribution of the parameter of interest.

The use of an exponential model has several attractive features. It will ensure that all parameters are strictly positive and avoids so the estimation of negative non-physiological individual values. Also ω_k^2 becomes dimensionless and the square root of the estimate is approximately the coefficient of variation in the model parameters.

The *interoccasion variability* (IOV) or intraindividual variability [11] arises when a parameter of the model, e.g. CL, varies within a subject between study occasions. The term "occasion" can be defined arbitrarily, usually logical intervals for an occasion are chosen, e.g. each dosing interval in multiple dose studies or each treatment period of a cross-over study can be defined as an occasion. To assess the IOV of a specific parameter more than one measurement per individual has to be available per occasion. The IOV can be implemented in the random effect model as described in the following:

$$P_{kiq} = \theta_k \cdot e^{\eta_{ki} + \kappa_{kiq}}$$

where P_{kiq} denotes the ith individual value of parameter P_k at study occasion q. It is commonly assumed that κ_{kiq} are independently and symmetrically distributed with zero mean and variance π^2.

The *residual variability* describes the extent of deviation between the observed and the model predicted value, including IIV and IOV. Deviations might be caused by errors in the documentation of the dosing and blood sampling times, analytical errors, misspecification in the models and other factors. The most common error models describing the residual variability are presented in the following.

The simplest one is the additive residual variability model:

$$Y_{ij} = f\left(\phi_i, x_{ij}\right) + \varepsilon_{ij}$$

Y_{ij} denotes the measured observation from the ith individual at time point j. By the function $f(\phi_i, x_{ij})$ the individual prediction is made using the structural model including the interindividual and the interoccasional variability. ε_{ij} denotes the random deviation between the individual prediction and the observed measurement for each individual i at time point j. This model applies if a constant variance over the whole measurement range is probable.

Another common residual variability model is the proportional residual variability model:

$$Y_{ij} = f\left(\phi_i, x_{ij}\right) \cdot \left(1 + \varepsilon_{ij}\right)$$

This error model applies if the deviations from the individual predictions are increasing proportionally with increasing observations (i.e. the relative deviations stay constant).

The combined residual variability model is another widely used residual variability model for the population approach. This residual variability model contains a proportional and an additive component:

$$Y_{ij} = f\left(\phi_i, x_{ij}\right) \cdot \left(1 + \varepsilon_{1ij}\right) + \varepsilon_{2ij}$$

This residual variability model behaves at small observation values like the additive residual variability model, for higher observation values the proportional component is dominating.

For the determination of the residual variability model it is assumed that ε_{ij} is a zero mean random variable with variance σ^2, that is multivariately symmetrically distributed.

17.4.1.3 Covariate Model

The covariate model describes the relationship between covariates and model parameters. Covariates are individual-specific variables that describe intrinsic factors like, e.g. sex, age, weight, creatinine clearance or extrinsic factors like, e.g. concomitant medication, alcohol consumption. The parameter–covariate relation should explain the variability in PK or PD model parameters to a certain extent.

The relationship between a covariate and individual model parameters should be exploratively investigated by a graphical presentation which often suggests a mathematical function that might best describe the relationship. The most commonly used functions are linear functions, so called "hockey stick" functions as well as exponential functions. The linear relationship can be implemented according the following equation:

$$\theta_{P-Cov} = \theta_P \cdot \left[1 + \theta_{Cov} \cdot \left(Cov - Cov_{median}\right)\right]$$

where θ_{P-Cov} describes the typical population parameter value for an individual with a certain covariate Cov. θ_P denotes the typical population value for an individual with a median covariate value. θ_{Cov} describes the influence of the covariate as a percentage change from θ_P per change of one covariate unit from the median covariate value Cov_{median}.

If the covariate covers a large range of values the data may not be satisfactorily described by the above presented linear model. An alternative model for those situations is the "hockey stick" model described in the following equation.

$$\theta_{P-Cov} = \begin{cases} \theta_P \cdot [1 + \theta_{Cov} \cdot (Cov - Cov_{median})] \text{ if covariate } \leq \text{median} \\ \theta_P \text{ if covariate } > \text{median} \end{cases}$$

The hockey stick function assumes a linear relationship until a node point is reached; afterwards another relationship can be used. In the above equation, the median was chosen as the node point and for all individuals having a larger covariate value than the median value no influence of the covariate was assumed.

Categorical covariates are most often implemented as a step function, where an example for a sex difference is given in the following equation:

$$\theta_{P-Cov} = \begin{cases} \theta_{P,male} & \text{if sex} = \text{male} \\ \theta_{P,female} & \text{if sex} = \text{female} \end{cases}$$

17.4.1.4 Population Model

The combination of all submodels together results in a population model which can be described by the following equation, where exemplarily an additive error model is used:

$$Y_{ij} = f[g(\Theta, z_i) + \eta_i + \kappa_{iq}, x_{ij}] + \varepsilon_{ij}$$

The equation describes each measured observation y from the ith individual at a certain time point j at occasion q by the individual prediction of the function $f()$ and the residual variability ε_{ij}. The function f characterizes the relationship between all investigated data and contains the documented and given independent variables x_{ij} (e.g. dose) and z_i (covariate, e.g. age), the vector Θ of all individual fixed effect parameters θ (PK and PD parameters, covariate influence) and the vectors (or scalars if one-dimensional) of the random effect parameters η_i, κ_{iq}, and ε_{ij}.

17.4.2
Parameter Estimation

Estimation of nonlinear mixed effects models has been implemented in a number of software packages and includes different estimation methods [12]. As NONMEM is the most commonly used software to estimate population parameters this program is base for the following description.

The aim of parameter estimation is an adaptation of the model function to the observations made to gain model parameters which describe the observed data best. In NONMEM this is done by the minimization of the extended least square objective O_{ELS} function, which provides maximum likelihood estimates under Gaussian conditions [13]. The equation calculating the O_{ELS} function is given in the following:

$$O_{ELS} = \sum_{i=1}^{n} \left[\frac{(y_i - E(y_i))^2}{var(y_i)} + \ln |var(y_i)| \right]$$

where var(y_i) is the variance–covariance of y_i, i.e. the ith individual vector of observations, and $E(y_i)$ is the expectation of y_i, the observation of the ith individual.

The objective function value is up to a constant, equal to minus twice the log-likelihood of the fit. Thus, a minimum objective function value reflects the maximum likelihood of the model parameters to describe the data best. The standard errors of the parameter estimates are also calculated by the maximum likelihood method.

In most models developed for pharmacokinetic and pharmacodynamic data it is not possible to obtain a closed form solution of $E(y_i)$ and var(y_i). The simplest algorithm available in NONMEM, the first-order estimation method (FO), overcomes this by providing an approximate solution through a first-order Taylor series expansion with respect to the random variables η_i, κ_{iq}, and ε_{ij}, where it is assumed that these random effect parameters are independently multivariately normally distributed with mean zero. During an iterative process the best estimates for the fixed and random effects are estimated. The individual parameters (conditional estimates) are calculated *a posteriori* based on the fixed effects, the random effects, and the individual observations using the maximum *a posteriori* Bayesian estimation method implemented as the *"post hoc"* option in NONMEM [10].

The FO method was the first algorithm available in NONMEM and has been evaluated by simulation and used for PK and PD analysis [9]. Overall, the FO method showed a good performance in sparse data situations. However, there are situations where the FO method does not yield adequate results, especially in data rich situations. For these situations improved approximation methods such as the first-order conditional estimation (FOCE) and the Laplacian method became available in NONMEM. The difference between both methods and the FO method lies in the way the linearization is done.

The FOCE method uses a first-order Taylor series expansion around the conditional estimates of the η values. This means that for each iteration step where population estimates are obtained the respective individual parameter estimates are obtained by the FOCE estimation method. Thus, this method involves minimizations within each minimization step. The interaction option available in FOCE considers the dependency of the residual variability on the interindividual variability. The Laplacian estimation method is similar to the FOCE estimation method but uses a second-order Taylor series expansion around the conditional estimates of the η values. This method is especially useful when a high degree of nonlinearity occurs in the model [10].

Due to the improved and complex mechanism both methods are more precise but also considerably slower than the FO method.

17.4.3
Building Population Models

For model building two different approaches are applicable: the top down and the bottom up approaches. The top down approach starts with the most complex model allowed by the data and this full model is reduced sequentially to include only the relevant features. This approach has the advantage that any change is evaluated in a model that includes as many true terms as possible, that means it is closer to the true model. Disadvantages are numerical instabilities of the full model and sometimes it is difficult to define the most complex model upfront. In addition, in most cases the top down approach is extremely time-consuming with respect to the computation times. The bottom up approach starts from the minimal model and expands this model until no more terms can be justified. As the data situation seldom allows the application of the top down approach and the computation times are so tremendous, the top down approach is not very common, while the bottom up approach is widely used for the building of population models [10].

To compare and evaluate different kinds of models created during the model development process graphical and statistical methods should be applied. A good description of the model building process can be found elsewhere [14].

17.5
Pharmacokinetic Models

Pharmacokinetics describe and predict the time-course of drug concentrations in body fluids. Pharmacokinetics answer the question "what does the body do to the drug?" The following processes occur after administration of a drug: absorption, distribution, metabolism, and excretion (ADME). PK models are quite common and well known in clinical drug development. In contrast to the PD models, the PK models can be clearly and easily classified into empirical and mechanistic models. In general, they are applied for the following situations:

- extrapolation of animal PK to human PK;
- simulation of the variability of plasma concentration–time profiles (mainly empirical models);
- simulations of multiple dose administrations from single dose data (mainly empirical models);
- simulation of sub-populations, e.g. different ages (pediatrics), weight, disease status, etc.;
- simulation of the impact of the release characteristics of different dosage forms on the PK profile;
- simulation of drug absorption after different routes of administration;
- simulation of drug concentrations in tissues and organs which are not accessible for sampling (mainly physiologically based pharmacokinetic models).

The following provides a brief overview of the empirical and mechanistic PK models.

17.5.1
Empirical Pharmacokinetic Models

In pharmacokinetics empirical models are for example "compartment models" where the body is sub-divided into one or more compartments and the drug is assumed to distribute and be eliminated with first-order rate constants. Typical model parameters are the rate constants and the volumes of the compartments. The compartmental models reflect the physiological reality only to a very limited degree. Despite this limitation compartment models are essential in drug development and have received considerable attention and showed huge utility and impact on the labeling of drugs on the market [4].

A full empirical PK model using the population approach contains, as described in Section 17.4, three sub-models: the structural, the statistical, and the covariate.

The structure of a compartmental PK model is given by the number of compartments being used and the way the compartments are connected. For most drugs the plasma concentration–time profiles can be sufficiently described by one-, two- or three-compartment models. A one-compartment model assumes that no time is necessary for the distribution of the drug and the whole distribution occurs within this one compartment. The two-compartment model implements in addition to a central compartment a peripheral compartment which allows the description of distribution processes of the compound in, e.g. tissues with different physicochemical properties. The three-compartment model provides an additional compartment for distribution processes. The use of more than three compartments is quite rare, but there are some situations where the model can get quite difficult, e.g. if the concentration–time profile of metabolites are also considered within the model. For more information regarding compartment models, refer to Rowland and Tozer [15].

The choice of the compartment model is mainly driven by the data quality and the frequency of the plasma sampling, e.g. if no samples are taken shortly after the administration of the compound, where most of the distribution processes occur, it might be possible that a simpler model (i.e. one with less compartments) is sufficient to describe the data.

The statistical model is incorporated as described in Section 17.4. The covariate model allows a description of relationships between covariates and model parameters, explaining parts of the intersubject variability and identifying sub-populations at risk for concentrations below or above the therapeutic range.

As described in Section 17.2, the model building is mainly data-driven and drug-dependent, model parameters and their variability are estimated based on the data available.

17.5.1.1 Example: NS2330 (Tesofensine)
There are many good application examples published. The following briefly summarizes one interesting example [16].

Background NS2330 is a new central nervous system (CNS)-active drug in clinical development for Alzheimer's disease and Parkinson's disease [17]. NS2330 shows a high volume of distribution of about 600 l and a low oral clearance of 30–40 ml/min. The high volume of distribution and low clearance result in a long half-life of NS2330 of about 200 h. M1 is the only metabolite found in human plasma and shows a longer half-life than the parent compound of about 300–400 h. The trough concentrations of M1 in humans after oral administration of NS2330 at steady state are about one-third of the trough steady-state concentrations of the parent compound. The metabolite M1 shows the same *in vitro* pharmacological profile as the parent compound with higher *in vitro* potency compared with NS2330.

Objective The objective of this analysis was to develop a population pharmacokinetic model for NS2330 and its major metabolite M1, based on data from a 14-week proof of concept study in Alzheimer's disease patients, including a screening for covariates that might influence the pharmacokinetic characteristics of the drug and/or its metabolite. Subsequently, several simulations should be performed to assess the influence of the covariates on the plasma concentration–time profiles of NS2330 and its metabolite.

Methods and Patients Plasma concentration–time profiles of 201 female and 119 male subjects after 0.25, 0.5 or 1.0 mg oral NS2330 (consisting of 1969 NS2330 and 1714 metabolite concentrations) were included in the population PK analysis.

The structural model was developed in a stepwise manner, starting with a one-compartment model for parent compound and metabolite, respectively. Analyses were performed using NONMEM V, ADVAN 5, and the FOCE INTERACTION estimation method.

17.5.1.2 Results

Modeling Plasma concentration–time profiles of NS2330 and M1 were best described by one-compartment models with first-order elimination processes for both compounds. Absorption of NS2330 was best modeled by a first-order absorption process. Setting the typical volume of distribution of the metabolite M1 (V3/F) to 0.768-fold of the typical volume of distribution of NS2330 (V2/F) described the data best. This value was based on prior knowledge gained in preclinical studies. A schematic illustration of the PK model is shown in Fig. 17.3. Low apparent total clearances (1.54 l/h for NS2330, 0.936 l/h for M1) in combination with the large apparent volumes of distribution (653 l for NS2330, 502 l for M1) resulted in long half-lives of 295 h (NS2330) and 371 h (M1).

The covariate analysis identified weight, sex, creatinine clearance, body mass index (BMI), and age as having an influence on PK parameters of NS2330 and/or M1, respectively. The overall clearance of NS2330 is reduced by 20.5% in females compared to males and the NS2330 clearance which accounted for any elimination pathways except the formation of M1 from NS2330 ($CL_{non\text{-}met}$/F) is reduced in patients with a creatinine clearance lower than 62.5 ml/min ($\sim$ 1.2% reduction per 1 ml/min reduction). The volume of distribution of NS2330 is influenced by body

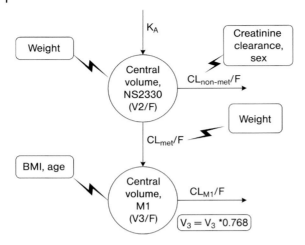

Fig. 17.3 Schematic population PK model.

weight ($\sim$ 1% increase per 1 kg increase) and the volume of distribution of M1 is influenced by BMI ($\sim$ 2% increase per 1 kg/m^2 increase).

Evaluation The robustness and predictive performance of the population model developed was demonstrated as it successfully predicted the observations obtained in an other independently performed study (14-week proof of concept study in 202 Parkinson patients treated with 0.125, 0.25, 0.5, and 1.0 mg NS2330 orally once daily – data not shown). The assumption was that the indication investigated did not have an impact on the PK characteristics.

Simulation As only clearance determines the extent of exposure, the influence of the covariates that influence CL/F were illustrated by simulation of plasma concentration–time profiles following once daily chronic dosing (1 mg). The exposure in a renally impaired female (CRCL=35.6 ml/min) was increased by 62% compared with a male with normal renal function (Fig. 17.4).

Conclusion A population PK model for the new compound NS2330 and its major metabolite could successfully be developed describing the plasma concentration–time profiles of the parent compound and its metabolite in Alzheimer's disease patients.

The covariate analysis revealed physiologically plausible covariate effects, partly explaining the variability observed in the pharmacokinetic parameters. A high interindividual variability on the CL/F not responsible for the generation of M1 still remained even after the incorporation of the covariates sex and creatinine clearance. This indicates that there still might be yet undiscovered covariates additionally influencing the elimination of NS2330.

The covariates sex and creatinine clearance had an influence on the plasma concentration–time profiles. The combination of both covariates resulted in an

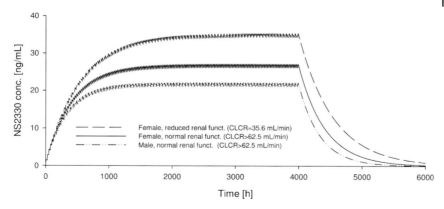

Fig. 17.4 Impact of sex and creatinine clearance on the plasma concentration–time profiles of NS2330 after oral administration of 1mg NS2330 once daily for 166 days (4000 h). Typical profiles of a male subject with normal renal function, a female subject with normal and reduced renal function.

62% increased exposure in a renally impaired female compared to a male with a normal renal function.

17.5.2
Mechanism-based Pharmacokinetic Models

A more mechanism-based approach to pharmacokinetic modeling is offered by physiology-based PK (PBPK) models, where the drug concentration–time profiles are described in the respective organs and body fluids considering the physiological system (= human body) and the physicochemical properties of the drug to a great extent [18]. The basic principle of PBPK modeling is to segment the mammalian body into physiological relevant compartments and develop mathematical expressions of physical and physiological processes for each compartment describing the fate of the compound within the respective compartment. The compartments principally reflect important organs, like liver, heart, lung, muscle, brain, etc. The structure, the complexity, and the level of detail of a PBPK model may vary tremendously depending on the aim of the model and the information available. Overall, there are two different types of PBPK models used in drug development [19]. First, whole-body PBPK models with a closed loop similar to Fig. 17.5; second, partial PBPK models of selected body systems, like the gastrointestinal tract describing the absorption/gut wall metabolism or the liver describing the hepatic elimination/metabolism.

To decrease the complexity of a PBPK model some organs can be lumped together if the information about the behavior of the single organ is not important for the aim of the model.

Once the model structure is defined, the model of the particular tissue or organ needs to be specified. The basic passive processes which determine the behavior of

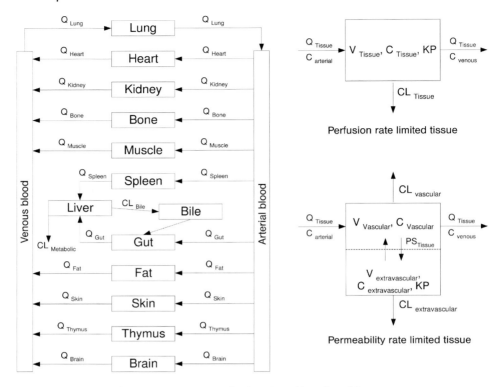

Fig. 17.5 Schematic representation of a physiological based model. Left figure shows the physiological structure, upper right figure shows a model for a perfusion rate limited tissue, and lower right figure shows a model for a permeability rate-limited tissue. Q denotes the blood flow, CL the excretion rate, KP the tissue:plasma distribution coefficient, and PS the permeability surface area coefficient.

the substance in an organ are mass transport by the blood flow, permeation from the vascular space into the organ tissue, and partitioning between blood/plasma and organ tissue. In principle, two different models are mainly used for the description of these passive processes (see Fig. 17.5). First, a perfusion rate-limited model and second a permeability rate-limited model. The perfusion-limited model represents each tissue or organ as a single-well stirred compartment and assumes that the drug distributes instantaneously in the whole volume of the tissue from the incoming blood. This represents a simplification of the biological situation, but often sufficiently describes the data. A more complex description of the organ or tissue is the permeability rate-limited model. This model consists of two-well stirred compartments with rate-limited transfer between both of them. There are also more complex models proposed, e.g. several models include the presence of drug transporters or metabolizing enzymes.

Table 17.2 Important PBPK parameters.

System parameters (drug-independent)	Drug-dependent parameters
Organ volumes	Lipophilicity
Blood flow rates	Protein binding
Effective surface areas for permeation processes	Molecular weight
pH values at different locations	Clearance
	Solubility
	pK_a/pK_b value

Overall, the development of a PBPK can be quite challenging if the whole model is newly developed. However, there are several PBPK software packages available which have predefined structures already implemented and need only the adaptations for the properties of the compound [19]. Those packages are more user-friendly and less time-consuming in development, but they are also less flexible than other software packages where the whole model needs to be implemented by the modeler. A summary of the PBPK software packages can be found in Section 17.10.

The parameter set of a PBPK model includes in general two sets of model parameters, the compound-independent physiological parameters ("systems parameter") and the compound-specific physicochemical parameters ("drug-dependent parameters") [20]. Some of the most important parameters are listed in Table 17.2.

17.5.2.1 System Parameters
Physiological parameters used for PBPK modeling are drug-independent and need to be collected from different sources for a successful development of a reliable model. As mentioned in Section 17.2, this might be challenging, as the quality of the public data is difficult to assess and rarely is information given about the methods of the measurements, standard errors, and variability. There are a few published papers summarizing important physiological parameters [21–24]. In addition, most pharmaceutical companies have internal sources of parameters which they might use for model development. There are also several off-the-shelf PBPK software products on the market, which have already an incorporated set of physiological parameters (see Section 17.10).

17.5.2.2 Drug-dependent Parameters
Physicochemical parameters determine several characteristics of the compound, like binding, partition, permeability or excretion. When building a PBPK model the level of information available for a drug may vary depending on the status of the development and the experiments performed. For model development the unknown parameters might be assumed and refined when the experiment is performed or some of the parameters can be estimated by the model, for example the different clearance values describing renal, hepatic and other excretions.

17.5.2.3 **Examples**

Currently, PBPK modeling is often applied in drug discovery, toxicology, and for allometric scaling from animals to humans. Examples within clinical drug development are rare. Examples for the application of PBPK are often shown for the therapeutic areas of oncology [25–28] and antibiotics [29–32]. Especially for these kinds of drugs the concentration within the target organ plays a major role. The PBPK modeling approach allows now the simulation of the plasma concentration–time profiles within the target organs. For oncology compounds the respective tumor compartments can be added to the model and adapted according to the type of tumor.

A good resource for other examples and also for future directions can be found on www.PBPK.org.

17.6
Pharmacodynamic Models

Pharmacodynamics (PD) is the study of the time course and intensity of drug effects and tries to answer the question "what does the drug do to the body?" Pharmacodynamic modeling has undergone major advances during recent years, due to technological advances, increased understanding of mechanism-based drug effects and increased possibilities to measure drug effects. In general, they are applied for the following situations:

- simulation of the variability of plasma concentration–effect time profiles (mainly empirical and semi-mechanistic models);
- simulation of the effect after multiple dosing from single-dose data (mainly empirical models and semi-mechanistic models);
- simulation of sub-populations, e.g. different ages (pediatrics), weight, disease status, etc.;
- simulation of the impact of the release characteristics of different dosage forms on the PD profile;
- knowledge gathering about PD processes (mainly semi-mechanistic and mechanistic models).

Overall, PD models can be quite heterogeneous, depending on the level of mechanism reflected; and a clear classification is almost impossible. However, the following gives a brief overview of the basic PD models.

17.6.1
Empirical Pharmacodynamic Models

Empirical PD models describe the relationship between drug concentrations in body fluids and the observed drug effect (efficacy, safety, biomarker measurements). Similar to empirical PK models, PD models are relatively simple and a variety of drug effects can be described with a few PD models. Empirical PD models have also been proven to be of tremendous value in drug development and are an integral part of PK/PD models. A full empirical PD model using the population approach includes, besides the structural model, also the statistical and covariate model to account for the observed variability.

A good review of the history of PD models is given, e.g. by Csajka and Verotta [33].

The easiest models to describe concentration–response relationships are linear models which postulate that the effect and concentration are direct proportional, e.g.:

$$E = E_0 + m * C$$

Where E_0 denotes the effect without medication (= baseline), m is the slope and C the drug (active metabolite) concentration. A more frequently used model is the log-linear model where a linear relationship between the effect and the logarithm of the concentration is assumed:

$$E = m * \log C + \text{Intercept}$$

Both models have the disadvantage that the effect increases with increasing concentrations without an upper limit which is a highly unphysiologic behavior. Therefore, such models should be used with caution or better should not be used for extrapolation outside the observed range. The "gold standard" of the empirical PD models is the sigmoid E_{max} model with or without the Hill coefficient:

$$E = \frac{E_{max} \cdot C^n}{EC_{50} + C^n}$$

where E_{max} is the maximum effect attributable to the compound and EC_{50} denotes the concentration producing 50% of the maximum effect, n is the Hill factor affecting the shape of the curve (a high value of n makes the curve steeper, in an extreme case a step function is obtained), and C reflects the concentration of the compound (Fig. 17.6). This E_{max} model can be easily modified to describe different kinds of situations, e.g. to describe the competitive interaction between parent compound and metabolite [34] and many other situations.

17.6.1.1 Linking Pharmacokinetics and Pharmacodynamics

An important aspect of PD models is the link between the pharmacokinetics and the pharmacodynamic model, i.e. which concentration drives the drug effect. For empirical models there are many different PK/PD link approaches described and the theory is presented in several review papers [33–35]. The two most popular approaches are described in the following and are also illustrated in Fig. 17.7. The direct link approach assumes that a change in the measured concentration is directly reflected in a change in the measured PD. This is most often observed if the site of PK measurement and the site of PD measurement are identical (e.g. PK measurement in plasma and clotting time as PD measurement).

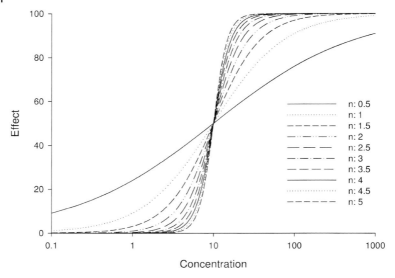

Fig. 17.6 E_{max} model with a given $E_{max} = 100$, $EC_{50} = 10$ and ten
different n values, ranging from 0.5 up to 5.0.

Often the drug effect lags behind the measured plasma concentration–time profile ("hysteresis"), i.e. the same PD measurement corresponds to different measured plasma concentrations. In this case often a hypothetical effect compartment
is introduced into the model, where a rate constant k_{eo} determines the elimination
out of this effect compartment. The concentration–time profile within this effect
compartment is now linked to the PD effect model. This allows the description of a
time delay between the plasma concentration and the PD (see Fig. 17.7) and again
a unique relation between concentration and the observed PD is established.

17.6.2
Mechanism-based Pharmacodynamic Models

Mechanism-based models try to incorporate preferably a highly realistic representation of the biological system. The level of detail may vary from genetic level up
to the semi-mechanism-based level. The following section refers to models which
are also known as systems biology models. Meanwhile a few pharmaceutical companies have actually established own systems biology groups [36] or computational
biology groups with the aim to develop mathematical models for drug research and
development.

The modeling building process is iterative and is guided by the knowledge available in literature, developed in house or on hypothesis. The model parameters are
in general taken from public databases (KEGG, etc.), from literature or from own
investigations performed. This mixture of sources makes the standardization and
the quality assurance of the chosen parameter value a challenge. Datasets in public databases can be incomplete, they are rarely standardized, or properly anno-

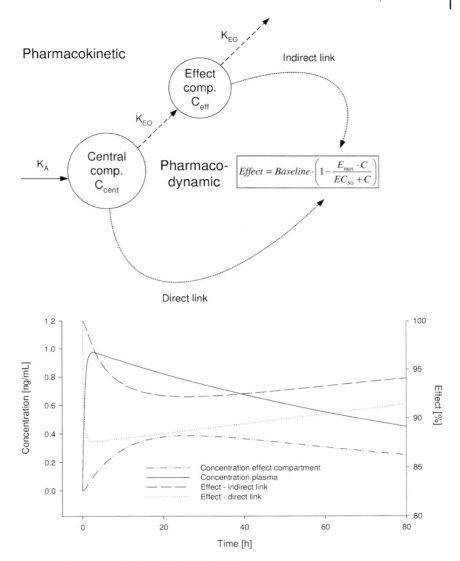

Fig. 17.7 Direct versus indirect link of PK and PD.

tated with the method used for estimation [37]. In addition, standard errors and variability of the parameters are infrequently provided. This makes it difficult to implement uncertainty and variability in those models when used for simulation and is therefore rarely done. Another challenge in model development is the fact that some biological processes are currently very difficult to identify and quantify. Mechanism-based models require also a high interdisciplinary effort. Different kinds of scientists are necessary for a successful model development. Continuous communication to the whole team is a prerequisite. Overall, the model development process is a complicated and both time- and research-intensive procedure. The time of a model development process can also vary, but is much more intensive compared to the empirical models. In addition, the model might be modified over the time when new knowledge of specific processes is revealed.

Another challenge in the development of mechanism-based models is data ownership and information distribution. As the pharmaceutical industry is a highly competitive area, the information-sharing philosophy of the pharmaceutical companies is improvable. Hopefully, future liaisons and partnerships will improve the data-sharing between various institutions like companies, regulatory agencies, and academia.

Once the model is developed it has the big advantage that it can be applied generically to different situations, e.g. different drugs. This makes those models superior over the empirical models and very interesting for the drug development process, especially when follow-up compounds or compounds with similar properties need to be considered.

Overall, the implementation of complex mechanism-based models in drug development progresses slowly. There might be several reasons for this. One is that the number of success stories is small and, consequently, the investment in this technology is low. Second, the complexity of these models requires a new type of modeling expert which is currently very rare. Third, currently most mechanism-based models do not consider variability in their model parameters.

Overall, there is a need and an interest in the implementation of this new technology and the future will show whether these models will become a permanent part of the clinical drug development process.

17.6.2.1 Examples

The number of published mechanism-based PD models having an impact on clinical drug development is quite low. A good collection of different types of biological models can be found in the previous chapters of this book. Additionally, the BioModels database [38] (http://www.ebi.ac.uk/biomodels/) presents a good collection of different kind of models, some of which might also be applicable to situations in drug development.

Examples for interesting mechanistic platform models are presented by Entelos with their commercial PhysioLab systems (www.entelos.com). Entelos provides platforms in the areas of asthma, obesity, type 1 and type 2 diabetes, rheumatoid arthritis, and cholesterol metabolism. The models are developed by a top down approach using data from literature. Each PhysioLab platform represents a dynamic whole-body physiology containing a graphical disease map, mathematical equations, and a reference database. The PhysioLab technology might have some benefit also for the clinical drug development. The product allows the simulation of clinical trials using virtual patients that represent real patient populations. This approach might be able to successfully identify optimal dose, dose frequency, responder/nonresponder populations, and surrogate markers.

Another interesting example of platform modeling was presented by Eddy and Schlesinger with their Archimedes model. Initially, they created a model of the anatomy, pathophysiology, tests, treatments, and outcomes pertaining to diabetes [39]. In addition, the model contains aspects of the health care system, like health care personnel, facilities, procedures, regulations, and others. The model can be used to simulate, e.g. the outcome of clinical trials. The predictive performance of the model has already been shown [40–43]. Meanwhile this promising model is being extended with additional diseases, like coronary artery disease, congestive heart failure, asthma, stroke, hypertension, dyslipidemia, and obesity (www.archimedesmodel.com).

17.6.3
Semi-mechanistic Models

The empirical and fully mechanistic models reflect the extreme of both situations, either a huge reductionism or a complex description of the biological system. As each approach has its strengths and weaknesses, a hybrid of both (the so-called "semi-mechanistic" approach) was introduced into the field of clinical modeling and simulation. Within the area of M&S these models are also often just referred to as "mechanistic models".

Overall, these models are closer to the empirical PD models, but major elements of the biological system are implemented. Semi-mechanistic models are mostly developed using the population approach and consequently they are data-driven and parameters are estimated from the data available. Parameters which cannot be estimated might be either fixed to biologically meaningful values or they are explored by other studies, including *in vitro* or preclinical *in vivo* studies. Overall, the number of parameters is still small, compared with mechanistic PD models; and the majority of the parameters are estimated.

A recently published review [44] summarizes the importance and development of this approach. The major application is seen for the description of target site equilibration, transduction, target interaction and activation, homeostatic feedback and disease process. Especially, the search and implementation of new biomarkers makes this approach feasible and promising for the future application of PD models in clinical drug development.

17.6.3.1 Example: BIBN4096
Several applications have been published. The example by Troconiz et al. [45] is used in the following to illustrate the principles.

Background BIBN4096 is a nonpeptide CGRP receptor antagonist that based on *in vitro* and *in vivo* preclinical studies shows very high affinity and specificity for the human CGRP receptor. The compound is in the clinical development as a new target for the treatment of migraine.

In a proof of concept study where BIBN 4096 BS was administered intravenously by a 10-min infusion to patients with migraine attacks as rated as moderate to severe, the investigators found response rates similar to the efficacy rates reported for triptans, together with good safety and tolerability profiles. The pharmacokinetics of BIBN4096 were characterized by means of noncompartmental and population pharmacokinetic analyses. BIBN4096 showed a linear pharmacokinetic behavior that was not mainly affected by standard demographic characteristics.

Objective The objective of this analysis was to develop a population PK/PD model based on the data after intravenous administration that allows to explore *in silico* which absorption characteristics would be required for other administration routes and/or which receptor binding properties are crucial for backup compounds.

Methods and Patients Overall, 126 patients with an acute moderate to severe migraine attack lasting not more than 6 h were enrolled in this study. BIBN 4096 was given as a single intravenous 10-min infusion at different dose levels ranging from 0.25 mg to 10.0 mg. Severity of headache as a PD measurement was measured up to 24 h. Patients who did not show pain relief by 2 h were allowed to take rescue medication. Severity of headache and time to rescue medication measurements were fitted simultaneously using logistic regression and time-to-event analysis with NONMEM.

17.6.3.2 Results

Modeling Plasma concentration–time profiles of BIBN4096 were best described by a three-compartment model with first-order elimination processes. The drug effect was described using a logistic regression model, where the fraction of receptors that is blocked by BIBN 4096 determines the probability of a certain severity of headache. The link between the blocked receptors and PK is established using a drug–receptor interaction model:

$$\frac{dR*}{dt} = k_{on} \cdot C \cdot R - k_{off} \cdot R*$$

where $dR*/dt$ is the rate of change of $R*$ (fraction of blocked receptors), C corresponds to the predicted plasma concentrations of BIBN 4096 BS that were obtained from the pharmacokinetic model developed previously, R is the concentration of unblocked receptors, and k_{on} and k_{off} represent the second- and first-order rate constants representing the onset and offset of the anti-migraine effect elicited through the respective amount of CGRP receptors blocked by BIBN 4096 BS. Fig. 17.8 shows the schematic PK/PD model. The model predicted a slow rate of offset of the anti-migraine effect (half-life of $k_{off} = 21$ h). The model developed described the data well and was validated properly.

Simulation Simulations exploring the effect of the rate of absorption, bioavailability after an extravascular administration, and the rate of activation/inactivation of the anti-migraine effect were performed. To achieve a response rate of 60% at 2 h, the rate of absorption seems to play a minor role; if bioavailability fractions of at least 0.2–0.3 can be achieved and k_{on} is >0.081 ml/ng. At later times after administration, higher values of k_{off} are associated with faster offset of the response.

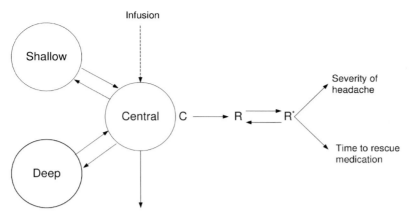

Fig. 17.8 Schematic representation of the PK/PD model. C = model
predicted drug concentrations in plasma; R = the free form of the
calcitonine gene-related peptide (CGRP) receptor; R* = the blocked
form of the CGRP receptor, which has been related to the severity of
headache and time to rescue medication using logistic regression and
time-to-event analysis.

Conclusion A population PK/PD modeling for BIBN4096 was successfully developed describing the plasma concentration–time profiles of BIBN4096 and its influence on the severity of the headache and the time to rescue medication.

Simulations gained important information for further pharmaceutical development with regard to the absorption and bioavailability prerequisites.

The simulations showed in addition that molecules with high k_{on} and low k_{off} values are the most promising which should be considered for further drug development of compounds with a similar mode of action.

17.7
Disease Progression Models

Disease progression can be defined as the change in disease status over time. For the simulation of long-term administration of drugs intended to treat degenerative diseases, progression models can be very helpful and should be considered for model building and interpretation of the results.

Especially for special therapeutic areas where it is unethical to treat patients for a longer time with a placebo, the development of disease progression models is of high interest because they allow a better discernment of the true treatment effect, for example, for these therapeutic areas: Alzheimer's disease, Parkinson's disease, Osteoporosis, HIV, diabetes, and cancer. Furthermore, such models can help to differentiate whether the drug has only a symptomatic effect or a disease-modifying effect. The implementation of disease progression models also improves the reliability and acceptance of simulations.

Disease progression can be modeled in different ways. This depends mainly on the level of detail that is provided within the pharmacodynamic model (see Section 17.6). Within a mechanism-based PD model the disease progression can easily be implemented if the mechanism of the progression of the disease is elucidated. For most of the above-listed diseases there is still a lack of knowledge that hopefully will be filled in the near future.

If the empirical approach is chosen, an empirical disease progression model can also be developed based on data collected during clinical trials or from public databases. A general empirical disease progression model has the following components:

$$DP(t) = \text{Baseline} + \text{Progression}/t$$

In case of a symptomatic effect of the drug the curve resulting from the above equation is just shifted in parallel, i.e. the same disease status is achieved at a later time point, whereas in case of a disease-modifying drug the Progression/t should be decreased by the drug.

The models are built similar to the descriptive mechanism-based PD models. Most of them are also estimated by the nonlinear mixed effects modeling approach considering interindividual and residual variability. In addition, covariates influencing the disease progression can also be investigated.

The model development also contains some challenges. The available data often lack information for untreated patients. For most of the above-listed diseases it is ethically not justifiable to leave these patients untreated for a long time period. Furthermore, it is time-consuming to collect longitudinal data; and within the clinical drug development this might be a critical hurdle. In addition, there is a large variability within the data which might require a large number of subjects to determine parameters accurately.

Despite the complexity of data collection and model generation, there are quite a few examples published in the literature describing the disease progression of various diseases (e.g. [46, 47]). Most of them are quite simple with just a few parameters and describe the disease progression just by a linear relationship. Nevertheless, they show a benefit of the understanding of drugs treating degenerative diseases [46]. In addition to the "simple" linear empirical models, there is also a huge trend towards a more mechanism-based description of the disease progression (e.g. [47]).

Overall, the disease progression models allow a visualization of the time course of diseases under treated and untreated conditions and allow one to investigate *in silico* the impact of different therapeutic interventions.

17.8
Patient Models

For a successful simulation the distribution of interesting intrinsic and extrinsic factors of the patient population should be available. Furthermore, the compliance of the patient population and the probability that a patient discontinued his/her treatment should be known. These features can also be described by models, mainly empirical models, as detailed in the following.

17.8.1
Covariate Distribution Model

The covariate distribution model defines the distribution and correlation of covariates in the population to be studied. The aim of a covariate distribution model is to create a virtual patient population that reflects the target population for simulations including patients' covariates. This model is of great importance for the realistic simulation of clinical trials.

In general, a covariate distribution model considers only the covariates influencing the PK and/or PD of the compound of interest. For example, if the covariates age, sex, and weight are identified as important covariates than correlated covariates like height, body mass index, and others might not be incorporated.

Single or independent covariates can be described by univariate distribution models, like a normal distribution. Multivariate distribution models need to be considered if two or more covariates are correlated like weight and height. This assures that "realistic" patients are created with a plausible set of covariates. For the creation of categorical covariates, like sex or smoking status, discrete functions are used. Most commonly the Bernoulli distribution is used. Another option is the use of a uniform distribution with simple cut-offs, e.g. if random number <0.5 then male, else female. However there are several other functions available to describe discrete functions. A comprehensive review of the mathematical background of covariate distribution models is given by Kimko and Dufful [48].

For the appropriate development of covariate distribution models, the pharmaceutical industry has huge amount of data in their clinical databases. In addition, there are also public databases available which can be used, like the *Congestive Heart Failure Database* (http://www.physionet.org/) derived from patients undergoing cardiac catheterization at Duke Medical Centre during 1990–1996 (about 4000 patients, data on demographics, risk factors histories, cardiac catheterization, EKG, cardiac scores, follow-up data).

Overall, it should be considered how the target population might look. For example, the constitution of patients of phase I trials used to develop a model might be completely different from the target population for the simulated phase III trial, e.g. geriatric or obese patients. In addition, it needs to be considered that some covariates might change over time, e.g. caused by treatment effect or by the progression of the disease.

17.8.2
Compliance Model

Noncompliance to medication is an important issue that occurs within ambulatory patients and within patients included in clinical trials. Noncompliance should be considered if simulations over longer periods are performed where multiple dosing is the basis of the simulation.

The most frequent noncompliance behaviors are deviations from the amount of dose (this includes leaving out doses), the timing of administration, and the duration of treatment. Thus, noncompliances range from short-time deviation from the administration time over randomly omitted doses up to so-called "drug holidays" where the patient "takes a break" from medication. The worst case is discontinuation of the medication, which is discussed in Section 17.8.3.

Deviations from dosing time can be easily implemented in the model by distributions around the recorded dosing time and the actual dosing time. Randomly omitted drug administrations and drug holidays can be comprehensively described by hierarchical Markov models. A good explanation is given by Kimko and Dufful [48].

17.8.3
Drop-out Model

Discontinuation of self-treatment by the patient (= drop-out) is a critical point for the evaluation of studies. There are two different types of drop-outs: informative and non-informative. Non-informative drop-out means that no reason for the discontinuation is given; this means that the patient stops taking medication by chance. Informative drop-out means that discontinuation occurs due to facts that give feedback for evaluation of the treatment, e.g. discontinuation due to drug-related side-effects. A comprehensive description of drop-out models can be found elsewhere [48].

17.9
Outlook/Future Trends

Modeling and simulation are important aspects of the current clinical drug development program in large pharmaceutical companies. The population approach for example has already shown its value for the drug development process. But there is still an urgent need for a more efficient drug development processes and the application of modeling and simulation might be a cornerstone to achieve this goal.

Overall, there seems to be a paradigm shift in pharmaceutical industry. The necessity to systematically integrate data across disciplines to gain knowledge is realized. Models represent the ideal tools to combine data and test hypothesis in a quantitative manner. Pharmacokinetic, pharmacodynamic, disease progression, patient characteristics, and competitor information should be assimilated in models. These models and appropriate simulations using these models will allow the design of informative development programs. More and more mechanistic components are used in the models to improve the predictability. To facilitate building of these models, (patho)-physiological and disease progression models within therapeutic areas of interest have been established which can be used across different projects.

Within the PK area, the mechanistic PBPK models is also becoming more acknowledged and seems to be more frequently applied in the early stages of drug development. They have shown major improvements in the recent years within their prediction and with the further improvement and application they might also play a role in the drug development process.

The mechanistic PD models, also known as systems biology models, are currently rarely applied in clinical drug development. The various reasons for that are explained in the respective section. However, the pharmaceutical companies also pay some attention to the development of these models. These models have a high potential to be applied in drug research, e.g. for discovering new targets or for the explanation of different behaviors of the compound. In clinical drug development these models might also become more relevant; and also examples like Entelos or Archimedes show some promising results. However, to become a major part of the clinical drug development process, these models need to present a face value, e.g. by the publication of some success stories within pharmaceutical companies.

Despite the hurdles and challenges on the way to a major improvement of the drug development process by M&S there are also some organizational issues that should be considered. To create new and complex models requires a rethinking in the behavior of all relevant parties in drug development. It will be important to establish cooperation between academia, industry, and regulatory agencies, working together and sharing data and knowledge. Only joint action by all parties might be the key for successful model development in the future.

17.10
Software

There are several software packages available to implement the models described in this chapter. The following presents the most prominent.

When software is used for regulatory relevant analyses the FDA's 21 CFR Part 11 (http://www.fda.gov/ora/compliance_ref/part11/) should be considered.

17.10.1
General Simulation Packages

There are several general simulation packages on the market like S-Plus, R, SAS, Matlab. The listed packages have the big advantage that they are extremely flexible and very powerful. Nearly every kind of model can be implemented and estimation algorithms are also implemented, including nonlinear mixed effect modeling technique. The challenge with these packages might be that the models need to be implemented manually into a program and consequently the modeler needs to be familiar with the syntax of the software.

However, there are some graphical add-on tool boxes available for, e.g. Matlab, making the implementation of complex models more convenient (http://www.sbtoolbox.org/, http://www.mathworks.com/products/simulink/). The biggest challenge is often to integrate the dosing history for PK models.

17.10.2
PBPK Software

There are several commercial software packages available for the simulation of PBPK models, e.g. PK-Sim (Bayer Technology Services, http://www.pk-sim.com/), GastroPlus (SimulationPlus, http://www.simulationplus.com), SimCyp (Simcyp Ltd, http://www.simcyp.com/) or MEDICI-PK (CIT, http://www.cit-wulkow.de/tbgmed.htm). In addition, PBPK can also be implemented in the above-mentioned general simulation packages and NONMEM can be used for modeling and simulation. However, the commercial PBPK software packages have several advantages compared with the general simulation packages. The most important aspect is their user-friendliness. They are equipped with a convenient graphical user interface. In addition (as mentioned in Section 17.5.2), either they have physiological parameters already implemented in a software-related database or they allow easy implementation. Only a few compound specific parameters need to be specified. The kind and number of parameters to be specified are dependent on the software package.

17.10.3
Population Approach Software

Meanwhile there are several software packages and options available for estimating population parameters [49], based on nonlinear mixed effects modeling. NONMEM [10] was the first software package available for population analysis. In later years, several other software packages became available (e.g. NLME procedure in S-Plus, NLMIX procedure in SAS, MONOLIX add-on for MATLAB, WIN-NONMIX), implementing comparable estimation methods. WINNONMIX shows, e.g. a graphical user interface, the remaining packages are line command-based. Nevertheless, NONMEM is still the most widely used [50] software package.

Implementation of models in all of the above-mentioned packages is not an easy task and requires some experience. With the exception of NONMEM and WIN-NONMIX, it is also very difficult to implement complete dosing histories into the programs. In addition, the packages are good for parameter estimation. A flexible and straightforward simulation afterwards is difficult and also requires some experience.

17.10.4
Clinical Trial Simulators

The *Trial Simulator* (Pharsight Corp., http://www.pharsight.com) is a comprehensive and powerful tool for the simulation of clinical trials. Population PK/PD models developed with tools mentioned in Section 17.10.3 can be implemented in a *Trial Simulator*. In addition, treatment protocols, inclusion criteria, and observations can be specified. Also covariate distribution models, compliance models, and drop-out models can be specified. All of these models can be implemented via a graphical user interface. For the analysis of simulation results a special version of S-Plus is implemented and results can also be exported in different formats, like SAS.

References

1 Sheiner, L. B., Rosenberg, B., Marathe, V. V.: Estimation of population characteristics of pharmacokinetic parameters from routine clinical data. *J Pharmacokinet Biopharm* **1977**, 5:445–479.

2 FDA: *Guidance for Industry: Population Pharmacokinetics*, **1999**.

3 FDA: *Administration. Guidance for Industry: Exposure–Response Relationships – Study Design, Data Analysis, and Regulatory Applications*, **2003**.

4 Bhattaram, V. A., Booth, B. P., Ramchandani, R. P., et al.: Impact of pharmacometrics on drug approval and labeling decisions: a survey of 42 new drug applications. *AAPS J* **2005**, 2005:7.

5 DiMasi, J. A., Hansen, R. W., Grabowski, H. G.: The price of innovation: new estimates of drug development costs. *J Health Econ* **2003**, 22:151–185.

6 FDA: *Challenge and Opportunity on the Critical Path to New Medical Products*, **2004**.

7 Chien, J. Y., Friedrich, S., Heathman, M. A., de Alwis, D. P., Sinha, V.: Pharmacokinetics/pharmacodynamics and the stages of drug development: role of modeling and simulation. *AAPS J* **2005**, 2005:7.

8 Aarons, L.: Population pharmacokinetics: theory and practice. *Br J Clin Pharmacol* **1991**, 32:669–670.

9 Mandema, J. W.: *Population Pharmacokinetics and Pharmacodynamics*, in Peter Welling (ed): *Pharmacokinetics: Regulatory,*

Industrial, Academic Perspectives. Marcel Dekker, **1995**, 411–450.

10 Beal, S. L., Sheiner, L. B.: *NONMEM Users Guides*. Universtity of California, San Francisco, **1998**.

11 Karlsson, M. O., Sheiner, L. B.: The importance of modeling interoccasion variability in population pharmacokinetic analyses. *J Pharmacokinet Biopharm* **1993**, 21:735–750.

12 Aarons, L., Balant, l. P., Mentre, F., et al.: Population approaches in drug development. Report on an expert meeting to discuss population pharmacokinetic/pharmacodynamic software. *Eur J Clin Pharmacol* **1994**, 46:389–391.

13 Beal, S. L.: Population pharmacokinetic data and parameter estimation based on their first two statistical moments. *Drug Metab Rev* **1984**, 15:173–193.

14 Bonate, P. L.: *Pharmacokinetic–Pharmacodynamic Modeling and Simulation*. Springer, **2006**.

15 Rowland, M. T. T.: *Clinical Pharmacokinetics: Concepts and Applications, 3rd edn*. Williams & Wilkins, **1995**.

16 Lehr, T., Staab, A., Tillmann, C., Trommeshauser, T., Schäfer, H. G., Kloft, C.: Population pharmacokinetic modelling of NS2330 and its major metabolite in patients with Alzheimer's disease. *British Journal of Clinical Pharmacology* **2007**, 64:36–48.

17 Thatte, U.: NS-2330 NeuroSearch. *Curr Opin Invest Drugs* **2001**, 2:1592–1594.

18 Reddy, M., Yang, R. S., Andersen, M., Clewell, H.: *Physiologically Based Pharmacokinetic Modelling: Science and Applications,* 2006.

19 Schmitt, W., Willmann, S.: Erratum (Physiology-based pharmacokinetic modeling: ready to be used. *Drug Discov Today Tech*, DOI 10.1016/j.ddtec.2004.09.006). *Drug Discov Today Tech* **2005**, 2:125–132.

20 Nestorov, I.: Whole body pharmacokinetic models. *Clin Pharmacokinet* **2003**, 42:883–908.

21 Brown, R. P., Delp, M. D., Lindstedt, S. L., Rhomberg, L. R., Beliles, R. P.: Physiological parameter values for physiologically based pharmacokinetic models. *Toxicol Ind Health* **1997**, 13:407–484.

22 Leggett, R. W., Williams, L. R.: Suggested reference values for regional blood volumes in humans. *Health Phys* **1991**, 60:139–154.

23 Kuwahira, I., Gonzalez, N. C., Heisler, N., Piiper, J.: Regional blood flow in conscious resting rats determined by microsphere distribution. *J Appl Physiol* **1993**, 74:203–210.

24 Davies, B., Morris, T.: Physiological parameters in laboratory animals and humans. *Pharm Res* **1993**, 10:1093–1095.

25 Kirman, C. R., Hays, S. M., Kedderis, G. L., Gargas, M. L., Strother, D. E.: Improving cancer dose-response characterization by using physiologically based pharmacokinetic modeling: An analysis of pooled data for acrylonitrile-induced brain tumors to assess cancer potency in the rat. *Risk Anal* **2000**, 20:135–151.

26 Zhu, H., Jain, R. K., Baxter, L. T.: Tumor pretargeting for radioimmunodetection and radioimmunotherapy. *J Nuclear Med* **1998**, 39:65–76.

27 Zhu, H., Baxter, L. T., Jain, R. K.: Potential and limitations of radioimmunodetection and radioimmunotherapy with monoclonal antibodies. *J Nuclear Med* **1997**, 38:731–741.

28 Devineni, D., Klein-Szanto, A., Gallo, J. M.: In vivo microdialysis to characterize drug transport in brain tumors: analysis of methotrexate uptake in rat glioma-2 (RG-2)-bearing rats. *Cancer Chemother Pharmacol* **1996**, 38:499–507.

29 Kawai, R., Mathew, D., Tanaka, C., Rowland, M.: Physiologically based pharmacokinetics of cyclosporine A: extension to tissue distribution kinetics in rats and scale-up to human. *J Pharmacol Exp Ther* **1998**, 287:457–468.

30 Nakajima, Y., Hattori, K., Shinsei, M., et al.: Physiologically-based pharmacokinetic analysis of grepafloxacin. *Biol Pharm Bull* **2000**, 23:1077–1083.

31 Manuilov, K. K., Navashin, S. M., Kuleshov, S. E.: Use of gentamicin physiologically-based model for individual dosing. *Int J Clin Pharmacol Res* **1993**, 13:59–63.

32 Manuilov, K. K.: Use of a physiologically-based pharmacokinetics model for analysis of antibiotic distribution in tissue. *Int J Clin Pharmacol Ther Toxicol* **1992**, 30:548–549.

33 Csajka, C., Verotta, D.: Pharmacokinetic–pharmacodynamic modelling: history and perspectives. *J Pharmacokinet Pharmacodyn* **2006**, 33:227–279.

34 Holford, N. H., Sheiner, L. B.: Kinetics of pharmacologic response. *Pharmacol Ther* **1982**, 16:143–166.

35 Holford, N. H. G., Sheiner, L. B.: Understanding the dose–effect relationship: clinical application of pharmacokinetic–pharmacodynamic models. *Clin Pharmacokinet* **1981**, 6:429–453.

36 Mack, G. S.: Can complexity be commercialized? *Nat Biotechnol* **2004**, 22:1223–1229.

37 Aderem, A.: Systems biology: its practice and challenges. *Cell* **2005**, 121:511–513.

38 Le Novere, N., Bornstein, B., Broicher, A., et al.: BioModels database: a free, centralized database of curated, published, quantitative kinetic models of biochemical and cellular systems. *Nucleic Acids Res.* **2006**, 34:D689–D691.

39 Eddy, D. M., Schlessinger, L., Kahn, R.: Archimedes: a trial-validated model of diabetes dealing with complexity in clinical diabetes: the value of Archimedes. *Diabetes Care* **2003**, 26:3093–3101.

40 Eddy, D. M., Schlessinger, L., Kahn, R.: Clinical outcomes and cost-effectiveness of strategies for managing people at high risk for diabetes. *Ann Intern Med* **2005**, 143:251–322.

41 Kahn, R.: Dealing with complexity in clinical diabetes: the value of Archimedes. *Diabetes Care* **2003**, 26:3168–3171.

42 Eddy, D. M., Schlessinger, L., Eddy, D. M., Schlessinger, L., Kahn, R.: Validation of the Archimedes diabetes model: a trial-validated model of diabetes dealing with complexity in clinical diabetes: the value of Archimedes. *Diabetes Care* **2003**, 26:3102–3110.

43 Brandeau, M. L., Eddy, D. M., Schlessinger, L., Eddy, D. M., Schlessinger, L., Kahn, R.: Modeling complex medical decision problems with the Archimedes model. A trial-validated model of diabetes dealing with complexity in clinical diabetes: the value of Archimedes. *Ann Intern Med* **2005**, 143:303–304.

44 Danhof, M., Alvan, G., Dahl, S. G., Kuhlmann, J., Paintaud, G.: Mechanism-based pharmacokinetic–pharmacodynamic modeling – a new classification of biomarkers. *Pharm Res* **2005**, 22:1432–1437.

45 Troconiz, I. F., Wolters, J. M., Tillmann, C., Schaefer, H. G., Roth, W.: Modelling of the anti-migraine effects of BIBN 4096 BS. *Clin Pharmacokinet* **2006**, 45:715–728.

46 Holford, N. H. G., Chan, P. L. S., Nutt, J. G., Kieburtz, K., Shoulson, I.: Disease progression and pharmacodynamics in Parkinson disease – evidence for functional protection with levodopa and other treatments. *J Pharmacokinet Pharmacodyn* **2006**, 33:281–311.

47 de Winter, W., Post, T., DeJongh, J., et al.: A mechanistic disease progression model for type 2 diabetes mellitus and pioglitazone treatment effects. *PAGE* **2006**, 2006:13.

48 Kimko, H. C., Duffull, S. B.: *Simulation for Designing Clinical Trials.* Marcel Dekker, **2003**.

49 Aaron,s L.: Software for population pharmacokinetics and pharmacodynamics. *Clin Pharmacokinet* **1999**, 36:255–264.

50 Mentre, F.: History and new developments in estimation methods in non-linear mixed-effects models. *PAGE* **2005**, 2005:14.

18
Biosimulation and Its Contribution to the Three Rs
Hanne Gürtler

Abstract

The use of experimental animals and human subjects in the development of new medicines raises a number of ethical issues and is a source of concern to the European public in general. Animal experimentation and testing in human subjects are today a necessary part of the development of new medicines. New and reliable methods for predicting efficacy and safety are needed before the use of experimental animals and human subjects in drug development can be effectively reduced and eventually replaced. The Three Rs (replacement, reduction, refinement) is an internationally recognized principle governing the use of experimental animals and is the basic tenet of European Union (EU) research and other policies concerning the use of animals in scientific testing and experimentation. A similar principle relating to the use of human subjects has not yet been established. Biosimulation holds great hopes for the future in the reduction, refinement, and replacement of animal and human experimentation in drug development and thus can contribute to the implementation of the Three Rs in both animal and human research.

18.1
Ethical Considerations in Drug Development

Today there is a greater awareness of the ethical dilemmas arising during drug development and use, and a shared commitment among politicians, scientists, industry, and NGOs to ensure that such ethical issues are properly addressed. Public expectations of greater accountability and transparency are rising, with companies expected to show how they ensure ethical considerations in the process of bringing products to market [1].

The use of experimental animals and human subjects in the development of new medicines raises a number of ethical issues and is a source of concern to the European public in general. Animal experimentation and testing in human subjects are today a necessary part of the development of new medicines. New and

Biosimulation in Drug Development. Edited by Martin Bertau, Erik Mosekilde, and Hans V. Westerhoff
Copyright © 2008 WILEY-VCH Verlag GmbH & Co. KGaA, Weinheim
ISBN: 978-3-527-31699-1

reliable methods for predicting efficacy and safety are needed before the use of experimental animals and human subjects in drug development can be effectively reduced and eventually replaced. There is an increased focus in the EU on better ways of conducting science while contributing to the solving of ethical issues. This includes both improving drug development and finding alternatives to animal and human experimentation.

While regulatory authorities monitor research to ensure it is conducted in accordance with relevant laws and universal principles, stakeholders also seek reassurance that companies consider any ethical concerns that emerge. In particular, this is a matter of being respectful of the integrity of people participating in clinical trials, animal welfare, and culturally founded objections to certain types of research. This condition is an obvious motivator for the adoption of alternative methods, such as biosimulation.

Ethics can be defined as the system of ethical values and principles that guide human behavior. In this sense, ethics is about what determines the right choice – in relation both to ourselves as humans, and to the world we live in. Different branches of ethics cover different areas of ethical reflection. Bioethics is a branch of ethics that, broadly speaking, addresses ethical issues related to medicine, life sciences, and biotechnology as applied to humans, animals, and the environment. There are different approaches to ethics. While so-called utilitarian ethics focus on the consequences of human action with the objective of maximizing benefits, other approaches attach greater importance to other values such as, e.g. integrity. It can be argued that regardless of the benefits of utilizing new technology, the integrity of living entities needs always to be respected. Today many ethical issues are approached following the utilitarian approach, by making risk assessments and weighing up costs and benefits. While this often makes good sense, it is widely recognized that many ethical issues cannot be approached solely by weighing up costs and benefits.

18.2
The Three Rs – An Ethical Approach to Animal Experimentation

The Three Rs were defined by Russell and Burch in 1959 in the classic publication *The Principles of Humane Experimental Technique* [2]. The Three Rs refer to methods or a modification of methods which reduce, refine, or replace animal experimentation. Since their introduction, these principles have become widely accepted internationally as the basic principles guiding animal use in research, teaching, and testing. The Three Rs provide a strategy for a rational and stepwise approach to minimizing animal use and the suffering caused by such use, without compromising the quality of the scientific work being undertaken.

The *Three Rs Declaration of Bologna* which was adopted in 1999 by the *Third World Congress on Alternatives and Animal Use in the Life Sciences* strongly endorsed and reaffirmed the principles of the Three Rs [3]. In the following year, 2000, the Three

Rs were officially endorsed by the *European Science Foundation* [4], and in 2001, they were recommended in an editorial of the *British Medical Journal*.

18.3
The Three Rs Alternatives

18.3.1
Replacement Alternatives

These refer to methods which permit a given purpose to be achieved without conducting experiments or other scientific procedures on animals [5]. A scientist's long-term aim should be to replace all experiments on living animals through the development and use of suitable non-animal methods. Replacement alternatives include:

- the use of *in vitro* systems, e.g. cell and tissue cultures, whole organs, parts of cells, and tissue slides;
- the use of lower organisms such as bacteria, fungi, and plants;
- the use of vertebrates at early stages of development;
- the use of mathematical and computer models;
- the use of imaging techniques;
- the use of physical and chemical techniques;
- the collation and use of information already gained.

18.3.2
Reduction Alternatives

These refer to methods for obtaining comparable levels of information from the use of fewer animals in scientific procedures, or for obtaining more information from the same number of animals [5]. Scientists should try to reduce the number of animals used in experiments to the minimum possible. The proper design of experiments enables the fewest animals needed to give meaningful results to be used. Reduction alternatives include:

- the application of better research strategies;
- the use of *in vitro* systems for screening;
- appropriate experimental design and size;
- careful planning and evaluation of experiments;
- maximizing the data from each procedure;
- proper choice of animal models;
- proper selection of experimental conditions and parameters measured;
- better control of variation;
- harmonization of regulatory guidelines for testing;
- regularly review of design of regulatory testing procedures;

- the use of computer-aided learning of experimental design and statistics;
- better information exchange to avoid repeat testing.

18.3.3
Refinement Alternatives

These refer to methods which alleviate or minimize potential pain, suffering, and distress, and which enhance animal well-being [5]. Scientists should try to refine their experiments, improving the care and treatment of laboratory animals to minimize any possible pain and suffering experienced by the animals which have to be used. Refinement alternatives include:

- the use of more humane endpoints;
- improving the health and care of test animals;
- improving experimental procedures;
- the use of non-invasive methods;
- improving housing conditions;
- appropriate training and socialization of animals;
- the use of appropriate anaesthetics and analgesics if surgical techniques are applied;
- the use of telemetric methods;
- appropriate training and education of personnel.

18.4
The EU and the Three Rs

The Three Rs alternatives are the basic tenets of EU research and other policies concerning the use of animals in scientific testing and experimentation, and the present review by the European Commission (Directive 86/609/EEC) on the *Protection of Animals Used for Experimental and other Scientific Purposes* focuses on the implementation of the Three Rs [6].

The establishment in 2002 of the European Consensus Platform for Alternatives (ECOPA) brings together representatives of animal welfare, industry, government, and academia; and the inclusion of the Three Rs in the *Community Action Plan on the Protection and Welfare of Animals 2006–2010* further demonstrates the commitment of all stakeholders in EU to the Three Rs alternatives [7, 8].

In 2005, a European Partnership between the EU Commission and industry to promote alternative approaches to animal testing was announced, clearly demonstrating the EU focus on the Three Rs alternatives [9].

18.4.1
European Partnership

The European Partnership for Alternative Approaches to Animal Testing (EPAA) is a joint initiative from the European Commission, European trade associations from seven industry sectors, and individual companies. The partners have agreed to a *Three Rs Declaration* [10]. Based on that declaration, an action program was established which focuses on identifying barriers to progress and proposing appropriate solutions in order to promote the development, validation, and regulatory acceptance of alternative approaches, such as e.g.:

- mapping of research activities and current strategies;
- cooperation in research to strengthen and enlarge current activities between the partners and other relevant stakeholders;
- development of alternative approaches, including intelligent testing strategies;
- practical mechanisms to improve the validation process using available knowledge;
- practical mechanisms to facilitate the regulatory acceptance process of alternative approaches;
- widening stakeholders' dialogue;
- practical mechanisms to foster innovation in the areas of alternative approaches.

An annual report from the partnership on the implementation of the action program will be published for the attention of the Council, the European Parliament, and other relevant stakeholders. The first progress report was presented at a conference in December 2006 [10].

18.4.2
European Centre for the Validation of Alternatives

A European Centre for the Validation of Alternative Methods (ECVAM) was created by a Communication from the Commission to the Council and the Parliament in 1991, pointing to a requirement in Directive 86/609/EEC for the protection of animals used for experimental and other scientific purposes, which requires that the Commission and the Member States should actively support the development, validation, and acceptance of methods which could reduce, refine, or replace the use of laboratory animals [11]. As defined in the Communication of the European Commission to Council and the European Parliament, the duties of ECVAM are:

- to coordinate the validation of alternative test methods at the EU level;
- to act as a focal point for the exchange of information on the development of alternative test methods;
- to set up, maintain, and manage a data base on alternative procedures;
- to promote dialogue between legislators, industries, biomedical scientists, consumer organizations, and animal welfare groups, with a view to the development, validation, and international recognition of alternative test methods.

ECVAM seeks to promote the scientific and regulatory acceptance of alternative methods which are of importance to the biosciences, through research, new test development and validation, and the establishment of specialized databases, with the aim of contributing to the replacement, reduction, and refinement of laboratory animal procedures in accordance with the Three Rs concept of Russell and Burch [2, 11].

18.4.3
European Consensus Platform for Alternatives

ECOPA brings together representatives of animal welfare, industry, government, and academia and further demonstrates the commitment of all stakeholders in EU to the Three Rs alternatives [7]. The concept of consensus between the parties concerned, i.e. animal welfare, industry, academia, and governmental institutions has been accepted in various countries as an efficient way to stimulate research into alternatives to animal experiments and to enforce the acceptance of alternatives in experimental practice. The goal of ECOPA is to respond to the need for the creation of a pan-European platform. Besides the fact that a link is needed between the different national platforms, consensus discussions with all relevant groups will maximize results and minimize conflicts within the Three Rs strategy. Jointly accepted opinions transformed into a strong plea have a substantial impact on issues centring on alternative methods.

18.5
Applying the Three Rs to Human Experimentation

The ethical implications of conducting clinical trials are wide-ranging and much debate has been generated on best practice. In order to prove that a new medicine or treatment is safe and effective, it has to be tested on human volunteers.

There is currently a focus on ethical practices in the conduct of clinical trials. An increasing need for new participants in clinical studies raises a number of ethical issues that companies must address now and in the years to come. These include, among others: choosing the appropriate research question and design, ensuring a prior scientific and ethical review of the proposed protocol, selecting equitable participants, obtaining informed consent, minimizing risks to research participants, maintaining an equal regard for all participants, and providing appropriate treatment to participants during and after the trial [12–14].

Indeed, ethical behavior is not only an essential ingredient in sustaining public support for research, it is an integral part of the process of planning, designing, implementing, and monitoring research involving human beings.

Just as good science requires appropriate research design, consideration of statistical factors, and a plan for data analysis; it must also be based on sound ethical principles. Only then can research succeed in being efficient and cost-effective, while at the same time embodying appropriate protections for the rights and welfare of human participants.

Principles for the conduct of clinical research are set forth in internationally recognized documents, such as the *Declaration of Helsinki* and the *Guideline for Good Clinical Practice of the International Conference on Harmonization* (ICH) [14].

In 2000 the *Declaration of Helsinki* on biomedical research was amended and now takes into account the Three Rs. It now opens possibilities for testing in humans based, among other sources of information, on data obtained from validated *in vitro* analysis without using live animals. In light of this development more focus should be given to alternative approaches in relation to human experimentation.

Reduction, Replacement, and *Refinement* alternatives are also important in human research and the Three Rs should be considered as a principle guiding this area too. The possibilities of extending the principle of the Three Rs to human experiments are presently being evaluated as part of the objectives of BioSim – an EU Network of Excellence in Biosimulation [15].

The Three Rs principle has been and still is a useful tool for engaging all stakeholders in the development of a common strategy in relation to animal experimentation. The establishment of a Three Rs principle pertaining to human experiments and thereby the introduction of a new and useful tool for engaging different stakeholders in developing a common strategy in relation to the future use of human subjects may help to solve some of the key ethical issues relating to clinical trials.

18.6
Biosimulation and its Contribution to the Three Rs

18.6.1
Biosimulation – A New Tool in Drug Development

Biosimulation is a potential new tool in drug development and health care that, by use of advanced *in silico* models (computer models), could make it possible in future to simulate the behavior of complex biological systems.

A simulation model describes the temporal variation of a system in terms of the processes and interactions that are presumed to be at work. In connection with the development of new medicines, the model combines a pharmacokinetic description of the absorption, distribution, metabolism, and excretion of the medicine, with a detailed representation of the mechanisms responsible for its function and for the development of side-effects or possible synergetic interactions with other medicines [16].

By continuously comparing model predictions with experimental results, biosimulation can provide better insights into the behavior of biological systems like the human body, the progression of diseases, and a better prediction of the function and effects of new medicines by integrating experimental results into a coherent structure.

18.6.2
The Challenges in Drug Development

As the drug development process is conducted today there appears to be an enormous need for improvement.

The discovery and development of a new medicine is long, complex, and requires extensive resources, including the use of experimental animals and humans in clinical trials. Despite technical progress in drug discovery technologies, there is no concomitant increase in understanding the clinical basis of disease and therefore the development of novel effective therapies. Today the rate of failure of new medicines poses a difficult challenge to the industry.

There is currently ongoing discussion on the challenges facing the European biopharmaceutical industry in relation to drug development. In an attempt to address these challenges, the European Innovative Medicine Initiative (IMI) was launched in 2004. The results are described in the final IMI draft report [17].

The IMI stresses the importance of improving the predictability of safety testing, integrating new technologies such as *in silico* predictive techniques. One of the main recommendations concerning safety is the need to develop *in silico* methods for predicting toxicity. The IMI stresses that there are overarching needs, common to all disease areas, which illustrate the challenges of improving efficacy, such as developing a better understanding of disease mechanisms and *in silico* simulations of disease pathology.

The IMI draft report [17] explicitly stresses that the greatest need for the pharmaceutical industry is to detect the possibility of failure at the earliest stage possible, and it is in this context that advances in basic biomedical science within the European research community could make the greatest contribution. A major key to reducing the failure of new medicines is the development and use of preclinical models that are more predictive of efficacy and safety in clinical trials. To succeed, the IMI stresses the importance of integrating a variety of different new technologies and innovative approaches, such as improvements in predictive biology, systems biology, and *in silico* predictive techniques. Accordingly, employing innovative approaches such as these potentially not only decreases the cost of drug development and speeds up the delivery of innovative medicines to patients; it also revolutionizes and completely changes the process by which new medicines are developed.

The *Innovative Medicine Initiative Strategic Research Agenda* explicitly concludes that scientists are urgently needed within specific areas such as biosimulation, systems biology, and *in silico* modeling. The current approach to drug discovery and development can be significantly improved through the development of efficient means to:

- extract the information available in the individual trial and validate it in terms of current knowledge in cell biology, medicine, etc.;
- apply an adaptive trial process where information acquired in a given trial is continuously assessed and immediately used to improve the procedure of the trial;
- extrapolate results obtained from experiments on cell cultures and from animal experiments for application to human patients;
- predict the variation of drug efficacy and the occurrence of side-effects in dependence of genetic modifications;

- predict the likelihood that a particular chemical compound functions as a drug on the basis of knowledge about related compounds.

This situation may be viewed as a consequence of the extraordinary complexity of living systems. However, the more drug development resembles a trial-and-error process, and the less information one extracts from each individual test, the more tests must be performed before the drug can be approved.

18.6.3
Biosimulation's Contribution to Drug Development

Millions of people's lives and health conditions depend on getting the right treatment. Often the only treatment is to resort to medicine. Not all medicines, however, have the same level of efficacy, and not all people respond in the same way to the same medicine. Unfortunately, certain types of medicines may even have serious side-effects, which often become apparent only after introduction to the market.

The EU strives to ensure that medicinal products for human use help maintain a high level of protection for public health. Biosimulation underpins this objective as a potential future tool which may contribute to the development of safer and more effective medicines better adapted to patients needs and by improving today's knowledge about how patients need to administer medicine. By providing better predictions of the effects of new medicine as well as better insights into the behavior of biological systems like the human body, biosimulation could effectively contribute to the improvement of today's drug development [15, 16].

By integrating experimental results into a coherent structure, biosimulation can provide better predictions of the function and effects of new medicines. By providing better predictions of the effects of new medicines, biosimulation could contribute to the likelihood of fewer failures during drug development, the development of safer and more effective medicines, and faster development of medicine at lower costs, allowing more focus on medicine for rare diseases and individualized medicines. By continuously comparing model predictions with experimental results, biosimulation can provide better insights into the behavior of biological systems like the human body and the progression of diseases. By providing better insights into the behavior of living systems, particularly human, biosimulation helps contribute to:

- a more quantitative and systems oriented understanding of the function of the human body in health and disease;
- a better understanding of the role of the complex dynamic phenomena involved, for instance, in cellular communication and hormonal control.

Increased use of modeling and simulation will help the understanding of exposure/response relationships with regard to both safety and efficacy. It will help to understand the drug metabolism and also the target biology in humans. Modeling can also be used to define disease biomarker. Clinical trial simulation is a valuable tool to test trial design factors, identifying non-robust co-variants likely to jeopardize trial outcome. The resulting study design will be more robust, execute faster with fewer subjects, and fewer non-responders. The resultant clinical programs will be cheaper and provide more informed decisions [17].

In silico modeling and simulation can be applied at every stage of the drug development process, from the virtual simulation of a cellular function, e.g. the whole network of molecular interactions involved in cell biology, to modeling virtual populations.

During the initial stage of the drug development process, one can use the simulation model to test any hypothesis one might have regarding the function of the drug vis-à-vis the established biological understanding [15].

During the trial phase, the simulation model can be used as a vehicle to define an effective test protocol. As stressed in the *American Food and Drug Administration Critical Path Opportunities List*, clinical trial simulation can predict efficient designs for development programs that reduce the number of trials and patients, improve decisions on dosing, and increase informativeness [18].

During the approval process, the regulatory agencies can use models to check that the tests are adequately performed. The predictive modeling approach is strongly recommended by the American Food and Drug Administration that already uses mathematical models in its evaluation of applications for drug approval.

18.6.4
Biosimulation's Contribution to the Three Rs

Biosimulation holds great promise for the future in solving key ethical issues relating to drug development and drug use. Biosimulation allows for a better planning of experiments, a more rational exploitation of the information acquired in each test, and a better prediction of the effect of new medicines before administering them in animals and humans and may contribute to:

- the reduction, refinement, and replacement of animal experimentation (the Three Rs);
- the reduction, refinement, and replacement of human experimentation;
- more ethically acceptable drug evaluation in particular at risk patient groups, such as children, pregnant women, and those with specific gene modifications.

The IMI only implicitly addresses both biosimulation and the Three Rs. As a response to the absence of an explicit focus on the Three Rs, an article on the IMI's implications for the Three Rs was later published [19]. Applying biosimulation to drug development helps contribute to the Three Rs by:

- enabling better designs and planning of experiments;
- enabling better exploitation of the information acquired in each test;
- combining *in vitro* with *in silico* simulation models to predict what happens *in vivo*/in man;
- enabling better use of animal data to predict what happens in man;

- providing better predictions of efficacy and toxicity;
- providing a more sophisticated understanding of the function of living systems.

Biosimulation may also contribute to better animal models and improved animal data. A better understanding of the basis of disease together with *in silico* simulations of disease processes and/or drug intervention could lead to better animal models. Developing better animal models includes the need to select appropriate species. In this context, biosimulation could also help find appropriate models down the so-called 'evolutionary line'.

In relation to human experimentation the development of virtual population models can, through a better delineation of patient groups, reduce the number of phase-3 trials and thus the number of human subjects needed for clinical trials [15].

Evaluating the specific contribution of biosimulation is, however, not an easy task. Specific contributions may vary depending on disease area, type of research, kind of medicine in question, specific stage in drug development, etc. Building *in silico* simulation models requires experimental information from animal work before a reduction is possible. Building better animal models of disease also might require increased animal work initially before a reduction is possible. While in some areas one will definitely see an initial reduction in the use of animal experimentation, in other areas the possibility of an initial increase before reduction is a realistic scenario. The long-term outcome, however, is that of a very positive contribution to the Three Rs in terms of reduction, refinement, and partial replacement.

To quote The European Federation of Pharmaceutical Industries and Associations' (EFPIA) Policy Statement on the use of animals in research and development, "as our biological knowledge increases, so too does the usefulness of non-animal methods. There are still enormous gaps in our biological knowledge that limit the usefulness of cell culture and computer based research".

Biosimulation is in its infancy and its future success depends on:

- collaboration and communication among those in various academic disciplines ranging from the life sciences to physics and mathematics;
- training and education programs in biosimulation, thereby providing an adequate cohort of professional staff with the necessary expertise in modeling biological systems;
- strong links between industry, academia, and regulatory authorities;
- recognition of the benefits of biosimulation not only to science and industry but also to patients and society.

References

1 Bioethics: *Novo Nordisk Annual Report,* **2006**, available at: www.novonordisk.com.

2 Russell, W.M.S., Burch, R.L.: *The Principles of Humane Experimental Technique, Special Edition,* UFAW, London, 1959 reprinted **1992**.

3 UFAW: *The Three Rs Declaration of Bologna* (3rd World Congress on Alternatives and Animal Use in the Life Sciences), **1999**, available at: www.ufaw.org.uk/Annual_report_2000/Bologna.htm.

4 European Science Foundation: *European Science Foundation Policy Briefing N° 9 – Use of Animals in Research,* **2000**, available at: www.esf.org/esf_genericpage.php?section=4&language=0&genericpage=213.

5 FRAME: *Fund for the Replacement of Animals in Medical Experiments,* available at: www.frame.org.uk.

6 EEC: *Revision of Directive 86/609/EEC on the Protection of Animals Used for Experimental and Other Scientific Purposes,* available at: europa.eu.int/comm/environment/ chemicals/lab_animals/home_en.htm.

7 ECOPA: *European Consensus-Platform for Alternatives,* available at: ecopa.vub.ac.be/.

8 EU: *Community Action Plan on Animal Welfare 2006–2010,* available at: ec.europa.eu/food/animal/welfare/ actionplan/actionplan_en.htm.

9 EU: *Europe Goes Alternative* (Conference on Alternative Approaches to Animal Testing, Brussels), **2005**, available at: europa.eu.int/comm/enterprise/events/ animal_tests/index_en.htm.

10 EPAA: *European Partnership on Alternative Approaches to Animal Testing,* available at: ec.europa.eu/enterprise/epaa/ index_en.htm.

11 ECVAM: *European Centre for the Validation of Alternative Methods,* available at: ecvam.jrc.it/index.htm.

12 Novo Nordisk: *Clinical Trials,* in: *Novo Nordisk Annual Report,* **2006**, available at: www.novonordisk.com.

13 PhRMA: *Principles of Conduct of Clinical Trials and Communication of Clinical Trial Results,* The Pharmaceutical Research and Manufacturers of America (PhRMA), available at: www.phrma.org/.

14 World Medical Association: *Helsinki Declaration,* available at: www.wma.net/e/.

15 BioSim: *BioSim – an EU Network of Excellence in Biosimulation,* available at: www.biosim-network.net/.

16 Mosekilde, E., Colding-Jørgensen, M.: Biosimulation – a new tool in drug development, SCREENING. *Trends in Drug Discovery,* **2004**.

17 The Innovative Medicines Initiative: *Strategic Research Agenda – Creating Biomedical R&D Leadership for Europe to Benefit Patients and Society, Final Draft,* **2005**, available at: europa.eu.int/comm/research/fp6/ index_en.cfm?p=1_innomed.

18 FDA: *Critical Path Opportunities List,* available at: www.fda.gov/oc/initiatives/ criticalpath/reports/opp_list.pdf.

19 Ragan, I.: *The Innovative Medicines Initiative: Implications for the 3Rs,* **2006**, available at: www.nc3rs.org.uk/news.asp?id=16.

Index

Cell-free Protein Synthesis. Edited by Alexander Spirin and James Swartz
Copyright © 2008 WILEY-VCH Verlag GmbH & Co. KGaA, Weinheim
ISBN: 978-3-527-31649-6